DEVELOP
STUDIES

A READER

Editor
Stuart Corbridge
University Lecturer in South Asian Geography and
Fellow of Sidney Sussex College, Cambridge

A member of the Hodder Headline Group
LONDON • NEW YORK • SYDNEY • AUCKLAND

First published in Great Britain 1995 by Arnold,
a member of the Hodder Headline Group
338 Euston Road, London NW1 3BH

Co-published in the United States of America by
Oxford University Press Inc.,
198 Madison Avenue, New York, NY10016

British Library Cataloguing in Publication Data
A catalogue record for this book is available from the British Library

Library of Congress Cataloging-in-Publication Data
A catalog record for this book is available from the Library of Congress

ISBN 0 340 61452 8

6 7 8 9 10

Typeset by Phoenix Photosetting, Chatham, Kent
Printed and bound in Great Britain by J W Arrowsmith Ltd, Bristol

CONTENTS

ACKNOWLEDGEMENTS

I am grateful to Laura McKelvie for commissioning this book and seeing it into production with an unfailing mixture of good humour and patient prompting. Sophie Oliver and Patrick Bonham were models of efficiency in preparing the manuscript for the printers, and Ian Agnew produced the line diagrams in short time. I would also like to thank three referees for Edward Arnold who pushed me to rethink some aspects of the organization of the Reader. Above all, I would like to thank John Agnew, David Booth and Michael Watts for their continuing support and inspiration, and the authors of the Readings collected here for allowing me to reprint their work. The book is dedicated to a group of research scholars from each of whom I have learned a great deal about development studies: Amitabh, Jim Bentall, Tim Conway, Alan Hudson, Sarah Jewitt, Emma Mawdsley, Julian Richards, Babar Sobhan, Jody Solow, Glyn Williams and Jock Wills.

The authors and publishers would like to thank the following for permission to use copyright material in this book:

Ashgate Publishing Ltd for the excerpts from *Why Poor People Stay Poor* by Michael Lipton (Temple Smith, 1977); Butterworth-Heinemann for Alan Gilbert, 'Latin America's Urban Poor: Shanty Dwellers or Renters of Rooms?' from *Cities*, Vol. 4, No. 1 (1987), pp. 43–51; University of California Press for the excerpts from *Death Without Weeping: Mother Love and Child Death in Northeast Brazil* (1992), and for the extract from *Toward a Political Economy of Development: A Rational Perspective* by Robert Bates (1988); Frank Cass Publishers for the excerpts from John Harriss, 'Does the Depressor Still Work? Agrarian Structure and Development in India' from *Journal of Peasant Studies*, Vol. 19, No. 2 (1992), pp. 189–222, © Frank Cass & Company; University of Chicago Press for the excerpts from Bert Hoselitz, 'Non-Economic Barriers to Economic Development' from *Economic Development and Cultural Change*, Vol. 1, No. 1 (1952); Edinburgh University Press for the excerpts from Judith Carney and Michael Watts, 'Manufacturing Dissent: Work, Gender and the Politics of Meaning in a Peasant Society' from *Africa*, Vol. 60 (1990), pp. 207–40; Elsevier Science Ltd for the excerpts from David Booth, 'Marxism and Development Sociology: Interpreting the Impasse' from *World Development*, Vol. 13 (1985), pp. 761–87, A.K. Zen, 'Food and Freedom' from *World Development*, Vol. 17 (1989), pp. 769–81, for the excerpts from Paul Streeten, 'Structural Adjustment: Issues and Option' from *World Development*, Vol. 15 (1988), pp. 1469–82, and the excerpts from Susanna Hecht, 'Environment, Development and Politics: Accumulation and the Livestock Sector in eastern Amazonia' from *World Development*, Vol. 13 (1985), pp. 663–84; Harper Collins Publishers, Inc. for the foreword by Mario Vargas Llosa from *The Other Path* by Hernando de Soto (Harper and Rowe, 1989), © 1989 Hernando de Soto; Harvard University Press and Harvester Wheatsheaf for the excerpts from *The Development Frontier: Essays in Applied Economics* by Peter Bauer (1991), © Peter Bauer; IMF Publication Services for Deepak Lal, 'The Misconceptions of "Development Economics"' from *Finance and Development*, Vol. 22 (1985), pp. 10–13; Longman Group for the excerpts from *Rural Development: Putting the Last First* by Robert Chambers (1982); Monthly Review Foundation for A.G. Frank, 'The Development of Under-Development' from *Monthly Review*, Vol. 18 (September 1966), © 1966 Monthly Review Inc; The Open University for excerpts from 'TNCs, Dominance and Dependency in the World Economy' from B. Crow and M. Thorpe *et al* (eds), *Survival and Change in the World Economy* (1988); Oxford University Press for the extract from *Of Peasants, Migrants and Paupers: Rural Labour Circulation and Capitalist Production in West India* by Jan Breman (1985), the excerpts from *World Development Report 1994* by The World Bank, and for the excerpts from *'Unimodal' and 'Bimodal' Strategies of Agrarian Change* by B.F. Johnston and P. Kilby (1975). A much more recent statement of Johnston and Kilby's views on these topics will be found in *Transforming Agrarian Economies: Opportunities Seized, Opportunities Missed* by T.P. Tomich, P. Kilby, and B.F. Johnston (Cornell University Press, 1995). Population Investigation Committee for Mead Cain, 'The Consequences of Reproductive Failure' from *Population Studies*, Vol. 40 (1986), pp. 375–88; Princeton University Press for the excerpts from *Lessons from East Asia* by Robert Wade (1990); Sage Publications, Inc. for Cathy Schneider, 'Radical Opposition Parties and Squatters Movements in Pinochet's Chile' from *Latin American Perspectives*, Vol. 18 (1991), pp. 92–112; Sage Publications for the excerpts from James Boyce, 'Of Coconuts and Kings: The Political Economy of an Export Crop' from *Development and Change*, Vol. 23 (1992), pp. 1–25; Royal Holloway and Bedford New College Geography Department for the extracts from Adrian Leftwich, 'Governance, Democracy and Development in the Third World', *Third World Quarterly*, Vol. 14 (1993), pp. 605–24; Zed Books for Arturo Escobar, 'Planning' from W. Sachs (ed.), *The Development Dictionary* (1992), pp. 132–45, and the excerpts from Chapter 1 of *Cold Hearths and Barren Slopes* by Bina Agarwal (1986).

PREFACE AND PROSPECTUS

In putting together this Reader on Development Studies I have drawn extensively on my own experiences as a teacher of development issues, and on criticisms, questions and suggestions I have received from students at Huddersfield, London, Syracuse, Cambridge and Jawaharlal Nehru Universities. One question commonly put to me by cash-strapped students has been: 'which book do you recommend for your course on development studies?'. Far too often my first response has been to deny that any one book can serve such a purpose, which is true enough but which also betrays a common academic arrogance and is less than helpful. When pushed I have cited a few of my favourite books, which have changed over the years, and pointed students to the family of textbooks on development put together by the UK's Open University (Bernstein, Crow and Johnson, 1992; Hewitt, Johnson and Wield, 1992; Wuyts, Mackintosh and Hewitt, 1991). When pushed further, I usually mention Michael Todaro's textbook, *Economic Development*, which is now in its fifth edition (Todaro, 1994). But even here I enter a caveat. I commend Todaro's book to students as a broad-based introduction to development economics, but point out that Todaro has little to say about some 'non-economic' debates about development. Given that many courses on development now cover issues like Orientalism, representation, and anti-development, and pay rather more attention to issues of gender, class, culture, the environment, citizenship and development ethics than do some courses in development economics, even Todaro will only take students so far along the route to a more catholic account of development studies.[1]

[1] I do not want this comment to be misconstrued or to sound ungenerous. The fifth edition of Todaro's textbook is an achievement many of us would be proud of and it is far more than just a textbook on the economics of development. Todaro is particularly adept at linking theories, models and policies.

Amongst existing Readers on development and underdevelopment, one that has stood the test of time, and which has been regularly updated, is that edited by Wilber and Jameson (1992, fifth edition). Rather like the Reader put together by Henry Bernstein (1976), however, the Wilber and Jameson collection leans strongly towards political economy perspectives on development and underdevelopment, where political economy is understood to be left-of-centre in political terms

Not surprisingly, perhaps, there is no one text that can do justice to the huge and growing field that is development studies – that pays close and proper attention to 'Third World' cultures and literatures, as well as to states, quasi-states and politics in the developing world, economic growth and employment issues, entitlements and distributive justice, and so on. Nor can any one text pay due attention to the diverse experiences, problems and achievements of communities in Africa, Asia, Latin America and the Pacific, or to issues like migration, demography, labour markets, housing, the informal sector, corruption, land reform, new social movements, non-governmental organizations (NGOs), survival strategies, trade, aid, debt, geopolitics and the new international economic order. There is just *too much to know* now (as, indeed, there always was). Development studies as an intellectual discipline has expanded so much in the past twenty or thirty years that it is in danger of losing its claim to distinctiveness. The journals that first grew up in the West[2] to serve the student of development were generalist journals like *Economic Development and Cultural Change* (first published in 1952), the *Journal of Development Studies* (1965), *Development and Change* (1970) and *World Development* (1973). Now, the keen student of development must also keep up with journals like *Development Policy Review*, the *European Journal of Development Research, Research in Rural Sociology and Development, Population and Development Review*, the *Journal of Peasant Studies* and the *World Bank Research Observer*, as well as journals like *Alternatives, Dialectical Anthropology, Gender and Society* and *Transition*, all of which deal regularly with wider development themes and issues. If we add in certain well-established area studies journals like *Africa*, the *Journal of Modern African Studies, Economic and Political Weekly (Bombay)*, the *Journal of Contemporary Asia, Latin American Perspectives, Latin American Research Review* and *Pacific Viewpoint*, we begin to see what a catch-all 'development studies' has become. If development studies means anything it presumably means the study of those countries which have a past in common, if not a common past – I refer to a history of colonialism and post-colonialism[3] –

and generally opposed to the accounts of neo-classical economics. Mention should also be made of the well-used Reader edited by Meier (1989) and of the two volumes of *A Handbook of Development Economics* edited by Chenery and Srinivasan (1988). The fact that these two volumes retailed at close to £100 ($150) a piece when they first came out meant that they were not readily adopted as student texts!

[2] This Reader draws heavily on extracts from books and papers published in Anglo-American journals. Although there are good prosaic reasons for this bias, and although many academics in developing countries consult these journals and publish in them, it has been a source of concern to the Editor and I advertise it to the reader as such.

[3] One reason for referring to a history of colonialism/post-colonialism is to distinguish the different pasts of the so-called Second and Third Worlds. But this distinction is becoming harder to draw. South Korea is clearly more prosperous than most parts of the ex-Soviet Union, and the very concept of a Third World only took shape with reference to the Cold War battles between the

and which can only reasonably be understood in relation to the so-called developed countries. No wonder that no-one has yet written the Great Development Studies text!

For many of us, though, the explosion and fragmentation of development studies has been at least as liberating as it has been frustrating. With questions of democratization, citizenship, liberalization, institution-building and the environment coming to the fore in the 1980s and the 1990s the need for an inter-disciplinary view of development – for a development studies, as opposed to a development economics/geography/sociology – has become ever more pressing. By the same token, the fact that no one text exists for most courses on development studies does not mean that teachers and students of development have given up on the idea of imposing some order on the fields they are covering in lectures, seminars, tutorials and practicals. I have talked to a number of colleagues about this and most are agreed that we are hemmed in by two main problems: (1) a lack of class time;[4] and (2) a lack of expertise across the board. We are forced to be partial in our coverage. These points acknowledged, most colleagues also maintain that they do have a *general vision* of development studies that they want to explore with students, and most have a mental checklist of the qualities they deem to be important in the make-up of a competent student of development. It is this vision and checklist that underpins the structure of any particular course on development – colleagues and timetables permitting – and which calls to mind for most teachers something like an ideal textbook for their course.

I share this perspective on development studies and the teaching of it. I first started teaching 'development studies' in 1981–82 and the content and style of the courses I have taught in successive years since then have changed quite radically in some respects, particularly so in regard to questions of gender, representation, the environment, and the role of the state. The rise of the counter-revolution in development theory and policy (or the New Right) has left a particular mark on the form and content of my classes. In still other respects I find myself returning to old

'First' and 'Second Worlds'. It may be of significance that some 'transitional' countries in the ex-Soviet Union, including Russia, have recently been advised by 'development economists' like Jeffrey Sachs (who had previously worked extensively in Latin America). The World Bank is not alone in seeking a way around the terminological minefield of First, Second and Third Worlds, when it refers to low-income economies, middle-income economies, upper-middle-income economies and so on. Because this Reader presents extracts from books and papers written over a forty-year period it would be churlish to object strongly to a phrase like the Third World. Even so, it is used here with some reluctance, and as little more than a synonym for the developing world.

[4] The time constraint will clearly not apply where students are following a full-time course in development studies – as at the University of East Anglia in the UK. By and large, however, development studies (or development economics/geography/sociology) is taught as part of a 'traditional' degree course. The time constraint is real enough in such cases.

themes, which might be a cause for comfort or discomfort, or possibly both. This Reader necessarily reflects my own views of what a course in development studies should try to do, suitably tempered by the views of many colleagues and students, and by the helpful written comments of three referees for Edward Arnold. My own checklist for a course/text in development studies would comprise at least the six points that follow.

(1) The course/text should give students a sense of the development of development studies as a discipline, in the process pointing out how it has changed partly in response to certain changes in the structures and problems of development that it is concerned to describe, understand, promote and possibly change. The objective is not to present a history of development studies for its own sake, or at great length, or even to suggest that neo-Marxist perspectives on development (say) continue to command wide support in the 1990s (they do not); it is rather to make clear how, and to what extent, contemporary empirical debates – for example, about food policies in sub-Saharan Africa – are conditioned by certain theoretical assumptions that often have a long and uneven history, and which need to be rendered transparent before they can be retired, amended, or properly acted upon.

(2) The course/text should give students a sense of the continuing and vital debates that lie at the heart of development studies and policy, while at the same time alerting them to those themes and policies that command wide support in the mainstream of the subject. That development studies is about debate – and differing, even opposing, policy options – can be illustrated well enough by reference to the success or otherwise of socialist development strategies, or the significance and activities of transnational corporations, or the relative merits of Green Revolution and 'indigenous' agricultural development strategies, or even the definition of development itself and the question of who is defining what for whom. That development studies is also about convergence and the rigorous empirical testing of ideas and policies can be illustrated with reference to the growing consensus that exists on the importance of infrastructure and institutions for development, or on the quality (as opposed to quantity) of state interventions to support market-led developments. A mainstream does exist in development studies and it is as well to signal this (see Williamson, 1993). At the same time, some of the most interesting work done under the rubric of development studies will always come from mavericks who challenge the conventional wisdom. It is important to hold these two strands together in fruitful tension.

(3) The course/text should give students a sense of the strengths and weaknesses of different 'traditional' disciplines, like economics, geography, sociology, anthropology and political science, insofar as they have

engaged with development issues, while also making the case for interdisciplinary studies.

(4) The course/text should introduce students to development issues at all spatial scales, running from the individual, household or village levels to the seemingly more abstract international level. The aim here is to establish that local development strategies must in large part depend upon the monetary and fiscal policies of a given country and on global trading and monetary regimes, just as abstract demands for trade liberalization should have regard to the specific economic and political structures through which these demands or policies must be operationalized. Some years ago Pranab Bardhan called for an extended conversation between anthropologists and economists which would marry the dominantly micro-sociological concerns of the anthropologists and the dominantly macro-sociological concerns of the economists (Bardhan, 1989). I support this call. Development economics is surely unhelpful to the extent that its models and landscapes are peopled by the social morons so famously lampooned by Amartya Sen (Sen, 1978); by the same token, a development studies which disregards some quite firmly established economic principles is liable to lapse into an unhelpful romanticism. Students also need to take on board questions of time and temporality and sequencing when it comes to the evaluation of development models and policies. Structural adjustment policies, for example, are often commended as being in the medium and long-term interest of a particular developing country, but what of the short term? What does the 'long term' mean exactly, and what relation does it bear to the lives of real men and women who 'in the long run are all dead'? Questions of spatial and inter-generational equity are at the heart of development studies, and are rarely susceptible to simple and one-sided answers. The hard part is learning to ask the right questions.

(5) The course/text should convey to students a sense of the often fragile means by which knowledge about 'development' or 'Other' societies is produced and circulated (be it by anthropological fieldwork, national sample surveys, World Bank forecasts, or whatever). This need not be a case of putting development economics on trial by mounting an anthropological case for the prosecution (Hill, 1986), but it should encourage a certain humility about what we know about often distant places and peoples.[5]

[5] Distance can be measured in absolute and relative terms, and with reference to social proximity as well as spatial propinquity. Rural areas of a Third World country can often be remote from the concerns and daily activities of urban-based policy makers in the same country. Distance also has to do with communication, or 'knowing'. It is bound up with problems of language, and the difficulties and opportunities imposed by status, age, race, ethnicity and gender. For an excellent discussion of the difficulties of understanding our so-called Others, which is nonetheless committed to the possibility and necessity of such an understanding/conversation, see Comaroff and Comaroff (1992, Ch. 1).

(6) The course/text should emphasize the links between development theories and policies such that students come to recognize the truth of Keynes's dictum (slightly amended) that 'the ideas of development workers, both when they are right and when they are wrong, are more powerful than is commonly understood' (after Keynes, 1973, p. 383).

Putting all this in slightly different terms, I would say: (1) that students of development (including teachers of development, of course) should be warned against a false distinction which opposes theories to policies, or ideas to actions; and (2) that students should be encouraged to reflect critically on ideas and policies relating to development that are put forward as self-evident truths. The aim should be to think counter-factually as well as associationally: to note not only that 'development' has been associated with many instances of environmental degradation (for example), but also to note that some environmental problems have been mitigated by development and to ask whether and how things might have been otherwise under different circumstances.

John Toye hints at just this point when he writes of the dilemmas that define the processes of development (Toye, 1987). Toye also notes that: 'When economic thinking is connected up with political movements of the right or the left, it seems almost impossible to avoid the ill-consequences of over-simplification' (Toye, 1987, p. viii). I share this concern about 'the violence of abstraction' (Sayer, 1987), even if I would slightly re-phrase it. I still cling to the view that our academic writings should have some political (or practical) purchase, but I would resist the command to write in support of a pre-ordained political end. If we are to avoid the 'ill-consequences of over-simplification' we must allow our politics to be driven by critical reflection rather than vice versa. Likewise, if there is a single principle that underpins my vision of development studies, it is the principle (adapted from Max Weber) that the true function of social science is to render problematic that which is conventionally regarded as self-evident. The challenge is to nurture an attitude of mind that can tolerate provisional truths and which encourages us to believe that we might be wrong. Our most precious beliefs should be tested continually against the reasoned beliefs of others who do not share our views. When exploring the 'debt crisis' we might well want to maintain that the commercial bank debts of some Latin American countries should have been written down in the 1980s, but we also need to ask what the consequences of this action would have been for different social groups, and what signals it would have sent to countries that were struggling to meet their debt repayments. In so doing we raise important questions about rights and responsibilities in the pursuit of different 'development policies'. An instinctively left-of-centre approach to debt and development ethics would here be set against the views of Buiter and Srinivasan, say, who have condemned commercial debt write-downs as a reward for profligate behaviour in Latin America, and as a punishment visited on those countries that behaved prudently like South Korea, or which were too poor to go into debt in the first place (Buiter and Srinivasan, 1987). Real issues are raised for students in this manner, even if real choices still need to be made in the world outside the classroom (often, it has to be said, on the basis of *Realpolitik* as much as informed opinion). By the same token, while an argument for a 'big bang' approach to the problems of transitional economies can be advanced by economists, we should also ponder the likely political consequences of such an approach over the short, medium and long terms. Economies rarely flourish amidst social dislocation and political chaos. Once again, space (in the first example) and time are at the heart of our intellectual *and practical* concerns, and they need to be clearly signalled to the student as such.

In my view, it is *asking* questions like these that makes development studies exciting and demanding. Complex economic and social problems are unlikely to call forth plausible simple answers, and we should be on our guard against various panaceas for development that are thrown up, from time to time, by committed advocates of modernization theory, or neo-Marxism, or a populist anti-developmentalism (see Section One). Development studies needs good theory, but it depends even more on a willingness to engage in reasoned argument and due consideration of other points of view.[6] The many pathways to development are indeed marked by dilemmas, contradictions and trade-offs, at different temporal and spatial scales, and suggestions to the contrary should be looked upon with suspicion. Most of the Readings in this book reflect this conviction. The authors of the Readings have their own views on development, of course, some of which are startlingly at odds with the views of other contributors, but most of them are also committed to a spirit of inquiry that takes seriously the views of others.

Such a prospectus for development studies is bound to seem like a tall order, and like all such checklists it is open to objections from students and colleagues and is by no means inclusive. We live and learn – the more so, as teachers, when we teach as part of a team. But tall orders can also concentrate the mind. They give us all a sense of what a course in development studies *might* be like, and of what an ideal textbook might look like to go with this course. What follows is something akin to my own ideal textbook on 'development studies', subject to the usual constraints of incomplete coverage, imperfect knowledge, lack of space and so on. Included here are many classic extracts from books and papers that have helped to define

[6] This should not be read as an invitation to intellectual relativism (as opposed to cultural pluralism). There is still a need to judge between competing arguments (and policies), and this might be done by reference to such things as empirical accuracy, completeness, and logical consistency. In still wider terms, it is important to distinguish between the very reasonable injunction to listen to the views of the victims/beneficiaries of development projects, and the less reasonable injunction, made in some quarters, to identify with these voices to the extent that they are assumed to be 'correct' as of right, or *a priori*.

the changing fields of development studies. Also included are many extracts from books and papers that are in the process of becoming classics, at least in my view. Taken together, they go some way to presenting a broad overview of development studies. They also foster a sense of development studies as a vital set of **debates** about the problems (and achievements) of poorer countries and peoples and about the different policies that might be adopted to tackle these problems. No single line on development is pushed. The Reader sets out several distinctive, if often overlapping, perspectives on particular development issues, and encourages the reader to think critically about the strengths and weaknesses of each approach.

The six sections that make up the book also cover a number of key subject areas in development studies, although many more are missing. The censor's axe has to fall somewhere and I have reluctantly had to leave out a number of exciting readings on 'Third World' cultures and politics, and on the increasing role played by non-governmental organizations in promoting development alternatives, to name but three areas.[7] Publishing, too, is riven by dilemmas and trade-offs – most notably, in this case, between a fuller text and a sensible cover price. The Reader also draws on the writings of men and women who have worked in different parts of the developing world, and who come from different disciplinary and geographical backgrounds.

As for the ordering of the six sections, I am sure that some colleagues would have done things differently. It did occur to me to begin with a theories and models section, and then to move down the hierarchy of spatial scales from global to national to household issues. In the end, though, my own experiences as a teacher counselled me against this ordering. In my judgement, it *is* logical to begin with the history of development thinking (Section One), and I think few would dispute this. We are all prisoners of ideas, but it helps to have a sense of the prison walls we are variously forced to live within. The Reader then moves on to rural development issues for the simple reason that most people in Africa and Asia still live in rural areas, whether or not they are primarily engaged in agricultural work. Vital debates continue to rage on the best

[7] The reader might find it helpful to consult some of the following books and papers on Third World cultures and politics: Alavi and Harriss (1989); Allen and Williams (1989); Appadurai (1990); Baddeley and Fraser (1989); Bayart (1993); Cammack (1991); Clapham (1985); Clifford (1988); Dornbusch and Edwards (1991); Dunkerley (1988); Flora and Torres-Rivas (1989); Gupta and Ferguson (1992); Kelly (1991); Kohli (1990); Manor (1991); Manuel (1994); Nkrumah (1962); Parry (1994); Rowe and Schelling (1991); Rushdie (1991); Said (1993); and Watts (1992). One important reason for including the study of Third World arts and literatures in a broad-based study of development is to challenge the all too common assumption that development studies is about the 'problems' of the developing world, to the extent that the developing world and development are both defined as problems (and only as problems). On non-governmental organizations, see Farrington *et al.* (1993), and Friedmann (1992).

means by which to increase poor peoples' entitlements to food in low-income countries. These debates have recourse to empirical research on agrarian structures, pricing policies and the creation of on-farm and off-farm employment opportunities (Section Two). Most people also have to survive in the rural environments of the developing world, making of their bodies (a gendered demography that explodes conventional notions of *household* survival strategies) and their environments what they can to subsist and to improve their life-chances or capabilities. Many people also migrate (and not just to urban areas) and some participate in local political movements that seek to empower poor people by making use of the 'weapons of the weak' (Scott, 1985): these themes and others are picked up in Section Three. Migration then leads us on, just as simply, to urban and industrial issues (including the role and significance of transnational corporations, and the lessons of the East Asian 'miracle': Section Four). This in turn folds back into more global issues relating to trade, aid, money and structural adjustment (Section Five). Closing the Reader with a Section (Six) on New Directions in Development Studies will surely raise few eyebrows, even if here, as throughout the Reader, the reader or reviewer might have chosen other themes and different extracts. Gender issues have deliberately not been stranded in a separate Reading: they are explored and taken seriously by several authors. Where gender issues do not figure prominently the reader might like to ask why not and to judge what the consequences of this 'absence' might be. Texts can be read for their absences as well as their presences.[8]

Let me close with one last note of caution. When Henry Bernstein published the second edition of his Reader on *Underdevelopment and Development* in 1976 (Bernstein, 1976) he remarked ruefully in the Preface that 'Several reviewers of this collection appeared bewildered by its eclecticism' (Bernstein, 1976, p. 9). I suspect that Bernstein was bemused and not a little hurt; after all, his Reader was at least dedicated to the goal (if not the party line) of a 'radical social science' (*ibid.*). This Reader cannot claim even this. What holds this Reader together is a catholic view of development studies as an academic discipline, and a deep regard for the spirit of constructive debate – and critical inquiry – that should mark the study of development and development issues/policies in the 1990s. I hope that students who approach the book in this spirit will find it of some value. The book is very clearly about *development*, but it is also about the *study* of development: about certain differences (and similarities)

[8] To say that texts can and should be read for absences as well as presences is hardly novel. Even so, it is important to stress that reading texts for what they do not say is a perilous activity, which needs to be undertaken with great care. It is very easy – too easy – to criticize an author for what he or she does not say, or to assume that an absence is always significant, or could easily have been attended to. Salman Rushdie is right to maintain that 'every story one chooses to tell is a kind of censorship, it prevents the telling of other tales' (Rushdie, 1983, p. 71).

in approaches to development issues that have real effects on the formulation of development policies. Particular Readings are put in context by an editorial introduction to each of the six sections. It goes without saying that the Reader is not presented as a substitute for further reading, but precisely as a spur to further study. Guides to Further Reading are provided at the end of the Editor's Introduction to each section.

References

Alavi, H. and Harriss, J. (eds) 1989: *Sociology of 'developing countries': South Asia*. Basingstoke: Macmillan.

Allen, C. and Williams, G. (eds) 1989: *Sociology of 'developing countries': Sub-Saharan Africa*. Basingstoke: Macmillan.

Appadurai, A. 1990: Disjuncture and difference in the global cultural economy. *Public Culture* 2 (2), 1–24.

Baddeley, O. and Fraser, V. 1989: *Drawing the line: Art and cultural identity in contemporary Latin America*. London: Verso.

Bardhan, P. (ed.) 1989: *Conversations between economists and anthropologists*. Delhi: OUP.

Bayart, J.-F. 1993: *The state in Africa: The politics of the belly*. Harlow: Longman.

Bernstein, H. (ed.) 1976: *Underdevelopment and development: The Third World today*. Harmondsworth: Penguin.

Bernstein, H., Crow, B. and Johnson, H. (eds) 1992: *Rural livelihoods: Crises and responses*. Oxford: OUP/Open University.

Buiter, W. and Srinivasan, T. 1987: Rewarding the profligate and punishing the prudent and poor: some recent proposals for debt relief. *World Development* 15, 411–17.

Cammack, P. 1991: Brazil: the long march to the new republic. *New Left Review* 190, 21–58.

Chenery, H. and Srinivasan, T. (eds) 1988: *A handbook of development economics* (2 volumes). Amsterdam: North Holland.

Clapham, C. 1985: *Third World politics: An introduction*. London: Croom Helm.

Clifford, J. 1988: *The predicament of culture: Twentieth century ethnography, literature and art*. Cambridge, MA: Harvard UP.

Comaroff, J. and Comaroff, J. 1992: *Ethnography and the historical imagination*. Boulder, CO: Westview.

Dornbusch, R. and Edwards, S. (eds) 1991: *The macroeconomics of populism in Latin America*. Chicago: Chicago UP/NBER.

Dunkerley, J. 1988: *Power in the isthmus: A political history of modern Central America*. London: Verso.

Farrington, J. and Bebbington, A. with Wellard, K. and Lewis, D. 1993: *Reluctant partners? Non-governmental organizations, the state and sustainable agricultural development*. London: Routledge.

Flora, J. and Torres-Rivas, E. (eds) 1989: *Sociology of 'developing countries': Central America*. Basingstoke: Macmillan.

Friedmann, J. 1992: *Empowerment: The politics of alternative development*. Oxford: Blackwell.

Gupta, A. and Ferguson, J. 1992: Beyond culture: space, identity and the politics of difference. *Cultural Anthropology* 7 (1), 6–23.
Hewitt, T., Johnson, H. and Wield, D. (eds) 1992: *Industrialisation and development*. Oxford: OUP/Open University.
Hill, P. 1986: *Development economics on trial: An anthropological case for a prosecution*. Cambridge: CUP.
Kelly, J. 1991: *A politics of virtue: Hinduism, sexuality and countercolonial discourse in Fiji*. Chicago: Chicago UP.
Keynes, J. M. 1973: *The general theory of employment, interest and money*. London: Macmillan.
Kohli, A. 1990: *Democracy and discontent: India's growing crisis of governability*. Cambridge: CUP.
Manor, J. (ed.) 1991: *Rethinking Third World politics*. Harlow: Longman.
Manuel, P. 1994: *Cassette culture: Popular music and technology in North India*. Chicago: Chicago UP.
Meier, G. (ed.) 1989: *Leading issues in economic development*. Oxford: OUP.
Nkrumah, K. 1962: *Towards colonial freedom*. London: Heinemann.
Parry, J. 1994: *Death in Banaras*. Cambridge: CUP.
Rowe, W. and Schelling, V. 1991: *Memory and modernity: Popular culture in Latin America*. London: Verso.
Rushdie, S. 1983: *Shame*. London: Jonathan Cape.
Rushdie, S. 1991: *Imaginary homelands*. London: Granta.
Said, E. 1993: *Culture and imperialism*. London: Chatto and Windus.
Sayer, D. 1987: *The violence of abstraction: The analytic foundations of historical materialism*. Oxford: Blackwell.
Scott, J. 1985: *Weapons of the weak: Everyday forms of peasant resistance*. New Haven: Yale UP.
Sen, A. 1978: Rational fools: a critique of the behavioural foundations of economic theory. In Harris, H. (ed.), *Scientific models and men*. London: OUP.
Todaro, M. 1994: *Economic development* (5th edition). Harlow: Longman.
Toye, J. 1987: *Dilemmas of development: Reflections on the counter-revolution in development theory and policy*. Oxford: Blackwell.
Watts, M. 1992: The shock of modernity: petroleum, protest and fast capitalism in an industrializing society. Chapter 2 of Pred, A. and Watts, M., *Reworking modernity: Capitalisms and symbolic discontent*. New Brunswick: Rutgers UP.
Wilber, C. and Jameson, K. 1992: *The political economy of development and underdevelopment* (5th edition). New York: McGraw-Hill.
Williamson, J. 1993: Democracy and the 'Washington consensus'. *World Development* 21, 1329–36.
Wuyts, M., Mackintosh, M. and Hewitt, T. 1991: *Development policy and public action*. Oxford: OUP/Open University.

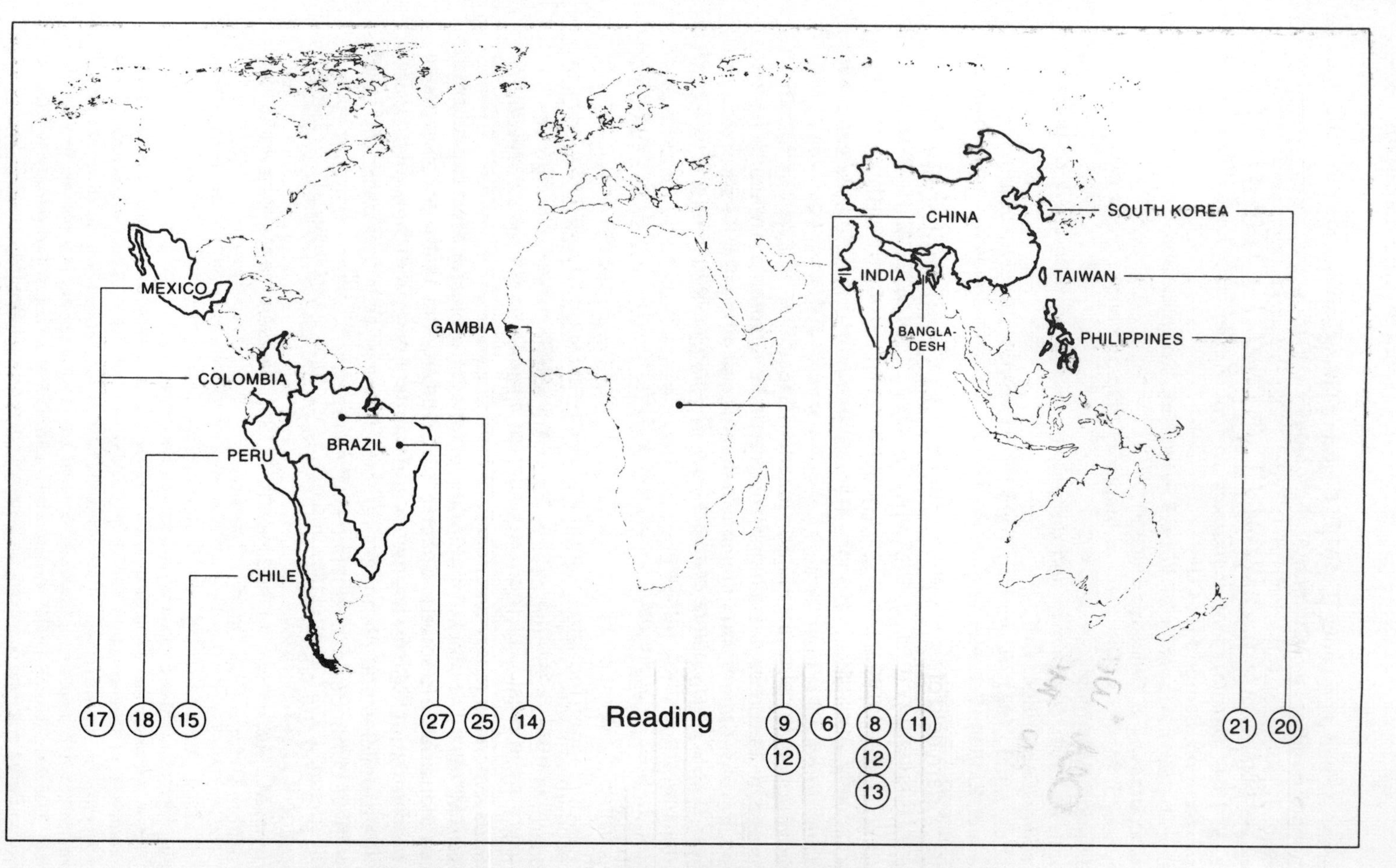

Specific regions or countries referred to in selected Readings.

SECTION ONE

THINKING ABOUT DEVELOPMENT

Editor's Introduction

The modern field of development studies is barely fifty years old. In the first part of this century most discussions of the world's poorer regions and countries were structured by an enduring mixture of environmental determinism and social Darwinism. Poor countries were said to be poor because their peoples were the prisoners of tropical climates that discouraged effort and innovation, or because they lacked the mental faculties that could be taken for granted by Caucasians (Haller, 1975; Pratt, 1992). Colonialism was a necessary consequence of this division between rich and poor, temperate and tropical, able and less able. It was the white *man's* burden to look after – and slowly 'improve' – the native populations of Africa, Asia, Latin America and the Pacific. This was the West's duty to the rest of the world, or to what Said has called the West's 'Others' (Said, 1979).[1] But development as such, in the sense of planned industrialization and sustained economic growth, was not yet considered a serious option by mainstream academics and policy-makers in the West.[2] The modern notion of development is very much a product of the 1940s and 1950s, when ostensibly European–American ideas of progress were extended to what soon would be Europe's erstwhile colonies, and when a battle loomed between the USA (a largely non-territorial imperial power) and the USSR for the hearts and minds of the rest of the world. Development policies and planning first took shape against this backdrop, and they have continued to be shaped and reshaped with reference to East–West tensions and the changing organization of the global political economy (Pletsch, 1981).

[1] The pre-history of 'development studies' is not properly covered in this volume. Useful guides would include Arndt (1981; 1987) and Dorfman (1991). See also Hall (1992), Mitchell (1991) and Young (1990) for further discussions of Orientalism and Othering.

[2] This is not to say that intellectuals in the colonies accepted Western accounts of the necessary backwardness of their countries and peoples: most did not. Latin American accounts of the development or underdevelopment are discussed later in this Introduction. In India, the period from 1880 to 1905 witnessed a significant rise and growth of economic nationalism: see Chandra (1966). Dadabhai Naoroji (1901) was especially influential in linking Indian poverty to what he called 'unBritish' rule, and demanding an industrial future for a modern India.

When development studies emerged in the 1940s a majority of its practitioners (including scholars, development planners in the Third World, and aid workers) subscribed to a conception of development-as-modernization that is well to the fore in **Reading 1** by **Bert F. Hoselitz**.[3] Modernization theorists held, first, that developed societies were distinguished by their economic, social, cultural and political *modernity*, which contrasted sharply with the *traditional* values of underdeveloped, developing, backward, Third World or latecomer societies. Development was thus about the modernization of traditional societies, and the first major journal of development studies was fittingly called *Economic Development and Cultural Change*. (The journal was first published in 1952 and a version of the first paper in the first issue is reprinted here as **Reading 1**.)[4] Economic development would clearly revolve around industrialization and the transfer of an under-employed rural labour force to more productive occupations in the urban–industrial sector. This much was agreed by Ragnar Nurske (1953) and W. Arthur Lewis (1954, 1955), to name just two noted exponents of development as structural transformation. But development could not be reduced to industrialization and economic growth alone: modern workers had to be created to work in modern industries and this would entail massive investment in education systems and the introduction of modern forms of time-management (McClelland, 1961). The size of a typical Third World family would also have to be reduced, or made more like 'modern/Western' families. Not surprisingly, perhaps, clocks and condoms were among the most tangible forms of official development assistance in the 1950s.

Other propositions were then developed with reference to this leading

[3] Again, it is important to note that the existence of a mainstream in development studies at any one time, or in social science more generally, does not mean that several important scholars and activists were not even then challenging the conventional wisdom. The general dominance of modernizing ideas in the 1950s certainly was challenged early on by Marxist-inspired scholarship: most notably by Paul Baran, Paul Sweezy and others in the *Monthly Review* school of American Marxism. The 1950s and early 1960s also saw the publication of Mannoni's trenchant critique of the pyschology of colonization (Mannoni, 1950), and Frantz Fanon's classic account of *The Wretched of the Earth* (Fanon, 1992; first published in 1961).

[4] The second article to be published in the first issue of *Economic Development and Cultural Change* was an account of 'The appeal of communism to the peoples of underdeveloped areas' by Morris Watnick (Watnick, 1952). The publication of this article makes the point very clearly that First World discussions of the Third World were then (and for many years after) bound up with First World readings of the Second World and its intentions. Watnick begins his discussion as follows: 'If time is a power dimension in any political strategy, the odds facing the West in the underdeveloped areas of the world today are heavily weighted against it. The effort to capture the imagination and loyalties of the populations of these areas did not begin with the West in President Truman's plea for a "bold new program" of technical aid to backward areas. It began more than a generation earlier when the Communist International at its second world congress in 1920, flung out the challenge of revolution to the peoples of the colonial and dependent countries, and proceeded to chart a course of action calculated to hasten the end of Western overlordship' (Watnick, 1952, p. 22).

proposition. In terms of the economics of growth it was widely assumed in the 1950s and 1960s that market failure was the norm in developing countries. The state would have to mobilize domestic and foreign savings to create an investment pool from which it could finance a programme of directed industrial development. Direction usually came in the form of state development plans and by means of the controls that most development states exercise over scarce foreign exchange reserves. The state would also protect infant industries by means of tariff and quota policies (hence, import-substitution industrialization), and would encourage investment in heavy industries like steel and engineering – the commanding heights of the economy according to India's Second and Third Five Year Plans (1956–66) – by running such industries themselves. The state alone was credited with the ability to think and act in the long-term interest of all its citizens. Markets would play a supporting role in most capitalist developing countries, and the state would embody all that was most modern and disinterested in the changing cultural and political make-up of a developing country.[5]

It was also widely accepted that developing countries could and should learn from the development experiences of the already developed, or pioneer countries. (They could also learn from the experiences of the late-industrializing countries of Eastern Europe: Rosenstein-Rodan, 1943.) Walt Whitman Rostow has perhaps been most closely associated with the view that there is a common pathway to development which has to be trod by all countries intent on becoming modern (Rostow, 1960). Latecomer societies have to 'take-off' into development – a suitably space-conquering metaphor for the late 1950s/1960s when the USA and the USSR were competing with each other above the Earth as well as on it – by mobilizing a critical mass of savings and investing it carefully in private and public programmes of industrialization. Happily, the fact that latecomer societies could draw on the experiences and resources of the pioneer countries meant that the transition from a state of tradition to a state of modernity could be accomplished quite rapidly. The only danger was that some countries would choose the deviant route of socialism–communism.[6]

[5] The counter-revolution in development studies has sometimes seemed to suggest that an endorsement of the public sector was a defining feature of the 'development economics' that dominated in the 1950s and 1960s, and that most development theorists and planners at this time were strongly biased against the private sector. I doubt this was the case. Rostow's work is now more often cited than read, but his discussion of the stages of economic growth (Rostow, 1960) argues that development will best be served by a dynamic (and free) private sector that is suitably supported in the short run by an efficient public sector. This aspect of Rostow's work resonates more closely with contemporary concerns for developmental institutions and public–private sector partnerships than does some of the more fundamentalist work emerging from the counter-revolution.

[6] The subtitle of Rostow's account of *The Stages of Economic Growth* is, of course, *A Non-Communist Manifesto* (Rostow, 1960: see also footnote 5). Communist countries were considered by Rostow to be deviant insofar as they had departed from the *normal* road of capitalist development. The pairing of normal and deviant is instructive here.

Thirty or forty years later on – and in the wake of continuing famines in Africa, a lost decade of development in Latin America, and the tragedy in Rwanda – it can be difficult to credit the spirit of optimism that marked the Golden Years of the 1950s and 1960s. But these were decades when policy-makers in the West (and the East: Murphey, 1967) believed that 'Man' could overcome 'Nature', and the boom–bust cycles of capitalism could be tamed by means of Keynesian demand-management policies. The corporatist state was then a matter of pride and not a source of dismay and distrust. President Kennedy is well remembered for his bold claim that the United States would put a man on the Moon by the end of the 1960s, ahead of the USSR. It is less often recalled that the Kennedy Administration also helped usher in the United Nations Development Decade (now called the First Development Decade since we are living through the fourth!); a decade when, or so the story ran, development would cease to be a problem as the entire capitalist world made confident strides towards a state of development/modernity.

The extreme optimism of the 1950s and early 1960s gave way in the late 1960s and 1970s to more measured accounts of economic growth and development, even within the 'modernization' camp. The World Bank began to talk about redistribution with growth and moved anti-poverty programmes to the centre of its discourses on development (Chenery and associates, 1974). Growth and development ceased to be used as synonyms (even if the former was still widely considered a condition of the latter), and the urban–industrial biases of 1950s-style modernization theories/policies began to be challenged. By the mid-1970s it made little sense to refer to modernization theories or policies in the way that supporters and critics had referred to them previously. A new mainstream was emerging in developing studies that was less committed to development as Americanization, and which had less faith in developmental panaceas or rapid development progress. Even so, this mainstream was still committed to forms of development built around industrialization (albeit in significantly different forms: see Section Four), and was and is committed to a strong role being played by the state in managing local processes of development. In these respects, a modernizing mainstream has shared much in common with the first major tradition of dissent which faced it in the 1950s and 1960s; a tradition that we can broadly refer to as the *dependencia* tradition.

Reading 2 by **Andre Gunder Frank** is a classic statement of what might be called the hard-line position within the *dependencia* camp; or the development of underdevelopment position. In this reading we see Frank taking issue with the modernization accounts that dominated the development industry in the mid-1960s. Frank argues that it is absurd to look to Western capitalism to sponsor the industrial development of the periphery of the world-system. In his view, development in the core, and underdevelopment in the periphery, are two sides of the same global capitalist coin. Metropolitan capitalism depends on the exploitation and active underdevelopment of an already capitalist periphery. According to Frank, old style colonialism has simply given way to a neo-colonialism dominated by the IMF and the multinationals, and enforced by transfer pricing and unequal exchange in world trade. The choice for the Third World lies between the continuing barbarism of capitalism or the great promise of socialism. (It is worth recalling that Frank's essay was written shortly after the Cuban Revolution and at a time when China and the USSR were still considered by many in the West to be successful sponsors of economic growth.)

Several writers in the broad *dependencia* tradition have advanced the neo-Marxist views propounded by Frank: Samir Amin (1976), Paul Baran (1957), Arghiri Emmanuel (1972), Walter Rodney (1972) and Immanuel Wallerstein (1974, 1979) all come to mind. But other authors have used the term dependency in a more nuanced and arguably more empirical fashion (Palma, 1978). Fernando Henrique Cardoso has often been lauded for updating the work of Prebisch and the Cepalinos,[7] and for producing an account of dependent development that emphasizes the injustice and irrationality of the many asymmetrical relationships of production, trade and power which make the Third World dependent on the First World long after the Age of Empire (Cardoso, 1977; Cardoso and Faletto, 1979). Such accounts of dependent development are at odds with the early-Frankian position that peripheral industrialization is impossible within the capitalist world system, and the view that socialism (and autarkic socialism at that) can save the Third World from continuing underdevelopment. In 1977 Cardoso famously declared that history, in the form of the newly industrialized countries (NICs), was preparing a trap for the pessimists.

The rise of the NICs and the recent collapse of many state socialist countries has done a lot of damage to Marxist social science, and to Marxissant development theories specifically. This does not mean that such theories and models are irrelevant, however, or that some of their propositions and assumptions do not command wide support. Marxist-inspired accounts of colonialism and the formation of rich and poor countries continue to be widely regarded (Mintz, 1985; Wolf, 1982), and many popular views about the activities of Western multinationals in the Third World are subtly conditioned by neo-Marxist accounts of exploitation and unequal exchange (see also **Reading 19**). It is also important to note that Marxism is not a unitary discourse, any more than is modernization theory or the counter-revolution in development theory and policy.

[7] CEPAL is the Spanish-language acronym for the UN's Economic Commission for Latin America [and the Caribbean]; Raul Prebisch is the Argentinian economist who helped to forge a CEPAL line on the problems facing dependent peripheral countries as a result of the quasi-monopolistic powers exercised by core countries in global labour and trading markets. For a full discussion, see Kay (1989).

In the late 1970s and early 1980s the main challenge to the neo-Marxist views of Amin, Baran, Frank and Wallerstein came from scholars who claimed to draw on a more authentic, or 'classical', tradition of Marxism. Bill Warren made himself unpopular with many of his comrades on the Left by arguing that a post-imperial capitalism would promote development in the Third World just as it had done in the First World (Warren, 1980; see also Brewer, 1980). Warren suggested that capitalism is always and everywhere dynamic and developmental (in terms of economic growth), even if it is irrational and unfair in the manner in which it distributes the fruits of this growth. (Warren looked to a more rational 'world socialism' to take the place of a truly global capitalism.)

Other scholars joining this seemingly esoteric debate on the nature of capitalist development were armed with certain influential French texts that had taken shape with reference to 'structural Marxism' (Meillassoux, 1972; Rey, 1971; Terray, 1972). This gave rise to a body of work on the articulation of capitalist and pre-capitalist modes of production, before a more general attack was launched on the economism and reductionism of all Marxist development sociology by **David Booth** (see **Reading 3**). The modes of production literature arguably saw Marxist development theory at its most rigorous, notwithstanding certain flaws noted by Booth. It attempted to account for the fracturing of the ex-Third World into NICs and low-income countries in terms of the ability of an autocentric capitalism to take hold in the former regions but not in the latter. It suggested that in most parts of the Third World small pockets of urban–industrial capitalism were being supported, or reproduced, by regular infusions of poor and exploited labour from local pre-capitalist peripheries (as in white–urban and black–rural South Africa: Wolpe, 1980). There was an articulation, or interaction, of two modes of production that was functional for the capitalist mode of production, and which made local capitalists loath to promote true capitalism in their rural hinterlands. By such means did Marxism attempt to come to terms with the different experiences of developing countries and regions; a developing world that was also being remade by a new international division of labour that saw First World companies moving to some Third World countries to exploit cheap labour and new markets (Frobel, Heinrichs and Kreye, 1980).[8]

Whatever the academic merits of Marxist development theories, they have largely failed to punch their weight because of the lack of plausible policy advice to which they have given rise. The same charge cannot be so easily levelled against a counter-revolution in development theory and policy that took shape in the 1960s and 1970s and which came into its own

[8] References to a new international division of labour and autocentric capitalism are also suggestive of a later 'post-Marxist' literature on development that draws, variously, on French regulation theories and Sklar's work on post-imperialism. See: Becker and Sklar (1987); Booth (1994); Corbridge (1990); Lipietz (1987); Vandergeest and Buttel (1988).

in the 1980s. **Deepak Lal (Reading 4)** has been closely associated with this counter-revolution, along with writers and policy-makers like Bela Balassa (1982), Peter Bauer (1972), Jagdish Bhagwati (1993), Gottfried Haberler (1987), Harry Johnson (1971), Anne Krueger (1974) and Ian Little (1982). In subtly different terms these writers have all attacked what they see as the damaging consensus on development economics that took shape soon after World War II: a consensus which, or so Lal suggests, misreads the true legacy of Keynes, fails to see that most states are likely to be predatory, rent-seeking economic actors, and fails to recognize that state failures are at least as likely as market failures in the developing world, and far more damaging. Drawing variously on the libertarian creeds of Hayek and Friedman, and on monetarist, supply-side and neo-classical economics, the counter-revolution has hailed the price-responsive behaviour of farmers (when they are free to respond: Schultz, 1964), and the dynamism of entrepreneurs and small businesses. The counter-revolution has also been associated with calls for 'trade not aid' and 'privatization not nationalization'. To some extent, too, it has been associated with a shift in the policy direction of development institutions like the World Bank, such that a general principle of 'the user pays' is becoming widely accepted. For their part, critics maintain that the counter-revolution is yet one more example of a metanarrative (or jumbo theory) of development that attempts to squeeze a diverse social reality into the straitjacket of a singular way of seeing or way of thinking (see Schuurman, 1993). The counter-revolution in development theory and policy has also been attacked for its simplistic accounts of the nature of 'real' markets in many developing countries and for its one-dimensional accounts of what motivates apparently isolated economic actors (Killick, 1986; Stewart, 1985). According to one prominent critic, 'The magnificent vision of the 1980s is of a world developing its resources and capacities in response *only* to the ups and downs of relative prices and the self-imposed stasis of limited government' (Toye, 1987, p. viii; emphasis added).

The shift away from so-called metanarratives has also been apparent in certain New Directions in Development Studies that are signalled in Section Six of this Reader. In part this echoes the break-up of *the* 'Third World', but even more so it reflects a disenchantment on the part of many development activists with both the 'successes' and failures of capitalist (and socialist) development strategies. Section Six is mainly concerned with the environmental costs of development, and with the continuing threat to (gendered) human bodies that often lies behind a screen of development and democratization. **Readings 25–27** make a case for sustainable forms of development that increase the life-chances and well-being of men, women and children, and which respect their rights to individual and group differences. For some authors this means putting a premium upon development actions rather than development theories, so that Michael Edwards writes pointedly of the 'irrelevance of development studies' (Edwards, 1989, 1994). For others (and indeed for Edwards) it means making a space for development that relies much more heavily on the skills and voices of the people that are being 'developed', often with the support of non-governmental organizations (Edwards and Hulme, 1992; Farrington and Bebbington, 1993).

Some others go much further still and challenge the very idea of development. In a sense this takes us back to where we began this Introduction: with long-standing ideas of a world divided into the West and the Rest (Hall, 1992), where the West stands ready to colonize, *or develop*, its Others. Some of the most challenging recent work on development has drawn on Said's accounts of Orientalism, and the work of Foucault and some post-modernists, to suggest that the modern concept of 'development' is itself beset by all sorts of colonizing ambitions, whether or not these are explicitly recognized, and regardless of the formal ending of empires and territorial imperialism. Colonialism, it suggests, is about the production of 'Other' societies and peoples as part of a discourse which defines the West in opposition to the rest of the world: as the singular source of reason, progress and enlightenment. Thus defined, colonialism can easily survive the physical fact of decolonization. Development might then be read as an attempt by the West to produce Other societies in its own image: to erase/(colonize) the specific identities of these Others, where once they would have been preserved as reminders of their incapacity to become modern or 'Western'.

Arturo Escobar (Reading 5) is a leading member of the loosely-knit group of 'anti-development' scholars and activists to whom I am referring. Escobar challenges the humanist assumptions that are conventionally built into development studies and the project of development. Anti-developmentists turn the discourse of Orientalism back on itself. Some go so far as to suggest that the West is a source of 'disease' which threatens to infect the more authentic and sustainable livelihood systems of 'less-developed' countries (Shucking and Anderson, 1991). More commonly, post-war efforts at development are condemned for imposing on non-Western populations a peculiarly Western conception of industrial progress and economic growth that has proved difficult to promote in practice, and which is less than desirable as an end-state. According to Escobar, development planning is a form of technical knowledge that has lent legitimacy to this notion of (anti)-development. His Reading attempts to show how Western conceptions of development and development planning seek to normalize non-Western populations, or make the latter more like the former by command. This process of development/anti-development obviously works to the advantage of Western companies and enjoys the support of elite groups and the state in developing countries. Indeed, for Escobar and other anti-developmentists, the state in the Third World is usually an instrument of neo-colonialism which arrogates to itself the ability to speak and write *with authority* about the so-called problems of developing societies. Knowledge is power, Escobar maintains, and the state's knowledge is expressed in the form of development plans. By contrast, the voices of the 'victims of development' (Seabrook, 1993), the peoples and communities whose lives are being re-made in its name, are denied a status equivalent to the 'scientific' edicts of the modern state. Anti-developmentists join with some populists (who promote the interests of the rural poor above all else) in demanding either an end to 'development' as it is presently conceived, and/or a much greater attentiveness to the knowledge, skills and needs of the peoples who are to be 'developed' (or whose lives are to be improved/empowered). Escobar's Reading traces the development of this and other modes of development thinking, and notes the role played by feminist critics of development, in particular, in providing a 'knowledge in opposition' to the conventional wisdom about development. (For more on gendered hierarchies in development thought, see the important recent book by Naila Kabeer: Kabeer, 1994.)

Of course, there have always been dissenting groups who have challenged Western conceptions of development and scientific progress. Said points out that Orientalism itself is by no means a singular discourse of domination, but is rather a multiple set of discourses through which the West seeks to understand itself by reference to certain qualities that are projected onto its Others. Even in the nineteenth century a minor key of Orientalism held that the West had fallen from Grace and was being made inauthentic and unwholesome by virtue of economic growth, materialism, progress and science. According to this account, which is well represented in the paintings of Gauguin, the 'exotic' and more 'natural' sexual and other mores of Other societies were to be prized and preserved and not attacked and dismantled. In the twentieth century, Gandhi (1966) and Schumacher (1973), and more recently some proponents of indigenous ecological/agricultural development (Brokensha, Warren and Werner, 1980), have breathed fresh life into this perspective. When anti-development writers develop similar themes they are tapping into a long-established body of thought. Not surprisingly, perhaps, they join with populists and neo-populists in acclaiming the virtues of rural life and indigenous forms of resource management. They are also accused by their critics of rehearsing and expanding some of the common 'failings' of populism, namely: (1) of romanticizing the lifestyles of indigenous peoples and inventing an account of their histories which does not always survive rigorous scrutiny; (2) of failing to acknowledge certain very real improvements in people's life chances and capabilities in the 'age of development' (as measured by infant mortality rates, life expectancy, educational achievement, access to clean water, etc.); and (3) perversely, of attributing to local people a mistrust of 'development' that they might not share in full: many such people recognize that development is about costs and benefits, and they are more interested in the balance of these items than in the possibility of a painless development or non-development. (For a stunning critique of some aspects of anti-developmentism and the 'populist orthodoxy', see Nanda, 1991.)

Anti-development ideas are at one end of a spectrum of ideas that challenge conventional perspectives on the purposes and practicalities of development. They are proving attractive to some on the Left who have lost faith in 'Socialism' as a panacea for the ills of poverty. Aspects of anti-developmentalism are also finding favour with social scientists attentive to the problems of writing about Other cultures (the ethnographic turn: Clifford and Marcus, 1986), and with activists anxious to listen to the voices of local subaltern populations (and to women as much as men). Far more widely accepted, however, is the proposition, floated long ago by Dudley Seers (Seers, 1969), that development and economic growth are not one and the same thing. Compared with just ten years ago, there is less concern now with the merits and nature of capitalist growth strategies *per se*, and far more concern for the nature of local relationships between development and the environment, citizenship, gender issues, justice, institutions and democracy. History may or may not be dead (Fukuyama, 1991), but development theory continues to move on.

It also bears saying that the five Readings in this Section are exemplars of particular ways of thinking about development. They give us a sense of what is at stake in talking about development and in promoting different forms of development and different development policies or projects. They also make us think about certain alternatives, like socialism or 'non-development', and they remind us that development studies has a history which survives into the present: even in the mid-1990s it is not difficult to find individuals who closely identify with only one of these traditions of development thought. But most people who think about development, or who are engaged in development work, do not subscribe to all of the views expressed in just one of these paradigms, or super-models, and probably never have done. Most of us think about development in a manner which hybridizes two or more of these traditions. Often the hybrid emerges because of the debates that rage between protagonists of different traditions. The same holds true for development policy. It is true that pro-market policies for development have become more important over the past twenty years, and became especially influential in the World Bank in the mid-1980s when Anne Krueger was Vice-President of Economics and Research. But it would not be true to say that we live in an age of uncontested market idolatry.[9] The

[9] At least, not in the academy, and probably not any longer at the level of official US or World Bank policies.

mainstream in development studies and policy today is probably best characterized by its commitment to making effective states and effective markets work in tandem by means of efficient local institutions (see **Reading 24**). The new mainstream is also more attentive to gender and environmental issues than it was ten years ago. It can fairly be argued that development studies does learn from its past, from its mistakes and from its successes.

Guide to further reading

On the history of development thinking

Adas, M. 1989: *Machines as the measures of men: Science, technology, and ideologies of western dominance*. Ithaca: Cornell UP.

Arndt, H. 1981: Economic development; a semantic history. *Economic Development and Cultural Change* 29 (3).

Arndt, H. 1987: *Economic development: The history of an idea*. Chicago: University of Chicago Press.

Asad, T. (ed.) 1973: *Anthropology and the colonial encounter*. London: Ithaca Press.

Booth, D. (ed.) 1994: *Rethinking social development: Theory, research and practice*. Harlow: Longman.

Boserup, E. 1970: *Women's role in economic development*. New York: St Martin's Press.

Brookfield, H. 1975: *Interdependent development*. London: Methuen.

Dorfman, R. 1991: Economic development from the beginning to Rostow. *Journal of Economic Literature* 29, 573–91.

Edwards, E. (ed.) 1992: *Anthropology and photography, 1860–1920*. London: Yale UP/The Royal Anthropological Institute.

Escobar, A. 1992: Reflections on development. *Futures* 24, 411–36.

Esteva, G. 1992: Development. In Sachs, W. (ed.), *The development dictionary*. London: Zed Press, 6–25.

Hall, S. 1992: The West and the Rest: discourse and power. In Hall, S. and Gieben, B. (eds), *Formations of modernity*. Cambridge: Polity/Open University, 275–320.

Haller, J. 1975: *Outcasts from evolution: Scientific attitudes of racial inferiority, 1859–1900*. New York: McGraw-Hill.

Hettne, B. 1990: *Development theory and the three worlds*. London: Routledge.

Hunt, D. 1987: *Economic theories of development*. Hemel Hempstead: Harvester Wheatsheaf.

Kabeer, N. 1994: *Reversed realities: Gender hierarchies in development thought*. London: Verso.

Mitchell, T. 1991: *Colonising Egypt*. Berkeley: University of California Press.

Moore, H. 1988: *Feminism and anthropology*. Cambridge: Polity.

Pletsch, C. 1981: The three worlds, or the division of social scientific labour, circa 1950–1975. *Comparative Studies in Society and History* 23, 565–90.

Pratt, M. L. 1992: *Imperial eyes: Travel writing and transculturation*. London: Routledge.

Said, E. 1979: *Orientalism*. New York: Pantheon.
Sen, A. K. 1983: Development: which way now? *Economic Journal* 93, 745–62.
Stern, N. 1989: The economics of development: a survey. *Economic Journal* 99, 573–91.
Young, R. 1990: *White Mythologies: Writing history and the West*. London: Routledge.

On modernization theories (including critiques)

Bell, C. 1989: Development economics. In Eatwell, J. Milgate, M. and Newman, P. (eds), *The new Palgrave: Economic development*. London: Macmillan, 1–17.
Bernstein, H. 1979: Sociology of underdevelopment versus sociology of development. In Lehmann, D. (ed.), *Development theory*. London: Frank Cass, 77–106.
Chandra, B. 1966: *The rise and growth of economic nationalism in India, 1880–1905*. Delhi: OUP.
Chenery, H. Ahluwalia, M. Bell, C. Duloy, J. and Jolly, R. 1974: *Redistribution with growth: An approach to policy*. Oxford: OUP/World Bank.
Friedmann, J. 1966: *Regional development policy: A case study of Venezuela*. Cambridge, MA: MIT Press.
Hirschman, A. 1958: *Strategy of economic development*. New Haven: Yale UP.
Lewis, W. A. 1954: Economic development with unlimited supplies of labour. *Manchester School* 22, 139–91.
Lewis, W. A. 1955: *The theory of economic growth*. London: George Allen and Unwin.
Mannoni, O. 1950: *Psychologie de la colonisation*. Paris: Editions du Seuil.
McClelland, D. 1961: *The achieving society*. New York: Random House.
Murphey, R. 1967: Man and nature in China. *Modern Asian Studies* 1, 313–33.
Naoroji, D. 1901: *Poverty and un-British rule in India*. London: Macmillan.
Nurske, R. 1953: *Problems of capital formation in underdeveloped countries*. New York: OUP.
Rosenstein-Radan, P. 1943: Problems of industrialization in Eastern and South-Eastern Europe. *Economic Journal* 53, 202–11.
Rostow, W. 1960: *The stages of economic growth: A non-communist manifesto*. London: CUP.
Seers, D. 1969: The meaning of development. *International Development Review* 11, 2–6.
Slater, D. 1973: Geography and underdevelopment. *Antipode* 5, 21–53.
Taylor, J. 1979: *From modernisation to modes of production*. London: Macmillan.
Watnick, M. 1952: The appeal of communism to the peoples of underdeveloped areas. *Economic Development and Cultural Change* 1, 22–36.

On dependency theories (including critiques)

Amin, S. 1976: *Imperialism and unequal development*. Hassocks: Harvester.
Baran, P. 1957: *The political economy of growth*. New York: Monthly Review.
Brenner, R. 1977: The origins of capitalist development: a critique of neo-Smithian Marxism. *New Left Review* 104, 25–92.

Cardoso, F. H. 1977: The originality of the copy: CEPAL and the idea of development. *CEPAL Review* 2, 17–24.

Cardoso, F. H. and Faletto, E. 1979: *Dependency and development in Latin America*. Berkeley: University of California Press.

Corbridge, S. 1986: *Capitalist world development: A critique of radical development geography*. London: Macmillan.

Emmanuel, A. 1972: *Unequal exchange*. New York: Monthly Review.

Frank, A. G. 1969: *Latin America: Underdevelopment or revolution*. New York: Monthly Review.

Griffin, K. and Gurley, J. 1985: Radical analysis of imperialism, the Third World, and the transition to socialism: a survey article. *Journal of Economic Literature* 23, 1089–143.

Hout, W. 1993: *Capitalism and the Third World: Development, dependence and the world system*. Aldershot: Edward Elgar.

Kay, C. 1989: *Latin American theories of development and underdevelopment*. London: Routledge.

Larrain, J. 1989: *Theories of development*. Cambridge: Polity.

Palma, G. 1978: Dependency: a formal theory of underdevelopment or a methodology for the analysis of concrete situations of underdevelopment? *World Development* 6, 881–924.

Rodney, W. 1972: *How Europe underdeveloped Africa*. London: Bogle l'Ouverture.

Ruccio, D. and Simon, L. 1988: Perspectives on underdevelopment: Frank, the modes of production school, and Amin. In Wilber, C. and Jameson, K. (eds), *The political economy of development and underdevelopment* (5th edn). New York: McGraw-Hill, 119–50.

Wallerstein, I. 1974: *The modern world system: Capitalist agriculture and the origins of the European world-economy in the sixteenth century*. New York: Academic Press.

Wallerstein, I. 1979: *The capitalist world economy*. Cambridge: CUP.

On modes of production [articulation] theories, post-Marxism and the radical impasse (including critiques)

Becker, D. and Sklar, R. (eds) 1987: *Postimperialism: International capitalism and development in the late twentieth century*. Boulder: Lynne Reinner.

Booth, D. 1985: Marxism and development sociology: interpreting the impasse. *World Development* 13, 761–87.

Brewer, A. 1980: *Marxist theories of imperialism*. London: Routledge and Kegan Paul.

Corbridge, S. 1990: Post-Marxism and development studies: beyond the impasse. *World Development* 18, 623–39.

Frobel, F., Heinrichs, J. and Kreye, O. 1980: *The new international division of labour*. Cambridge: CUP.

Lipietz, A. 1987: *Mirages and miracles: The crises of global Fordism*. London: Verso.

Meillassoux, C. 1972: From reproduction to production. *Economy and Society* 1, 93–105.

Mintz, S. 1985: *Sweetness and power: The place of sugar in modern history*. New York: Viking.
Mouzelis, N. 1988: Sociology of development: reflections on the present crisis. *Sociology* 22, 23–44.
Rey, P.-P. 1971: *Colonialisme, néo-colonialisme et transition au capitalisme*. Paris: Maspero.
Sklair, L. 1991: *Sociology of the global system*. Hemel Hempstead: Harvester Wheatsheaf.
Terray, E. 1972: *Marxism and 'primitive' societies*. London: Monthly Review.
Vandergeest, P. and Buttel, F. 1988: Marx, Weber and development sociology. *World Development* 16, 677–97.
Warren, B. 1980: *Imperialism: Pioneer of capitalism*. London: Verso.
Wolf, E. 1982: *Europe and the people without history*. Berkeley: University of California Press.
Wolpe, H. (ed.) 1980: *The articulation of modes of production*. London: Routledge and Kegan Paul.

On market-led development and the counter-revolution in development theory and policy (including critiques)

Balassa, B. and associates 1982: *Development strategies in semi-industrial economies*. Baltimore: Johns Hopkins UP.
Bates, R. (ed.) 1988: *Toward a political economy of development: A rational choice perspective*. Berkeley: University of California Press.
Bauer, P. 1972: *Dissent on development*. London: Weidenfeld and Nicolson.
Bauer, P. 1991: *The development frontier*. Hemel Hempstead: Harvester Wheatsheaf.
Beenstock, M. 1984: *The world economy in transition*. London: George Allen and Unwin.
Bhagwati, J. 1993: *India in transition: Freeing the economy*. Oxford: OUP.
Bierstaker, T. 1992: The 'triumph' of neo-classical economics in the developing world: policy convergence and bases of governance in the international economic order. In Rosenau, J. and Czempiel, E.-O. (eds), *Governance without government: Order and change in world politics*. Cambridge: CUP, 102–31.
Chaudhry, K. 1993: The myths of the market and the common history of late developers. *Politics and Society* 21, 245–74.
Colclough, M. and Manor, J. (eds) 1991: *States or markets? Neoliberalism and the development policy debate*. Oxford: Clarendon Press.
Haberler, G. 1987: Liberal and illiberal development policy. In Meier, G. (ed.), *Pioneers in development: Second series*. Oxford: OUP/World Bank, 51–83.
Johnson, H. 1971: A word to the Third World: a western economist's frank advice. *Encounter* 37.
Kabeer, N. and Humphrey, J. 1991: Neo-liberalism, gender, and the limits of the market. In Colclough, C. and Manor, J. (eds), *States or markets? Neo-liberalism and the development policy debate*. Oxford: Clarendon Press, 78–100.
Killick, A. 1986: Twenty-five years of development: the rise and impending decline of market solutions. *Development Policy Review* 4.

Krueger, A. 1974: The political economy of the rent-seeking society. *American Economic Review* 64, 291–303.
Lal, D. 1983: *The poverty of 'development economics'*. London: IEA.
Little, I. 1982: *Economic development: Theory, policy and international relations.* New York: Basic Books.
Schultz, T. W. 1964: *Transforming traditional agriculture.* New Haven: Yale UP.
Sender, J. and Smith, S. 1985: What's right with the Berg report and what's left of its critics? *Capital and Class* 24, 125–46.
Srinivasan, T. 1985: Neoclassical political economy, the state and economic development. *Asian Development Review* 3, 38–58.
Stewart, F. 1985: The fragile foundations of the neoclassical approach to development. *Journal of Development Studies* 21.
Toye, J. 1987: *Dilemmas of development: Reflections on the counter-revolution in development theory and policy.* Oxford: Blackwell – 2nd edition, 1993.
Watts, M. 1991: Visions of excess: African development in an age of market idolatry. *Transition* 51, 124–51.

On the new populism in development studies, and anti-developmentalism (and critiques)

Apter, D. 1987: *Rethinking development: Modernization, dependency and post-modern politics.* London: Sage.
Bhabha, H. 1983: The Other question. *Screen* 24, 18–35.
Brokensha, D. Warren, D. and Werner, O. (eds) 1980: *Indigenous knowledge systems and development.* Lanham, MD: University Press of America.
Chambers, R. 1983: Putting last thinking first: a professional revolution. *Third World Affairs* 4, 78–94.
Clifford, J. and Marcus, G. (eds) 1986: *Writing culture.* Berkeley: University of California Press.
Davis, D. 1992: Unlearning languages of development. *Latin American Research Review* 27, 151–68.
Development and Change 1992: Special issue on 'Emancipations – Modern and Post-Modern', Volume 23 (3).
Edwards, M. 1989: The irrelevance of development studies. *Third World Quarterly* 11, 116–35.
Edwards, M. 1994: Rethinking social development: the search for 'relevance'. In Booth, D. (ed.), *Rethinking social development: Theory, research and practice.* Harlow: Longman, 279–97.
Edwards, M. and Hulme, D. (eds) 1992: *Making a difference: NGOs and development in a changing world.* London: Earthscan.
Escobar, A. 1988: Power and visibility: development and the invention and management of the Third World. *Cultural Anthropology* 3.
Fanon, F. 1992: *The wretched of the earth.* Harmondsworth: Penguin.
Farrington, J. and Bebbington, A. with Wellard, K. and Lewis, D. 1993: *Reluctant partners? Non-governmental organizations, the state and sustainable agricultural development.* London: Routledge.
Fukuyama, F. 1991: *The end of history and the last man.* London: Hamish Hamilton.

Gandhi, M. K. 1966: *An autobiography: The story of my experiments with truth*. Harmondsworth: Penguin.

Marglin, F. and Marglin, S. (eds) 1990: *Dominating knowledge: Development, culture and resistance*. Oxford: Clarendon/WIDER.

Nanda, M. 1991: Is modern science a western, patriarchal myth? *South Asia Bulletin* XI, 32–61.

Richards, P. 1985: *Indigenous agricultural revolution*. London: Hutchinson.

Sachs, W. (ed.) 1992: *The development dictionary*. London: Zed.

Schumacher, E. 1973: *Small is beautiful*. London: Blond and Briggs.

Schuurman, F. (ed.) 1993: *Beyond the impasse: New directions in development theory*. London: Zed.

Seabrook, J. 1993: *Victims of development: Resistance and alternatives*. London: Verso.

Shiva, V. 1989: *Staying alive: Women, ecology and development*. London: Zed.

Shucking, H. and Anderson, P. 1991: Voices heeded and unheeded. In Shiva, V. (ed.), *Biodiversity: Social and ecological perspectives*. London: Zed, 13–41.

Slater, D. 1992: On the borders of social theory: learning from other regions. *Society and Space* 10, 307–27.

Spivak, G. C. 1990: *The post-colonial critic*. London: Routledge.

Trinh, T. M. 1989: *Woman, native, other: Writing postcoloniality and feminism*. Bloomington: Indiana UP.

1 Bert F. Hoselitz, 'Non-Economic Barriers to Economic Development'

Excerpts from: *Economic Development and Cultural Change* 1 (1), 8–21 (1952)

Although discussion of economic development is currently widespread there is little conscious agreement of what are its implications. Above all, there is confusion over whether the process of economic development can be regarded as a slow evolution of new forms of economic activity, or whether (and to what extent) it consists in a sharp break with the past and the sudden introduction of new forms. But in addition to ambiguity surrounding the notions of the speed with which economic change occurs, there is confusion with regard to the depth to which it penetrates. Does economic development mean only a change in certain aspects of overt behavior, notably the acquisition of new skills or the exercise of new forms of productive activity, or is it accompanied by or contingent upon more basic changes in social relations and even the structure of values and beliefs of a culture?

These problems are, essentially, questions of definition and theory and rather than discussing them at this point, I shall assume that economic development, especially if it involves industrialization, implies a rapid, and in a sense, revolutionary process which, if it is to take root in a society, must penetrate widely and deeply and hence affects the social structural and cultural facets of a society. In other words, economic development consists not merely in a change of production techniques, but also, in the last resort, in a reorientation of social norms and values. Any analysis of economic development which is to be fruitful and complete must include a set of propositions relating changes in production techniques to changes in values.

Most past attempts at bringing about economic development can be viewed as proceeding from either one of these extremes. Current proposals, especially those made by some publicists who are well-intentioned but often ill-informed, and even those made by some spokesmen for government agencies or international organizations, appear to be based on notions very close to a theory of economic determinism. If the underdeveloped countries were only supplied with capital in appropriate form valued at several billion dollars annually – so the argument runs – their economy and presumably their society would be changed drastically. Even a conservative interpretation of this view comes to the result that economic changes, notably the introduction of new techniques and new capital instruments, are a necessary prerequisite, and indeed the most likely path by which social behavior patterns and cultural norms can be changed.[1] However it is doubtful whether the transformation of a society can be explained in such simple fashion and there is no doubt that the obstinacy with which people hold to traditional values, even in the face of a rapidly changing technology and economic organization may impose obstacles of formidable proportions.

It may be asked whether a more fruitful procedure may not be the attempt to alter values first and to expect that this will create a climate favorable for new economic forms and processes. But it appears, from theoretical reflection and historical experience, that this method has little chance of success. We have the testimony of our most distinguished anthropologists who argue that a diffusion of values or value systems is impossible.[2] The historical experiences, notwithstanding the success of some individual missionaries, confirms this. In those instances in which religious conversions of whole societies were attempted, as in the case of the Spanish and Portuguese colonies, the long-run effects on social structure and the economy have been negligible; we also note failure of attempts to remodel only secular values, while leaving the religion undisturbed, as was the case in upper Burma, where the British tried to replace traditional quasi-tribal social relations by a social system based on the free market and the rule of law. The result was negative; Burma experienced social disorganization on a large scale, culminating in gang warfare and a formidable increase of violent crimes; the expected positive results were not forthcoming. Although there was great improvement in output, there was little improvement in the level of living of the average Burman and the allocation of developmentally most significant functions continued to be influenced strongly by status considerations rather than considerations of equity and efficiency.[3]

If we try to interpret the aspirations of the presently economically less advanced countries, we find there also a strange ambiguity which appears to be the result of partial unawareness of the close interconnectedness of economic advancement and cultural change. For the spokesmen of the poorer countries most emphatically favor economic progress resulting in an elevation of general levels of living, and blame their poverty on previous colonial status or quasi-colonial imperialistic exploitation. At the same time their rejection of colonialism and imperialism manifests itself in a heightened sense of nationalism, the

[1] This view is expressed in its most simple form by Benjamin A. Javits, *Peace by Investment*, New York, 1950, and Willard R. Espy, *Bold New Program*, New York, 1950. Although he does not adhere to a simple theory of economic determinism such a functional relationship between variables is assumed by Kingsley Davis, when he says in speaking of the Indian population problem that 'any policy that would rapidly industrialize Pakistan and India would be a far greater shock to the basic social institutions than would any policy that attacked fertility directly. Fast industrialization would sweep both the *ryot* and the *zamindar* from their moorings, transforming them into workers in a collectivized mechanized agriculture utterly foreign to their habits.' (The economic demography of India and Pakistan, in Phillips Talbot (ed.), *South Asia in the World Today*, Chicago, 1950, pp. 104–5.)

[2] See for example, Ralph Linton, *The Study of Man*, New York, 1936, p. 339.

[3] For an analysis of the Burmese experience see J. S. Furnivall, *Colonial Policy and Practice*, Cambridge, 1948.

symbolic expression of which consists in the repudiation of foreign philosophies and external behavior patterns and the reaffirmation of traditionally honored ways of acting and thinking. For example, the nationalism in Gandhi's independence movement was associated with the return to highly inefficient methods of traditional Indian activity, and in present-day Burma independence is not only accompanied by a resumption of traditional names and dress, but a strengthening of Buddhism, a religion which reflects an ideology totally opposed to efficient, progressive economic activity. The realization of economic advancement meets thus with numerous obstacles and impediments. Many of these obstacles are in the realm of economic relations: there is scarcity of capital, there is a demand for new skills and new techniques, there is a need for better roads and improved systems of communication, for public utilities and new sources of power. But some of the impediments to economic progress are beyond the area of economic relations. If the observation is made that among the prerequisites of economic development is the growth of a middle class, or the evolution of a spirit of venturesomeness, or the elimination of corruption among officialdom, we are confronted with changes in the social organization and culture of a people rather than in its economy. I propose to discuss in the remainder of this paper some of these non-economic 'factors' which are yet too little explored, but which appear to exercise a strong negative and positive influence on the attainment of economic betterment.

If we ask how technological or economic innovations are introduced in a society we immediately encounter two problems. One is the question of which innovations will be adopted with different degrees of ease and which will be rejected, and the other is the question of what person or group of persons performs the tasks of adopting and further spreading the new techniques in a society. Within the context of this paper the first question is of subordinated importance, since we are not dealing with specific innovations but the general problem of development and all underdeveloped countries are eager to accept more modern forms of economic activity, although for diverse reasons some of them may reject one or the other type. For example, though India may reject or hesitate to adopt modern methods of meat packing, it is eager to introduce other industries. But the second question, who carries the main burden in the process of innovation, is of great interest to us; evidence for this is the fact that in discussing economic development emphasis has so often been placed on the presence or absence of certain social groups exhibiting particular attitudes (e.g. venturesomeness) or performing special roles (e.g. bureaucracy, 'middle class'). In somewhat different terms, we may say that economic development requires the formation of a social group which constitutes the spearhead of different kinds of innovations.

It is plain from these considerations that one of the prerequisites of economic and technical advancement is a high degree of social mobility. If, for whatever reasons, social advancement of people in less privileged ranks in society is made difficult, or if the cleavages in status, power, and wealth between different social groups is great, an impediment to economic development of considerable magnitude is thereby in existence. Very primitive societies apart, the main status-determining factors are wealth, political power, and education. The ideal form of the liberal state is based on the assumption that each of these factors will be in the hands of a different social group or class and that in addition to the separation of powers in the political field, there will be a 'balance' of social power and status. It will be remembered that an important aspect of the Marxian criticism of 'bourgeois capitalism' was based on the assertion that this separation did not exist, or existed only in appearance and that, in Engels' words, 'the modern state no matter what its form is essentially a capitalist machine, the state of the capitalists.'[4]

Now whether or not this statement was true of the nations of nineteenth century Europe, the social situation on which it is based is true of many underdeveloped countries today. In many of these countries, wealth, political power, and education are concentrated in a small group of people, and not infrequently the very individuals who control political power are also the richest and best educated men in the society. But this very monopoly of status-conferring factors is an impediment to economic development. The gap between the privileged and the masses, between the rulers and the ruled, is immense and there is nothing to fill it. But even to the extent to which this gap is being filled by an incipient middle class consisting chiefly of educators, government officials, and members of the intelligentsia, this group must, in order to assure its maintenance, either align itself with the ruling group or suffer being pushed into positions of harsh antagonism to that group. Hence intellectuals often attain positions of leadership among the discontented, the unprivileged, the poor; hence the appeal the communist ideology exerts on intellectuals in underdeveloped countries; hence also the enhanced social cleavage which becomes little, if at all, mitigated by the rise of the middle class. The cleavage of the world into two antagonistic camps becomes reflected in the political and ideological issues in a developing country and the possibility of evolutionary development toward higher levels of living disappears more and more as a practical third alternative between either the maintenance of social status quo or the revolution which threatens to throw the country into the arms of communism. If the issues are seen in this light the rigidities of the class structure in many underdeveloped countries become understandable. But the very sharpening of the issues, the fact that many enlightened people see as the only alternative to the maintenance of the existing class structure a communist revulsion makes them reluctant to advocate rapid and decisive innovations which, if they are to take root, inevitably affect the social status quo. Hence technical and economic innovations, if sponsored at all, are received with the greatest caution, are severely limited in application, and are often permitted only as token or symbolic performances.

Another obstacle to economic development which is located in conditions in

[4] Friedrich Engels, *Socialism, Utopian and Scientific*, Chicago (Charles H. Kerr Co.), 1902, p. 123.

underdeveloped countries is the nature of their aspirations and the form in which the realization of these aspirations is pictured. In more concrete terms this may be stated by saying that economic development plans are often unrealistic and divorced from the more immediate needs and productive capacities of these countries. I have drawn attention earlier to the ambiguity of simultaneously aiming at economic progress and the preservation of national and cultural traditions. But there is also an ambiguity in the thinking of many leaders of underdeveloped countries between the objectives of developmental efforts and practicable attainments. The point is sometimes stated in a rather drastic form by emphasizing the fact that many development plans fostered by underdeveloped governments give a high priority to the establishment of steel mills and other forms of heavy industry even though such plants may have little justification on the basis of considerations of efficiency and rational allocation of resources. We may look at this matter from two points of view. We may either regard the wish for a steel mill as a childish, irrational desire which only merits ridicule. But we may also regard it as a symbolic expression of the wish for industrialization, and the implicit acknowledgement of the fact that little is known about the priorities and time sequence of such a process. I would regard it as evidence of the latter alternative, and here the obstacle to fruitful development is founded on defective knowledge and consequent inability to make rational workable plans.

In spite of numerous surveys of natural resources, soil types, and other environmental factors, knowledge of the natural endowment as well as the human resources of most underdeveloped countries is very imperfect. The United Nations and its specialized agencies have often been confronted by this fact. A mission of experts sent to Haiti produced a voluminous report on developmental possibilities in that country. Yet the chief impression one gains from reading the report of that mission is the frequent repetition of the statement that fruitful recommendations cannot be made because of the absence of reliable information.[5] Similarly, the International Bank for Reconstruction and Development and the Food and Agriculture Organization have been hamstrung in actually carrying through developmental projects for which funds would have been available, simply because workable projects which could withstand the careful scrutiny of experts were not forthcoming.

[5] United Nations, *Mission to Haiti*, New York, 1949, *passim*, but esp. pp. 12–22. This situation is by no means limited to Haiti, but all missions of the United Nations and its specialized agencies in underdeveloped countries encounter not only ignorance of technical procedures, but also ignorance of the most elementary facts.

The situation is not better in countries which publish statistics. As a rule these statistics are wrong and hence misleading. Inspection by the writer of the statistical and information services maintained in El Salvador, for example, reveal that with the exception of a few series (notably foreign trade and monetary and banking statistics) published figures are unreliable and Salvadoran administrators do not believe in them. At the same time large numbers of reports by industria inspectors and other petty administrators on wages, prices, and other economically significant data are filed in government offices without ever being used as raw material for the compilation of figures published by the statistical office.

Ignorance is always an obstacle to rational action. But in the case of economic development it is doubly fatal, because in this case action cannot be postponed for political or morale reasons. The consequence is that on the one hand programs are undertaken in fields where obvious needs for improvement exist (such as public health, for example) which, however, cannot maintain themselves because the necessary concomitant adjustments in the economy lag behind, and on the other hand that short-run programs are initiated which tend to lead to such allocation of resources (and hence to certain new rigidities and vested interests) as to make the attainment of the long-run objective more difficult. Evidence for both contingencies is frequent. As concerns public health programs, the attempts at control of tropical diseases are very instructive. For example, yaws and malaria are dreaded diseases in Haiti. A campaign to control yaws in the Marbial valley failed to have lasting results, although inoculation with antibiotics led to temporary relief, because the economic status of the mass of the population was not elevated enough to permit them to meet the most elementary standards of cleanliness. A swamp drainage project designed to eliminate carriers of malaria fell into disrepair after a few years, the drainage canals became stopped up and large expenditures turned out to have been in vain, because owing to indifference, corruption, and mismanagement the project was not kept going properly after its foreign initiators had left.[6]

Examples for the conflict between short-run and long-run aims are also frequent. For many 'one-crop' countries it presents a real dilemma. The long-run objective of economic development programs for these countries is greater diversification of production, to make them less dependent on one or two or three staple exports, the prices of which are determined on the world market, subjecting, in this way, the international accounts of the one-crop country to great uncertainties and often violent fluctuations. At the same time the major export industry deserves full support in the short-run because it is the chief asset producing foreign exchange which, in the absence of generous loans or foreign aid, provides the wherewithal for economic development. The experience with coffee planters in some Latin American countries and rubber planters in some countries of South East Asia shows the restraining influence on long-run plans of economic diversification exercised by vested interests in an important export industry.

Another instance of conflict between short-run and long-run objectives of economic development plans is reported by Wilbert E. Moore. Moore believes that the *Ejido*, although it 'alleviated the immediate economic ills of the Mexican rural population . . . it did . . . make possible a re-establishment of the partially isolated village, agricultural underemployment, and all the conservative traditions of a land-hungry peasantry . . . All indirect evidence indicates that in terms of long-run developments the *Ejido* was a strongly conservative move in the strict sense; that is, the possible increase in market

[6] *Ibid.*, p. 69.

orientation, improved education, and productive technique seems to have been offset by re-establishing the traditional village.'[7]

Professor Moore here again points to the fact that the implementation of short-run objectives creates vested interests which impede the full realization of the long-run developmental goals. But the nature of the vested interests in this case is very different from that of the vested interests fostered by the extension of an export crop in a 'one-crop' country. In the latter case these vested interests are based chiefly on the expectation of economic gain, in the former case they consist in the rejuvenation of a traditional way of life which, in many of its aspects, is opposed to economic progress.

I believe that the dilemma found by Professor Moore in Mexico poses a general problem, notably for areas in which an extension of agricultural settlement is still possible. The fact that some underdeveloped countries, in spite of great rural population density in certain localities, have still considerable areas of uncultivated arable land, is often regarded as fortunate. In a country like India or Egypt, where further horizontal expansion of agriculture is virtually impossible, economic development is pushed necessarily into the channels of industrialization accompanied by intensification of agriculture, i.e., the application of policies resulting in higher yields per acre. This process is accompanied in all likelihood by a rapid increase of the population, and, since industrialization is associated with urbanization, by an increase in the required quantity of real output per family or productively employed individual. It is probable that under these conditions increase in agricultural productivity will not be commensurate with increase in demand for products grown on the land (foods, fibers, hides, and skins, chemical raw materials, such as oils and lumber), and the scarcity of economically usable land becomes a serious bottleneck to development. On the other hand, countries in which substantial areas of unused land are still available can syphon off part of the developing population surplus by settling it on new lands, and can simultaneously expand the output of agricultural raw materials with the increasing demand for such materials by developing industry.

Looked at purely from the viewpoint of the strategy of resource allocation planning in the short run this group of countries (among which belong most countries of South East Asia, the Middle East, and Latin America) is therefore in a position in which rational planning may mitigate the economic sacrifices involved in industrialization. But the existence of an agricultural frontier and the knowledge that such a frontier exists exerts an influence on policies actually made. As Professor Moore has shown in his book referred to earlier, recruitment of large masses of peasants and primitives for industry is a hard task. Resettlement may not be much easier, but it may be more acceptable to some native peoples than induction into the industrial labor force. To the extent, therefore, to which local population pressure is mitigated by resettlement – as for example by moving people in the Philippines from Luzon to Davao, or in

[7] Wilbert E. Moore, *Industrialization and Labor*, Ithaca, 1951, pp. 237–8.

Indonesia from Java to the outer islands – the traditional agricultural way of life with its pre-industrial and non-rational aspects (in Max Weber's sense) is given a new lease of life and industrial progress made more difficult.

In essence the conflict between the two ways of life is a conflict of values. Just as Hinayana Buddhism in Burma, with its other-worldly orientation, calls forth a philosophy of life which is not conducive to economic advancement, so the strengthening of traditional methods of small scale agriculture reinforces the system of values which may have flowered into great cultural achievements in the past, but which it has been impossible to adapt to rational, efficient, economic activity. This conflict in values has sometimes been expressed as the antagonism between city and countryside, the mutual estrangement between the urban and the rural inhabitant. To a contemporary American, and perhaps also to a contemporary European, this distinction may appear spurious. But it is a difference which is obvious to a student of the social structure of oriental societies and it crops up every now and then in the sociological literature of the eighteenth and early nineteenth century.

Among earlier writers in the field of social science most persistent attention to this problem was given by Karl Marx. References to the cleavage between the city and the countryside run through his works from the early 1840's to the end of his life. The tersest formulation which he ever gave to this problem is a passage in *Capital*. He says:

> 'The foundation of every division of labor that is well developed and brought about by the exchange of commodities is the separation between town and country. It may be said that the whole economical history of society is summed up in the movement of this antithesis.'[8]

Although these words refer primarily to the difference in the forms of economic organization between medieval and capitalist Europe, they imply, and other writings plainly show, his recognition of this difference in the entire way of life on the land and in the cities in countries with a non-capitalistic, small-scale primitive agriculture. In Western Europe the transition from one way of life to the other – and the values implied in each of them – took place during three or four hundred years, and was aided by enclosure acts, 'Bauern legen', pauperization, the adaptation of the Calvinist ideology to the objectives of the commercial and industrial middle class, and other measures which turned the rolling green hills of Warwickshire and the Tyneside [sic] into the 'Black Country' and the fields and wastes of Lancashire into the cotton center of the world.

I am not expressing nostalgic regrets over the passing of the European middle ages, but I am trying to indicate that parallel with the external change in the landscape the minds and aspirations of men, the things they valued and were taught to value, changed; and with this in mind it is perhaps not quite wrong to say that the England or France of the thirteenth century resembled

[8] Karl Marx, *Capital*, Chicago (Charles H. Kerr Co.), 1903, vol. I, p. 387.

more the present Middle East or South East Asia, than the England or France of our day.

As has been pointed out earlier value systems offer special resistances to change, but without wishing to be dogmatic, I believe, it may be stated that their change is facilitated if the material economic environment in which they can flourish is destroyed or weakened. This seems to be the experience from the history of Western European economic development, and it seems to be confirmed by the findings of students of colonial policy and administration, and research results on the impact of industrialization in underdeveloped countries. Economic development plans which combine industrialization with an extension of traditional or near-traditional forms of agriculture are thus creating a dilemma which in the long run may present serious repercussions on the speed and facility with which ultimate objectives can be reached. This does not mean that, wherever this is possible, extension of agricultural production should not be undertaken in combination with industrialization. But rural resettlement should be regarded as a form of industrialization rather than an extension of traditional methods of farming. In view of existing pressures and the absence in almost all underdeveloped countries of an efficient administrative apparatus the difficulties which such a program confronts are obvious.

Reference to the experience of the transition from medieval to modern economic organization in Western Europe brings to mind another important factor which may prove to be a serious obstacle to technological advancement. This obstacle is found in the changes required in methods of work and levels of skill which necessarily accompany technical change and alterations in the scope and form of economic activity. Little needs to be said here about these two points since much of the relevant evidence has been collected by Professor Moore in his book.[9] From these remarks it appears that these obstacles, although real, are less significant than those opposing economic development because of vested interests of an elite, or the vigor of a non-industrial system of values. To a certain extent resistances against new modes of technical and economic processes and changing kinds and levels of skill are specific aspects of the two last-named factors. But since from the socio-psychological point of view economic development may be interpreted above all as a change in the division of social labor, some special attention to skills and modes of work appears to be in order.

Confining ourselves, at first, to a discussion of skills, the first question which might be raised is whether economic development requires a transition from less complex to more complex skills or vice versa. This question is impossible to answer because there exists no generally agreed upon classification of skills in terms of their complexity, and even if it existed the answer would be ambiguous. Certainly the manual skill of a hand-loom weaver is superior to that of a man who runs a power-loom, but the skill of the mechanic who tends the power-loom and keeps it in repair is probably superior to that of the hand-loom

[9] Moore, *op. cit.*, pp. 44–7, 55–9, 90–4, 114–26, 274–8, and 308–10.

weaver. In general, it may be said, that mechanization, by 'putting the skill into the machine' has two opposite effects. It simplifies many manual operations and makes possible the rapid training of large numbers of unskilled or semi-skilled workers. It thus creates a large demand for people whose skill level is indifferent and who can acquire the necessary manipulatory accomplishments by a process of on-the-job training. At the same time it requires the development of a group of men, foremen, engineers, technical maintenance men, petty administrators and others capable of rendering services which often require not only a high level of dexterity but a considerable variety of aptitudes. Now, I suppose that you have all heard of the African native mechanic who, in some relatively isolated place in the Sudan, repairs and keeps going with some wire and pieces of sheet metal a model T Ford, which in this country would be considered fit only for the junk heap. These men exist, and I am far from denying their dexterity and ingenuity. But they belong in the same class as the Chinese ivory worker who produced the most delicately carved decorations of a cigarette-case, and the anonymous medieval stone-cutter who chiseled the capitals of the Gothic cathedrals of France. It is granted that human capacity for the exercise of highly skilled tasks, that human ingenuity, that human intelligence is fairly evenly distributed over the globe. The problem of developing a group of skilled technicians is not a psychological question of the capacity of persons in underdeveloped countries to learn, but a social problem: the creation of attitudes and material and psychological compensations which will make the choice of such careers attractive. In other words, the question we must ask is not: 'How can the people of technically less advanced countries learn the modern techniques?', but: 'Will they learn them, and how can they be induced to want to learn them?'.

. . . .

A final word of caution remains to be added. I have discussed in this paper obstacles and impediments to economic advancement, and have not tried to indicate means by which they could be overcome. This does not imply that such means do not exist or could not be devised. Their careful examination would require, however, another paper at least as long as this. But these impediments had to be discussed because they are important factors intervening in development plans. As has been stated earlier, there exists a strong desire for development in all poorer countries. From the top government officials to the most menial peasants and laborers, aspirations are inculcated and find expression which point to this. Economic development, although endowed with different meanings to members of different societies and different classes of the same society has become a slogan with formidable powers of attraction. The execution of actual development projects, the attainment of sensibly higher living standards for large masses of people, is a very great task and certain to lead to many disappointments and disillusionments. In the enthusiasm of regarding economic development as a cure-all one is wont to overlook or belittle difficulties which might stand in the way of the easy attainment of too frequently all too ambitious targets. An honest and critical evaluation of economic[10] and non-economic barriers to such development, may therefore have the wholesome effect of inducing the drawing up of plans which are capable of actual realization and will avoid the emergence of unforeseen by-products which may jeopardize the attainment of the objectives of developmental efforts.

2 Andre Gunder Frank, 'The Development of Underdevelopment'

Reprinted in full from: *Monthly Review* (September) (1966)

We cannot hope to formulate adequate development theory and policy for the majority of the world's population who suffer from underdevelopment without first learning how their past economic and social history gave rise to their present underdevelopment. Yet most historians study only the developed metropolitan countries and pay scant attention to the colonial and underdeveloped lands. For this reason most of our theoretical categories and guides to development policy have been distilled exclusively from the historical experience of the European and North American advanced capitalist nations.

Since the historical experience of the colonial and underdeveloped countries has demonstrably been quite different, available theory therefore fails to reflect the past of the underdeveloped part of the world entirely, and reflects the past of the world as a whole only in part. More important, our ignorance of the underdeveloped countries' history leads us to assume that their past and indeed their present resembles earlier stages of the history of the now developed countries. This ignorance and this assumption lead us into serious misconceptions about contemporary underdevelopment and development. Further, most studies of development and underdevelopment fail to take account of the economic and other relations between the metropolis and its economic colonies throughout the history of the worldwide expansion and development of the mercantilist and capitalist system. Consequently, most of our theory fails to explain the structure and development of the capitalist

[10] I have not stressed economic barriers to economic development in this paper because they have been summarized in a brilliant fashion by Professor Jacob Viner in Chapter VI of his *Lectures on the Theory of International Trade*, (published in Portuguese translation in *Revista Brasileira de Economia*, Ano V, Numero 2, June, 1951).

system as a whole and to account for its simultaneous generation of underdevelopment in some of its parts and of economic development in others.

It is generally held that economic development occurs in a succession of capitalist stages and that today's underdeveloped countries are still in a stage, sometimes depicted as an original stage, of history through which the now developed countries passed long ago. Yet even a modest acquaintance with history shows that underdevelopment is not original or traditional and that neither the past nor the present of the underdeveloped countries resembles in any important respect the past of the now developed countries. The now developed countries were never *under*developed, though they may have been *un*developed. It is also widely believed that the contemporary underdevelopment of a country can be understood as the product or reflection solely of its own economic, political, social, and cultural characteristics or structure. Yet historical research demonstrates that contemporary underdevelopment is in large part the historical product of past and continuing economic and other relations between the satellite underdeveloped and the now developed metropolitan countries. Furthermore, these relations are an essential part of the structure and development of the capitalist system on a world scale as a whole. A related and also largely erroneous view is that the development of these underdeveloped countries, and within them of their most underdeveloped domestic areas, must and will be generated or stimulated by diffusing capital, institutions, values, etc., to them from the international and national capitalist metropoles. Historical perspective based on the underdeveloped countries' past experience suggests that on the contrary, economic development in the underdeveloped countries can now occur only independently of most of these relations of diffusion.

Evident inequalities of income and differences in culture have led many observers to see 'dual' societies and economies in the underdeveloped countries. Each of the two parts is supposed to have a history of its own, a structure, and a contemporary dynamic largely independent of the other. Supposedly only one part of the economy and society has been importantly affected by intimate economic relations with the 'outside' capitalist world; and that part, it is held, became modern, capitalist, and relatively developed precisely because of this contact. The other part is widely regarded as variously isolated, subsistence-based, feudal, or pre-capitalist, and therefore more underdeveloped.

I believe on the contrary that the entire 'dual' society thesis is false and that the policy recommendations to which it leads will, if acted upon, serve only to intensify and perpetuate the very conditions of underdevelopment they are supposedly designed to remedy.

A mounting body of evidence suggests, and I am confident that future historical research will confirm, that the expansion of the capitalist system over the past centuries effectively and entirely penetrated even the apparently most isolated sectors of the underdeveloped world. Therefore the economic, political, social, and cultural institutions and relations we now observe there are the products of the historical development of the capitalist system no less than are the seemingly more modern or capitalist features of the national metropoles of these underdeveloped countries. Analogous to the relations between development and underdevelopment on the international level, the contemporary underdeveloped institutions of the so-called backward or feudal domestic areas of an underdeveloped country are no less the product of the single historical process of capitalist development than are the so-called capitalist institutions of the supposedly more progressive areas. I should like to sketch the kinds of evidence which support this thesis and at the same time indicate lines along which further study and research could fruitfully proceed.

The Secretary General of the Latin American Center for Research in the Social Sciences writes in that Center's journal: 'The privileged position of the city has its origin in the colonial period. It was founded by the Conqueror to serve the same ends that it still serves today; to incorporate the indigenous population into the economy brought and developed by that Conqueror and his descendants. The regional city was an instrument of conquest and is still today an instrument of domination.'[1] The Instituto Nacional Indigenista (National Indian Institute) of Mexico confirms this observation when it notes that 'the mestizo population, in fact, always lives in a city, a center of an intercultural region, which acts as the metropolis of a zone of indigenous population and which maintains with the underdeveloped communities an intimate relation which links the center with the satellite communities.'[2] The Institute goes on to point out that 'between the mestizos who live in the nuclear city of the region and the Indians who live in the peasant hinterland there is in reality a closer economic and social interdependence than might at first glance appear' and that the provincial metropoles 'by being centers of intercourse are also centers of exploitation.'[3]

Thus these metropolis–satellite relations are not limited to the imperial or international level but penetrate and structure the very economic, political, and social life of the Latin American colonies and countries. Just as the colonial and national capital and its export sector become the satellite of the Iberian (and later of other) metropoles of the world economic system, this satellite immediately becomes a colonial and then a national metropolis with respect to the productive sectors and population of the interior. Furthermore, the provincial capitals which thus are themselves satellites of the national metropolis – and through the latter of the world metropolis – are in turn provincial centers around which their own local satellites orbit. Thus, a whole chain of constellations of metropoles and satellites relates all parts of the whole system from its metropolitan center in Europe or the United States to the farthest outpost in the Latin American countryside.

[1] *América Latina*, Año 6, No. 4 (October–December 1963), p. 8.

[2] Instituto Nacional Indigenista, *Los centros coordinadores indigenistas* (Mexico, 1962), p. 34.

[3] *Ibid.*, pp. 33–34, 88.

When we examine this metropolis-satellite structure, we find that each of the satellites, including now underdeveloped Spain and Portugal, serves as an instrument to suck capital or economic surplus out of its own satellites and to channel part of this surplus to the world metropolis of which all are satellites. Moreover, each national and local metropolis serves to impose and maintain the monopolistic structure and exploitative relationship of this system (as the Instituto Nacional Indigenista of Mexico calls it) as long as it serves the interests of the metropoles which take advantage of this global, national, and local structure to promote their own development and the enrichment of their ruling classes.

These are the principal and still surviving structural characteristics which were implanted in Latin America by the Conquest. Beyond examining the establishment of this colonial structure in its historical context, the proposed approach calls for study of the development – and underdevelopment – of these metropoles and satellites of Latin America throughout the following and still continuing historical process. In this way we can understand why there were and still are tendencies in the Latin American and world capitalist structure which seem to lead to the development of the metropolis and the underdevelopment of the satellite and why, particularly, the satellized national, regional, and local metropoles in Latin America find that their economic development is at best a limited or underdeveloped development.

That present underdevelopment of Latin America is the result of its centuries-long participation in the process of world capitalist development, I believe I have shown in my case studies of the economic and social histories of Chile and Brazil.[4] My study of Chilean history suggests that the Conquest not only incorporated this country fully into the expansion and development of the world mercantile and later industrial capitalist system but that it also introduced the monopolistic metropolis-satellite structure and development of capitalism into the Chilean domestic economy and society itself. This structure then penetrated and permeated all of Chile very quickly. Since that time and in the course of world and Chilean history during the epochs of colonialism, free trade, imperialism, and the present, Chile has become increasingly marked by the economic, social, and political structure of satellite underdevelopment. This development of underdevelopment continues today, both in Chile's still increasing satellization by the world metropolis and through the ever more acute polarization of Chile's domestic economy.

The history of Brazil is perhaps the clearest case of both national and regional development of underdevelopment. The expansion of the world economy since the beginning of the sixteenth century successively converted

[4] 'Capitalist Development of Underdevelopment in Chile' and 'Capitalist Development of Underdevelopment in Brazil' in *Capitalism and Underdevelopment in Latin America* (New York & London: Monthly Review Press, 1967 and 1969).

the Northeast, the Minas Gerais interior, the North, and the Center–South (Rio de Janeiro, São Paulo, and Paraná) into export economies and incorporated them into the structure and development of the world capitalist system. Each of these regions experienced what may have appeared as economic development during the period of its golden age. But it was a satellite development which was neither self-generating nor self-perpetuating. As the market or the productivity of the first three regions declined, foreign and domestic economic interest in them waned and they were left to develop the underdevelopment they live today. In the fourth region, the coffee economy experienced a similar though not yet quite as serious fate (though the development of a synthetic coffee substitute promises to deal it a mortal blow in the not too distant future). All of this historical evidence contradicts the generally accepted theses that Latin America suffers from a dual society or from the survival of feudal institutions and that these are important obstacles to its economic development.

During the First World War, however, and even more during the Great Depression and the Second World War, São Paulo began to build up an industrial establishment which is the largest in Latin America today. The question arises whether this industrial development did or can break Brazil out of the cycle of satellite development and underdevelopment which has characterized its other regions and national history within the capitalist system so far. I believe that the answer is no. Domestically the evidence so far is fairly clear. The development of industry in São Paulo has not brought greater riches to the other regions of Brazil. Instead, it has converted them into internal colonial satellites, de-capitalized them further, and consolidated or even deepened their underdevelopment. There is little evidence to suggest that this process is likely to be reversed in the foreseeable future except insofar as the provincial poor migrate and become the poor of the metropolitan cities. Externally, the evidence is that although the initial development of São Paulo's industry was relatively autonomous it is being increasingly satellized by the world capitalist metropolis and its future development possibilities are increasingly restricted.[5] This development, my studies lead me to believe, also appears destined to limited or underdeveloped development as long as it takes place in the present economic, political, and social framework.

We must conclude, in short, that underdevelopment is not due to the survival of archaic institutions and the existence of capital shortage in regions that have remained isolated from the stream of world history. On the contrary, underdevelopment was and still is generated by the very same historical process which also generated economic development: the development of capitalism itself. This view, I am glad to say, is gaining adherents among students of Latin America and is proving its worth in shedding new light on the problems of the

[5] Also see, 'The Growth and Decline of Import Substitution,' *Economic Bulletin for Latin America*, IX, No. 1 (March 1964); and Celso Furtado, *Dialectica do Desenvolvimiento* (Rio de Janeiro: Fundo de Cultura, 1964).

area and in affording a better perspective for the formulation of theory and policy.[6]

The same historical and structural approach can also lead to better development theory and policy by generating a series of hypotheses about development and underdevelopment such as those I am testing in my current research. The hypotheses are derived from the empirical observation and theoretical assumption that within this world-embracing metropolis–satellite structure the metropoles tend to develop and the satellites to underdevelop. The first hypothesis has already been mentioned above: that in contrast to the development of the world metropolis which is no one's satellite, the development of the national and other subordinate metropoles is limited by their satellite status. It is perhaps more difficult to test this hypothesis than the following ones because part of its confirmation depends on the test of the other hypotheses. Nonetheless, this hypothesis appears to be generally confirmed by the non-autonomous and unsatisfactory economic and especially industrial development of Latin America's national metropoles, as documented in the studies already cited. The most important and at the same time most confirmatory examples are the metropolitan regions of Buenos Aires and São Paulo whose growth only began in the nineteenth century, was therefore largely untrammeled by any colonial heritage, but was and remains a satellite development largely dependent on the outside metropolis, first of Britain and then of the United States.

A second hypothesis is that the satellites experience their greatest economic development and especially their most classically capitalist industrial development if and when their ties to their metropolis are weakest. This hypothesis is almost diametrically opposed to the generally accepted thesis that development in the underdeveloped countries follows from the greatest degree of contact with and diffusion from the metropolitan developed countries. This hypothesis seems to be confirmed by two kinds of relative isolation that Latin America has experienced in the course of its history. One is the temporary isolation caused by the crises of war or depression in the world metropolis. Apart from minor ones, five periods of such major crises stand out and are seen to confirm the hypothesis. These are: the European (and especially Spanish) depression of the seventeenth century, the Napoleonic Wars, the First World War, the Depression of the 1930's, and the Second World War. It is clearly established and generally recognized that the most important recent industrial development – especially of Argentina, Brazil, and Mexico, but also of other countries such as Chile – has taken place precisely during the periods of the two

[6] Others who use a similar approach, though their ideologies do not permit them to derive the logically following conclusions, are Aníbal Pinto, *Chile: Un caso de desarrollo frustrado* (Santiago: Editorial Universitaria, 1957); Celso Furtado, *A formaçao económica do Brasil* (Rio de Janeiro: Fundo de Cultura, 1959) which was recently translated into English and published as *The Economic Growth of Brazil* by the University of California Press; and Caio Prado Junior, *Historia Económica do Brasil* (7th ed., São Paulo: Editora Brasiliense, 1962).

world wars and the intervening Depression. Thanks to the consequent loosening of trade and investment ties during these periods, the satellites initiated marked autonomous industrialization and growth. Historical research demonstrates that the same thing happened in Latin America during Europe's seventeenth-century depression. Manufacturing grew in the Latin American countries, and several, such as Chile, became exporters of manufactured goods. The Napoleonic Wars gave rise to independence movements in Latin America, and these should perhaps also be interpreted as in part confirming the development hypothesis.

The other kind of isolation which tends to confirm the second hypothesis is the geographic and economic isolation of regions which at one time were relatively weakly tied to and poorly integrated into the mercantilist and capitalist system. My preliminary research suggests that in Latin America it was these regions which initiated and experienced the most promising self-generating economic development of the classical industrial capitalist type. The most important regional cases probably are Tucumán and Asunción, as well as other cities, such as Mendoza and Rosario, in the interior of Argentina and Paraguay during the end of the eighteenth and the beginning of the nineteenth centuries. Seventeenth- and eighteenth-century São Paulo, long before coffee was grown there, is another example. Perhaps Antioquia in Colombia and Puebla and Querétaro in Mexico are other examples. In its own way, Chile was also an example since before the sea route around the Horn was opened this country was relatively isolated at the end of a long voyage from Europe via Panama. All of these regions became manufacturing centers and even exporters, usually of textiles, during the periods preceding their effective incorporation as satellites into the colonial, national, and world capitalist system.

Internationally, of course, the classic case of industrialization through non-participation as a satellite in the capitalist world system is obviously that of Japan after the Meiji Restoration. Why, one may ask, was resource-poor but unsatellized Japan able to industrialize so quickly at the end of the century while resource-rich Latin American countries and Russia were not able to do so and the latter was easily beaten by Japan in the War of 1904 after the same forty years of development efforts? The second hypothesis suggests that the fundamental reason is that Japan was not satellized either during the Tokugawa or the Meiji period and therefore did not have its development structurally limited as did the countries which were so satellized.

A corollary of the second hypothesis is that when the metropolis recovers from its crisis and re-establishes the trade and investment ties which fully re-incorporate the satellites into the system, or when the metropolis expands to incorporate previously isolated regions into the worldwide system, the previous development and industrialization of these regions is choked off or channeled into directions which are not self-perpetuating and promising. This happened after each of the five crises cited above. The renewed expansion of trade and the spread of economic liberalism in the eighteenth and nineteenth centuries choked off and reversed the manufacturing development which Latin America had experienced during the seventeenth century, and in some places at the beginning of the nineteenth. After the First World War, the new national industry of Brazil suffered serious consequences from American economic invasion. The increase in the growth rate of Gross National Product and particularly of industrialization throughout Latin America was again reversed and industry became increasingly satellized after the Second World War and especially after the post-Korean War recovery and expansion of the metropolis. Far from having become more developed since then, industrial sectors of Brazil and most conspicuously of Argentina have become structurally more and more underdeveloped and less and less able to generate continued industrialization and/or sustain development of the economy. This process, from which India also suffers, is reflected in a whole gamut of balance-of-payments, inflationary, and other economic and political difficulties, and promises to yield to no solution short of far-reaching structural change.

Our hypothesis suggests that fundamentally the same process occurred even more dramatically with the incorporation into the system of previously unsatellized regions. The expansion of Buenos Aires as a satellite of Great Britain and the introduction of free trade in the interest of the ruling groups of both metropoles destroyed the manufacturing and much of the remainder of the economic base of the previously relatively prosperous interior almost entirely. Manufacturing was destroyed by foreign competition, lands were taken and concentrated into latifundia by the rapaciously growing export economy, intra-regional distribution of income became much more unequal, and the previously developing regions became simple satellites of Buenos Aires and through it of London. The provincial centers did not yield to satellization without a struggle. This metropolis-satellite conflict was much of the cause of the long political and armed struggle between the Unitarists in Buenos Aires and the Federalists in the provinces, and it may be said to have been the sole important cause of the War of the Triple Alliance in which Buenos Aires, Montevideo, and Rio de Janeiro, encouraged and helped by London, destroyed not only the autonomously developing economy of Paraguay but killed off nearly all of its population unwilling to give in. Though this is no doubt the most spectacular example which tends to confirm the hypothesis, I believe that historical research on the satellization of previously relatively independent yeoman-farming and incipient manufacturing regions such as the Caribbean islands will confirm it further.[7] These regions did not have a chance against the forces of expanding and developing capitalism, and their own development had to be sacrificed to that of others. The economy and industry of Argentina, Brazil, and other countries which have experienced the effects of

[7] See for instance Ramiro Guerra y Sánchez, *Azúcar y Problación en las Antillas*, 2nd ed. (Havana, 1942), also published as *Sugar and Society in the Caribbean* (New Haven: Yale University Press, 1964).

metropolitan recovery since the Second World War are today suffering much the same fate, if fortunately still in lesser degree.

A third major hypothesis derived from the metropolis-satellite structure is that the regions which are the most underdeveloped and feudal-seeming today are the ones which had the closest ties to the metropolis in the past. They are the regions which were the greatest exporters of primary products to and the biggest sources of capital for the world metropolis and were abandoned by the metropolis when for one reason or another business fell off. This hypothesis also contradicts the generally held thesis that the source of a region's underdevelopment is its isolation and its pre-capitalist institutions.

This hypothesis seems to be amply confirmed by the former super-satellite development and present ultra-underdevelopment of the once sugar-exporting West Indies, Northeastern Brazil, the ex-mining districts of Minas Gerais in Brazil, highland Peru, and Bolivia, and the central Mexican states of Guanajuato, Zacatecas, and others whose names were made world famous centuries ago by their silver. There surely are no major regions in Latin America which are today more cursed by underdevelopment and poverty; yet all of these regions, like Bengal in India, once provided the life blood of mercantile and industrial capitalist development – in the metropolis. These regions' participation in the development of the world capitalist system gave them, already in their golden age, the typical structure of underdevelopment of a capitalist export economy. When the market for their sugar or the wealth of their mines disappeared and the metropolis abandoned them to their own devices, the already existing economic, political, and social structure of these regions prohibited autonomous generation of economic development and left them no alternative but to turn in upon themselves and to degenerate into the ultra-underdevelopment we find there today.

These considerations suggest two further and related hypotheses. One is that the latifundium, irrespective of whether it appears today as a plantation or a hacienda, was typically born as a commercial enterprise which created for itself the institutions which permitted it to respond to increased demand in the world or national market by expanding the amount of its land, capital, and labor and to increase the supply of its products. The fifth hypothesis is that the latifundia which appear isolated, subsistence-based, and semi-feudal today saw the demand for their products or their productive capacity decline and that they are to be found principally in the above-named former agricultural and mining export regions whose economic activity declined in general. These two hypotheses run counter to the notions of most people, and even to the opinions of some historians and other students of the subject, according to whom the historical roots and socioeconomic causes of Latin American latifundia and agrarian institutions are to be found in the transfer of feudal institutions from Europe and/or in economic depression.

The evidence to test these hypotheses is not open to easy general inspection and requires detailed analyses of many cases. Nonetheless, some important confirming evidence is available. The growth of the latifundium in nineteenth-century Argentina and Cuba is a clear case in support of the fourth hypothesis and can in no way be attributed to the transfer of feudal institutions during colonial times. The same is evidently the case of the post-revolutionary and contemporary resurgence of latifundia, particularly in the north of Mexico, which produce for the American market, and of similar ones on the coast of Peru and the new coffee regions of Brazil. The conversion of previously yeoman-farming Caribbean islands, such as Barbados, into sugar-exporting economies at various times between the seventeenth and twentieth centuries and the resulting rise of the latifundia in these islands would seem to confirm the fourth hypothesis as well. In Chile, the rise of the latifundium and the creation of the institutions of servitude which later came to be called feudal occurred in the eighteenth century and have been conclusively shown to be the result of and response to the opening of a market for Chilean wheat in Lima.[8] Even the growth and consolidation of the latifundium in seventeenth-century Mexico – which most expert students have attributed to a depression of the economy caused by the decline of mining and a shortage of Indian labor and to a consequent turning in upon itself and ruralization of the economy – occurred at a time when urban population and demand were growing, food shortages were acute, food prices skyrocketing, and the profitability of other economic activities such as mining and foreign trade declining.[9] All of these and other factors rendered hacienda agriculture more profitable. Thus, even this case would seem to confirm the hypothesis that the growth of the latifundium and its feudal-seeming conditions of servitude in Latin America has always been and is still the commercial response to increased demand and that it does not represent the transfer or survival of alien institutions that have remained beyond the reach of capitalist development. The emergence of latifundia, which today really are more or less (though not entirely) isolated, might then be attributed to the causes advanced in the fifth hypothesis – i.e., the decline of previously profitable agricultural enterprises whose capital was, and whose currently produced economic surplus still is, transferred elsewhere by owners and merchants who frequently are the same persons or families. Testing this hypothesis requires still more detailed analysis, some of which I have undertaken in a study on Brazilian agriculture.[10]

[8] Mario Góngora, *Origen de los 'inquilinos' de Chile central* (Santiago: Editorial Universitaria, 1960); Jean Borde and Mario Góngora, *Evolución de la propiedad rural en el Valle del Puango* (Santiago: Instituto de Sociología de la Universidad de Chile); Sergio Sepúlveda, *El trigo chileno en el mercado mundial* (Santiago: Editorial Universitaria, 1959).

[9] Woodrow Borah makes depression the centerpiece of his explanation in 'New Spain's Century of Depression,' *Ibero-Americana*, No. 35 (Berkeley, 1951). François Chevalier speaks of turning in upon itself in the most authoritative study of the subject, 'La formación de los grandes latifundios en México,' *Problemas Agrícolas e Industriales de México*, VIII, No. 1, 1956 (translated from the original French and recently published by the University of California Press). The data which provide the basis for my contrary interpretation are supplied by these authors themselves. This problem is discussed in my '¿Con qué modo de producción convierte la gallina maíz en huevos de oro?' and it is further analyzed in a study of Mexican agriculture under preparation by the author.

[10] Capitalism and the Myth of Feudalism in Brazilian Agriculture,' in *Capitalism and Underdevelopment in Latin America*.

All of these hypotheses and studies suggest that the global extension and unity of the capitalist system, its monopoly structure and uneven development throughout its history, and the resulting persistence of commercial rather than industrial capitalism in the underdeveloped world (including its most industrially advanced countries) deserve much more attention in the study of economic development and cultural change than they have hitherto received. Though science and truth know no national boundaries it is probably new generations of scientists from the underdeveloped countries themselves who most need to, and best can, devote the necessary attention to these problems and clarify the process of underdevelopment and development. It is their people who in the last analysis face the task of changing this no longer acceptable process and eliminating this miserable reality.

They will not be able to accomplish these goals by importing sterile stereotypes from the metropolis which do not correspond to their satellite economic reality and do not respond to their liberating political needs. To change their reality they must understand it. For this reason, I hope that better confirmation of these hypotheses and further pursuit of the proposed historical, holistic, and structural approach may help the peoples of the underdeveloped countries to understand the causes and eliminate the reality of their development of underdevelopment and their underdevelopment of development.

3 David Booth, 'Marxism and Development Sociology: Interpreting the Impasse'

Excerpts from: *World Development* 13 (7), 761–87 (1985)

. . . .

1. Dependency theory and the sociology of development: fifteen years after

The dominant feature on the horizon of radical development theory today is undoubtedly the decline and threatened, but never quite realized, disappearance of the dependency perspective as a widely accepted approach. I refer here to the general belief, influential in research on a number of parts of the world, that the development problems and hence the social structures and politics of less developed countries are to be understood primarily in terms of the particular nature of their insertion into the international capitalist system – rather than in terms of largely domestic considerations. Different writers in the dependency tradition have, of course, assigned different weights to the several constituent properties of 'dependent' or 'peripheral' status, emphasizing respectively trade, finance, ownership of productive assets, technology or ideology and culture. There have also been sharp disagreements about what precisely dependency theory is supposed to explain. Cutting across these differences, however, is a shared conviction that in the analysis of underdevelopment and patterns of change in Third World social formations, external relations determine the role of 'domestic' structural properties, not vice versa.

Although perhaps they were never as generally influential as some of us once imagined, dependency . . . conceptions of Third World reality have today, quite rightly, lost much of their pull. The views of Frank, Amin and Wallerstein, and to a lesser extent Sunkel, Cardoso and Quijano, have been vigorously criticized from a number of angles over many years. It is now commonplace to observe that radical development theory has generated rather less cumulative empirical research and more sterile controversy than might have been hoped, and it is increasingly usual to lay this at the door of the dependency perspective. I was comparatively slow to come around to this point of view but would now go somewhat further than most in drawing a balance sheet of the dependency phase of the new development sociology. It is now reasonably clear, I would argue, that the dependency position is vitiated by a variable combination of circular reasoning, fallacious inferences from empirical observation and a weak base in deductive theory.

The circular logic of crucial arguments in the dependency armoury was recognized at an early date but has perhaps never been given sufficient prominence. The dependency view of development is in fact fatally flawed on logical grounds at precisely the point where it seems strongest empirically – for example, in Frank's early (1966, 1967) historical sketch of the economic history of Latin America. Over the years Frank's early studies have been much maligned as representing an unusually vulgar or simplistic variant of the dependency position; yet they contain one of the very few attempts in the primary dependency literature to formulate a theory in the sense of a logically interrelated set of general propositions, and to derive and test empirical generalizations on the basis of research. Thus it is not perverse to take as a leading case in point here Frank's 1966 validation of the dependency position, the more so since in broad terms it exemplifies perfectly a form of argument still in common use.

Frank's main theoretical proposition asserted: 'contemporary underdevelopment is in large part the historical product of past and continuing economic and other relations between the satellite underdeveloped and the now developed metropolitan countries.' This generated the hypothesis that 'the satellites experience their greatest economic development and especially their *most classically capitalist industrial development* if and when their ties to their metropolis are weakest.' (Frank, 1966, pp. 4, 9–13). Several sorts of historical and comparative evidence were invoked in support of this proposition.

Today there are grounds for questioning some of the factual elements in Frank's proof, notably improved evidence on the timing of the first stages of industrialization in some of the larger Latin American economies. Nevertheless the crucial flaw remains the definition of 'development' (and hence 'underdevelopment') that was smuggled into the statement of the hypothesis, with the result that the proposition becomes tautologically true, rendering the historical material illustrative rather than corroborative. In the hypothesis and the accompanying text, satisfactory development is *identified* as 'self-sustaining' or 'autonomous' and (hence?) industrial growth; this is the meaning attached to 'classically capitalist' development. But this amounts to saying that it is non-satellite (or to use the language subsequently adopted, non-dependent) development. Hence the empirical demonstration that in historical fact satisfactory development occurs only to the extent that metropolis-satellite, or dependency, links are broken or weakened, is nothing of the kind, but an exercise in tautology. This argument or something rather similar has been forcefully advanced by Smith (1980, p. 13) with reference to the work of Samir Amin, and Warren (1980, pp. 160, 165–6) has pointed to the element of logical circularity in Dos Santos's famous attempt to define 'dependence.' The basic critique applies well to Rodney's (1972) essay on African underdevelopment and to several other influential studies of dependency and alternative development paths for Third World regions.

Circular, that is tautological, reasoning is central to dependency theory and to those variants of Marxist development theory which operate within a dependency framework; but it is perhaps not an intrinsic characteristic, and at any rate there is more to the tradition than circular arguments about autonomous and non-autonomous economic growth. Many Latin American dependency writers were concerned from the beginning not with structural underdevelopment or dependency conceived in these broad and problematic terms, but with more specific economic and social problems held to be characteristic of the latest phase in Latin America's relations with international capitalism. Focusing on the one hand on patterns of deteriorating income distribution, social 'marginalization' and authoritarian politics, and on the other on the role of multinationals, inappropriate technology and/or cultural alienation, this type of dependency proposition lends itself less readily to tautologous presentation. At least in principle, it is capable of being formulated as a set of substantive hypotheses linking proposed causal factors to independently identified effects.

To recognize the existence of such hypotheses is, however, by no means to accept them. There remains the very live issue of whether they can be validated empirically, that is whether marginalization and related processes can be shown to be the result of factors of 'dependence' such as the colonization of the most dynamic sectors of the economy by transnational capital, as opposed to other factors suggested by other kinds of development thinking. In the original dependency literature (e.g. in the work of Cardoso, Sunkel and Quijano) the answer to this question tended to be presented as self-evident on the basis of a small range of broadly similar national experiences, frequently those of the larger Latin American countries. In other words no systematic effort was made to distinguish the effects to transnationalization *per se* from those of the local social and political context, the prevailing economic policy regime and so on. More recently there has been a trend towards the use of cross-national statistical testing to vindicate dependency propositions. However, despite protestations to the contrary the bulk of this work has not escaped the empiricism of the original dependency formulations, so that the conditions for drawing valid causal inferences from statistical analyses have not usually been met. Systematic and theoretically informed comparative analysis has in contrast, seldom been employed by those most sympathetic to the dependency viewpoint, and the results of such exercises have typically proved unfavorable to the standard dependency claims.[1]

The required evidence is of course patchy. Nevertheless the essays by Lall (1975) and Weisskopf (1976) have made effective use of comparative analysis to distinguish the effects of dependence (variously defined) from those of capitalism in general and/or from those of particular policy patterns or institutional arrangements. Drawing on a very wide range of empirical sources, Morawetz (1977, pp. 40–1) has concluded that 'fast-growing, market-oriented countries,' that is successful 'dependent developers,' include *both* cases where the income share of the poorest has declined *and* a group of countries (conspicuously excluding Latin American ones) whose record by this criterion has not been at all bad. Morawetz suggests the working hypothesis (which would seem to be strongly supported by the historical record of certain East Asian countries) that what determines the trend is the degree of inequality at the outset.

Dependency Marxism would seem, then, to be logically flawed where it appears at first sight to be supported by important historical evidence, and incapable of withstanding serious comparative analysis even where it is logically sound. This ought not to be surprising, because, additionally, writers in the dependency tradition have typically worked from an extraordinarily weak base in deductive economic theory. A number of critics (Leys, 1977; Phillips, 1977; Bernstein, 1979; Taylor, 1979; Weeks, 1981) have made much of the fact that dependency and underdevelopment theory is not rooted in a rigorous application of Marxist economic theory, but unless one lives in a thoroughly Manichean world of proletarian truth and bourgeois error, the point is surely that it is not rooted in *any* rigorous body of deductive-type theory. This is not to say there are no economic ideas behind dependency theory, but dependency writers tend to be either almost literally the slaves of defunct economists or amateurish and uncritical consumers of economic literature.

Dependency theory was the child of its time in both a passive and an active

[1] Rhys Jenkins's study of the Latin American auto industry (1977) might seem to be an exception on both counts; but both the conclusion, that there are important long-run, dynamic benefits from retaining domestic ownership and control, and the analysis, which considers foreign ownership in the context of several other explanatory factors, place this book more in the reputable tradition of Latin American structuralism – or alternatively in a 'post-dependency, post-neoclassical' phase – than in the dependency tradition as considered in this paper. A recent work of somewhat similar scope by a sociologist (Gereffi, 1983) contains some particularly lucid remarks about what is wrong with standard dependency evaluations of the impact of transnationals (e.g., pp. 60–1) and would similarly seem to qualify as something other than dependency analysis.

sense. Dependency writers assumed unquestioningly either one or both of the theories influential for a short time in the 1960s to the effect that participation in world trade was likely to be secularly impoverishing for less developed countries (qua primary producers, or as low-wage areas). This remained the case even after these theories were subjected to devastating criticism and academic support for them in their original form was reduced to a rump inside certain international bureaucracies.[2] Other concepts which passed rapidly out of vogue after the mid-1960s but were permanently absorbed into the dependency lexicon, include the target notion of 'rapid and self-sustaining growth' and the belief that shortages of local savings and capital (or 'surplus') are among the critical obstacles to development in the typical Third World country.

Dependency also took over and transformed in a more active way the trend in nationalist thinking – not only in Latin America – in the late 1950s and the 1960s towards the view that the causes of the apparently multiplying difficulties of the national development process were located 'outside' rather than 'inside' the national society. The crucial juncture came with the realization, first in Latin America and East Asia, later elsewhere, that industrialization by import substitution was deepening rather than resolving serious social inequities and balance-of-payments problems. Careful studies of various theoretical and ideological complexions were eventually to appear that documented analytically and empirically the ways in which these problems could be seen to be inherent in the ISI policy package. These studies agreed in indicting ISI policies, or their more extreme manifestations, as sufficient causes of the regressive income-distribution trends and external vulnerability (and in this special sense dependency) observed in many semi-industrialized countries. The odd unusually sophisticated defender of the dependency approach (e.g. Seers, 1977) has tried, not very convincingly, to show that dependency theory can subsume such insights. But there really is not much doubt that mainstream 1960s dependency theorists were taken in by a perfectly ordinary spurious correlation. Advanced import substitution was associated with an invasion of manufacturing multinationals, growing external vulnerability and regressive trends in employment and income distribution. Plausibly but wrongly the last two things were laid at the door of the first. For a variety of reasons this was a mistake most easily committed in Latin America, and later easily transported to Africa.

[2] This refers to the 'Prebisch thesis' and the original, neo-Marxist formulation of unequal exchange theory by Arghiri Emmanuel (1972). When dependency theorists explain that what was novel about their contribution in the late 1960s was not a stress on 'external' factors but the conceptualization of dependence in terms of the internal structure of societies (e.g., Cardoso, 1977a, pp. 12–13, or Quijano, 1977, pp. 99–103) they implicitly underline this uncritical assimilation of the ECLA/UNCTAD view of international trade. Cardoso (1977b) illustrates the point in another way. Immanuel Wallerstein has to my knowledge never explained the theory of unequal trade on which his enormously influential world-system concept (Wallerstein, 1974, 1979, 1980) crucially depends. For an up-to-date professional view on the Prebisch thesis, see Spraos (1983).

We may reasonably say, in sum, that the dependency position has been shown to be untenable on a combination of logical, analytical and theoretical grounds. Something of the kind is increasingly recognized even, and in some respects particularly, among scholars who contributed significantly to the dependency inspired critiques of the older development literature. Vigorous rebuttals of the dependency approach have come from scholars whose previous work placed them squarely within that tradition. And yet the change in the intellectual scene is in some ways more apparent than real.

. . . .

Major foci of empirical work in the sociology of development such as the formation of national bourgeoisies and working classes have continued to be dominated by false problems and schematic tendencies derived from dependency Marxism, as is now beginning to be recognized. In alliance with a form of Marxian state theory that scarcely anybody now defends, dependency assumptions are a pervasive influence on studies of politics – including the politics of meaningful reform – in major developing countries, contributing to the obfuscation of real political choices in a number of areas. These contradictions between the ground that has been covered in theory and the actual practice of development sociology is a striking and disturbing feature of the present state of the field.

2. Warren's challenge: strengths and weaknesses

Similarly unsatisfactory is the way the literature has responded, or failed to respond, to the revival of radically anti-dependency Marxism in the work of Bill Warren and others. Warren has provided what is arguably the most thorough and courageous critique of the dependency Marxist viewpoint. He also goes some way towards identifying the political 'uses' of dependency Marxism and thus explaining in part its persistent attractions in some quarters. But although his position is beginning to be accorded a grudging respect on this basis, the full implications are far from having been grasped.

This should not be interpreted as an endorsement of Warren's approach. On the contrary I think there are solid reasons why even those like myself who think Warren's main theses are supported by the evidence, should be unenthusiastic about embracing them as an alternative framework. Interest here focuses on the arguments in the second half of Warren's book regarding the effects of colonialism and of the post-colonial international system on the development of the productive forces in Third World countries. These are summarized as follows:

(i) 'Contrary to current Marxist views, empirical evidence suggests that the prospects for successful capitalist development in many underdeveloped countries are quite favourable . . .'

(ii) '[T]he period since the end of the Second World War has witnessed a major surge in capitalist social relations and productive forces in the Third World.'

(iii) 'Direct colonialism, far from having retarded or distorted indigenous capitalist development that might otherwise have occurred, acted as a powerful engine of progressive social change . . .'

(iv) 'Insofar as there are obstacles to [capitalist] development, they originate not in current relationships between imperialism and the Third World, but in the internal contradictions of the Third World itself.'

(v) The overall, net effect of the policy of 'imperialist' countries and the general economic relations of these countries with the underdeveloped countries actually favours the industrialization and general economic development of the latter.'

(vi) 'Within a context of growing economic interdependence, the ties of '*dependence*' (or subordination) binding the Third World and the imperialist world have been and are being markedly loosened with the rise of indigenous capitalisms. . . .' (Warren, 1980, pp. 9–10)

I do not suggest that these propositions, even when elaborated as they are in the original text, are unproblematic. However, Warren's main assertions are well informed and well supported, and not open to the kinds of objections that have been most commonly raised. Although perhaps propositions (iii) and (iv) remain controversial, much of what Warren has to say is rather standard stuff in development economics, and the fact that it was necessary to insist so much on these points says something about the way left-tending social scientists and activists have seen fit to close their minds to pertinent mainstream literature. Overall, it would not be easy to establish that any of his carefully worded propositions are actually wrong.

Even while stressing this aspect, however, it is hard not to be struck by a certain perversity in the way Warren's theses organize the evidence, and it is difficult to dismiss the feeling that while what he has said is true, it is in a certain sense not the whole truth. Specifically, I would argue, his views on the current prospects and problems of Third World development are subject to important *non*-dependency objections centering on the level and mode of abstraction employed. The basic feature is a single-minded and unremitting concentration on the general, intrinsic and mainly economic qualities and effects of development understood as the unfolding or diffusion of the capitalist mode of production. This theoretically determined approach has several consequences or corollaries.

First, in examining the postwar economic experience of the less developed world, Warren frequently acknowledges variations in the general pattern of development progress, but consistently downplays them. There is an almost explicit theoretical denial of the possibility of important systematic variations within the general pattern. The book does not claim that the surge in the forces of production since the War has affected all areas of the Third World equally, and prospects are said to be favorable only in 'many' underdeveloped countries. This is probably correct, but certainly insufficient. The past performance of different regions and countries of the less developed world has been notoriously uneven, and by and large the poorest countries have done least well. It is tragically the case that the prospects for the next decades vary between countries from excellent (maintenance of GDP growth rates of 8–12% per year) to disastrous (negative growth in per capita if not absolute terms). Warren's method of argument takes us from the indiscriminate pessimism of the dependency view of the world to a barely less misleading generalized optimism. Worse, it distracts attention from some of the most deplorable aspects of the contemporary situation, and hence from the exploration of the underlying causes.[3]

As a rule, secondly, specific national policy regimes and institutional arrangements get extremely short shrift in Warren's treatment of the variations in performance between countries. This is a feature of his discussion of income distribution (the interesting experience of postwar Taiwan in mentioned only to subsume it within a general vision in which institutions and policies are entirely subordinate to the stage of the development process that has been reached), and it is also true more generally. Obviously, there are various both positive and negative features of the Third World development experience that can be said to be inherent in capitalist (if not all) development. But equally clearly, there are achievements and failures that must be laid squarely at the door of governments and their policies. Warren recognizes this but is most unwilling to accept its implications. On the last three pages of the book, he tells us of 'major policy blunders' that resulted in a 'squandering of many of the benefits of Third World postwar economic development;' a different set of policies 'would have permitted the promotion of a more efficient and humane capitalist development' than actually occurred. But this, we are assured, 'has failed to halt the gathering momentum of capitalist advance,' so the discussion of policy errors appears merely as a gloss on the main theme, an illustration of the silliness of failing to recognize capitalist development when it occurs (1980, pp. 253–235; cf. pp. 177, 236).

Warren's approach, I am suggesting, is unhelpful on the role of national policy regimes and institutional arrangements. For related reasons, thirdly and fourthly, it has difficulty dealing with direct colonialism and its aftermath, and generates a very unsatisfactory treatment of nationalism and 'the national question.' In both areas, the problems arise from the attempt to understand the

[3] An important instance is the treatment of trends in income distribution. Major variations in country performance, including between countries with similar rates of economic growth, are first acknowledged and then set aside in the interests of the general argument. Warren also commits the textbook fallacy of moving directly from current cross-section data to conclusions about the March of Progress, espousing an evolutionary view of the relation of inequality to growth that was fashionable among development economists twenty or more years ago but has lost ground for very good reasons in recent times (pp. 199–211).

development of national social formations relying entirely on a general concept of capitalism and its dynamics.

. . .

In sum, Warren's approach is limiting in a series of connected ways, not because the authors and positions he criticizes are after all right, but because of the nature of the theoretical framework he offers in their place. The capitalist mode of production and its dynamics (including 'uneven development') are offered as sufficiently explaining both the ugly and the attractive faces of development in the Third World today and in the recent past: both aspects are simply a function of normal capitalist development. But important variations within the general pattern are certainly not sufficiently explained by differences in the evolution or spread of capitalist social relations. Moreover, there is much that is of interest and importance in the contemporary experience of the less developed countries which escapes the attention of this framework more or less entirely. This makes it virtually unuseable as a framework for social science research, let alone politics or policy formation.

In light of the above comments, the fact that the demise of dependency theory has not led to the emergence of a flourishing Warrenite or 'classical Marxist' development sociology, is neither surprising nor particularly regrettable. What is to be regretted is that we have not yet entered a post-Warren era in sociological development research. There is a widespread sense of the futility of pursuing the dependency-Warren debate any further. Yet one feels the sources of Warren's 'extremism' have not been properly grasped or the implications thought through; rather as in the previously considered case of the critique of dependency theory, the response to Warren has been in the first place overpersonalized and then stronger on perceiving theoretical weaknesses than on grasping the nettle of metatheory – the reasons why a given intellectual tradition articulates problems *for* theory in the way that it does. In summary, while the need to move on from the type of generalized controversy represented by Warren *vs.* dependency is now obvious and widely accepted, there is as yet little real understanding of what this might entail.

3. The mode of production debate: impasse within an impasse

It may well be thought that the necessary theoretical basis for a development sociology which cuts across the simplistic polarities of the 1970s already exists in the form of the large and now quite diverse literature on subordinate modes of production, 'forms of exploitation' and the formal subsumption of labor under capital. After all it has been over a decade since the debate prompted by Laclau's (1971) seminal article on 'Feudalism and Capitalism in Latin America,' which was itself widely regarded as an important step beyond the dependency problematic. Since then, research on subordinate forms of production has been one of the most vigorous areas of work in the sociology of development. Arguably Frank's transitive underdevelopment theory and Warren's evolutionary mode of production theory are unrepresentative polar positions about which it is no longer necessary to concern ourselves.

There is some truth in this, in my view, but as a comment on the empirical strengths of the work in this area, not on its theoretical basis. The growing aridity and final inconclusiveness of the 'mode of production debate' is now widely acknowledged. Careful observers (e.g., Foster-Carter, 1978; Goodman and Redclift, 1981) also admit that the attempt to resuscitate the concept of mode of production and its associated categories has produced a crop of seemingly unresolvable conceptual problems. What the debate has served to establish most sharply, I would argue, however, is the impossibility of steering a middle course between dependency and the so-called 'classical' (Warren) position using the concepts of theoretical Marxism.

From a theoretical point of view, there have been three outstanding contributions to the mode of production debate, each of which has contributed important clarifications, so that the level of general understanding of the issues involved is now very high by the standards of the recent past. Each presents a facet of the truth about the rigorous application of classical Marxist concepts to the purposes of development theory. But because the positions adopted have also proved to be mutually exclusive, the controversy has not led to a progressive resolution of the outstanding issues. What has been established instead is that the concept of mode of production is subject to multiple and in practice contradictory theoretical requirements which make it incapable of consistent application to the task of illuminating world development since the sixteenth century.

Laclau (1971) argued convincingly that the 'world capitalist system' of Frank's early studies confused the realms of production and exchange, giving in effect causal primacy to the latter. This was not only at variance with Marx's method in which circulation is subordinate to the process of production, but it produced a theory of underdevelopment that could not account for certain important aspects of modern world history, and was in a general sense lacking in explanatory power. Recognizing that capitalism and feudalism are distinguished by their different relations of production, rather than in the sphere of exchange, we can admit the fundamental concept that the world economic system has been marked since the seventeenth century (not the sixteenth) by the combination (or as later writers were to say, articulation) of elements belonging to different modes of production – with pre-capitalist social relations of production being restructured but not eliminated by the intrusion of merchant capital.

Laclau's analysis, however, was quite weak where its claims were boldest. The propensity of capitalist expansion in certain times and places to promote the reinforcement and even restoration of precapitalist relations in agriculture was attributed not to such factors as the local dynamics of labor supply and class struggle, but to the needs of capital in the metropolitan countries understood in terms of a variant of the Marx/Bukharin/Lenin theory of capital exports to backward regions (pp. 35–7). Leaving aside the scientific status of the parts of Marx's economic theory to which the latter belongs, this account was subject to the well-worn objection that it did not and probably could not specify the mechanisms by which what capital 'needed' was translated into reality at the local level. In this sense, the explanatory power of the alternative theory of underdevelopment rested largely on bluff. The claim that the theory was rooted in production relations, understood in terms of groups of producers and non-producers and their struggles was also illusory.

On the other hand, Laclau undoubtedly detected a major theme in Marx and drew attention once again to the point made in the 1950s by Dobb (Sweezy *et al.*, 1976) that comparative economic and social history provides many suggestions as to the correctness of a 'productionist' approach to conceptualizing transitions to capitalism. In opposition to this view, there remain a few Marxist scholars who take an unrepentantly 'circulationist' position in the tradition of Sweezy's 1950s contributions (*ibid.*). In general, however, the basic point has been well taken, and what has drawn the attention of critics has been the 'formalism' of Laclau's critique of Frank, and his only partial grasp of Marx's theory of modes of production and the transitions between them. The writings of Alavi (1975, 1982) and Banaji (1972, 1977) in particular have drawn attention to another aspect of the role of capitalism in the periphery, and to another major theme in Marx.

The common starting point in this second stage of the debate is that it is formalistic and wrong to characterize the social relations resulting from colonial imposition in Asia, Latin America and elsewhere as feudal, and in this sense non-capitalist. In Marx, it is argued, the feudal mode of production is a system of localized production and appropriation with a specific place in a whole epoch of European history. This is in contrast with the experiences of areas like north-eastern Europe in the sixteenth century and after, and parts of Latin America in the late nineteenth century, where feudal-type tenancies were introduced in response to the stimulus of long-distance trade. The latter forms are typical of what happened in the territories of direct colonialism, which represents a distinct epochal reality (Alavi, 1975; Banaji, 1972; also C. F. S. Cardoso, 1976).

What makes it impossible to dismiss this argument as a simple resurgence of un-Marxist 'circulationism' is the accompanying critique of Laclau's understanding of the concept of mode of production. Alavi (1975, p. 182) stresses 'the inadequacy of any conception of the "mode of production" that is premised narrowly on sets of relationships that are arbitrarily assigned to the "structure" ignoring the totality'; for Marx the totality was fundamental. To put it another way, in Marx the capitalist mode of production (the only mode about which he wrote extensively) was not just 'an articulated combination of relations and forces of production,' but also, or instead, 'a definite totality of historical laws of motion.' This being the case, 'relations of production . . .

become a *function of the given mode of production.*' 'The character of any definite type of production relations, is . . . impossible to determine until these laws of motion are themselves determined' (Banaji, 1977, p. 10). This accounts for some references in Marx and in subsequent Marxist analysis to the presence of various 'relations of exploitation' other than wage labor in production systems dominated by capital. It also explains why Marx leaves us in no doubt that crises of the transition apart, only *one* mode of production can prevail in any one place at any single time.

To the extent that this interpretation is correct, it becomes absurd to talk about combinations of (plural) *modes* of production as a permanent feature of Third World social formations. The issue becomes a double one: first, which are the forms of exploitation of labor that are being employed; and second and above all, what are the laws of motion under which these are subsumed? Most of the writers who have posed the matter in this way have had recourse, at least in the first instance, to a 'colonial mode of production,' a construction designed to admit *both* that a variety of relations other than that of wage-labor typified the periphery in the colonial era *and* that these 'forms of exploitation' had their *raison d'être* in a totality embracing the metropolitan economy. However, the colonial mode idea was short-lived.

The authors of the concept were among the first to admit that even at a common-sense level it is full of holes. Alavi concedes that there is a 'highly problematic area' relating to whether the 'colonial mode of production' leaves any room for capitalism proper, since on the criteria established by this literature the metropolitan economy cannot be involved in both. An additional question-mark hangs over whether and why a new (post- or neo-colonial) mode should be said to intervene as a consequence of the achievement of administrative Independence by ex-colonial territories (Alavi, 1975, pp. 190–3; Foster-Carter, 1978, p. 72).

At a more fundamental level, the colonial mode concept has proved unstable. Despite the strong emphasis placed on the specification of laws of motion as the means of conceptualizing a mode of production, what the 'laws' of the colonial mode are, and what if anything distinguishes them from the 'laws' established by Marx for capitalism, has not been explained beyond the assertion that whereas the latter rapidly expand the forces of production, the former do not (Banaji, 1972, p. 2500; Alavi, 1975, p. 187). At the same time it has been impossible to ignore two aspects of Marx's method in this area: on the one hand; the insistence on the necessity of uncovering laws of motion; and, on the other hand; the belief that the laws of motion of a specific mode arise from and are in some sense determined by the character of the prevailing combination of relations and forces of production. In this regard, different authors have moved in different, and indeed opposite, directions.

Alavi represents a tendency to allow a strict understanding of these terms to go by the board in favor of the empirical investigation of the circuits of capital and forms of labor recruitment of what comes to be called colonial capitalism or peripheral capitalism. This tendency now occupies the murky middle-ground between classical Frank and Laclau, and is barely distinguishable in theoretical terms from what is today the position of Frank himself and Samir Amin. In this literature Marx's now famous distinction (1976, pp. 1019ff.) between the merely formal subsumption of the labor process under capital and the 'real' subsumption characteristic of 'the specifically capitalist mode of production in its developed form' is increasingly used to buttress the concept of peripheral capitalism. But this idea, too, is unstable; it is also applied (following Marx literally) in a more evolutionary spirit, so that formal subsumption represents a more or less extended *moment* in the transition to capitalism.

Some have interpreted this sort of conceptual indeterminacy as a manifestation of the 'ideological' character of dependency/underdevelopment theory (Bernstein, 1979, p. 88). At least in part, however, this is to confuse cause and consequence. I believe Alavi and company have cast themselves back into the arms of Frank, so to speak, because only one other step seems conceptually allowable: it is either this or take the course apparently preferred by Banaji and revert to the conception of the capitalist mode of production and its relation to pre-capitalism that is expressed with different emphases and on different terrains by Warren and Brenner.

Brenner's article (1977) is deservedly famous not only for its fascinating demolition of the circulationist interpretation of the history of uneven development in the early modern world, but also for its lucid exposition of the Marxian view of capitalism. The link between the emergence of a system of free wage labor and competition between capitals on the one hand, and the dynamic transformative characteristics of modern capitalism on the other, is well explained and convincingly supported on the broad canvas of modern world history. There is much to be said for this 'classic' view of matters. But for all its brilliance, Brenner's essay does not give us what many people have looked for in the mode of production literature, namely, a genuine third position in the debate over colonial and contemporary development in the Third World.

Brenner's theory is limited in one rather obvious way: it tells us nothing about the contemporary, post-Independence Third World, or indeed about anything much subsequent to the decline of feudalism and slavery (the dating of which is in any case unclear). The vision of capitalist development, once that process is under way, is broadly optimistic, at least as far as the forces of production and the improvement of the productivity of human labor are concerned, but it is also highly unspecific. In this sense, it does not by itself offer a comprehensive alternative to dependency and colonial-mode theories. Brenner's contribution is best treated as an interesting, and across a narrow historical field, thoroughly convincing footnote to the only real alternative to dependency within Marxism: the one espoused by Warren.[4]

[4] The relevance of Brenner's thesis is actually even narrower, and its critical impact on dependency/world-systems work weaker, than appears at first sight: Brenner barely discusses the phenomenon of colonialism. The issue is taken to be between those who emphasize external, trade-related causes of development and underdevelopment, and those who see that the decisive

To sum up the discussion in this section, it has not proved possible to bring the mode of production controversy to a satisfactory close. Wolpe's perceptive article (1980) notwithstanding, the obstacles to an adequate theory of articulation are not so insubstantial as to yield to some simple conceptual innovation (such as the distinction between 'restricted' and 'extended' concepts of mode of production) useful as such things may be for expositional purposes. The truth of the matter is that the debate has defined several positions which are for different reasons irreconcilable with Marxist theoretical norms, and none which are genuinely independent of the theoretical poles of dependency and Warrenism. All roads lead to one or other of the basic variants of Marxist development theory, and the mode of production concept as such is no guarantee against either.

. . . .

4. Interpreting the impasse

Until now, the development debate as it has affected sociology has taken the form, almost exclusively, of exchanges between different poles of opinion within Marxism: circulationism vs 'productionism,' dependence (in its Marxist or *marxisant* versions) vs 'classical Marxism,' and so forth. I would argue that it is now time to stand back a bit from the controversy and allow the light of scrutiny to fall on some features which are *common* to all of the major contributions. We also need to move on from purely theoretical to *meta*theoretical considerations of a certain type. While previous writers (Leys, 1978; Bienefeld, 1980; Bernstein, 1982; Browett, 1983) have in different degrees indicated an awareness that, for example, the positions of Frank and Warren mirror one another's weaknesses, this insight has not been taken very far. To my mind it ought to be pressed further, so that we are asking not just in what senses both Frank *et al.* and Warren were wrong but also, in a deeper sense, *why* they were wrong – what it was that led them to advance and persist in false or theoretically limiting positions. This type of questioning, which can be applied also to the theoretical and empirical modes-of-production/subsumption-under-capital literatures, should also shed light on the inability of the standard critiques to prevent the reproduction of the old errors in new guises.

determinants are internal social relations, class struggles and systems of power. In fact, of course, the case for viewing underdevelopment as an externally imposed condition rests in part on a particular view of the effects of direct colonialism and other non-economic relations between nations and territories. In *this* context, it is often perfectly correct to see domestic class structure as derivative of inter-societal relations: colonial powers dramatically transformed the social and economic fabrics of the places they colonized, and where in conformity with Brenner's thesis, mere commerce proved insufficient to break down pre-capitalist social relations, they took whatever other steps were necessary. To the extent that they have been among those drawing attention to this aspect of the history of the Third World, dependency theorists can hardly be accused of economic determinism and 'neo-Smithianism;' in fact, the boot of economism would seem to be on the other foot – in a sense which recalls our discussion of Warren above. Thus, Brenner's critique seems to be doubly limited, to a particular historical period, and to the historically rather special case of pure trade.

The conclusion I have reached is that there is a basic problem with Marxist theory as an input to development sociology that transcends the particular forms in which it has been manifested. This is its metatheoretical commitment to demonstrating that what happens in societies in the era of capitalism is not only explicable, but also in some stronger sense *necessary*. This is what is most fundamentally wrong about the 'dependency debate' as it has usually been conducted over the last decade; it is because they share this underlying commitment that both sides in the debate are in different degrees myopic and one-dimensional. The same urge to establish that prevailing patterns of exploitation are not only explicable but necessary also accounts for the aridity and repetition in the modes/subsumption literature. Overall, this is what explains the inability of the radical literature genuinely to go beyond itself even years after the need for some decisive advance has been recognized.

The Marxist commitment to the 'necessity' of socio-economic patterns under capitalism is expressed in two main forms in the development literature. The first operates through the way in which it is usual to conceive of the relation between the theoretical concept of the capitalist mode of production and the national or international economies, polities, and social formations under analysis. The other – if anything more persistent and fundamental – involves a form of system teleology or functionalism. As components of Marxist theory in general, both forms have been usefully discussed by a number of recent writers, the most important of whom are Hindess and Hirst and their collaborators (Hindess and Hirst, 1977; Cutler *et al.*, 1977–78). However, the implications of the new critical insights for development studies in particular have not been delineated.

. . . .

5. Conclusion

This paper has sought to explain the nature and examine the causes of the current impasse in the 'new,' Marxist-influenced development sociology. The bulk of the discussion has been devoted to familiar themes – the demise of dependency theory, the 'classical' alternative expounded by Bill Warren, the mode of production controversy, and high and low points in recent sociological development research. At this level, my contention has been that the impasse is indeed a general one: not the product of the weaknesses of one particular theoretical perspective (dependency) or even of a mutually contradictory pair of perspectives (dependency vs Warren) but the result of a generalized theoretical disorientation affecting in different degrees all of the main positions in the radical development debate.

The once-dominant dependency approach, I argued, has been subjected to what ought to have been fatally damaging criticism on logical, analytical and theoretical grounds. But partly because in various ways the radical literature has conspired to weaken the impact of this attack, the dependency perspective has refused to die, continuing to influence much of the published work in the field and contributing to the lack of a serious body of cumulative research in such areas as the sociology of class. On the other hand, no significant 'Warrenite' development sociology has emerged in response to the crisis of dependency Marxism, for the good reason that Warren's theory is systematically limiting: the complex and challenging issues of development in the Third World today cannot be sufficiently grasped in terms of the dynamics and differential spread of the capitalist mode of production, the theoretical primacy of which is the hallmark of the 'classical-Marxist' approach. This much is perhaps beginning to be accepted, but what lies *behind* the respective theoretical limitations of the dependency and Warrenite views has not been explored very far until now.

This is partly because the notion still persists that there are promising avenues of research employing fundamental Marxist concepts that occupy a sort of middle ground between the polarized paradigms of the 1970s. However, a re-examination of the mode of production controversy does not suggest that there is any such middle position, or indeed that the central concepts involved are capable of consistent application to the subject matter of development studies. The modes/subsumption literature has produced some useful empirical research – cumulative at a certain level – on peasantries and the urban poor, but despite rather than because of the input from theory. It has also left some notable lacunae. Although committed to a political economy approach, the new development sociology has tended to shun actual collaboration with economists and has failed to contribute to the extent that one might have hoped to the illumination of current development issues.

The reasons for these serious limitations, I have argued, are theoretical, but what accounts for the persistence in false or theoretically limiting positions and the repeated reproduction of old fallacies in new guises, is a matter of metatheory. Behind the distinctive preoccupations, blind spots and contradictions of the new development sociology there lies a metatheoretical commitment to demonstrating that the structures and processes that we find in the less developed world are not only explicable but necessary under capitalism.

This general formula covers two variants: the type of necessity entailed by the Marxian insistence that the salient features of capitalist national economies and social formations can be derived or 'read off' from the concept of the capitalist mode of production and its laws; and another, also inspired in Marx's theory, that involves a system teleology or functionalism. The critique of 'reading off' would seem to add something significant to our appreciation of the one-dimensional character of Warren's theory and the inconclusiveness of the debate started by Laclau on the analysis of modes of production. On the other hand the discussion of functionalism seems to cast light on the persistent weaknesses, and the persistence in weakness, of those forms of dependency and articulation analysis less directly descended from the Marxism of *Capital*.

Jointly, I suggest, the two modes of necessitist metatheoretical commitment constitute the basic underlying cause of the current impasse of the new development sociology and the main obstacle to be removed if we are to do better in the future.

It would not be appropriate for this paper to conclude with an attempt to map out in detail the problems of research and analysis to which development sociology should turn its attention during the later 1980s and beyond. The paper has tried to uncover the causes of a general malaise in a certain broad field of enquiry: it has not attempted the sort of survey that can result in the construction of a new research agenda. Substantive problems and new lines of theoretical advance will no doubt emerge from the usual encounter between the crises and sufferings of the world and the intellectual creativity of us all. What is more uncertain is how far this enterprise will continue to be dogged by the sorts of problems discussed at length in the paper.

A final clarification is in order. If the analysis presented here is correct, the main idea to be taken on board is not that a particular tradition of theory, Marxism, has proved to contain some central difficulties and needs to be looked at in a more critical light in the future. This is certainly an implication of what has been said, but the essential point is in one sense more specific, and in another more general. First, the objection is not to the Marxist tradition as a whole (either in the name of some alternative tradition or on empiricist grounds), but to theoretical formulations of a particular type that are central to Marxism and happen to have been influential in recent years primarily in a Marxist form. Development sociology does not need to be purged wholesale of questions and lower-order concepts derived from Marx, but specifically of abstract entities conceived as having 'necessary effects inscribed in their structure' or as being endowed with the capacity to shape socio-economic relations in accordance with their 'needs.' Curiosity about why the world is the way it is, and how it may be changed, must be freed not from Marxism but from Marxism's ulterior interest in proving that within given limits the world *has* to be the way it is.

The second important idea is that there is a case for a relative (and hopefully temporary) shift of emphasis in the sociological development debate from theory to metatheory. Along with a revitalized interest in the real-world problems of development policy and practice, an enhanced sensitivity to questions of this type seems essential if we are to get out, and stay out, of the impasse discussed in this paper.

Selected References

Alavi, H. 1975: India and the colonial mode of production, in *The Socialist Register, 1975*, Ralph Miliband and John Saville (eds). London: Merlin Press.

Alavi, H. 1982: The structure of peripheral capitalism, in Alavi and Shanin.

Alavi, H. and Shanin, T. (eds) 1982: *Introduction to the Sociology of Developing Societies*. London: Macmillan.

Amin, S. 1976: *Unequal Development: An Essay on the Social Formations of Peripheral Capitalism*. New York: Monthly Review Press.
Banaji, J. 1972: For a theory of colonial modes of production. *Economic and Political Weekly* 52 (23 Dec.).
—— 1977: Modes of production in a materialist conception of history. *Capital and Class* 3 (Autumn).
Bernstein, H. 1979: Sociology of underdevelopment vs. sociology of development?, in Lehmann.
—— 1982: Industrialization, development and dependence, in Alavi and Shanin.
Bienefeld, M. 1980: Dependency in the eighties, in Bienefeld and Godfrey.
—— and Godfrey, M. (eds) 1980: Is dependency dead?, special issue of *IDS Bulletin* (Sussex), 12 (1) (Dec.).
Brenner, R. 1977: The origins of capitalist development: A critique of neo-Smithian Marxism. *New Left Review* 104 (July–Aug.).
Browett, J. 1983: Out of the dependency perspective, in Limqueco and McFarlane.
Cardoso, Ciro F. S. 1976: Los modos de producción coloniales: Estado de la cuestión y perspectiva téorica, in Roger Bartra *et al.*, *Modos de Producción en América Latina* (Lima: Delva).
Cardoso, F. H. 1972: Dependency and development in Latin America. *New Left Review* 74 (July–Aug.).
—— 1977a: The consumption of dependency theory in the United States. *Latin American Research Review* 12 (3).
—— 1977b: The originality of a copy: CEPAL and the idea of development. *Cepal Review*, (2).
—— and Faletto, E. 1979: *Dependency and Development in Latin America* (Berkeley: University of California Press).
Cutler, A., Hindess, B., Hirst, P. and Hussein, A. 1977–78: *Marx's 'Capital' and Capitalism Today*, 2 vols. (London: Routledge & Kegan Paul).
Dos Santos, T. 1970: The structure of dependence. *American Economic Review* 60 (2) (May).
Emmanuel, A. 1972: *Unequal Exchange: A Study of the Imperialism of Trade* (New York: Monthly Review Press).
Foster-Carter, A. 1978: The modes of production controversy. *New Left Review* 107 (Jan–Feb.).
Frank, A. G. 1966: The development of underdevelopment. *Monthly Review* 18 (4) (Sept.).
—— 1967: *Capitalism and Underdevelopment in Latin America* (New York: Monthly Review Press).
Gereffi, G., 1983: *The Pharmaceutical Industry and Dependency in the Third World* (Princeton: Princeton UP).
Goodman, D. and Redclift, M. 1981: *From Peasant to Proletarian: Capitalist Development and Agrarian Transitions*. Oxford: Basil Blackwell.
Hindess, B. and Hirst, P. 1977: *Mode of Production and Social Formation: An Auto-Critique of 'Pre-Capitalist Modes of Production'* (London: Macmillan).
Jenkins, R. 1977: *Dependent Industrialization in Latin America: The Automotive Industry in Argentina, Chile, and Mexico* (New York: Praeger).
Laclau, E. 1971: Feudalism and capitalism in Latin America. *New Left Review* 67 (May–June 1971); reprinted with a postscript in his *Politics and Ideology in Marxist Theory* (London: New Left Books 1977).

Lall, S. 1975: Is 'Dependence' a useful concept in analysing underdevelopment? *World Development* 3, (11/12) (Nov.–Dec.).
Lehmann, D. (ed.) 1979: *Development Theory: Four Critical Studies* (London: Frank Cass).
Leys, C. 1977: Underdevelopment and dependency: Critical notes. *Journal of Contemporary Asia* 7 (1).
—— 1978: Accumulation, class formation and dependency: Kenya, in Ralph Miliband and John Saville (eds), *The Socialist Register, 1978* (London Merlin Press).
Limqueco, P. and McFarlane, B. (eds) 1983: *Neo-Marxist Theories of Development* (London: Croom Helm).
Marx, K. 1976: Results of the immediate process of production. Appendix to the 1976 Penguin edition of *Capital, Volume I* (Harmondsworth: Penguin/New Left Review).
Morawetz, D. 1977: *Twenty-Five Years of Economic Development, 1950–75* (Washington, D.C.: World Bank).
Phillips, A. 1977: The concept of 'development'. *Review of African Political Economy* 8 (Jan.–April).
Quijano, A. 1977: *Dependencia, Urbanización y Cambio Social en América Latina* (Lima: Mosca Azul).
Rodney, W. 1972: *How Europe Underdeveloped Africa* (London: Bogle-l'Ouverture).
Seers, D. 1977: Indian bias?, in *Urban Bias – Seers versus Lipton*, Institute of Development Studies, University of Sussex, Discussion Paper 116, (Aug.).
Smith, S. 1980: The ideas of Samir Amin: Theory or tautology? *Journal of Development Studies* 17 (1) (Oct.).
Spraos, J. 1983: *Inequalising Trade? A Study of Traditional North/South Specialisation in the Context of Terms of Trade Concepts* (Oxford: Clarendon Press with UNCTAD).
Sunkel, O. 1973: Transnational capitalism and national disintegration in Latin America. *Social and Economic Studies* 22 (1).
Sweezy, Paul *et al.* 1976: *The Transition from Feudalism to Capitalism* (London: New Left Books).
Taylor, J. G. 1979: *From Modernization to Modes of Production: A Critique of the Sociologies of Development and Underdevelopment* (London: Macmillan).
Wallerstein, I. 1974: *The Modern World-System: Capitalist Agriculture and the Origins of the European World-Economy in the Sixteenth Century* (New York: Academic Press).
—— 1979: *The Capitalist World-Economy* (Cambridge: Cambridge University Press/ Maison des Sciences de l'Homme).
—— 1980: *The Modern World-System II: Mercantilism and the Consolidation of the European World-Economy, 1600–1750* (New York: Academic Press).
Warren, B. 1980: *Imperialism: Pioneer of Capitalism* (London: New Left Books).
Weeks, J. 1981: The differences between materialist theory and dependency theory and why they matter, in *LAP*.
Weisskopf, T. E. 1976: Dependence as an explanation of underdevelopment: A critique (mimeo.) (University of Michigan, Center for Research on Economic Development, March).
Wolpe, H. 1980: Introduction, in Wolpe (ed.) *The Articulation of Modes of Production* (London: Routledge & Kegan Paul).

4 Deepak Lal, 'The Misconceptions of "Development Economics"'

Reprinted in full from: *Finance and Development* (June), 10–13 (1985)

Ideas have consequences. The body of thought that has evolved since World War II and is called 'development economics' (to be distinguished from the orthodox 'economics of developing countries') has, for good or ill, shaped policies for, as well as beliefs about, economic development in the Third World. Viewing the interwar experience of the world economy as evidence of the intellectual deficiencies of conventional economics (embodied, for instance, in the tradition of Marshall, Pigou, and Robertson) and seeking to emulate Keynes' iconoclasm (and hopefully renown), numerous economists set to work in the 1950s to devise a new unorthodox economics particularly suited to developing countries (most prominently, Nurkse, Myrdal, Rosenstein-Rodan, Balogh, Prebisch, and Singer). In the subsequent decades numerous specific theories and panaceas for solving the economic problems of the Third World have come to form the corpus of a 'development economics.' These include: the dual economy, labor surplus, low level equilibrium trap, unbalanced growth, vicious circles of poverty, big push industrialization, foreign exchange bottlenecks, unequal exchange, 'dependencia,' redistribution with growth, and a basic needs strategy – to name just the most influential in various times and climes.

Those who sought a new economics claimed that orthodox economics was (1) unrealistic because of its behavioral, technological, and institutional assumptions and (2) irrelevant because it was concerned primarily with the efficient allocation of given resources, and hence could deal neither with the so-called dynamic aspects of growth nor with various ethical aspects of the alleviation of poverty or the distribution of income. The twists and turns that the unorthodox theories have subsequently taken may be traced in four major areas: (1) the role of foreign trade and official or private capital flows in promoting economic development, (2) the role and appropriate form of industrialization in developing countries, (3) the relationship between the reduction of inequality, the alleviation of poverty, and the so-called different 'strategies of development,' and (4) the role of the price mechanism in promoting development.

The last is, in fact the major debate that in a sense subsumes most of the rest and it is the main concern of this article; for the major thrust of much of 'development economics' has been to justify massive government intervention through forms of direct control usually intended to supplant rather than to improve the functioning of, or supplement, the price mechanism. This is what I label the *dirigiste dogma*, which supports forms and areas of *dirigisme* well beyond those justifiable on orthodox economic grounds.

The empirical assumptions on which this unwarranted *dirigisme* was based have been repudiated by the experience of numerous countries in the postwar period. This article briefly reviews these central misconceptions of 'development economics.' References to the evidence as well as an elucidation of the arguments underlying the analysis (together with various qualifications) can be found in the author's work cited in the accompanying box.

Denial of 'economic principle'

The most basic misconception underlying much of development economics has been a rejection (to varying extents) of the behavioral assumption that, either as producers or consumers, people, as Hicks said, 'would act *economically*; when the opportunity of an advantage was presented to them they would take it.' Against these supposedly myopic and ignorant private agents (that is, individuals or groups of people), development economists have set some official entity (such as government planners, or policymakers) which is both knowledgeable and compassionate. It can overcome the defects of private agents and compel them to raise their living standards through various *dirigiste* means.

Numerous empirical studies from different cultures and climates, however, show that uneducated private agents – be they peasants, rural-urban migrants, urban workers, private entrepreneurs, or housewives – act economically as producers and consumers. They respond to changes in relative prices much as neoclassical theory would predict. The 'economic principle' is not unrealistic in the Third World; poor people may, in fact, be pushed even harder to seek their advantage than rich people.

Nor are the preferences of Third World workers peculiar in that for them too (no matter how poor), the cost of 'sweat' rises the harder and longer they work. They do not have such peculiar preferences that when they become richer they will not also seek to increase their 'leisure' – an assumption that underlies the view that there are large pools of surplus labor in developing countries that can be employed at a low or zero social opportunity cost. They are unlikely to be in 'surplus' in any meaningful sense any more than their Western counterparts.

Nor are the institutional features of the Third World, such as their strange social and agrarian structures or their seemingly usurious informal credit systems, necessarily a handicap to growth. Recent applications of neoclassical theory show how, instead of inhibiting efficiency, these institutions – being second-best adaptations to the risks and uncertainties inherent in the relevant economic environment – are likely to enhance efficiency.

Finally, the neoclassical assumption about the possibilities of substituting different inputs in production has not been found unrealistic. The degree to which inputs of different factors and commodities can be substituted in the national product is not much different in developed or developing countries.

Changes in relative factor prices do influence the choice of technology at the micro level and the overall labor intensity of production in Third World economies.

Market vs bureaucratic failure

A second and major strand of the unwarranted *dirigisme* of much of development economics has been based on the intellectually valid arguments against *laissez-faire*. As is well known, *laissez-faire* will only provide optimal outcomes if perfect competition prevails; if there are universal markets for trading all commodities (including future 'contingent' commodities, that is commodities defined by future conditions, such as the impact of weather on energy prices); and if the distribution of income generated by the *laissez-faire* economy is considered equitable or, if not, could be made so through lump-sum taxes and subsidies. As elementary economics shows, the existence of externalities in production and consumption and increasing returns to scale in production, or either of them, will rule out the existence of a perfectly competitive utopia. While, clearly, universal markets for *all* (including contingent) commodities do not exist in the real world, to that extent market failure must be ubiquitous in the real world. This, even ignoring distributional considerations, provides a *prima facie* case for government intervention. But this in itself does not imply that any or most forms of government intervention will improve the outcomes of a necessarily imperfect market economy.

For the basic cause of market failure is the difficulty in establishing markets in commodities because of the costs of making transactions. These transaction costs are present in any market, or indeed any mode of resource allocation, and include the costs of excluding nonbuyers as well as those of acquiring and transmitting the relevant information about the demand and supply of a particular commodity to market participants. They drive a wedge, in effect, between the buyer's and the seller's price. The market for a particular good will cease to exist if the wedge is so large as to push the lowest price at which anyone is willing to sell above the highest price anyone is willing to pay. These transaction costs, however, are also involved in acquiring, processing, and transmitting the relevant information to design public policies, as well as in enforcing compliance. There may, consequently, be as many instances of bureaucratic as of market failure, making it impossible to attain a full welfare optimum. Hence, the best that can be expected in the real world of imperfect markets and imperfect bureaucrats is a second best. But judging between alternative second best outcomes involves a subtle application of second-best welfare economics, which provides no general rule to permit the deduction that, in a necessarily imperfect market economy, particular *dirigiste* policies will increase economic welfare. They may not; and they may even be worse than *laissez-faire*.

Foretelling the future

Behind most arguments for *dirigisme*, particularly those based on directly controlling quantities of goods demanded and supplied, is the implicit premise of an omniscient central authority. The authority must also be omnipotent (to prevent people from taking actions that controvert its diktat) and benevolent (to ensure it serves the common weal rather than its own), if it is to necessarily improve on the working of an imperfect market economy. While most people are willing to question the omnipotence or benevolence of governments, there is a considerable temptation to believe the latter have an omniscience that private agents know they themselves lack. This temptation is particularly large when it comes to foretelling the future.

Productive investment is the mainspring of growth. Nearly all investment involves giving hostages to fortune. Most investments yield their fruits over time and the expectations of investors at the time of investment may not be fulfilled. Planners attempting to direct investments and outputs have to take a view about future changes in prices, tastes, resources, and technology, much like private individuals. Even if the planners can acquire the necessary information about current tastes, technology, and resources in designing an investment program, they must also take a view about likely changes in the future demand and supply of myriad goods. Because in an uncertain world there can be no agreed or objective way of deciding whether a particular investment gamble is sounder than another, the planned outcomes will be better than those of a market system (in the sense of lower excess demand for or supply of different goods and services) only if the planners' forecasts are more accurate than the decentralized forecasts made by individual decision makers in a market economy. There is no reason to believe that planners, lacking perfect foresight, will be more successful at foretelling the future than individual investors.

Outcomes based on centralized forecasts may, indeed, turn out to be worse than those based on the decentralized forecasts of a large number of participants in a market economy, because imposing a single centralized forecast on the economy in an uncertain world is like putting all eggs in one basket. By contrast, the multitude of small bets, based on different forecasts, placed by a large number of decision makers in a market economy *may be* a sounder strategy. Also, bureaucrats, as opposed to private agents, are likely to take less care in placing their bets, as they do not stand to lose financially when they are wrong. This assumes, of course, that the government does not have better information about the future than private agents. If it does, it should obviously disseminate it, together with any of its own forecasts. On the whole, however, it may be best to leave private decision makers to take risks according to their own judgments.

This conclusion is strengthened by the fact, emphasized by Hayek, that most relevant information is likely to be held at the level of the individual firm and the household. A major role of the price mechanism in a market economy is to transmit this information to all interested parties. The 'planning without prices' favored in practice by some planners attempts to supersede and suppress the price mechanism. It thereby throws sand into one of the most useful and relatively low-cost social mechanisms for transmitting information, as well as for coordinating the actions of large numbers of interdependent market participants. The strongest argument against centralized planning, therefore, is that, even though omniscient planners might forecast the future more accurately than myopic private agents, there is no reason to believe that ordinary government officials can do any better – and some reason to believe they may do much worse.

It has nevertheless been maintained that planners in the Third World can and should directly control the pattern of industrialization. Some have put their faith in mathematical programming models based on the use of input–output tables developed by Leontief. But, partly for the reasons just discussed, little reliance can be placed upon either the realism or the usefulness of these models for deciding which industries will be losers and which will be winners in the future. There are many important and essential tasks for governments to perform (see below), and this irrational *dirigisme* detracts from their main effort.

Redressing inequality and poverty

Finally, egalitarianism is never far from the surface in most arguments supporting the *dirigiste dogma*. This is not surprising since there may be good theoretical reasons for government intervention, even in a perfectly functioning market economy, in order to promote a distribution of income desired on ethical grounds. Since the distribution resulting from market processes will depend upon the initial distribution of assets (land, capital, skills, and labor) of individuals and households, the desired distribution could, in principle, be attained either by redistributing the assets or by introducing lump-sum taxes and subsidies to achieve the desired result. If, however, lump-sum taxes and subsidies cannot be used in practice, the costs of distortion from using other fiscal devices (such as the income tax, which distorts the individual's choice between income and leisure) will have to be set against the benefits from any gain in equity. This is as much as theory can tell us, and it is fairly uncontroversial.

Problems arise because we lack a consensus about the ethical system for judging the desirability of a particular distribution of income. Even within Western ethical beliefs, the shallow utilitarianism that underlies many economists' views about the 'just' distribution of income and assets is not universally accepted. The possibility that all the variegated peoples of the world are utilitarians is fairly remote. Yet the moral fervor underlying many economic prescriptions assumes there is already a world society with a common set of ethical beliefs that technical economists can take for granted and use to make judgments encompassing both the efficiency and equity components of economic welfare. But casual empiricism is enough to show that there is no such world society; nor is there a common view, shared by mankind, about the content of social justice.

There is, therefore, likely to be little agreement about either the content of distributive justice or whether we should seek to achieve it through some form of coercive redistribution of incomes and assets when this would infringe other moral ends, which are equally valued. By contrast, most moral codes accept the view that, to the extent feasible, it is desirable to alleviate abject, absolute poverty or destitution. That alleviating poverty is not synonymous with reducing the inequality of income, as some seem still to believe, can be seen by considering a country with the following two options. The first option leads to a rise in the incomes of all groups, including the poor, but to larger relative increases for the rich, and hence a worsening of the distribution of income. The second leads to no income growth for the poor but to a reduction in the income of the rich; thus the distribution of income improves but the extent of poverty remains unchanged. Those concerned with inequality would favor the second option; those with poverty the first. Thus, while the pursuit of efficient growth may worsen some inequality index, there is no evidence that it will increase poverty.

Surplus labor and 'trickle down'

As the major asset of the poor in most developing (as well as developed) countries is their labor time, increasing the demand for unskilled labor relative to its supply could be expected to be the major means of reducing poverty in the Third World. However, the shadows of Malthus and Marx have haunted development economics, particularly in its discussion of equity and the alleviation of poverty. One of the major assertions of development economics, preoccupied with 'vicious circles' of poverty, was that the fruits of capitalist growth, with its reliance on the price mechanism, would not trickle down or spread to the poor. Various *dirigiste* arguments were then advocated to bring the poor into a growth process that would otherwise bypass them. The most influential, as well as the most famous of the models of development advanced in the 1950s to chart the likely course of outputs and incomes in an overpopulated country or region was that of Sir Arthur Lewis. It made an assumption of surplus labor that, in a capitalist growth process, entailed no increase in the income of laborers until the surplus had been absorbed.

It has been shown that the assumptions required for even under-employed rural laborers to be 'surplus,' in Lewis' sense of their being available to industry at a constant wage, are very stringent, and implausible. It was necessary to assume that, with the departure to the towns of their relatives, those rural workers who remained would work harder for an unchanged wage. This implied that the preferences of rural workers between leisure and income are perverse, for workers will not usually work harder without being offered a higher wage. Recent empirical research into the shape of the supply curve of rural labor at different wages has found that – at least for India, the country supposedly containing vast pools of surplus labor – the curve is upward-sloping (and not flat, as the surplus labor theory presupposes). Thus, for a given labor supply, increases in the demand for labor time, in both the industrial and the rural sectors, can be satisfied only by paying higher wages.

The fruits of growth, even in India, will therefore trickle down, in the sense either of raising labor incomes, whenever the demand for labor time increases by more than its supply, or of preventing the fall in real wages and thus labor incomes, which would otherwise occur if the supply of labor time outstripped the increase in demand for it. More direct evidence about movements in the rural and industrial real wages of unskilled labor in developing countries for which data are available has shown that the standard economic presumption that real wages will rise as the demand for labor grows, relative to its supply, is as valid for the Third World as for the First.

Administrative capacities

It is in the political and administrative aspects of *dirigisme* that powerful practical arguments can be advanced against the *dirigiste dogma*. The political and administrative assumptions underlying the feasibility of various forms of *dirigisme* derive from those of modern welfare states in the West. These, in turn, reflect the values of the eighteenth-century Enlightenment. It has taken nearly two centuries of political evolution for those values to be internalized and reflected (however imperfectly) in the political and administrative institutions of Western societies. In the Third World, an acceptance of the same values is at best confined to a small class of Westernized intellectuals. Despite their trappings of modernity, many developing countries are closer in their official workings to the inefficient nation states of seventeenth- or eighteenth-century Europe. It is instructive to recall that Keynes, whom so many *dirigistes* invoke as a founding father of their faith, noted in *The End of Laissez-Faire:*

> But above all, the ineptitude of public administrators strongly prejudiced the practical man in favor of *laissez-faire* – a sentiment which has by no means disappeared. Almost everything which the State did in the 18th century in excess of its minimum functions was, or seemed, injurious or unsuccessful.

It is in this context that anyone familiar with the actual administration and implementation of policies in many Third World countries, and not blinkered by the *dirigiste dogma*, should find that oft-neglected work, *The Wealth of Nations*, both so relevant and so modern.

For in most of our modern-day equivalents of the inefficient eighteenth-century state, not even the minimum governmental functions required for economic progress are always fulfilled. These include above all providing public goods of which law and order and a sound money remain paramount, and an economic environment where individual thrift, productivity, and enterprise is cherished and not thwarted. There are numerous essential tasks for *all* governments to perform. One of the most important is to establish and maintain the country's infrastructure, much of which requires large, indivisible lumps of capital before any output can be produced. Since the services provided also frequently have the characteristics of public goods, natural monopolies would emerge if they were privately produced. Some form of government regulation would be required to ensure that services were provided in adequate quantities at prices that reflected their real resource costs. Government intervention is therefore necessary. And, given the costs of regulation in terms of acquiring the relevant information, it may be second best to supply the infrastructure services publicly.

These factors justify one of the most important roles for government in the development process. It can be argued that the very large increase in infrastructure investment, coupled with higher savings rates, provides the major explanation of the marked expansion in the economic growth rates of most Third World countries during the postwar period, compared with both their own previous performance and that of today's developed countries during their emergence from underdevelopment.

Yet the *dirigistes* have been urging many additional tasks on Third World governments that go well beyond what Keynes, in the work quoted above, considered to be a sensible agenda for *mid-twentieth century* Western polities:

> the most important *Agenda* of the State relate not to those activities which private individuals are already fulfilling, but to those functions which fall outside the sphere of the individuals, to those decisions which are made by no one if the State does not make them. The important thing for governments is not to do things which individuals are doing already, and to do them a little better or a little worse; but to do those things which at present are not done at all.

From the experience of a large number of developing countries in the postwar period, it would be a fair professional judgment that most of the more serious distortions are due not to the inherent imperfections of the market mechanism but to irrational government interventions, of which foreign trade controls, industrial licensing, various forms of price controls, and means of inflationary financing of fiscal deficits are the most important. In seeking to improve upon the outcomes of an imperfect market economy, the *dirigisme* to which numerous development economists have lent intellectual support has led to policy-induced distortions that are more serious than, and indeed compound, the supposed distortions of the market economy they were designed to cure. It is these lessons from accumulated experience over the last three decades that have undermined development economics, so that its demise may now be conducive to the health of both the economics and economies of developing countries.

5 Arturo Escobar, 'Development Planning'

Reprinted in full from: W. Sachs (ed.), *The Development Dictionary: A Guide to Knowledge as Power*, pp. 132–45. London: Zed Books Ltd. (Originally published as 'Planning') (1992)

Planning techniques and practices have been central to development since its inception. As the application of scientific and technical knowledge to the public domain, planning lent legitimacy to, and fuelled hopes about, the development enterprise. Generally speaking, the concept of planning embodies the belief that social change can be engineered and directed, produced at will. Thus the idea that poor countries could move more or less smoothly along the path of progress through planning has always been held as an indubitable truth, an axiomatic belief in need of no demonstration, by development experts of most persuasions.

Perhaps no other concept has been so insidious, no other idea gone so unchallenged. This blind acceptance of planning is all the more striking given the pervasive effects it has had historically, not only in the Third World, but also in the West, where it has been linked to fundamental processes of domination and social control. For planning has been inextricably linked to the rise of Western modernity since the end of the 18th century. The planning conceptions and routines introduced in the Third World during the post-World War II period are the result of accumulated scholarly, economic and political action; they are not neutral frameworks through which 'reality' innocently shows itself. They thus bear the marks of the history and culture that produced them. When deployed in the Third World, planning not only carried with it this historical baggage, but also contributed greatly to the production of the socio-economic and cultural configuration that we describe today as underdevelopment.

Normalizing people in 19th century Europe

How did planning arise in the European experience? Very briefly, three major factors were essential to this process, beginning in the 19th century – the development of town planning as a way of dealing with the problems of the growing industrial cities; the rise of social planning, and increased intervention by professionals and the state in society, in the name of promoting people's welfare; and the invention of the modern economy, which crystallized with the institutionalization of the market and the formulation of classical political economy. These three factors, which today appear to us as normal, as natural parts of our world, have a relatively recent and even precarious history.

In the first half of the 19th century, capitalism and the industrial revolution brought drastic changes in the make-up of cities, especially in Northwestern

Europe. Ever more people flooded into old quarters, factories proliferated, and industrial fumes hovered over streets covered with sewage. Overcrowded and disordered, the 'diseased city', as the metaphor went, called for a new type of planning which would provide solutions to the rampant urban chaos. Indeed, it was those city officials and reformers who were chiefly concerned with health regulations, public works and sanitary interventions, who first laid down the foundations of comprehensive urban planning. The city began to be conceived of as an object, analysed scientifically, and transformed according to the two major requirements of traffic and hygiene. 'Respiration' and 'circulation' were supposed to be restored to the city organism, overpowered by sudden pressure. Cities (including the colonial chequerboards outside Europe) were designed or modified to ensure proper circulation of air and traffic, and philanthropists set out to eradicate the appalling slums and to bring the right morals to their inhabitants. The rich traditional meaning of cities and the more intimate relationship between city and dweller were thus eroded as the industrial–hygienic order became dominant. Reifying space and objectifying people, the practice of town planning, along with the science of urbanism, transformed the spatial and social make-up of the city, giving birth in the 20th century to what has been called 'the Taylorization of architecture' (McLeod, 1983).

Just like planners in the Third World today, the 19th century European bourgeoisie also had to deal with the question of poverty. The management of poverty actually opened up a whole realm of intervention, which some researchers have termed the social. Poverty, health, education, hygiene, unemployment, etc. were constructed as 'social problems', which in turn required detailed scientific knowledge about society and its population, and extensive social planning and intervention in everyday life. As the state emerged as the guarantor of progress, the objective of government became the efficient management and disciplining of the population so as to ensure its welfare and 'good order'. A body of laws and regulations was produced with the intention to regularize work conditions and deal with accidents, old age, the employment of women, and the protection and education of children. Factories, schools, hospitals, prisons became privileged places to shape experience and modes of thinking in terms of the social order. In sum, the rise of the social made possible the increasing socialization and subjection of people to dominant norms, as well as their insertion into the machinery of capitalist production. The end result of this process in the present day is the welfare state and the new professionalized activity known as social work.

Two points have to be emphasized in relation to this process. One, that these changes did not come about naturally, but required vast ideological and material operations, and often times plain coercion. People did not become accustomed to factory work or to living in crowded and inhospitable cities gladly and of their own volition; they had to be disciplined into it! And two, that those very operations and forms of social planning have produced 'governable' subjects. They have shaped not only social structures and institutions, but also the way in which people experience life and construct themselves as subjects. But development experts have been blind to these insidious aspects of planning in their proposals to replicate in the Third World similar forms of social planning. As Foucault said, 'the "Enlightment", which discovered the liberties, also invented the disciplines.' (Foucault, 1979). One cannot look on the bright side of planning, its modern achievements (if one were to accept them), without looking at the same time on its dark side of domination. The management of the social has produced modern subjects who are not only dependent on professionals for their needs, but also ordered into realities (cities, health and educational systems, economies, etc.) that can be governed by the state through planning. Planning inevitably requires the normalization and standardization of reality, which in turn entails injustice and the erasure of difference and diversity.

The third factor in European history that was of central importance to the development and success of planning was the invention of the 'economy'. The economy, as we know it today, did not even exist as late as the 18th century in Europe, much less in other parts of the world. The spread and institutionalization of the market, certain philosophical currents such as utilitarianism and individualism, and the birth of classical political economy at the end of the 18th century provided the elements and cement for the establishment of an independent domain, namely 'the economy', apparently separated from morality, politics and culture. Karl Polanyi refers to this process as the 'disembeddedness' of the economy for society, a process which was linked to the consolidation of capitalism and which entailed the commodification of land and labour. There were many consequences of this development, besides generalized commodification. Other forms of economic organization, those founded upon reciprocity or redistribution, for instance, were disqualified and increasingly marginalized. Subsistence activities became devalued or destroyed. And an instrumental attitude towards nature and people became the order of the day, which in turn led to unprecedented forms of exploitation of people and nature. Although today most of us take for granted the modern market economy, this notion and the reality of how it operates have not always existed. Despite its dominance, even today there persist in many parts of the Third World subsistence societies, 'informal' economies, and collective forms of economic organization.

In sum, the period 1800–1950 saw the progressive encroachment of those forms of administration and regulation of society, urban space and the economy that would result in the great edifice of planning in the early post-World War II period. Once normalized, regulated and ordered, individuals, societies and economies can be subjected to the scientific gaze and social engineering scalpel of the planner, who, like a surgeon operating on the human body, can then attempt to produce the desired type of social change. If social science and planning have had any success in predicting and engineering social change, it is precisely because certain economic, cultural and social regularities have already been attained which confer some systematic element and consistency with the real world on the planners' attempts. Once you organize factory work and discipline workers, or once you start growing trees in plantations, then you can predict industrial output or timber production. In the process, the exploitation of workers, the degradation of nature, and the elimination of other forms of knowledge – whether it be the skills of the craftsman or those who live off the forest – are also affected. These are the kind of processes that are at stake in the Third World when planning is introduced as the central technique of development. In short, planning redefines social and economic life in accordance with the criteria of rationality, efficiency and morality which are consonant with the history and needs of capitalist, industrial society, but not those of the Third World.

Dismantling and reassembling societies

Scientific planning came of age during the 1920s and '30s, when it emerged from rather heterogeneous origins – the mobilization of national production during World War I, Soviet Planning, the scientific management movement in the USA, and Keynesian economic policy. Planning techniques were refined during the Second World War and its aftermath. It was during this period, and in connection with the War, that operations research, systems analysis, human engineering, and views of planning as 'rational social action' became widespread. When the era of development in the Third World dawned in the late 1940s, the dream of designing society through planning found an even more fertile ground. In Latin America and Asia, the creation of a 'developing society', understood as an urban-based civilization characterized by growth, political stability and increasing standards of living, became an explicit goal, and ambitious plans were designed to bring it about with the eager assistance of international organizations and experts from the 'developed' world.

To plan in the Third World, however, certain structural and behavioural conditions had to be laid down, usually at the expense of people's existing concepts of social action and change. In the face of the imperatives of 'modern society', planning involved the overcoming or eradication of 'traditions', 'obstacles' and 'irrationalities', that is, the wholesale modification of existing human and social structures and their replacement with rational new ones. Given the nature of the post-war economic order, this amounted to creating the conditions for capitalist production and reproduction. Economic growth theories, which dominated development at the time, provided the theoretical orientation for the creation of the new order, and national development plans the means to achieve it. The first 'mission' – note its colonial, Christian missionary overtones – sent by the World Bank to an 'underdeveloped' country in 1949, for instance, had as its goal the formulation of a 'comprehensive program of development' for the country in question, Colombia. Staffed by experts in many fields, the mission saw its task as 'calling for a comprehensive and internally consistent program . . . Only through a generalized attack throughout the whole economy on education, health, housing, food and productivity can the vicious circle of poverty, ignorance, ill health and low productivity be decisively broken.' Moreover, it was clear to the mission that:

> One cannot escape the conclusion that reliance on natural forces has not produced the most happy results. Equally inescapable is the conclusion that with knowledge of the underlying facts and economic processes, good planning in setting objectives and allocating resources, and determination in carrying out a program for improvement and reforms, a great deal can be done to improve the economic environment by shaping economic policies to meet scientifically ascertained social requirements . . . In making such an effort, Colombia would not only accomplish its own salvation but would at the same time furnish an inspiring example to all other underdeveloped areas of the world. (International Bank for Reconstruction and Development, 1950)

That development was about 'salvation' – again the echoes of the colonial civilizing mission – comes out clearly in most of the literature of the period. Countries in Latin America, Africa and Asia were seen as 'relying on natural forces', which had not produced the 'most happy results'. Needless to say, the whole history of colonialism is effaced by this discursive way of putting it. What is emphasized instead is the introduction of poor countries to the 'enlightened' world of Western science and modern economics, while the conditions existing in these countries are constructed as being characterized by a 'vicious circle' of 'poverty', 'ignorance' and the like. Science and planning, on the other hand, are seen as neutral, desirable and universally applicable, while, in truth, an entire and particular rationality and civilizational experience was being transferred to the Third World through the process of 'development'. The Third World thus entered post-World War II Western consciousness as constituting the appropriate social and technical raw material for planning. This status of course depended, and still does, on an extractive neo-colonialism. Epistemologically and politically, the Third World is constructed as a natural–technical object that has to be normalized and moulded through planning to meet the 'scientifically ascertained' characteristics of a 'development society'.

By the end of the 1950s, most countries in the Third World were already engaged in planning activities. Launching the first 'Development Decade' at the beginning of the 1960s, the United Nations could thus state that:

> The ground has been cleared for a non-doctrinaire consideration of the real problems of development, namely saving, training and planning, and for action on them. In particular, the advantages in dealing with the various problems not piecemeal, but by a comprehensive approach through sound development planning, became more fully apparent. . . . Careful development planning can be a potent means of mobilizing . . . latent resources for a rational solution of the problems involved. (United Nations, 1962)

The same optimism – and, at the same time, blindness to the parochial and ethnocentric attitudes of the planners – was echoed by the Alliance for Progress. In President Kennedy's words:

> The world is very different now. For man (*sic*) holds in his mortal hands the power to abolish all forms of human poverty and all forms of human life. . . . To those people in the huts and villages of half the globe struggling to break the bonds of mass misery . . . we offer a special pledge – to convert our good words in good deeds – in a new alliance for progress – to assist free men and free governments in casting off the chains of poverty. (Kennedy, 1961)

Statements such as these reduce life in the Third World simply to conditions of 'misery', overlooking its rich traditions, different values and life styles, and long historical achievements. In the eyes of planners and developers, people's dwellings appear as no more than miserable 'huts', and their lives – often times, especially at this early point in the development era, still characterized by subsistence and self-sufficiency – as marked by unacceptable 'poverty'. In short, they are seen as no more than crude matter in urgent need of being transformed by planning. One does not need to romanticize tradition to realize that, what for the economist were indubitable signs of poverty and backwardness, for Third World people were often integral components of viable social and cultural systems, rooted in different, non-modern social relations and systems of knowledge. It was precisely these systems that came under attack first by colonialism and later on by development, although not without much resistance then as today. Even alternative conceptions of economic and social change held by Third World scholars and activists in the 1940s and '50s – the most notable being that of Mahatma Gandhi, but also, for instance, those of certain socialists in Latin America – were displaced by the enforced imposition of planning and development. For developers, what was at stake was a transition from a 'traditional society' to an 'economic culture', that is, the development of a type of society whose goals were linked to future-oriented, scientific-objective rationality and brought into existence through the mastering of certain techniques. 'So long as everyone played his part well,' planners believed, 'the system was fail-safe; the state would plan, the economy would produce, and working people would concentrate on their private agendas: raising families, enriching themselves, and consuming whatever came tumbling out from the cornucopia. (Friedmann, 1965, pp. 8–9).

As Third World elites appropriated the European ideal of progress – in the form of the construction of a prosperous, modern nation through economic development and planning; as other surviving concepts of change and social action became even more marginalized; finally, as traditional social systems were disrupted and the living conditions of most people worsened, the hold of planning grew ever stronger. Elites and, quite often, radical counter-elites found in planning a tool for social change which was in their eyes not only indispensable, but irrefutable because of its scientific nature. The history of development in the post-World War II period is, in many ways, the history of the institutionalization and ever more pervasive deployment of planning. The process was facilitated time after time by successive development 'strategies'. From the emphasis on growth and national planning in the 1950s, to the Green Revolution and sectoral and regional planning of the 1960s and '70s, including 'Basic Needs' and local level planning in the '70s and '80s, to environmental planning for 'sustainable development' and planning to 'incorporate' women, or the grassroots, into development in the '80s, the scope and vaulting ambitions of planning have not ceased to grow.

Perhaps no other concept has served so well to recast and spread planning as that of the Basic Human Needs strategy. Recognizing that the goals of reducing poverty and ensuring a decent living standard for most people were 'as distant as ever', development theorists – always keen on finding yet another gimmick which they could present as a 'new' paradigm or strategy – coined this notion with the aim of providing 'a coherent framework that can accommodate the increasingly refined sets of development objectives that have evolved over the past thirty years and can systematically relate these objectives to various types of policies' (Crosswell, 1981, p. 2), including growth. The key arenas of intervention were primary education, health, nutrition, housing, family planning, and rural development. Most of the interventions themselves were directed at the household. As in the case of the mapping of 'the social' in 19th century Europe, where society first became the target of systematic state intervention, Third World people's health, education, farming and reproduction practices all became the object of a vast array of programmes introduced in the name of increasing these countries' 'human capital' and ensuring a minimum level of welfare for their people. Once again, the epistemological and political boundaries of this kind of 'rational' approach – aimed at the modification of life conditions and inevitably marked by class, race, gender and cultural features – resulted in the construction of an artificially homogeneous monochrome, the 'Third World', an entity that was always deficient in relation to the West, and so always in need of imperialist projects of progress and development.

Rural development and health programmes during the 1970s and '80s can be cited as examples of this type of biopolitics. They also reveal the arbitrary mechanisms and fallacies of planning. Robert McNamara's famous Nairobi speech, delivered in 1973 before the boards of governors of the World Bank and the International Monetary Fund, launched the era of 'poverty-oriented' programmes in development, which evolved into the Basic Human Needs approach. Central to this conception were so-called national food and nutrition planning and integrated rural development. Most of these schemes were designed in the early 1970s at a handful of US and UK universities, at the World Bank, and at United Nations technical agencies, and implemented in many Third World countries from the mid 1970s until the late 1980s. Comprehensive food and nutrition planning was deemed necessary, given the magnitude and complexity of the problems of malnutrition and hunger. Typically, a national food and nutrition plan included projects in primary health care, nutrition education and food supplementation, school and family vegetable gardens, the promotion of the production and consumption of protein-rich foods, and integrated rural development generally. This latter component contemplated measures to increase the production of food crops by small farmers through the supply of credit, technical assistance and agricultural inputs, and basic infrastructure.

How did the World Bank define integrated rural development? 'Rural development', the World Bank's policy dictated:

> is a strategy designed to improve the economic and social life of a specific group of people – the rural poor. It involves extending the benefits of development to the poorest among those who seek a livelihood in rural areas. A strategy of rural development must recognize three points. Firstly, the rate of transfer of people out of low productivity agriculture into more rewarding pursuits has been slow. . . . Secondly, . . . their position is likely to get worse if population expands at unprecedented rates. . . . Thirdly, rural areas have labor, land and at least some capital which, if

> mobilized, could reduce poverty and improve the quality of life. . . . [Rural development] is clearly designed to increase production and raise productivity. It is concerned with the monetization and modernization of society, and with its transition from traditional isolation to integration with the national economy. (World Bank, 1975)

That most people in the 'modern' sector, namely those living under marginal conditions in the cities, did not enjoy 'the benefits of development' did not occur to these experts. Peasants – that 'specific group of people' which is in reality the majority of the Third World – are seen in purely economic terms, not as trying to make viable a whole way of life. That their 'rate of transfer into more rewarding pursuits' had to be accelerated, on the other hand, assumes that their lives are not satisfying – after all, they live in 'traditional isolation', even if surrounded by their communities and those they love. The approach also regards peasants as suitable for moving around like cattle or commodities. Since their labour has to be 'mobilized', they must surely have just been sitting about idly (subsistence farming does not involve 'labour' in this view), or perhaps having too many babies. All of these rhetorical devices that reflect the 'normal' perceptions of the planner contribute to obscure the fact that it is precisely the peasants' increasing integration into the modern economy that is at the root of many of their problems. Even more fundamentally, these statements, which become translated into reality through planning, reproduce the world as the developers know it – a world composed of production and markets, of 'traditional' and 'modern' or developed and underdeveloped sectors, of the need for aid and investment by multinationals, of capitalism versus communism, of material progress as happiness, and so forth. Here we have a prime example of the link between representation and power, and of the violence of seemingly neutral modes of representation.

In short, planning ensures a functioning of power that relies on, and helps to produce, a type of reality which is certainly not that of the peasants, while peasant cultures and struggles are rendered invisible. Indeed the peasants are rendered irrelevant even to their own rural communities. In its rural development discourse, the World Bank represents the lives of peasants in such a way that awareness of the mediation and history inevitably implicated in this construction is excluded from the consciousness of its economists and from that of many important actors – planners, Western readers, Third World elites, scientists, etc. This particular narrative of planning and development, deeply grounded in the post-World War II global political economy and cultural order, becomes essential to those actors. It actually becomes an important element in their insular construction as a developed, modern, civilized 'we', the 'we' of Western man. In this narrative, too, peasants, and Third World people generally, appear as the half-human, half-cultured benchmark against which the Euro-American world measures its own achievements.

Knowledge as power

As a system of representation, planning thus depends on making people forget the origins of its historical mediation. This invisibility of history and mediation is accomplished through a series of particular practices. Planning relies upon, and proceeds through, various practices regarded as rational or objective, but which are in fact highly ideological and political. First of all, as with other development domains, knowledge produced in the First World about the Third gives a certain visibility to specific realities in the latter, thus making them the targets of power. Programmes such as integrated rural development have to be seen in this light. Through these programmes, 'small farmers', 'landless peasants' and the like achieve a certain visibility, albeit only as a development 'problem', which makes them the object of powerful, even violent, bureaucratic interventions. And there are other important hidden or unproblematized mechanisms of planning; for instance, the demarcation of new fields and their assignment to experts, sometimes even the creation of a new sub-discipline (like food and nutrition planning). These operations not only assume the prior existence of discrete 'compartments', such as 'health', 'agriculture', and 'economy' – which in truth are no more than fictions created by the scientist – but impose this fragmentation on cultures which do not experience life in the same compartmentalized manner. And, of course, states, dominant institutions, and mainstream views are strengthened along the way as the domain of their action is inevitably multiplied.

Institutional practices such as project planning and implementation, on the other hand, give the impression that policy is the result of discrete, rational acts and not the process of coming to terms with conflicting interests, a process in which choices are made, exclusions effected, and worldviews imposed. There is an apparent neutrality in identifying people as 'problems', until one realizes first, that this definition of 'the problems' has already been put together in Washington or some capital city of the Third World, and second, that problems are presented in such a way that some kind of development programme has to be accepted as the legitimate solution. It is professional discourses which provide the categories in terms of which 'facts' can be identified and analysed. This effect is reinforced by the use of labels such as 'small farmer' or 'pregnant women', which reduces a person's life to a single trait and makes him/her into a 'case' to be treated or reformed. The use of labels also allows experts and elites to delink explanations of 'the problem' from themselves as the non-poor, and assign them purely to factors internal to the poor. Inevitably, people's lives at the local level are transcended and objectified when they are translated into the professional categories used by institutions. In short, local realities come to be greatly determined by these non-local institutional practices, which thus have to be seen as inherently political.

The results of this type of planning have been, for the most part, deleterious to Third World people and economies alike. In the case of rural development, for instance, the outcome was seen by experts in terms of two possibilities: '(a) the small producer may be able to technify his productive process, which entails his becoming an agrarian entrepreneur; and (b) the small producer is not prepared to assume such level of competitiveness, in which case he will be displaced from the market and perhaps even from production in that area altogether.' (Depto. Nacional De Planeación de Colombia, 1979, p. 47). In other words, 'produce (for the market) or perish'. Even in terms of increased production, rural development programmes have had dubious results at best. Most of the increase in food production in the Third World has taken place in the commercial capitalist sector, while a good part of the increase has been in cash or export crops. In fact, as has been amply shown, rural development programmes and development planning in general have contributed not only to growing pauperization of rural people, but also to aggravated problems of malnutrition and hunger. Planners thought that the agricultural economies of the Third World could be mechanically restructured to resemble the 'modernized' agriculture of the United States, overlooking completely not only the desires and aspirations of people, but the whole dynamics of economy, culture and society that circumscribe farming practices in the Third World. This type of management of life actually became a theatre of death (most strikingly in the case of the African famine), as increased production of food resulted, through a perverse shift, in more hunger.

The impact of many development programmes has been particularly negative on women and indigenous peoples, as development projects appropriate or destroy their basis for sustenance and survival. Historically, Western discourse has refused to recognize the productive and creative role of women and this refusal has contributed to propagating divisions of labour that keep women in positions of subordination. For planners and economists, women were not, until recently, 'economically active', despite the fact that a great share of the food consumed in the Third World is grown by women. Moreover, women's economic and gender position frequently deteriorated in the 1970s as a result of the participation in rural development programmes by male heads of household. It is not surprising that women have opposed much more actively than men these rural development programmes. With the 'technological packages', specialization in the production of certain crops, rigid lay-out of fields, pre-set cultivation routines, production for the market, and so forth, they contrast sharply with the more ecological and varied peasant farming defended by women in many parts of the Third World – in which production for subsistence and for the market are carefully balanced. Unfortunately, the recent trend towards incorporating women into development has resulted for the most part in their being targeted for what in all other respects remain conventional programmes. 'Target group categories are constructed to further development agency procedures to organize, manage, regulate, enumerate and rule the lives of ordinary women.' (Mueller, 1987). Thus the development industry's clientele has been conveniently doubled by this shift in representation.

Another important recent instance of planned development is the industrialization schemes in so-called free trade zones in the Third World, where multinational corporations are brought in under very favourable conditions (e.g., tax breaks, assurances of cheap, docile labour and a 'stable' political climate, lower pollution standards, etc). Like all other forms of planning, these industrialization projects involve much more than an economic transformation, and on an ever larger scale. What is at stake here is the rapid transformation of rural society and culture into the world of factory discipline and modern (Western) society.

Brought into Third World countries in the name of development, and actively promoted and mediated by Third World states, the free trade zones represent a microcosm in which households, villages, traditions, modern factories, governments and the world economy are all brought together in unequal relations of knowledge and power. It is no accident that most of the workers in the new factories are young women. The electronics industries in South East Asia, for instance, rely heavily on gender forms of subordination. The production of young women factory workers as 'docile bodies' through systematic forms of discipline in the factory and outside it, does not go, however, without resistance, as Aihwa Ong shows in her excellent study of Malaysian women factory workers. Women's forms of resistance in the factory (destruction of microchips, spirit possession, slow-downs etc.) have to be seen as idioms of protest against labour discipline and male control in the new industrial situation. Moreover, they remind us that, if it is true that 'new forms of domination are increasingly embodied in the social relations of science and technology which organize knowledge and production systems', it is equally true that 'the divergent voices and innovative practices of subjected peoples disrupt such cultural reconstructions of non-Western societies.' (Ong, 1987, p. 221).

Knowledge in opposition

Feminist critics of development and critics of development as discourse have begun to join forces, precisely through their examination of the dynamics of domination, creativity and resistance that circumscribe development. This hopeful trend is most visible in a type of grassroots activism and theorizing that is sensitive to the role of knowledge, culture and gender in supporting the enterprise of development and, conversely, in bringing about more pluralistic and egalitarian practices. As the links between development, which articulates the state with profits, patriarchy and objectivizing science and technology on the one hand, and the marginalization of people's lives and knowledge on the other, become more evident, the search for alternatives also deepens. The imaginary ideas of development and 'catching up' with the West are drained of their appeal as violence and recurrent crises – economic, ecological, political – become the order of the day. In sum, the attempt by states to set up totalizing systems of socio-economic and cultural engineering through development is running into a dead end. Practices and new spaces for thinking and acting are being created or reconstituted, most notably at the grassroots, in the vacuum left by the crisis of the colonizing mechanisms of development.

Speaking about ecology movements in India, many of them started by women at the grassroots level, Vandana Shiva, for instance, sees the emerging process as:

> a redefinition of growth and productivity as categories linked to the production, not the destruction, of life. It is thus simultaneously an ecological and a feminist political project that legitimizes the ways of knowing and being that create wealth by enhancing life and diversity, and which delegitimizes the knowledge and practice of a culture of death as the

> basis for capital accumulation. . . . In contemporary times, Third World women, whose minds have not yet been dispossessed or colonized, are in a privileged position to make visible the invisible oppositional categories that they are custodians of. (Shiva, 1989, pp. 13, 46)

One does not need to impute to Third World women, indigenous people, peasants and others a purity they do not have, to realize that important forms of resistance to the colonization of their life world have been maintained and even nurtured among them. And one does not need to be overly optimistic about the potential of grass-roots movements to transform the development order to visualize the promise that these movements hold, and the challenge they increasingly pose to conventional top-down, centralized approaches or even to those apparently decentralized, participatory strategies which are geared for the most part towards economic ends. ('Participatory' or local level planning, indeed, is most often conceived not in terms of a popular power that people could exercise, but as a bureaucratic problem that the development institution has to solve.) Shiva's argument that many groups of Third World people, especially rural women and indigenous peoples, possess knowledge and practices opposite to those that define the dominant nexus between reductionist science, patriarchy, violence and profits – forms of relating to people, knowledge and nature which are less exploitative and reifying, more localized, decentred and in harmony with the ecosystem – is echoed by observers in many parts of the world. These alternative forms, which are neither traditional nor modern, provide the basis for a slow but steady process of construction of different ways of thinking and acting, of conceiving of social change, of organizing economies and societies, of living and healing.

Thus Western rationality has to open up to the plurality of forms of knowledge and conceptions of change that exist in the world and recognize that objective, detached scientific knowledge is just one possible form among many. This much can be gleaned from an anthropology of Reason that looks critically at the basic discourses and practices of modern Western societies, and discovers in Reason and its key practices – such as planning – not universal truths but rather very specific, and even somewhat strange or at least peculiar, ways of being. This also entails, for those working within the Western tradition, recognizing – without overlooking the cultural content of science and technology – that:

> (1) The production of universal, totalizing theory is a major mistake that misses most of reality, probably always, but certainly now; (2) taking responsibility for the social relations of science and technology means refusing an anti-science metaphysics, a demonology of technology, and so means embracing the skilful task of reconstructing the boundaries of daily life, in partial connection with others, in communication with all of our parts. (Haraway, 1985, p. 100)

As we have shown, planning has been one of those totalizing universals. While social change has probably always been part of the human experience, it was only within European modernity that 'society', i.e. the whole way of life of a people, was open to empirical analysis and made the object of planned change. And while communities in the Third World may find that there is a need for some sort of organized or directed social change – in part to reverse the damage caused by development – this undoubtedly will not take the form of 'designing life' or social engineering. In the long run, this means that categories and meanings have to be redefined; through their innovative political practice, new social movements of various kinds are already embarked on this process of redefining the social, and knowledge itself.

The practices that still survive in the Third World despite development thus point the way to moving beyond social change and, in the long run, to entering a post-development, post-economic era. In the process, the plurality of meanings and practices that make up human history will again be made apparent, while planning itself will fade away from concern.

References

Crosswell, M. J. 1981: Basic Human Needs: A Development Planning Approach, in D. M. Leipziger and P. Streeten (eds), *Basic Needs and Development*. Cambridge, Mass: Oelgeschlager, Gunn and Hain Publishers Inc.

Depto. Nacional De Planeación de Colombia, Programa de Desarrollo Rural Integrado 1979: *El Subsector de Pequeña Producción y el Programa DRI*. Bogota: DNP, July 47.

Foucault, M. 1979: *Discipline and Punish*. New York: Pantheon Books.

Friedmann, J. 1965: *Venezuela: From Doctrine to Dialogue*. Syracuse: Syracuse UP.

Haraway, D. 1985: A Manifesto for Cyborgs: Science, Technology, and Socialist Feminism in the 1980s. *Socialist Review* 15 (2).

International Bank for Reconstruction and Development 1950: *The Basis of a Development Program for Colombia*. Baltimore: Johns Hopkins UP.

Kennedy, J. F. 1961: Presidential Address, January 20.

McLeod, M. 1983: 'Architecture or Revolution': Taylorism, Democracy, and Social Change. *Art Journal* (summer) 132–47.

Mueller, A. 1987: Power and Naming in the Development Institution: The 'Discovery' of 'Women in Peru' presented at the 14th Annual Third World Conference, Chicago, April 4.

Ong, A. 1987: *Spirits of Resistance and Capitalist Discipline*. Albany, New York: SUNY Press.

Shiva, V. 1989: *Staying Alive: Women, Ecology and Development*. London: Zed Books.

United Nations 1962: Dept. of Economic and Social Affairs. *The United Nations Development Decade: Proposals for Action*, New York: United Nations.

World Bank 1975: *Assault on World Poverty*. Baltimore: Johns Hopkins UP.

Bibliography

Edward Said's *Orientalism*, New York: Vintage Books, 1979, still constitutes the point of departure for examining European or Euro-American representations of non-Western peoples. The general orientation for the discursive critique of representations is provided by Foucault, especially in *The History of Sexuality*, Vol. I, New York: Vintage Books, 1980, and *Power/Knowledge*, New York: Pantheon Books, 1981. These works provide the general framework for analysing development as a discourse, i.e. as a Western form of social description. Extensions of these works in connection with development are I. Gendzier, *Managing*

Political Change: Social Scientists and the Third World, Boulder: Westview Press, 1985; P. Morandé, *Cultura y Modernización en América Latina*, Santiago: Pontificia Universidad Católica de Chile, 1984; V. Y. Mudimbe, *The Invention of Africa*, Bloomington: Indiana University Press, 1988; and A. Escobar, 'Power and Visibility: Development and the Invention and Management of the Third World', *Cultural Anthropology*, 3(4), November 1988.

On the origins of town planning, see L. Benevolo, *History of Modern Architecture*, Cambridge: MIT Press, 1971; and F. Choay, *The Modern City: Planning in the Nineteenth Century*, New York: George Bazillier, 1969. The rise of the social is documented in J. Donzelot, *The Policing of Families*, New York: Pantheon Books, 1979, and *L'Invention du Social*, Paris: Fayard, 1984. I. Illich discusses the professionalization of needs in *Toward a History of Needs*, Berkeley: Heyday Books, 1977. More recently, P. Rabinow has tackled the management of space and the normalization of the population in the context of French Colonial North Africa in *French Modern: Norms and Forms of the Social Environment*, Cambridge: MIT Press, 1989. The role of bio-politics and the narratives of science in the articulation of nature, gender and culture is examined in D. Haraway's *Primate Visions: Gender, Race and Nature in the World of Modern Science*, New York: Routledge, 1989. The two most insightful books on the origins of the modern economy, on the other hand, are K. Polanyi, *The Great Transformation*, Boston: Beacon Press, 1957 and L. Dumont, *From Mandeville to Marx: The Genesis and Triumph of Economic Ideology*, Chicago: The University of Chicago Press, 1977.

Perhaps the most comprehensive (retrospective and prospective) look at planning is J. Friedmann's *Planning in the Public Domain*, Princeton: Princeton University Press, 1987. The critical analysis of institutional practices has been pioneered by D. Smith, *The Everday World as Problematic: A Feminist Sociology*, Boston: Northeastern University Press, 1987, and extended by A. Mueller in her doctoral dissertation, *The Bureaucratization of Development Knowledge: The Case of Women in Development*, Ontario Institute for Studies in Education, University of Toronto, 1987. E. J. Clay and B. B. Schaffer provide a thorough analysis of the 'hidden' practices of development planning in *Room for Manoeuvre: An Exploration of Public Policy Planning in Agriculture and Rural Development*, Rutherford: Fairleigh Dickinson University Press, 1984, while G. Wood focuses on the relation between labels and power in his article, 'The Politics of Development Policy Labelling', *Development and Change*, Vol. 16, 1985. A. Ong provides a complex view of the manifold practices and effects of development as biopolitics in *Spirits of Resistance and Capitalist Discipline: Factory Women in Malaysia*, Albany: SUNY Press, 1987. An insightful general treatise on practices of domination and resistance is M. de Certau's *The Practice of Everyday Life*, Berkeley: University of California Press, 1984.

Important elements for redefining development, especially from the vantage point of grassroots alternatives, are found in D. L. Shet, 'Alternative Development as Political Practice', *Alternatives*, XII(2), 1987; V. Shiva, *Staying Alive: Women, Ecology and Development*, London: Zed Books, 1989; O. Fals Borda, *Knowledge and People's Power*, Delhi: Indian Social Institute, 1988; R. Kothari, 'Masses, Classes, and the State', *Alternatives*, XI(2), 1986; A. Nandy, *The Intimate Enemy*, Bombay: Oxford University Press, and *Traditions, Tyranny and Utopias*, Delhi: Oxford University Press, 1987; G. Esteva, 'Regenerating People's Space', *Alternatives*, XXI(1); and M. Rahnema, 'A New Variety of AIDS and Its Pathogens: Homo Economicus, Development and Aid', *Alternatives*, XIII(1), 1988. The role of social movements in articulating alternative visions of social and political change is explored in A. Escobar and S. Alvarez (eds), *New Social Movements in Latin America: Identity, Strategy and Democracy*, Boulder: Westview Press, 1991.

SECTION TWO

AGRARIAN CHANGE AND RURAL DEVELOPMENT

Editor's Introduction

Rural development issues were not always to the fore in the modernizing development studies that emerged in the 1940s and 1950s. Critics have complained that the urban–industrial biases of this time encouraged development planners to think of agriculture as a 'bargain basement' (a phrase associated with India's Prime Minister Nehru in the 1950s), and to propose largely urban solutions to essentially rural problems. The Lewis two-sector model seemed to embody this philosophy perfectly. In his classic work on economic development with unlimited supplies of labour (Lewis, 1954, 1955), Lewis wrote of massive disguised unemployment in the traditional, rural sectors of developing economies, and urged rural-to-urban migration as a costless way of transferring labour to more profitable enterprises in the modern sector.

There is clearly something to be said for this account of development studies in the heyday of modernization theories and policies, but much is also missing. Although rural development only became 'a distinctive field of policy and practice, and of research' in the 1970s (Harriss, 1982, p. 15), a broader concern for rural development and agricultural productivity was always present in the 1950s and 1960s and prevailing orthodoxies were often challenged. In the 1950s most attention was devoted to agrarian reform, and to questions of farm size–productivity relationships and the effects of different land tenure systems on farm output. A good deal of work maintained not just that rural areas of the Third World were burdened by 'traditional' agricultural practices and attitudes, but also that such traits were encouraged, or even produced, by local systems of land ownership and operation that vested land mainly in the hands of unenterprising 'feudal' or 'pre-capitalist' elites (Stavenhagen, 1970; Thorner and Thorner, 1962). These elites were content to live off land rents and an immiserated peasantry, and saw little need to play the part of modern, improving, or capitalist farmers. Potential farm output and rural incomes were depressed as a result. So too was the marketed surplus that was needed to support lower real wage rates in the 'modern' sector of the economy (Byres, 1974).

The 'depressor' argument is taken up in various ways by Amartya Sen, Bruce Johnston and Peter Kilby, and John Harriss in Readings 6, 7 and 8, although these Readings cover much else besides. In **Reading 6, Amartya Sen** addresses some fundamental questions relating to food and freedom. Drawing on Isaiah Berlin's distinction between negative and positive freedoms, Sen notes that the 'distinction may be quite central to different approaches to the idea of freedom and its implications'. In an interesting comparison of China and India from 1950 to 1980, Sen shows that rural development policies in communist China increased life expectancies there significantly more quickly than did the nominally socialist policies pursued in and by independent India. Positive freedoms were thereby increased in China, in the sense that most people were free from the threat of famine (save, terribly, between 1958 and 1961). Negative freedoms, however, or the freedom of the individual from interference by others, were not greatly valued by the communist rulers of China. Sen notes that it was the ability of the Chinese government to dictate to its citizens – to curtail their negative freedoms – that in part allowed the famine of 1958 to continue for three years and to claim 29.5 million lives. In India, by contrast, 'no government at the center – or at the state level – can get away without extreme political damage if it fails to take early action against famines'. Sen concludes his paper by noting that 'freedom is not a remote consideration in policy making', including food policy making. He might equally have said that development theories and policies are ever closely entwined (see Section One).

Sen's paper makes uncomfortable reading for those who would privilege positive over negative freedoms, but it should also be unsettling for anyone who believes that socialist development strategies have nothing to commend them. The radical redistribution of assets (or the means of production) in countries like China, Cuba and socialist Nicaragua, together with strong public investment in education and health-care systems, did usually improve the living conditions of a majority of the rural population in the early years following revolutionary change (Post and Wright, 1989). But some pressing practical and intellectual issues still remain when we come to assess the records of socialist developing countries (as Sen clearly shows us). In particular, it is important to ask whether an initial and effective redistribution of assets on a cooperative or collectivized basis is always compatible with sustained economic growth over a matter of decades. This is bound up with the incentive problem, and questions of peasant or worker motivation. It is also important to ask whether a more effective redistribution of assets might be effected within local systems of private property holding, or a 'mixed economy'.

Sen touches on this last issue when he offers an assessment of Sri Lanka's achievements since 1940. The issue is confronted more directly still in **Reading 7** by **Bruce Johnston and Peter Kilby**. Johnston and Kilby contrast unimodal and bimodal strategies of agrarian change in late-developing countries. Bimodal strategies have commonly been pursued in Latin America (de Janvry, 1982) and parts of South Asia (Herring, 1982). The bimodal option seeks 'a crash modernization strategy' in the countryside which concentrates private and public resources in 'a highly commercialized subsector'. A relatively small number of very large farming enterprises are encouraged to produce cash crops for the global market-place and to produce sizeable marketed surpluses. If there is any land redistribution going on it is usually from poorer to richer households, at least at the level of land operated (if not always land owned where ceilings legislation is on the statute books). Agribusiness links will flourish in this environment (Goodman and Redclift, 1991), along with all manner of 'plantation crops'. So too will expensive imports of pesticides, fertilizers and farm machinery.

Johnston and Kilby suggest that bimodal strategies are deficient in three key respects: (1) they fail to empower poorer farming households directly, with the result that an artificial land scarcity ensues; (2) they fail to maximize crop outputs on a per acre basis because large farm units are often managed extensively; and (3) they fail to induce the backward and forward linkages that create off-farm employment in surrounding rural and urban locations. It follows that a progressive modernization of the entire (hence unimodal) agricultural sector is much to be preferred, although not by means of a cloying collectivization. Johnston and Kilby point to the post-war land reforms in Japan and Taiwan as two examples of the initiation of unimodal strategies of agrarian change that redistributed land on a more egalitarian basis, but within a private property framework. The *Land to the Tiller Program* in Taiwan in the early 1950s forced Program landlords 'to sell their land in excess of a very small acreage ceiling' (Johnston and Kilby, 1975, p. 255), and set the scene for a rapid increase in per acre farm output on the basis of intensive peasant agriculture. Although they do not make such connections in the Reading published here, Johnston and Kilby are in this way lending their support to a neo-populist tradition of rural development thinking (and policy) which makes a virtue of the self-exploitation of peasant household labour (the Chayanovian tradition: see Chayanov, 1966). Their work also resonates with those accounts from the early 1970s which announced the compatibility, and even desirability, of redistribution with (or for) growth.

Johnston and Kilby's Reading concludes with a brief discussion of the political constraints on different strategies of agrarian change (see also Tomich, Kilby and Johnston, 1995). They join with Amartya Sen in insisting that so-called economic strategies for development cannot usefully be designed or propagated in the absence of a careful consideration of local political conditions and possibilities. It might be argued, though, that neither Reading 6 nor Reading 7 is properly alert to the possibility that agrarian reform is promoted for political reasons, as well as to advance economic ends. Fitzgerald points to this motive when he writes that: 'in capitalist (or "mixed") economies during the post-World War II years . . . there flourished a considerable enthusiasm for redistributive land reform, which was seen principally as constituting (or reconstituting) a prosperous small-farmer class on estates expropriated from the aristocracy or foreigners with a particular function of underwriting democracy, while responding to the millenarian demands of the peasantry for security' (Fitzgerald, 1989, p. 199). **John Harriss**, too, in **Reading 8**, is sensitive to this aspect of the politics of agrarian reform. In an extraordinarily wide ranging and up-to-date review of the evidence on agrarian change in India, Harriss moves well beyond the depressor arguments that dominated debate in the 1950s (without leaving these debates to one side) to consider also the agrarian consequences of the Green Revolution and an associated increase in state investment in rural development infrastructures.

The technological approach to rural development places a particular faith in the capacity of men and women, aided by the state, private industry and agricultural universities, to create a more productive, and in many respects less 'natural' environment for increases in farm output. Where agrarian reform places its faith in institutional changes in the countryside (and in the fact of there being 'surplus' land to redistribute to poorer households), the Green Revolution promises a scientific solution to the world's food problems. It is instructive that one of the plant scientists most closely associated with improved cereal varieties, Norman Borlaug, won the Nobel Prize for Peace. The Green Revolution was presented to the world as a peaceful revolution and not as a Red Revolution.

It is against this backdrop that Harriss uses his local knowledge of India to advance a prospectus for rural development studies that will be of much wider interest. Harriss notes that the Green Revolution in India has not been as polarizing as many observers once feared, in part because poorer farming households were in time empowered to partake of the Green Revolution 'package' of modern high-yielding varieties, chemical fertilizers, irrigation and pesticides. Their increased ease of access to improved varieties was aided by new credit arrangements in the countryside and by the growth of rural producer and marketing cooperatives. Harriss also highlights the extraordinary importance of rural non-agricultural activity as a means of guarding against rural poverty or improving rural livelihoods. In many areas of India the rural poor – the landless and marginal farming households – have been able to combine some agricultural activities with rural or urban non-agricultural work. Rural development studies in the academy have followed them out of the field and into interlinked land, labour and product markets (see Binswanger, 1990; Haggblade, Hazell and Brown, 1989; Lipton, 1983; Singh,

1990). Harriss further notes that the sluggish growth of off-farm employment is a cause for concern in India, and probably has something to do with local structures of land ownership and operation that are still skewed in favour of a rich minority of households (Lipton and Longhurst, 1989). The operation of local labour markets in turn will depend, in part, and in ways that can be difficult to control, on demographic trends, government pricing policies (for foodstuffs and non-foodstuffs), and on wider patterns of economic growth and stagnation. In India, too, where negative freedoms are still taken into account at the ballot box (notwithstanding rural vote banks and the capture of poll booths by political activists), peasant households have been encouraged to protect their perceived interests by organizing as a rural bloc. Organizations like the *Bharatiya Kisan Union* and *Shetkari Sanghatana* have sought to advance the claims of an 'authentic' rural India, or *Bharat*, against an allegedly 'colonial' urban India that perpetrates policies of urban bias (Rudolph and Rudolph, 1987; Bentall and Corbridge, 1995).

Harriss' paper joins with the others in this section in bringing a welcome degree of complexity into debates on rural development and agrarian change. All the papers move beyond a singular focus on the case for and against land reform, or the Green Revolution, or even the claims of socialism and capitalism. If the history of rural development strategies and policies suggests anything to us, it is that a complex set of relationships between labour markets, product markets, farm household decision-making, the environment, demography and farm output cannot satisfactorily be reduced to the independent operations of a single variable or relationship. Harriss' work also highlights the importance of government policies in creating a framework for rural development, and this theme is taken up forcefully by **Robert H. Bates** in **Reading 9**. Although Bates' work on governments and agricultural markets in Africa advances lessons that move beyond that region, it is no coincidence that Africa is the focus of his particular contribution to development studies.

The key issue for Bates concerns the decline in per capita food production in sub-Saharan Africa from about 1965, when comparable figures for Asia and Latin America have been rising. (In parts of Asia the rate of increase in per capita food production has slowed recently, prompting concern that the Green Revolution is stirring an ecological backlash in the form of depleted soils, salinization, eutrophication, a loss of biodiversity, and so on: see Adams, 1990; Shiva, 1991.) This same issue has also concerned the international donor and development communities, and was addressed by the World Bank, in 1981, in the form of Elliott Berg's report on *Accelerated Development in Sub-Saharan Africa* (World Bank, 1981). The gist of Berg's study, as of Bates' work, is that food production in Africa has suffered mainly because the incentive for most farmers to produce food has been eroded by government policies that favour urban–industrial interests. Farm output prices are depressed in many African countries by the monopsonistic activities of local marketing boards and by an overvaluation of the domestic currency. The result is a transfer of resources from ordinary farmers to urban–industrial interests and regions, and a concomitant decline in per capita food production. Such investment as there is in the African countryside tends to be directed to agricultural projects that benefit only a minority of wealthy farmers, who remain loyal to the government as a result. An overvaluation of the domestic currency also permits a subsidized supply of capital-intensive farm inputs to this constituency.

What makes the work of Robert Bates so interesting is his insistent examination of the political conditions of existence of these damaging macro-economic policies. Bates clearly draws on the work of Theodore Schultz (Schultz, 1964), and others as diverse as Elliott Berg and Michael Lipton (Lipton, 1977), all of whom have argued that 'given the right incentives, farmers in the developing world would "turn sand into gold."' But Bates (like Lipton) refuses to follow a true believer in the counter-revolution in development studies, like Schultz, in the view that all will be well in Africa if only governments refuse to introduce 'distortions' into agricultural markets. Bates accepts that African peasants are making rational choices about production and consumption, *given the situation they find themselves in*, but he has extended this same rational-choice framework to the actions of key interest groups in rural and urban Africa, and to the actions of governments themselves. Bates contends that distorted market prices and a project bias in agricultural policy are immediately rational for governments that depend upon the patronage of key interest groups in the countryside and the major cities. (The latter is surely more true of parts of Africa than Asia because of the often dispersed geography of settlement there, and a local disposition to change governments by coups that are targeted on the capital city.) If there is hope for Africa, it is that the political economy of the African state is always changing, and is continually reshaped by the food crises induced by present 'poor' policies and by pressures from the international community and agencies like the World Bank and the IMF (Bayart, 1993). It is one thing to assume that individuals and groups behave rationally, and quite another to specify what might be rational at any given time and place.

The final reading in this section is by **Robert Chambers (Reading 10).** Chambers would doubtless agree with much of the analysis of Robert Bates; indeed, the damaging effects of urban-biased policies are now widely recognized by neoliberals, Marxists, environmentalists and populists alike (although the question of how a surplus for industrial development is to be created remains). But Chambers is probably closer to Michael Lipton than Robert Bates in wanting to highlight the claims of rural dwellers as a means of addressing the curse of absolute poverty in the Third World (see also de Janvry and Sadoulet, 1989). The paper by

Chambers also sets out two further sets of claims that continue to make their mark on contemporary development studies. First, Chambers notes the difficulties that have marked outsiders' contacts with 'rural poverty in general, and with the deepest poverty in particular'. In a frank discussion of six sets of biases that impede our understanding of these issues, Chambers urges us – as students of development, as activists, as planners – to move beyond the tarmac road and the urban fringe, and to eschew research that is conducted only in the dry season, or in model or project villages, or where most contacts are with male members of the local elite. If we can do this, Chambers suggests secondly, we will come to perceive rural poverty in a way that is more akin to the perceptions of those trapped in rural poverty. We will then learn to act with more humility, taking the time to listen and to learn from local people. Chambers urges his readers to put poor people first. Like many other scholars and activists in the 1980s and 1990s, he has emphasized the need for local people to be empowered to improve their own lives. Development 'experts' must be careful about imposing their own professional biases and 'solutions' on people who often know what they want and need, even where they cannot access the resources required to improve their life- chances.

Taken together, the Readings in this section give a sense of some of the key issues and debates that continue to lie at the heart of rural development studies. Above all, they highlight the persistent tension that lies at the heart of rural development itself: the need to protect the interests of rural producers and the rural poor while also creating a surplus for use elsewhere in the economy. Several related issues are taken up in succeeding sections: notably on the environmental contexts for rural development (**Readings 12 and 25**), on gender issues and peasant struggles in rural societies (**Reading 14**), and on the exchanges of labour and commodities that bind rural areas into wider space-economies, both nationally and internationally (**Readings 13, 16 and 21**). Some key texts on issues that are not properly developed in this Reader – such as farmer decision making, the Green Revolution in Africa and Latin America, the merchant state, and the development of rural industries – are signposted in the Guide to Further Reading that follows.

Guide to further reading

General

Adams, W. 1990: *Green development: Environment and sustainability in the Third World*. London: Routledge.

Bayart, J.-F. 1993: *The state in Africa: The politics of the belly*. London: Longman.

Berry, R. and Cline, W. 1979: *Agrarian structure and productivity in developing countries*. Baltimore: Johns Hopkins UP.

Boserup, E. 1970: *Women's role in economic development*. New York: St Martin's Press.

Chayanov, A. V. 1966: *The theory of peasant economy*. Homewood, IL: Richard Irwin.

Colburn, F. 1982: Current studies of peasants and rural development; applications of the political economy approach. *World Development* 34, 437–49.

Friedmann, H. 1993: The political economy of food. *New Left Review* 197, 29–57.

Ghai, D., Khan, A. R., Lee, E. and Radwan, S. (eds) 1979: *Agrarian systems and rural development*. New York: Holmes and Meier.

Goodman, D. and Redclift, M. 1991: *Refashioning nature: Food, ecology and culture*. London: Routledge.

Grigg, D. 1993: *The world food problem* (2nd edn). Oxford: Blackwell.

Harriss, J. (ed.) 1982: *Rural development: Theories of peasant economy and agrarian change*. London: Hutchinson.

Hayami, Y. and Ruttan, V. 1985: *Agricultural development: An international perspective*. Baltimore: Johns Hopkins UP.

Johnston, B. F. and Kilby, P. 1975: *Agriculture and structural transformation: Economic strategy in late-developing countries*. Oxford: OUP.

Kandiyoti, D. 1990: Women and rural development policies: the changing agenda. *Development and Change* 21, 5–22.

Lewis, W. A. 1954: Economic development with unlimited supplies of labour. *Manchester School* 22, 139–91.

Lewis, W. A. 1955: *The theory of economic growth*. London: Allen and Unwin.

Lipton, M. 1977: *Why poor people stay poor: A study of urban bias in world development*. London: Temple Smith.

McMichael, P. 1992: Tensions between national and international control of the world food order. *Sociological Perspectives* 35, 343–65.

Raju, S. and Bagchi, D. (eds) 1993: *Women and work in South Asia*. London: Routledge.

Rao, M. and Caballero, R. 1990: Agricultural performance and development strategy. *World Development* 18, 899–913.

Sen, A. K. 1981: *Poverty and famines*. Oxford: Clarendon.

Singh, A. and Tabatabai, M. 1992: Agriculture and economic development in the 1990s. *International Labour Review* 131, 405–30.

Timmer, C., Falcon, W. and Pearson, S. 1983: *Food policy analysis*. Baltimore: Johns Hopkins UP.

Timmer, C. 1988: The agricultural transformation. In Chenery, H. and Srinivasan, T. (eds), *Handbook of development economics*, volume 1. Amsterdam: North Holland, 276–331.

Tomich, T., Kilby, P. and Johnston, B. 1995: *Transforming agrarian economies: Opportunities seized, opportunities missed*. Ithaca: Cornell UP.

Watts, M. 1983: *Silent violence: Food, famine and peasantry in Northern Nigeria*. Berkeley: University of California Press.

World Bank 1986: *World development report, 1986*. Oxford: OUP/World Bank.

Agrarian reform and socialist agricultural strategies

Bramall, C. 1993: *In praise of Maoist economic planning*. Oxford: Clarendon Press.

Bratton, M. 1987: The comrades and the countryside: the politics of agricultural policy in Zimbabwe. *World Development* 39, 174–202.
Byres, T. 1974: Land reform, industrialisation and the marketed surplus in India: an essay on the power of rural bias. In Lehmann, D. (ed.), *Agrarian reform and agrarian reformism*. London: Faber, 221–61.
Carney, J. 1992: Peasant women and economic transformation in the Gambia. *Development and Change* 23, 67–90.
Christadoulou, D. 1990: *The unpromised land: Agrarian reform and conflict worldwide*. London: Zed.
Cornia, G. 1985: Farm size, land yields and the agricultural production function: an analysis of fifteen developing countries. *World Development* 13, 513–34.
de Janvry, A. 1982: *The agrarian question and reformism in Latin America*. Baltimore: Johns Hopkins UP.
Dorner, P. 1982: *Land reform and economic development*. Harmondsworth: Penguin.
Fitzgerald, E. V. K. 1989: Land reform. In Eatwell, J., Milgate, M. and Newman, P. (eds), *The new Palgrave: Economic development*. Basingstoke: Macmillan.
Ghose, A. K. (ed.) 1983: *Agrarian reform in contemporary developing countries*. Beckenham: Croom Helm.
Griffin, K. 1974: *The political economy of agrarian change*. London: Macmillan.
Herring, R. 1982: *Land to the tiller: The political economy of agrarian reform in South Asia*. New Haven: Yale UP.
Hinton, W. 1968: *Fanshen: A documentary of revolution in a Chinese village*. New York: Monthly Review Press.
Jones, S., Joshi, P. and Murmis, M. (eds) 1983: *Rural poverty and agrarian reform*. New Delhi: Allied.
Lardy, N. 1983: *Agriculture in China's modern economic development*. Cambridge: CUP.
Lenin, V. I. 1956: *The development of capitalism in Russia*. Moscow: Progress.
Ling, Z. 1990: The transformation of the operating mechanisms in Chinese agriculture. *Journal of Development Studies* 26, 229–42.
Molyneaux, M. 1985: Mobilization without emancipation: women's interests, state and revolution. *Feminist Studies* 11, 227–54.
Nolan, P. 1988: *The political economy of collective farms*. Cambridge: Polity.
Post, K. and Wright, P. 1989: *Socialism and underdevelopment*. London: Routledge.
Rahmato, D. 1985: *Agrarian reform in Ethiopia*. Trenton: Red Sea Press.
Samatar, A. 1988: The state, agrarian change and crisis of hegemony in Somalia. *Review of African Political Economy* 43, 26–41.
Sklar, R. 1988: Beyond capitalism and socialism in Africa. *Journal of Modern African Studies* 26, 1–21.
Sobhan, R. 1993: *Agrarian reform and social transformation: Preconditions for development*. London: Zed.
Stavenhagen, R. 1970: *Agrarian problems and peasant movements in Latin America*. New York: Doubleday.
Thorner, D. and Thorner, A. 1962: *Land and labour in India*. London: Asia.
Warriner, D. 1969: *Land reform in principle and practice*. Oxford: Clarendon.
Yang, Y. and Tyers, R. 1989: The economic costs of food self-sufficiency in China. *World Development* 17, 81–98.

The Green Revolution and beyond: fields, farms, labour markets, linkages and struggles

Bebbington, A. and Carney, J. 1990: Geographers in the international agricultural research centers: theoretical and practical considerations. *Annals of the Association of American Geographers* 80, 34–48.

Bentall, J. and Corbridge, S. 1995: Urban–rural relations, demand politics and the 'new agrarianism' in north-west India: the Bharatiya Kisan Union. *Transactions of the Institute of British Geographers* (forthcoming).

Binswanger, H. 1990: The policy response of agriculture. In *Proceedings of the World Bank Annual Conference on Development Economics*. Washington, DC: World Bank.

Bhalla, S. 1987: Trends in employment in Indian agriculture. *Indian Journal of Agricultural Economics* 42, 537–60.

Bliss, C. and Stern, N. 1982: *Palanpur: The economy of an Indian village*. Oxford: Clarendon.

Boyce, J. 1987: *Agrarian impasse in Bengal: Institutional constraints to technological change*. Oxford: OUP.

Bray, F. 1987: *The rice economies*. Cambridge: CUP.

de Janvry, A. and Sadoulet, E. 1989: Investment strategies to combat rural poverty: a proposal for Latin America. *World Development* 17, 1203–21.

Evenson, R. and Kislev, Y. 1975: *Agricultural research and productivity*. New Haven: Yale UP.

Farmer, B. 1985: Perspectives on the 'green revolution' in South Asia. *Modern Asian Studies* 20, 175–99.

Glaeser, B. 1987: *The green revolution revisited: Critique and alternatives*. London: Macmillan.

Haggblade, S., Hazell, P. and Brown, J. 1989: Farm–nonfarm linkages in rural sub-Saharan Africa. *World Development* 17, 1173–201.

Harriss, B. 1985: *State and market*. Delhi: Concept.

Harriss, B. 1987: Regional growth linkages from agriculture: discussion. *Journal of Development Studies* 23, 275–89.

Harriss, J. 1982: *Capitalism and peasant farming: Agrarian structure and ideology in northern Tamil Nadu*. Bombay: OUP.

Hart, G. 1986: Exclusionary labour arrangements: interpreting evidence and employment trends in rural Java. *Journal of Development Studies* 22, 681–96.

Hazell, P. and Roell, A. 1983: Rural growth linkages: household expenditure patterns in Malaysia and Nigeria. *Research Report* 41. Washington, DC: International Food Policy Research Institute.

Islam, R. 1986: Non-farm employment in rural Asia: issues and evidence. In Shand, R. (ed.), *Off-farm employment in the development of rural Asia*. Canberra: Australian National University.

Kenney, M. and Buttel, F. 1985: Biotechnology: prospects and dilemmas for Third World development. *Economic Development and Cultural Change* 34, 17–52.

Lipton, M. 1983: *Labour and poverty*. Washington, DC: World Bank Staff Working Paper 616.

Lipton, M. with Longhurst, R. 1989: *New seeds and poor people*. London: Unwin Hyman.

Mellor, J. 1976: *The new economics of growth*. Ithaca: Cornell UP.

Rudolph, L. and Rudolph, S. H. 1987: *In pursuit of Lakshmi: The political economy of the Indian state* (Chapters 12 and 13). Chicago: University of Chicago Press.

Schultz, T. 1964: *Transforming traditional agriculture*. New Haven: Yale UP.

Shiva, V. 1991: *The violence of the green revolution*. London: Zed.

Singh, I. 1990: *The great ascent: The rural poor in South Asia*. Baltimore: Johns Hopkins UP.

Watts, M. 1989: The agrarian question in Africa: debating the crisis. *Progress in Human Geography* 13, 1–41.

Whitehead, A. 1990: Food crisis and gender conflict in the African countryside. In Bernstein, H. *et al.* (eds), *The food question: Profits versus people?* London: Earthscan.

Wilson, C. 1994: *How sustainable has the green revolution proved to be?* Unpublished M. Phil. dissertation: Cambridge University, Department of Geography.

Markets, states, politics and participation

Barham, B., Clark, M., Katz, E. and Schurman, R. 1992: Non-traditional agricultural exports in Latin America. *Latin American Research Review* 27, 43–82.

Bates, R. 1981: *Markets and states in tropical Africa: The political basis of agricultural policies*. Berkeley: University of California Press.

Bates, R. 1989: *Beyond the miracle of the market: The political economy of agrarian change in Kenya*. Cambridge: CUP.

Bauer, P. T. 1954: *West African trade*. Cambridge: CUP.

Bebbington, A. 1994: Theory and relevance in indigenous agriculture: knowledge, agency and organization. In Booth, D. (ed.), *Rethinking social development*. Harlow: Longman.

Cernea, M. (ed.) 1991: *Putting people first: Sociological variables in rural development* (2nd edn). Oxford: OUP/IBRD.

Chambers, R. 1983: *Rural development: Putting the last first*. Harlow: Longman.

Chambers, R., Pacey, A. and Thrupp, L. (eds) 1989: *Farmer first: Farmer innovation and agricultural research*. London: IT Publications.

Clayton, E. 1983: *Agriculture, poverty and freedom in developing countries*. London: Macmillan.

Crow, B. 1989: Plain tales from the rice trade. *Journal of Peasant Studies* 16, 198–229.

Demery, L. and Addison, T. 1987: Food security and adjustment policies in sub-Saharan Africa. *Development Policy Review* 5, 177–96.

Fox, J. 1992: Democratic rural development. *Development and Change* 23, 201–44.

Harriss, B. 1993: Markets, society and the state: problems of marketing under conditions of smallholder agriculture in West Bengal. *DPP Working Paper 26*. Open University: Faculty of Technology.

Lipton, M. 1968: The theory of the optimizing peasant. *Journal of Development Studies* 4, 327–51.

Lipton, M. 1991: Market relaxation and agricultural development. In Colclough, C. and Manor, J. (eds), *States or markets?* Oxford: Clarendon, 26–47.

Richards, P. 1985, *Indigenous agricultural revolution*. London: Hutchinson.
Schultz, T. (ed.) 1978: *Distortions of agricultural incentives*. Bloomington: Indiana UP.
Sender, J. and Smith, S. 1985: What's right with the Berg Report and what's left of its critics? *Capital and Class* 24, 125–46.
Wilken, G. 1987: *Good farmers: Traditional agricultural resource management in Mexico and Central America*. Berkeley: University of California Press.
World Bank 1981: *Accelerated development in sub-Saharan Africa*. Washington, DC: World Bank.

6 Amartya Sen, 'Food and Freedom'

Reprinted in full, save for the first paragraph, from: *World Development* 17, 769–81 (1989)

1. Introduction

The links between food and freedom may at first sight appear to be rather remote to policy making and far from central to practical concerns. I shall argue against that view, trying to discuss the various important connections that have to be recognized more fully as background to practical food policy. I shall not, of course, deny the fact that these connections are not typically taken to be straightforward preludes to practical policy making, but I will argue that we have to probe deeper for an adequate background to policy making.

2. Food for freedom and freedom for food

'Grub first, then ethics,' thus runs a much quoted aphorism of Bertolt Brecht. There is undoubtedly some sense in this phased gradation. Ethics may seem like a much more remote and much less immediate subject than the command over food that we need to survive. Freedom too – as an important concept in ethics – may seem to be far less immediate than the compelling demands of grabbing grub.

But this contrast is quite artificial. The provision of food is indeed a central issue in general social ethics, since so much in human life does depend on the ability to find enough to eat. In particular, the freedom that people enjoy to lead a decent life, including freedom from hunger, from avoidable morbidity, from premature mortality, etc., is quite centrally connected with the provision of food and related necessities. Also, the compulsion to acquire enough food may force vulnerable people to do things which they resent doing, and may make them accept lives with little freedom. The role of food in fostering freedom can be an extremely important one.

On the other side, freedom may also causally influence the success of the pursuit of food for all. One consideration that has received a great deal of attention recently relates to the role of freedom to make profits in providing

This paper was given as the Sir John Crawford Memorial Lecture, Washington, DC, 29 October 1987, and was published in *World Development*, 17 (1989). Minor modifications have been made on pp. 102–103 [here] to take account of more recently available data and publications.

incentives for the expansion of food production, thus helping to solve the food problem. This consideration has often cropped up in the critical evaluation of agricultural policies pursued in many countries in Africa and Asia. For example, the rapid expansion of agricultural output in China in the economic reforms carried out from 1979 onwards has, with much justice, been seen to be closely related to the freeing of markets and the unleashing of productive opportunities connected with profit incentives. These experiences invite attention and scrutiny.

Other types of freedom may also have important instrumental roles to play in the guaranteeing of food for all. Insofar as public policy to combat hunger and starvation – including rapid intervention against threatening famines – may depend on the existence and efficiency of political pressure groups to induce governments to act, political freedom too may have a close connection with the distribution of relief and food to vulnerable groups. There are other possible causal connections – operating in both directions – which may be worth investigating, and some of these I will indeed try to examine and assess in this lecture. Freedom to make profits is not the only freedom the causal influence of which would have to be considered.

Thus, what may superficially appear to be rather remote connections between food and freedom can be seen to be, in fact, central in importance and extremely rich in the variety of influences involved, operating in the two respective directions, viz., from food to freedom, and from freedom to food. I shall try to supplement the conceptual and theoretical discussions with illustrations from practical problems with empirical content. Freedom and ethics are indeed very practical matters in the determination of food policy.

3. Four concepts of freedom

In a justly famous essay called 'Two Concepts of Liberty', Isaiah Berlin (1969) made an important distinction between 'negative' and 'positive' theories of freedom. The negative view sees freedom exclusively in terms of the independence of the individual from interference by others, including governments, institutions and other persons. The positive view, which can be characterized in many different ways, sees freedom not in terms of the presence or absence of interference by others, but in terms of what a person is actually able to do or to be. The distinction may be quite central to different approaches to the idea of freedom and its implications (see Sen, 1985b). If a person is not free from hunger and lacks the means and the practical opportunities to feed himself or herself adequately, then that person's positive freedom must be seen as having been thoroughly compromised. On the other hand, his or her negative freedom may be completely unviolated, if this failure to acquire enough food is not a result of his or her having been stopped by interference from others.

There is another distinction which is quite central to the content and role of freedom, and this concerns the issue of *intrinsic* importance of freedom as such, in addition to its *instrumental* roles. That freedom must have instrumental importance as a *means* to other ends is obvious enough. Our freedom to choose one bundle of commodities rather than another may have an important effect on the living standards we can have, the happiness we can enjoy, the well-being we can achieve, and the various objectives of our lives we can fulfill. Similarly, the absence of interference by others may have important causal influence on various things that we can do and value doing. In the 'instrumental' view, freedom is taken to be important precisely because of its being a means to other ends, rather than being valuable in itself.

In contrast, the 'intrinsic' view of the importance of freedom asserts that freedom is valuable in itself, and not only because of what it permits us to achieve or do. The good life may be seen to be a life of freedom, and in that context freedom is not just a way of achieving a good life, it is *constitutive* of the good life itself. The 'intrinsic' view does not deny that freedom may *also* be instrumentally important, but does reject the view that its importance lies *entirely* on its instrumental function.

It is easy to see that the two ways of categorizing different approaches to freedom can be combined with each other, yielding four distinct categories. It is indeed possible to look through the history of ideas to see how different thinkers sharing a regard for freedom fall into different categories, related to the *positive-negative* distinction and to the *intrinsic-instrumental* classification. At the risk of oversimplification I might illustrate the distinctions involved by referring to some particular examples.

For example, Milton Friedman and James Buchanan have both tended to put considerable emphasis – indeed priority – on the negative view of freedom, related to non-interference by the state, institutions and other individuals.[1] This contrasts with the emphasis on the positive view of freedom that can be found in the writings of, say, Bentham or Marx. On the other hand, within the negative perspective, Friedman is much more concerned with the instrumental role of freedom rather than its intrinsic importance, while Buchanan constructs a 'non-instrumental' normative case in favor of giving priority to liberties and democratic rights. Whereas Friedman concentrates primarily on what he calls 'the fecundity of freedom,' Buchanan goes largely beyond this role of freedom as a means to other ends. Attaching intrinsic importance to negative freedom is seen also in the writings of John Rawls, Robert Nozick and other contemporary moral philosophers, and it is a position that was broadly shared also by John Stuart Mill (Mill, 1859; Rawls, 1971; Nozick, 1974).

Similarly, among the various theories concentrating on positive freedom, some have seen freedom to be intrinsically important, such as Adam Smith and Karl Marx, following a line of reasoning that goes back to Aristotle in *Nicomachean Ethics* and *Politics*. In fact, Aristotle had direct influence on Marx's writings on this subject. Marx's philosophical focus included giving a

[1] Buchanan (1986), and Friedman and Friedman (1980). I have discussed this contrast, among others, in Sen (1988).

foundational role to bringing 'the conditions for the free development and activity of individuals under their own control,' with a vision of a liberated society in the future that would make 'it possible for me to do one thing today and another tomorrow, to hunt in the morning, fish in the afternoon, rear cattle in the evening, criticize after dinner, just as I have in mind, without ever becoming hunter, fisherman, shepherd or critic.' (Marx and Engels, 1845–46, republished 1947, p. 22). While his urban middle-class origins may have influenced Marx's evident belief that evening is a good time to rear cattle (he was obviously on more familiar ground with 'criticize after dinner'), the placing of this general perspective of freedom in Marx's entire approach to economics, politics and society was altogether foundational.

While John Rawls' case for the 'priority of liberty' attaches overriding importance to *negative* freedom, his advocacy of the importance of 'primary goods' commanded by people reflects his basic concern for *positive* freedom as well. Primary goods include 'rights, liberties and opportunities, income and wealth, and the social bases of self-respect.' Possessing these things adequately makes a person positively more free to pursue his or her objectives and ends, and Rawls develops his political concept of social justice based on the efficiency and equity in the distributions of these freedoms.

In contrast, Jeremy Bentham's ultimate concern is with utility only, and positive freedom is regarded as important in the Benthamite system only because that freedom may be conducive to more happiness. This is, of course, an instrumental view. The Benthamite *instrumental-positive* view of freedom contrasts with Marx's *intrinsic-positive* view. And each in turn contrasts with the *instrumental-negative* view of Friedman, on the one hand, and the *intrinsic-negative* view of Buchanan and Nozick on the other. It is easy to find other examples to illustrate the contrasts, but perhaps the ones already mentioned will do. I ought to warn that these categories are often not very pure, and the same writers may have a certain amount of plurality within their overall theories (this was, of course, clearly seen in the case of Rawls in the preceding discussion).

4. Food policy and alternative approaches to freedom

This categorization is of crucial relevance even in understanding various demands on food policy, arising from different views of freedom. For example, the advocacy of greater freedom to earn profit in agriculture and of greater use of free markets without much interference by the state and other public institutions (an advocacy that can, incidentally, be found in many documents of the host for this lecture, the World Bank) usually reflects an *instrumental-negative* view of freedom, applied to food policy. Freedom to earn profits without interference is advocated not because it is typically taken to be foundationally important on its own, but because it is seen to be conducive to such things as greater productivity, larger income and enhanced food output. In general, the perspective of *incentives* constitutes an instrumental focus, related to what Milton Friedman calls 'the fecundity of freedom,' and in this particular case this is applied primarily to the negative view of freedom, seen in terms of non-interference.

In contrast, the writings of some authors, such as Peter Bauer, have tended to go beyond the instrumental view even in the context of agricultural development, emphasizing the importance of people having the right to enjoy the fruits of their own creation, without interference by the state or by other institutions or individuals (Bauer, 1981). Bauer has seen this as a central feature of a good agricultural policy. The instrumental consideration of incentives is not denied in this perspective (far from it), but the ethical argument goes well beyond that, to intrinsic importance as well.

On the other hand, economic approaches emphasizing the need to fulfill 'basic needs' for food and other essentials, or to pursue public policy to guarantee 'freedom from hunger,' and so on, take a positive view of freedom, concentrating on what people are able actually to do or be, rather than what they are prevented by others from doing or being (Streeten *et al*, 1981; Morris, 1979; Stewart, 1985). The focus of this literature has often tended to be on pragmatic rather than foundational issues. Concentration on 'freedom from hunger' and related objectives can indeed be defended either on grounds of their supposed intrinsic importance, or because of their instrumental role in serving other – allegedly more basic – goals, such as enhancement of happiness or welfare of individuals. The instrumental view can be seen clearly in the analysis presented by one of the earliest writers on 'basic needs' (though he did not use that expression), viz. A. C. Pigou, in *The Economics of Welfare* (Pigou, 1920). For an example on the other side, Paul Streeten's approach is perhaps best seen in terms of intrinsic value being attached to these respective freedoms to fulfill the various 'basic needs.' (Streeten, 1981).

The instrumental-intrinsic distinction relates to the foundational question as to what is regarded as valuable in itself, and what must be seen as important only as a contributor to other more basic goals. This is a question of deep philosophical interest, but it has pragmatic importance too, since instrumental arguments turn ultimately on the correctness of the cause-effect relationships postulated. For example, if it emerges that free markets and profit earnings do not provide much incentive for the expansion of production, or do not contribute to bettering living standards, the instrumental defense of these free market policies may well collapse, but this need not disestablish at all the view (e.g., Bauer's) that would see the right to earn these profits to be intrinsically important. In this sense, the intrinsic view is less vulnerable to empirical counter-argument, but it has, of course, greater need of foundational ethical defense.

The position is a little different as far as positive freedom is concerned. A policy of state intervention, e.g., in the distribution of food, is scarcely ever regarded as being of fundamental value of its own. The possibility of foundational valuation arises at a somewhat later stage (in this respect its contrast with the valuing of right-based *procedures*, as in the systems of Robert Nozick or Peter Bauer, is quite sharp), and valuing positive freedom has to be based on a good deal of instrumental analysis in moving from the means of state intervention to the realization of positive freedom.

The difference between the 'intrinsic' and 'instrumental' views of positive freedom lies, in this context, in the length to which the instrumental analysis has to be carried. In the broadly Aristotelian view, which sees the capability to achieve important functionings as being valuable in itself, the instrumental analysis can end at that point, but in those views in which positive freedom happens to be no more than means to other ends, e.g., in the pursuit of utility, the instrumental analyses have to go further into the translation of freedom into the fulfillment of other goals. In each case there is need to examine the effects of policies such as public distribution of food on the positive freedoms that individuals can actually obtain, and the difference arises only at a later stage, in moving from freedom to achievement. In this respect the positive freedom view is basically more instrument-dependent than the negative freedom approach is.

These considerations may, at first glance, appear to be rather distant from the nitty-gritty of practical policy making in the field of food and hunger. But foundational questions are ultimately quite central to the acceptability of particular policy analyses. While the tendency to avoid facing these foundational questions is quite common, it is more a reflection of escapism than a demonstration of uncanny wisdom. Ultimately policies have to be justified in terms of what is valuable and how various policies may respectively enhance these valuable things. There is no escape, therefore, from considering both the question of what is fundamentally valuable and the question of what instruments enhance these things best. It is indeed the combination of the intrinsic considerations and instrumental analyses that can lead the way to an adequate examination of what should be done and why.

While these conceptual and theoretical discussions can be carried further – I have tried to discuss some of these further issues elsewhere (Sen, 1985a, 1985b, 1987a) – I shall devote the rest of this lecture to rather practical matters, dealing with actual policy disputes in the field of food and hunger.

5. Opulence and living standard

A preliminary point first. The process of economic development is often seen in terms of the expansion of the material basis of well-being and freedom that people can enjoy. This approach has a rationale that is easy to understand, since the positive freedoms that we can enjoy and the well-being levels that we can achieve are both dependent on the commodity bundles over which we can establish command. This clearly is the sense behind assessing economic development in terms of the progress of real gross national product per head. On the other hand, freedom and well-being depend also on the *use* that is made of the opulence of the nation. Income distributions can vary. No less importantly, the command that people enjoy over essential food, health services, medical attention, etc., depends crucially on the delivery system for these commodities. A public distribution system geared to the needs of the vulnerable sections of the community can bring the essentials of livelihood within easy reach of people whose lives may remain otherwise relatively untouched by the progress of real national income.

Table 6.1 illustrates the point. Oman or South Africa may have a gross national product per head that is a great many times higher than that of China or Sri Lanka, but each of the former has under-five mortality rates (covering infants and children) that is two or three times higher than those prevailing in the poorer economies. The life expectancy at birth in Oman and South Africa lingers around the mid-50s, while China and Sri Lanka have achieved longevity rates reasonably close to those prevailing in Europe and America.

Table 6.1 Opulence, life and death

	GNP per head ($) 1985	*Life expectancy at birth 1985*	*Under-5 mortality rate (per thousand) 1985*
Oman	6,730	54	172
South Africa	2,010	55	104
Brazil	1,640	65	91
Sri Lanka	380	70	50
China	310	69	48

Sources: World Bank (1987); UNICEF (1987).

This is, of course, a well-known point, but it is worth emphasizing in the present context, since the demands of agricultural policy in general and food policy in particular are often seen primarily in terms of expanding the material bases of well-being and freedom. Indeed, as we shall presently see, there is an important policy issue related to this question even in terms of the recent economic reforms in China. The point to note here is that the positive freedom to lead a long life may well be typically enhanced by expansion of material prosperity, but the relationship is far from a tight one, and indeed it is quite possible for the freedom to live long to go down, while the level of economic opulence goes up. The shift of focus from the national product to the freedom enjoyed by members of the nation can bring about a major reexamination of the requirements of economic policy.

The freedom to live long is, of course, only one of the positive freedoms that may be thought to be important. It is a freedom that is particularly valued since our ability to do other things is, obviously, conditional on our being here, and it is not surprising that the option of living longer is very rarely refused. This is, of course, the reason why longevity, which is an *achievement*, can also be seen as an important indicator of the *freedom* to live long (we tend to exercise this freedom, in most cases, to the maximum extent we can), and the metric of life expectancy is, thus, a fairly basic indicator of a foundational positive freedom. There are, however, other important positive freedoms as well, e.g., freedom from hunger and undernutrition, freedom from escapable morbidity, freedom to read and write and communicate. Indeed, the list of important freedoms must be seen to be a long one in any accounting that aims at some degree of comprehensiveness. While any practical analysis may have to confine attention to only a few indicators, the need to have a wider informational base for a more definitive analysis has to be borne in mind.

Often these indicators move in the same direction (e.g., life expectancy, avoidance of morbidity, and literacy frequently tend to be highly correlated), but this is not invariably the case. For example, in the contrast between different states in India, Kerala comes out as having very much higher life expectancy and literacy than any other Indian state, but in terms of morbidity rates, Kerala does not seem to have this advantage. Indeed, measured in the metric of reported illnesses, Kerala's morbidity rate is much higher than that of many other Indian states (Panikar and Soman, 1986; Kumar, 1987). Some of that difference may undoubtedly be due to the fact that a more literate population, with access to medical attention and health care, is likely to report illnesses more thoroughly. But it is possible that even after these corrections are made, there is some dissonance between Kerala's performance in the fields of literacy and life expectancy and that in the prevention of morbidity.[2] The conflicts between different indicators may not, of course, always be serious, but the general possibility has to be kept in view in interpreting results of empirical analysis based on one or a few indicators. In this sense, analyses of the kind pursued in this paper must be seen to be tentative, even though it can be argued that even a preliminary move in the direction of indicators of certain basic capabilities and freedoms can bring out aspects of economic policy in general and food policy in particular that tend to be overlooked in the more traditional concentration on national income in general and food production in particular.

6. China and India

The comparison of the performances of China and India in dealing with problems of well-being and elementary freedoms has been one of the subjects of great interest in the field of comparative economics. In terms of achievements of GNP per head, China's performance would seem to have been better than India's, even though in terms of standard estimated figures, the Chinese GNP per head of $310 is only about 15% higher than India's $270, for 1985. Since Simon Kuznets' (1966, pp. 360–1) estimate of GNP per head for China and India were about comparable, with a 'product per capita' 20% higher in

[2] The relatively low nutritional intakes in Kerala may have some effect on the prevalence of some illnesses, even when mortality is prevented by an extensive system of medical care.

China, in 1958, it is tempting to think that China's and India's performances in terms of production have been roughly comparable. In fact these figures underestimate the relative performance of China *vis-à-vis* that of India, and if more comparable figures are used, China would seem to be further ahead than India in terms of national product and national income per head (Perkins, 1988; Swamy, 1986). Nevertheless, it would appear that judged in this perspective, while the Chinese have done noticeably better than what has happened in India, the Chinese performance in this field is not tremendously superior to that of India. Furthermore, some of the advantages that China now enjoys compared with India as far as national product is concerned relate to the high growth rate of the Chinese economy in very recent years, since the economic reforms of 1979. More on this later.

In terms of calorie consumption per head, the Chinese picture is considerably better than India's, as Table 6.2 reports. Here again, a big part of the difference has arisen only in recent years through the rapid expansion of agricultural output in general and food output in particular since the economic reforms.

Table 6.2 China and India

	China	*India*
GNP per head ($) 1985	310	270
Calorie consumption per head 1985	2,602	2,189
Life expectancy at birth (years) 1985	69	56
Under five mortality rate (per thousand) 1985	50	158
Famine mortality (millions) Chinese famines 1958–61	29.5	
Excess Indian 'normal' annual mortality (millions) 1985		3.8

Sources: World Bank (1987); UNICEF (1987); Ashton *et al.* (1984).

If we look, instead, at the indicators of basic freedom to avoid premature mortality, i.e., life expectancy at birth, China's performance would seem to be of a different order of magnitude altogether from that of India. Figure 6.1 presents the respective time series of life expectancy in the two countries. Beginning with life expectancy figures quite close to each other – not much above 40 years – in the early 1950s, the Chinese have been able to raise the life expectancy figure to close to European standards, while India lags behind by a big margin. The difference in the achievement of a life expectancy close to 70 years and that in the mid-50s is very large indeed, as we know from the history of life-expectancy changes in different parts of the world.

7. Famines and prevention

One of the interesting features in the comparison of life expectancy of China and India is the remarkably sharp drop that the Chinese figure has around

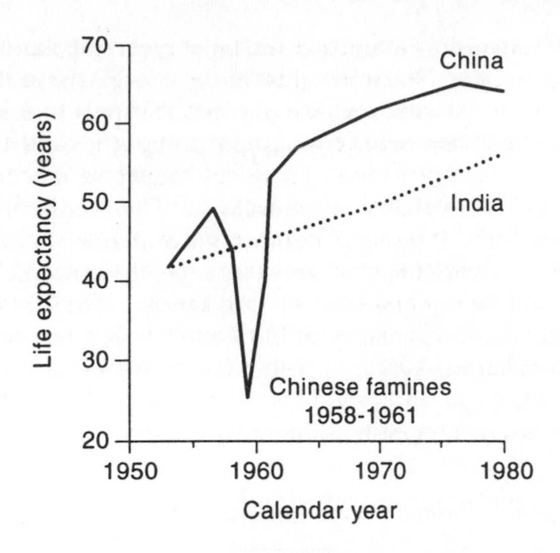

Figure 6.1

1958–61. These are the years of the Chinese famines following the failure of the Great Leap Forward. At one stage life expectancy had fallen to the mid-20s. While the Indian progress of life expectancy has been slow, it has not undergone fluctuations of this kind at all. Indeed, it must be recognized that in the field of famine prevention, India's record is distinctly superior to that of China.

I have tried to argue elsewhere (Sen 1982, 1983; see also Ram 1986) that India's success in eliminating famines since Independence is not primarily the result of raising food output per head, as it is often thought to be. Indeed the increase in availability of food per head in India has been fairly moderate (as it has also been in China up to the economic reforms: see Riskin, 1986, 1987) and the ratio of food to population has remained lower in the post-Independence period than it was in the late 19th century, when India had several famines. The main difference has been brought about by an administrative system which compensates the loss of entitlements as a result of such calamities as droughts and floods by providing employment – often at cash wages – giving the affected population renewed ability to command food in the market. The process is further helped by using substantial stocks held in the public distribution system which can be brought in, to supplement what the creation of income does in regenerating lost entitlements.

This administrative system does, in fact, have its roots in the Famine Codes formulated in British India in the 1880s. However, these Codes were often invoked too late, and intervention was often not a high priority item for the then government. In some cases, most notably in the notorious Bengal famine of 1943, no famine was ever officially 'declared', in order to avoid the necessity of taking actions required by the Famine Code (as Governor Rutherford of Bengal explicitly put it in an intra-government communication: see Sen,

1981b). The situation is now altogether different given the nature of politics in post-Independence India. No government at the center – or at the state level – can get away without extreme political damage if it fails to take early action against famines. The presence of active opposition parties and a relatively free news distribution system provide the political triggering mechanism that the Famine Codes in their original form lacked.[3] The availability of food in different parts of India has often fallen greatly *below* those prevailing in Ethiopia, Sudan, or the Sahel countries (see, for example, Table 6.3) at the time when they had their worst famines (see Drèze, 1986). Indeed, even the normal availability of food per head for India as a whole is not decisively higher than that of sub-Saharan Africa, as Table 6.4 shows (India coming halfway down the list of sub-Saharan economies, with less food availability per head than many countries with persistent famines).

Table 6.3 Famine, averted famine and cereal production:* Sahel and India

	Gross production per head			*Net availability per head*		
	Sahel	*India*	*Maharashtra*	*Sahel*	*India*	*Maharashtra*
1971	102	96	51	101	84	55
1972	75	92	46	76	84	57
1973	78	83	27	85	76	46
1974	115	88	62	120	82	73

* 100 = 182 kg per head per year.
Source: Drèze (1986).

The Chinese experience in this respect has been quite different. There was, of course, a very remarkable drop in food output per head after the Great Leap Forward (though not more than in some parts of India in different years, e.g., in Maharashtra in 1973), but there was no major revision of economic policy, no alert anti-famine relief operations, and not even an official recognition of the existence of famine for a number of years. The famine in China raged on for three years, and it is now estimated that the additional mortality because of the famine amounted to about 29.5 million.[4] It is quite remarkable that a famine of this magnitude could continue unredressed without bringing about a major policy shift, and this failure is certainly one connected closely with the absence of a relatively free press and the absence of opposition parties free to criticize

[3] On this see Sen (1982). Note that political and journalistic pressure is less effective in preventing high levels of 'normal' mortality than in countering deaths from open starvation which are more visible and easier material for news reporting and for political pressure. See footnote 4.

[4] Ashton *et al.* (1984). While this figure of famine mortality is exceptionally high, it should also be noted that normal mortality rates in China are now very low. Indeed, if India had the mortality rates prevailing in China, there would have been 3.8 million fewer deaths in India *each year* around the middle 1980s. That is, every eight years or so more people die in India in excess of Chinese normal mortality rates than died in China in the biggest famine of the century. India has no more reason to be smug than China has.

Table 6.4 Comparative food availability per head 1983: India and sub-Saharan Africa

Country	*Daily calorie supply per head*	*Country*	*Daily calorie supply per head*
Less than India		*More than India*	
Ghana	1,516	Sudan	2,122
Mali	1,597	Zaire	2,136
Chad	1,620	Botswana	2,152
Mozambique	1,668	Togo	2,156
Benin	1,907	Ethiopia	2,162
Kenya	1,919	Malawi	2,200
Zambia	1,929	Mauritania	2,252
Guinea	1,939	Niger	2,271
Zimbabwe	1,956	Tanzania	2,271
Burkina Faso	2,014	Rwanda	2,276
Nigeria	2,022	Uganda	2,351
Cameroon	2,031	Liberia	2,367
Angola	2,041	Lesotho	2,376
Central African Republic	2,048	Burundi	2,378
Somalia	2,063	Congo	2,425
Sierra Leone	2,082	Senegal	2,436
India	2,115	Ivory Coast	2,576

Source: World Bank (1986), Table 28.

and chastise the government in power. It may, thus, be argued that the massive deaths connected with starvation and famine during 1958–61 relate closely to the issue of freedom of information and criticism.

8. Chinese economic reforms

While the progress of food production in China was relatively moderate until the economic reforms, things have moved very fast indeed after 1979. Table 6.5 presents the gross value of agricultural output (including food output) between 1979 and 1986, as reported in the *Statistical Yearbooks of China*. It would seem that the agricultural output has doubled in the seven years since 1979, and the rate of growth of agriculture, which is typically much lower than the industrial growth rate, has in fact been exactly comparable. That the economic reforms permitting greater freedom to earn profits based on economic calculations have been a success from this point of view can scarcely be denied. It is possible to question some of the figures, and it has sometimes been argued that there were incentives for understating the agricultural output in the *pre*-reform period, but even when these corrections are made, the performance of Chinese agriculture since the economic reforms must be accepted to be altogether exceptional.

On the other hand, judged in terms of the freedom to avoid escapable mortality, the picture is much muddier. Even in Figure 6.1, one notices a slight tendency for the life expectancy to decline in the period following the economic

Table 6.5 China since 1979 reforms (indices)

	Index gross value of output		*Index death rate*	
	Industry	*Agriculture*	*National*	*Rural*
1979	100	100	100	100
1980	109	104	102	101
1981	113	111	102	102
1982	122	123	106	110
1983	135	135	114	120
1984	154	159	108	105
1985	181	181	106	104
1986	197	210	108	105

Sources: People's Republic of China (1986) and (1987).

reforms. This is brought out also in Table 6.5 in terms of the index of death rates, which goes up – rather than down – in the post-reform period. While the peak increase in death rate is now well past, the last reported death rates in China are still *higher* than that in the pre-reform period. Questions can be raised about the acceptability of these official mortality statistics, and it must also be recognized that the Chinese death rates were very low indeed by the time the economic reforms began.

The new policy package has included some radical changes in the distribution of health care in the rural areas, e.g., a withdrawal from the strategy of using 'barefoot doctors' (Table 6.6), and a general shortage of public funds for communal health care under the new privatized 'responsibility system.' Given the pro-male bias in Chinese society, this seems to have hit the female children hardest, reinforced by the population policy that came with the economic reforms (e.g., 'one child family' in many regions).[5]

It is remarkable that the enormous expansion of life expectancy in China, from a figure close to 40 years to one close to 70 years, took place in the pre-reform period with only a moderate increase in food availability per head but with a radical expansion in the delivery of health care and food to different sections of the population. Since the reforms, food availability per head has gone up radically, but the delivery system has undergone some changes, including contraction in some respects, and there seems to have been some decline from the previously achieved peak of high life expectancy and low death rate. While the Chinese economic reforms must be praised for what they have achieved – the increase in production has been altogether remarkable – there is need to reassess the policy lessons of the Chinese reforms, especially when attention is shifted from production, GNP and output per head, to the basic indicators of the freedom to live long and the related positive freedoms. While the Chinese experience of famines in 1958–61 raises one type of issue relating food to freedom (in that case concerning freedom of information and opposition), the

[5] This question is further discussed in my joint book with Jean Drèze, *Hunger and Public Action*, Oxford University Press, Chapter 11, (1989).

Table 6.6 Barefoot doctors in China

	Total		*Female*	
	Numbers (millions)	*Index (1975 = 100)*	*Numbers (millions)*	*Index (1975 = 100)*
1970	1.218	78	n.a.	n.a.
1975	1.559	100	0.502	100
1980	1.463	94	0.489	97
1981	1.396	90	0.443	88
1982	1.349	87	0.410	82
1983	1.279	82	0.371	76
1984	1.251	80	0.356	73

Sources: World Bank (1984); People's Republic of China (1985a) for 1983 and 1984; People's Republic of China (1985b) for 1981–83. *The Statistical Yearbooks of China* from 1986 onwards do not give the numbers of barefoot doctors any more.

post-reform experiences of China raise another type of question concerning that relation (involving in this case freedom to avoid premature mortality as an indicator of success, as opposed to the size of production and output)[6].

9. Assessment of Sri Lanka's achievements

Another country in which the enhancement of life expectancy has received much attention is Sri Lanka. As Table 6.1 indicated, Sri Lanka has a remarkably high life expectancy in comparison with its relatively low GNP per head. This achievement has been seen as being closely related to the policy of public intervention in Sri Lanka (Isenman, 1980; Sen, 1981a).

The role of public intervention in Sri Lanka in enhancing the positive freedom to live long has been questioned in a number of contributions in recent years, leading to some lively controversies. For example, based on comparing Sri Lanka's performance since 1960 with other countries, it has been argued that Sri Lanka has not been an exceptional performer.[7] Unfortunately, the period chosen for this comparative assessment, beginning with 1960, has made these comparative studies quite misleading. Extensive public intervention in Sri Lanka began in the early 1940s, and this was indeed accompanied by a sharp reduction in death rate, which went down from 20.6 per thousand in 1940 to 8.6 per thousand in 1960. By 1960, when the now-famous international comparative studies began, the death rate in Sri Lanka was within hitting distance of the

[6] That basic question remains, but later mortality data, including those emerging from the recent censuses, do not confirm the initial fear of an absolute increase in age-specific mortality rates. Progress in reducing death rates and raising life expectancies has, however, been relatively moderate in comparison with radical improvements in real national income per head and in industrial and agricultural production (Amartya Sen, 1995).

[7] Bhalla (1987); Bhalla and Glewwe (1986). See however, the rejoinder of Amartya Sen in the former volume, and those of Graham Pyatt and Paul Isenman in the latter journal, Vol. 1 (May 1987). See also Ravallion (1987); and Anand and Kanbur (1987).

more advanced countries in Europe and America. It is not surprising that the progress since then has been relatively slower, especially compared with other countries which had more scope for reduction in mortality rates. Also, as it happens, the period beyond 1960 has been one of some fluctuation of public intervention, and some of the major planks of public intervention used in Sri Lanka to enhance the quality of life have undergone, in fact, some decline in the 1970s. The policy of free or subsidized distribution of rice, which was introduced in 1942, has suffered from reductions in the later decades, and even the expansion of health services which was very fast during the 1940s and 1950s has slowed in the later periods, with a *reduction* in the number of doctors and other medical practitioners in the decade of the 1970s. Table 6.7 presents some of the relevant figures. The fact that the enhancement of life expectancy and related indicators has not been very fast since 1960 says very little about the alleged lack of effectiveness of public delivery systems in the expansion of life expectancy in Sri Lanka. By the time the comparative studies began much of the dramatic reduction in death rate in Sri Lanka had already taken place, and the comparisons also suffer from concentrating in a period in which there was nothing like the steady rise in public delivery arrangements for food and health care that had taken place in the earlier period, when mortality rates had indeed crumbled at a dramatic rate.

Table 6.7 Sri Lanka

	Public distribution of food	*Number of medical personnel per million people*	*Death rate per thousand*
1940	No (Introduced 1942)	271	20.6
1950	Yes	357	12.6
1960	Yes	557	8.6
1970	Yes (Reduced 1972, 1979)	693	7.5
1980	Yes	664	6.1

10. Periodization and British mortality decline

The issue of periodization, which proves to be central in appraising Sri Lanka's achievements, is in general an important question in assessing the effectiveness of different policies in the enhancement of life expectancy and in the decline of mortality. Even in Europe sharp reductions in premature mortality have been closely connected with expansion of public delivery of basic essentials of living, including health care and medical attention, and it is possible to move towards the identification of the relevant causal connections by distinguishing between different periods in which mortality reductions have been fast or slow.

Table 6.8 presents the extension of life expectancy at birth in England and Wales during the first six decades of this century. It can be seen that in every decade the life expectancy at birth went up moderately – by between one and four years – with two exceptions. In the decades between 1911 and 1921 and between 1940 and 1951 life expectancy increased by nearly seven years. These were, of course, the war years, and the improvement is to a great extent

Table 6.8 Extension of life expectancy at birth: England and Wales, 1901–60 (additional years)

Between	*Men*	*Women*
1901–11	4.1	4.0
1911–21	6.6	6.5
1921–31	2.3	2.4
1931–40	1.2	1.5
1940–51	6.5	7.0
1951–60	2.4	3.2

Source: Preston, Keyfitz, and Schoen (1972). See also Winter (1986).

recording the impact of public distribution systems that came in with protecting the general public from the possible effects of war. Public provision of food rationing and distribution, expansion of health services (including the introduction of the National Health Service just after the war), and other expansions of the involvement of the state in distributing food, health care, medical attention, etc., made a radical difference to the entitlements to these vital commodities enjoyed by the population at large, including its most vulnerable sections. The enhancement of life expectancy reflects these results of public policy, and it would be a mistake to think of the increase of life expectancy in Britain as the result entirely of enhanced overall economic opulence (or a general increase in GNP per head). Once the issue of periodization is appropriately faced, it is hard to escape the fact that even in the history of a country such as Britain, it is the delivery system of food and health care – over and above increases in economic opulence – that has played a strategic part in crucial periods of expansion in the elementary freedom to live long and live well.

11. Intrinsic and instrumental roles

In assessing the relevance of freedom in the making of food policy, both the intrinsic and the instrumental perspectives have to be kept very firmly in view. The instrumental perspective is often invoked in the context of emphasizing economic incentives in the expansion of national output in general and food production in particular, and there is undoubtedly much to be said for taking adequate note of this question, as the experiences of the Chinese economy in particular have sharply brought out in recent years. At the same time, the instrumental perspective has to be extended from the freedom to earn profits to freedoms of broader kinds, including political freedom in the form of freedom of opposition, freedom of information, and journalistic autonomy. We have seen that these freedoms can be quite crucial in the delivery and use of food.

The instrumental perspective is, however, inherently limited, since freedom can be seen as having intrinsic importance as well. In assessing economic development and social progress, it is natural to think of the enhancement of basic positive freedoms to avoid premature mortality, to escape morbidity, to eliminate undernutrition, and so on. While freedom is a complex notion, various aspects of it can be usefully studied in terms of statistical information of a kind that is frequently available and which can be made more easily accessible if the perspective of freedom is taken seriously by public policy makers.

The importance of this perspective arises partly from the fact that the metrics of gross national product, real income, etc., may often be quite misleading about the extents of freedom that people do enjoy and can build their lives on. Even in such elementary matters as avoiding premature mortality, the statistics of national products (including those of food output) can hide more than they reveal. It is possible for the national product per head and the food availability per person to go up sharply without reducing mortality rates, sometimes accompanied by increased mortality, as seems to have happened in China since the economic reforms of 1979. Once the process of economic development is reassessed in terms of the important indicators of elementary freedoms, a different light altogether may well be cast on economic policy changes that call for adequately broad evaluation. The Chinese economic reforms have been undoubtedly extremely successful in terms of raising production and enhancing income, but since the post-reform period has also seen an increase – rather than a reduction – in death rates, there is room for asking searching questions about the nature of the policy package that has gone with the economic reforms, and about variations to this package that can be considered from the economic point of view. The remarkable success of the Chinese economy in raising life expectancy at birth, from a figure close to 40 years just after the Revolution to a figure close enough to 70 years just prior to the economic reforms, was built on paying particular attention to public delivery systems involving food, health care, and related necessities. It is this aspect of the Chinese success that is in some danger of going out of focus – with possibly serious consequences – if the understandable concern with raising output and income distracts attention from the problem of delivery and public distribution.

Freedom is not a remote consideration in policy making. This applies just as much to the making and assessing of food policy as it does to many other fields of policy making in social and economic matters. Indeed, the inclusion of freedom as a consideration – both at the intrinsic and at the instrumental level – has the effect of appropriately broadening the concepts that must be invoked in the formulation and execution of food policies. The need for that broadening has been one of the main contentions I have tried to put across in this lecture.

The perspective of freedom, with its diverse elements, is much too important to be neglected in the making of food policy. Food and freedom are both central concerns in human life, and they have links that are both crucial and diverse. These links demand our attention. The elementary freedom to live long and live well for a great many million people is at stake.

References

Anand, S. and Kanbur, R. 1987: Public policy and basic needs provision: Intervention and achievement in Sri Lanka, WIDER Working Paper (Helsinki: WIDER). Also published in J. Drèze and A. Sen (eds), *The Political Economy of Hunger*. Oxford: Clarendon Press, and New York: OUP, 1990.

Ashton, B. *et al.* 1984: Famine in China 1958–61. *Population and Development Review* 10.

Bauer, P. 1981: *Equality, the Third World, and Economic Delusion*. Cambridge, MA: Harvard UP.

Berlin, I. 1969: *Four Essays on Liberty*. London and New York: OUP.

Bhalla, S. 1987: Is Sri Lanka an exception? A comparative study in living standards, in Srinivasan, T. N. and Bardhan, P. (eds), *Rural Poverty in South Asia*. New York: Columbia UP.

—— and Glewwe, P. 1986: Growth and equity in developing countries: A reinterpretation of Sri Lankan experience. *World Bank Development Review* 1 (September).

Buchanan, J. 1986: *Liberty, Market and the State*. Brighton: Wheatsheaf Books.

Drèze, J. 1986: Famine prevention in India. Helsinki: WIDER. Also published in Drèze, J. and Sen, A. (eds), *The Political Economy of Hunger*. Oxford: Clarendon Press, and New York: OUP, 1990.

Friedman, M. and Friedman, R. 1980: *Free to Choose*. London: Secker and Warburg.

Isenman, P. 1980: Basic needs: The case of Sri Lanka. *World Development* 8 (3).

Kumar, B. G. 1987: Poverty and public policy: Government intervention and levels of living in Kerala, India, D.Phil. dissertation. Oxford: Oxford University.

Kuznets, S. 1966: *Modern Economic Growth*. New Haven, CT: YUP.

Marx K. and Engels, F. 1947: *The German Ideology* (1845–46; republished, New York: International Publishers).

Mill, J. S. 1974: *On Liberty*. London: 1859; republished, Harmondsworth: Penguin).

Morris, M. D. 1979: *Measuring Conditions of the World's Poor*. Oxford: Pergamon Press.

Nozick, R. 1974: *Anarchy, State and Utopia*. Oxford: Blackwell, and New York: Basic Books.

Panikar, P. G. K., and Soman, C. R. 1986: *Health Status of Kerala*. Trivandrum: Center for Development Studies.

People's Republic of China 1987: *Zhongguo tongji nianjian 1987*. Beijing: Zhongguo tongji chuban she.

—— 1986: *Statistical Yearbook of China 1986*. Beijing: Zhongguo tongji chuban she.

—— 1985a: *Statistical Yearbook of China 1985*. Beijing: Zhongguo tongji chuban she.

—— 1985b: *Zhongguo tongji nianjian 1985*. Beijing: Zhongguo tongji chuban she.

Perkins, D. H. 1988: Reforming China's economic system. *Journal of Economic Literature* 26.

Pigou, A. C. 1952: *The Economics of Welfare*. London: Macmillan, 1920, sixth enlarged edition.
Preston, S., Keyfitz, N. and Schoen, R. 1972: *Causes of Death: Life Tables for National Populations*. New York: Academic Press.
Ram, N. 1986: An independent press and anti-hunger strategies – the Indian experience (Helsinki: WIDER, 1986). Also published in Drèze, J. and Sen, A. (eds). *The Political Economy of Hunger*. Oxford: Clarendon Press, and New York: OUP, 1990.
Ravallion, M. 1987: Growth and equity in Sri Lanka: A comment, mimeo. Canberra: The Australian National University.
Rawls, J. 1971: *A Theory of Justice*. Oxford: Clarendon Press, and Cambridge, MA: Harvard University Press.
Riskin, C. 1987: *China's Political Economy*. Oxford: Clarendon Press.
—— 1986: Feeding China: The experience since 1949. Helsinki: WIDER. Also published in Drèze, J. and Sen, A. (eds), *The Political Economy of Hunger*. Oxford: Clarendon Press, and New York: OUP, 1990.
Sen, A. 1988: Freedom of choice: Concept and content, Alfred Marshall Lecture to the European Economics Association, Copenhagen, August 1987. *European Economic Review* 32.
—— 1987a: *On Ethics and Economics*. Oxford: Blackwell.
—— 1987b: *Hunger and Entitlements*. Helsinki: WIDER.
—— 1985a: *Commodities and Capabilities*. Amsterdam: North Holland.
—— 1985b: Well-being, agency and freedom: The Dewey Lectures 1984. *Journal of Philosophy* 82 (April).
—— 1984: Development: Which way now? *Economic Journal* 93 (December 1983), reprinted in *Resources, Values and Development*. Oxford: Blackwell, and Cambridge, MA; Harvard UP.
—— 1986: How is India doing? *New York Review of Books* 21 (Christmas 1982), reprinted in Dilip Basu and Richard Sisson (eds), *Social and Economic Development in India: A Reassessment*. New Delhi, London, Beverly Hills, CA: Sage.
—— 1981a: Public action and the quality of life in developing countries. *Oxford Bulletin of Economics and Statistics* 43.
—— 1981b: *Poverty and Famines*. Oxford: Clarendon Press.
Stewart, F. 1985: *Planning to Meet Basic Needs*. London: Macmillan.
Streeten, P. 1981: *Development Perspectives*. London: Macmillan.
——, *et al.* 1981: *First Things First: Meeting Basic Needs in Developing Countries*. New York: OUP.
Swamy, S. 1986: Chinese price structure and comparative growth rates of China and India. Cambridge, MA: Harvard Institute of International Development.
UNICEF 1987: *The State of the World's Children*. New York: OUP.
Winter, J. M. 1986: *The Great War and the British People*. London: Macmillan.
World Bank 1987: *World Development Report 1987*. New York: OUP.
—— 1986: *World Development Report 1986*. New York: OUP.
—— 1984: *China: The Health Sector*. Washington, DC: World Bank.

7 Bruce Johnston and Peter Kilby, '"Unimodal" and "Bimodal" Strategies of Agrarian Change'

Excerpts from: B. F. Johnston and P. Kilby, *Agriculture and Structural Transformation*, Chapter 4. Oxford: OUP. (This version also published in J. Harriss (ed.), 1982, *Rural Development: Theories of Peasant Economy and Agrarian Change.* London: Hutchinson) (1975)

Introduction

Because of their structural and demographic characteristics, late-developing countries face a fundamental choice between a strategy aimed at the progressive modernization of the entire agricultural sector and a crash modernization strategy that concentrates resources in a highly commercialized subsector. We refer to the first alternative, well illustrated by the patterns of agricultural development in Japan and Taiwan, as a 'unimodal strategy'. The second alternative, which results in a development pattern based on a dualistic size structure of farm units, as in Mexico or Colombia, is labelled a 'bimodal strategy'.

We will be arguing that a unimodal strategy has significant advantages because it is consistent with maximum mobilization of a late-developing country's resources of labour and land. Because the non-agricultural sectors are so small in relation to the number of farm households, agriculture is subject to severe demand constraints. The resulting purchasing power constraint limits the extent to which expansion of the agricultural sectors' output can be based on increased use of purchased inputs, whether imported or manufactured domestically. These considerations underscore the importance of the dynamic forces that determine the rate and character of technical change, especially the process of generating a sequence of divisible innovations that leads to widespread increases in the productivity of land and labour. The success of individual farm units in allocating resources so as to minimize costs is clearly an essential ingredient of an efficient agricultural strategy. It is, however, the nature of technical innovations and their diffusion among farmers that are decisive in minimizing the cost of the sector-wide expansion of farm output and in determining the pattern of development. It will be suggested that the patterns of agricultural development associated with the unimodal and bimodal alternatives differ a great deal in the contributions that they make to achieving three major objectives of an agricultural strategy: advancing structural transformation, raising the welfare of the farm population, and fostering changes in rural attitudes and behaviour that will have beneficial effects on the process of modernization.

Although the concept of 'strategy' has become fashionable in development economics, few attempts have been made to define it. A useful general definition is that a strategy is a mix of policies and programmes that influences the pattern as well as the rate of growth. Particular attention is given here to the differential effects of the patterns of agricultural development associated with a unimodal as contrasted with a bimodal strategy. Any strategy for agricultural development will embrace some combination of (a) programmes of institution building related to such activities as agricultural research and rural education and farmer training, (b) programmes of investment in infrastructure, including irrigation and drainage facilities and rural roads, (c) programmes to improve product marketing and the distribution of inputs, and (d) policies related to prices, taxation, and land tenure. Its 'efficiency' will depend in part on promoting optimal use of available resources, and still more on modifying existing constraints.

In brief, the emphasis is on action to change the production possibilities available to farmers by modifying their institutional, technical, and economic environment. An underlying premise is that decentralized decision-making by individual producers has especially significant advantages in agriculture. The price mechanism performs a critical function in harmonizing decentralized decisions and in harnessing the powerful motive of profit. Although the role of market mechanisms in resource allocation is emphasized, we also stress the interactions between the activities of individual producers and government programmes and policies. Of special significance is the role of government in undertaking research and farmer-training programmes to favourably alter input-output relations while public investments in infrastructure enlarge the scope for applying profitable innovations. In addition, governments may find it desirable to adopt policies to make prices reflect more adequately the social costs and benefits of using resources in different types of productive activities.

Our stress on strategy differs sharply from the conventional approach of agricultural planning which has emphasized the setting of production targets for individual commodities. We would emphasize that a fundamental requirement of a suitable analytical approach to the design of an agricultural strategy is simultaneous consideration of the *objectives* to be furthered and the *means* (policies and programmes) by which those objectives are to be attained. It is also essential for the choice of objectives and of means to be guided by explicit recognition of certain *constraints* that can only be gradually eliminated, especially those imposed by the structural and demographic situation in a late-developing country. The critical factors limiting agricultural development and the pace of structural transformation are technological capabilities, availability of investable funds and foreign exchange, and the level of farm purchasing power. It is our argument that a unimodal agricultural strategy aimed at the progressive modernization of the bulk of a nation's cultivators, as contrasted to a bimodal crash modernization effort concentrated upon a small subsector of large-scale mechanized farms, minimizes the extent to which the above constraints impede the development of agriculture and the process of transformation. It does so through its effects on: (a) the disbursement pattern of farm cash receipts, (b) the allocation of investment resources, (c) the kinds of new technological knowledge that are produced, and (d) the proportion of the nation's producers that has access to the new modes of production.

It is clearly an oversimplification to concentrate on the polar extremes represented by unimodal and bimodal alternatives. There are good reasons, however, for focusing initially on the choice between those two extreme alternatives. Governments, like most bureaucratic organizations, are disposed to concentrate on coping with the agenda of pressing problems rather than on developing long-run strategies. Consequently, the 'choice' of an agricultural strategy will often be made by default. Moreover, there are often strong political pressures that tend to bias the outcome toward a bimodal strategy. For both reasons, it is especially important to arrive at a clear understanding of the nature of the alternatives and of their differential effects on the pattern of agricultural development and on overall economic growth.

Choice criteria: the multiple objectives of an agricultural strategy

What criteria should guide this choice between a unimodal and bimodal strategy? We propose that the efficiency of alternative strategies should be assessed in terms of their contributions to attaining three major objectives: first, facilitating the process of structural transformation and growth in national product; second, enhancing the welfare of the farm population; and third, promoting changes in attitudes and behaviour in rural communities that have a favourable impact on the process of social modernization.

Objectives of an agricultural strategy

Because agriculture and its inter-relations with other sectors bulk so large in late-developing countries, it is essential to assess alternative agricultural strategies in terms of their intersectoral effects as well as their direct effects on the expansion of farm output and incomes. Hence, the *first objective* of an agricultural strategy focuses on the need to achieve a rate and pattern of output expansion in the agricultural sector that will promote overall economic growth and structural transformation and take full advantage of positive interactions between agriculture and other sectors. This objective encompasses what has often been referred to as agriculture's 'contributions' to development: (a) providing increased supplies of food and raw materials to meet the needs of the expanding nonfarm sectors, (b) earning foreign exchange through production for export, and (c) providing a net flow of capital to finance a considerable part of the investment requirements for infrastructure and industrial growth.

The problems involved in achieving a net flow of resources from agriculture clearly represent the most difficult area of competitiveness between this first objective of fostering structural transformation and the second objective of improving the well-being of the rural population.

The expansion in the absolute and relative importance of commercial production in agriculture is, of course, an aspect of structural transformation and increasing sectoral interdependence. The growth of a marketable surplus of farm products, expansion of foreign exchange earnings, and increased availability of resources for capital formation are necessary conditions for the development of a diversified modern economy. At the same time the growth of farm cash income associated with structural transformation means increased rural demand for inputs and consumer goods that can provide an important stimulus to domestic industry. The strength of that stimulus and the associated feedback effects will be strongly influenced, however, by the composition of rural demand.

A broadly based expansion of farm cash income generating demand for low-cost and relatively simple inputs and consumer goods can be expected to foster efficient, evolutionary growth of domestic manufacturing that is characterized by relatively low import content and which leads to the strengthening and diffusion of entrepreneurial and technical competence. Basic to all of these inter-relations between agricultural development and overall economic growth is the creation of an integrated national economy characterized by increased specialization and growing interdependence among sectors. This requires the development of flexible and sensitive market networks and continuing improvement in transportation and other types of infrastructure.

The *second objective*, achieving broadly based improvement of the welfare of the rural population, is important simply because such a large fraction of the population of developing countries is destined to live and die in farming communities. Achievement of that objective depends in the long run on altering the predominantly agrarian structure of these economies. The possibility of enlarging the income of the agricultural sector, and still more the *average* income of farm households, is determined mainly by the rate and character of structural transformation, particularly as manifested in the decline of the relative and, eventually, absolute size of the farm workforce and the associated growth of commercial demand for agricultural products.

Inequality in income distribution is a conspicuous feature of most less-developed countries and a matter for particular concern because the poverty of the low-income groups is so extreme. The extent to which such inequality in income distribution will be either reduced or exacerbated will be determined mainly by whether the demand for labour increases more or less rapidly than the country's workforce. At issue is the growth of demand for labour in all sectors; but the increase in demand for labour in agriculture, including the employment opportunities available to family members working on their own or rented land, is of special significance. And the extent to which the expansion of farm output will lead to widespread increases in income-earning opportunities will, for reasons considered shortly, hinge on the development and diffusion of divisible innovations.

Certain dimensions of welfare can be furthered most effectively by direct action through government programmes, notably public health and related activities.

Rural works programmes can provide supplementary employment and income for some of the most disadvantaged elements of the rural population. But the indirect contribution of such programmes to the expansion of farm output and income, through the construction of roads, irrigation works, and other useful infrastructure, is likely to be more important. Considerable planning and technical supervision is required, however, to ensure the usefulness of employment-oriented projects of that nature. Because of those organizational problems and the fiscal constraints which limit their magnitude, such programmes apparently have not had a very large effect on under- and unemployment in rural areas. For the rural works programme in Bangladesh (then East Pakistan), which was one of the more ambitious undertakings of this type and one that was financed to a large extent by P. L. 480 grain imports, Walter Falcon reports an annual reduction of agricultural unemployment of only about 3.5 per cent.[1]

Judgements will differ concerning the importance of the *third objective*, that is, fostering a pattern of agricultural development that will have a favourable impact on social modernization as a result of inducing changes in rural attitudes, behaviour, and institutions. The evolutionary development of a variety of social institutions is clearly a significant feature of structural transformation. Salient examples pertaining to agriculture include the creation and strengthening of agricultural experiment stations; expansion of educational facilities and programmes for training farmers; establishing irrigation associations or other groups that enable farmers to concert their behaviour when group action is advantageous; and strengthening the organizations – private, public, or co-operative – that distribute credit and inputs and market farm products.

The need for 'institutional progress' is especially significant in countries undertaking a unimodal strategy of agricultural development. Hence, the interactions between technical and economic change at the farm level and institutional, attitudinal, and behavioural change merit attention in assessing the differential effects of alternative strategies. Broader participation in the modernization of agriculture implies a more widespread familiarity with calculations of costs and returns and with the evaluation and selective adoption of innovations. Such opportunities for 'learning by doing' foster the development and spread of managerial competence that facilitates the recruitment and training of the entrepreneurs and skilled workers required in a modernizing economy.

It is also to be expected that broad participation of the farm population in improved income-earning opportunities will influence the rural power structure and political institutions. This has obvious implications with respect to political and financial support for rural schools and other institutions to serve

[1] Walter P. Falcon's estimate of the employment impact of that programme is reported in Falcon (1973).

farming communities. There are, of course, reciprocal interactions between the effects of the pattern of rural development on the distribution of political power and the influence of the power structure on the choice of strategy for agricultural development.

There is one other area in which the inter-relations between the pattern of agricultural development and changes in rural attitudes and behaviour is potentially of very great significance. Many years ago John Stuart Mill asserted that an agricultural system based on peasant proprietorship would have a beneficial effect on the 'prudence' as well as the 'industry' of the rural population and would therefore 'discourage an improvident increase in their numbers. . . .'[2]

The key question concerns the way in which the modernization of agriculture will affect the spread of the knowledge, incentives, and motivation essential to the practice of family planning. It is certainly a reasonable hypothesis that conscious action to limit family size will take hold more readily if rural households are actively involved in a process of economic and technical change, whether as owner cultivators or as tenants, rather than being relegated to a 'surplus population-supporting sector' with slight opportunity to better their condition. The analysis of the relationships between various economic factors and fertility change in Taiwan seems to provide considerable support for that hypothesis. In her concluding comments, Eva Mueller declares that:

> Where agricultural improvement is confined to a minority of cultivators . . . the expansion of economic horizons will be more limited than in Taiwan. Only a minority will then experience the rising aspirations that in Taiwan seem to be contributing so importantly to acceptance of family planning in rural areas. The majority of farmers will have no experience with progress and no reason to raise their sights. They will continue to feel that yield-raising farm investments, a better education for their children, and modern consumer goods and services are not 'for them'. The transformation of household preferences which we observed in Taiwan will be much less extensive.[3]

Clearly, there is a marked contrast between the rapid reduction in birth-rates during the past two decades in Taiwan, and also South Korea, and the slight changes that have taken place in other developing countries.[4] This can probably

[2] Mill (1870), ch. 7, concluding paragraph.

[3] Mueller (1971), pp. 37–8. In a more recent paper, Professor Mueller (1973) examines additional evidence on linkages between agricultural and demographic change. She notes that much of the evidence is contradictory but continues to stress the role of rising aspirations.

[4] The contrast with Brazil and Mexico, where the pattern of agricultural development has been bimodal, is striking. In Taiwan there was a decline in the birth-rate from 41 to 36 per 1000 between 1947 and 1963 and a further decline to 26 per 1000 by 1970. In South Korea, the decline was from 45 to 30 per 1000 between 1950 and 1970. In contrast, the reduction in Brazil over the same twenty-year period was only from 41 to 38 per 1000 and the decline in Mexico was from 44 to 41 per 1000. See Kocher (1973), pp. 64–5. Clearly many factors contribute to those contrasts. Some would stress the absence of family planning programmes in Mexico and Brazil and the role of the Catholic Church and other cultural infuences; but the French peasantry that impressed Mill by its 'prudence' was predominantly Catholic yet it practised family planning on a significant scale at a time when birth control technology was very primitive.

be attributed in part to relatively well-organized family-planning programmes in those two countries. But the changes in rural attitudes and motivation resulting from broad participation in development and the widespread influence of education and other modernizing institutions and of the mass media undoubtedly strengthened the direct effects of family planning programmes. There are also indications that those factors and the rising aspirations which they have engendered have also contributed to more spontaneous reductions in fertility. One fact is beyond dispute. Given the structural-demographic characteristics of late-developing countries, there is no hope of bringing birth-rates into tolerable balance with the sharply reduced death-rates that now prevail unless conscious limitation of family size becomes widespread in rural areas.

Competitiveness and complementarity among objectives

To argue, as we have done in this section, that the choice of agricultural strategy should be guided by explicit attention to their effects on a set of objectives is a somewhat unorthodox approach – but one that is now receiving much attention. Richard Musgrave sums up the orthodox approach, which is in accord with the compensation principle of welfare economics, with the statement that policy makers should opt for 'the efficient choice' and then supplement that decision with 'the necessary distribution adjustment through a tax-transfer mechanism.' (Musgrave, 1969). But where poverty is widespread and tax revenues are severly limited, distributional adjustments through a tax-transfer mechanism cannot be carried out on a significant scale even if the political climate is favourable.

It has become a common practice to admonish policy makers in low-income countries to make development decisions by assigning appropriate weights to (a) growth of output, (b) employment expansion, and (c) income distribution goals. This is in accord with the predilection of economists to assume that there are invariably trade-offs between output and equity goals. This is, of course, true at the margin. Additional funds allocated to a programme of nutritional improvement will, at least indirectly, be at the expense of reduced allocations for some other programme such as agricultural research, farmer training, or investments in infrastructure.

Although it is necessary to consider trade-offs in making decisions with respect to particular policies or programmes, the trade-offs that arise in connection with the set of policies required to implement a unimodal strategy will be small compared to the situation under the bimodal alternative. In the latter case, for example, a truly massive (and usually politically not feasible) programme of rural public works would be required for unemployment relief to offset the labour-displacing effects of a capital-intensive expansion path. But within the framework of a unimodal strategy, the need to generate additional employment via a works programme would be much less. Moreover, the prospects for planning and financing rural public works projects that will have favourable effects on output are better in the context of a unimodal pattern of development; and in that context it is legitimate to give greater weight to output effects than employment creation.

The *crux* of our argument is that it is wrong to assume that the choice between unimodal and bimodal strategies necessarily involves a sacrifice with respect to the economic objective of increasing output in order to further the social objectives of expanding employment opportunities and reducing inequalities in income distribution. Given the economic constraints that condition the choice of means for promoting agricultural development, progressive modernization of the rural sector is, in general, the most efficient means of attaining the threefold objectives of an agricultural strategy.

Economic constraints and the choice of means

A major thesis of ours is that sequences of innovations *can* be generated and diffused that will foster the widespread increases in productivity that characterize a unimodal strategy and thus avoid the polarization of agriculture into subsectors using drastically different technologies.

A central element of a unimodal strategy is the development and diffusion of highly divisible innovations that promote output expansions within an agrarian structure made up of operational units relatively equal in size and necessarily small because of the large number of holdings relative to the cultivated area. The divisibility factor, by rendering new technology applicable to these small units, permits the progressive modernization of an increasing proportion of a country's farmers. There will, of course, be differences in the speed and efficiency with which farmers seize new opportunities depending on their initial resources, ability, and desire. Moreover, technical progress will inevitably have an uneven impact on different regions and types of farming. Such differences, however, are much less significant than those that result from the polarization of agriculture into modern and traditional sectors employing drastically different technologies.

Progressive modernization based on widespread adoption of a sequence of innovations compatible with the constraints imposed by structural-demographic characteristics makes it possible to exploit the large potential that exists for augmenting the productivity of the agricultural sector's internal resources of labour and land.

If purchased inputs are primarily divisible inputs such as seed and fertilizer, the new technologies can be widely adopted in spite of the purchasing power constraint. Spread of such inputs coupled with changes in farming practices and growth of farm cash income will generate demands for improved equipment to increase the precision as well as reduce the time required for various farming operations; but with progressive modernization this demand will be directed toward simple and inexpensive implements. And if output expansion results from widespread increases in productivity among the small farm units which necessarily predominate when the farm labour force is large relative to the total cultivated area, the capital requirements for investment in labour-saving farm equipment will be limited. The relative profitability – private and social – of outlays for fertilizer and other current inputs as compared to capital equipment will be influenced by the type of agricultural strategy pursued. For example, electric- or diesel-powered pumpsets in India and Pakistan have represented an innovation that has been essentially complementary to the internal resources of labour and land; and in spite of the lumpiness of the investment, the practice of selling tube-well water to near-by farmers makes the input provided quite divisible. The institution of contract ploughing can make the *services* of tractors divisible, and under certain circumstances, especially in semi-arid regions, mechanical cultivation can be complementary to yield-increasing inputs. And when the decision to hire tractor services is made by the farmer or tenant operating a unit of average size, that is, small, there is a good chance that the input will also be a complement to the labour of family members in contrast to the labour-displacing consequences when tractor mechanization is introduced in a large operational unit.

Our emphasis on framing a strategy aimed at promoting more productive utilization of the abundant supply of labour in the agricultural sector is not to be equated with the notion of a 'labour surplus' in the sense that there are individuals with zero marginal product. It is common to speak of a 'labour surplus' in traditional agriculture when the population of working age appears to be larger than the number of 'full-time workers' that would be required to maintain the level of production even with existing technologies. But those labour force concepts appropriate to a modern industrial economy are not really applicable. Particularly for the unpaid family labour that accounts for most of the rural workforce, there is no institutionally determined workday and no clear dichotomy between 'work' and 'leisure'.

For any given 'stock' of farm labour – in a household or in the sector – the actual 'flow' of labour inputs into agricultural production is determined by a 'subjective equilibrium' in the allocation of labour time. And the activities other than farming embrace pursuits such as handweaving and other types of cottage industry as well as leisure and a variety of 'non-economic' activities – litigation, ceremonies, hunting, and so on – many of which are readily *compressible* if altered circumstances make it attractive to increase the allocation of labour time to farming. In addition, agriculture is usually characterized by large seasonal variation in the demand for labour and farm work. Consequently, there is often considerable scope for increasing the number of hours worked per day, the number of work-days per year, and even the pace of work. The increase in time devoted to farming activities may also result from a reduction in the allocation of labour to cottage industries as the products of those industries are replaced by purchased goods. In brief, there is 'slack' that can be drawn into production if there is an increase in the marginal product attributable to additional inputs of labour or an increase in the marginal valuation that workers place on increments to income.

The prevalence of this labour slack in agriculture is a consequence of the economic structure and rapid growth of labour force in developing countries which make it inevitable that the farm workforce will continue to grow in absolute size for many years. In fact, 'labour slack' is merely a shorthand expression to describe a situation where the labour of a large part of the rural population of working age has a low opportunity cost. And because of rapid population growth the opportunity cost of labour is held down, or pushed even lower, unless a country's development strategy is generating new opportunities for productive employment at a pace more rapid than the rate of expansion of the labour force. The phenomenon is, of course, not confined to agriculture. Certain labour-intensive service trades characterized by easy entry also absorb part of the labour force that is unable to find jobs in firms that provide regular wage employment. But with the exception of some of the semi-industrialized countries in Latin America, agriculture is the dominant 'self-employment' sector and absorbs the majority of the annual additions to the labour force.

Because of the existence of labour slack, the impact on farm output of yield-increasing innovations goes beyond their effects on output per unit of labour input. Under a unimodal strategy, technical change and associated investments in infrastructure are likely to induce fuller use of farm labour and land as well as enlarged use of such inputs as fertilizer, because of the complementarity between the internal and external inputs. Taiwan's experience gives an indication of the quantitative importance of more productive seed–fertilizer combinations and investments in infrastructure which increased the returns to additional labour inputs. Investments in irrigation and drainage not only raised yields directly, they also facilitated multiple cropping, thereby raising the year-round utilization of both labour and land. Those who have argued that 'labour surplus' (in the sense of a zero marginal product) is an important feature of a traditional agriculture have often implied that the agricultural sector can be neglected in a development strategy. It should be clear from the foregoing that we are definitely not suggesting that agriculture can be neglected, but rather that a strategy of progressive modernization will have especially important advantages under those conditions.

If a strategy of progressive modernization has significant advantages in achieving both the economic and social goals of development, how are we to account for the fact that so many developing countries appear to be pursuing bimodal strategies?

In the absence of institutional arrangements to generate and diffuse innovations capable of raising yields on small farm units, a concentration of resources in a sub-sector of large and capital-intensive farms may appear to be the only feasible alternative. Tractors and their associated equipment constitute an innovation that is readily transferred. Their introduction is a powerful means of enlarging the area under cultivation, and it may also facilitate expansion of the planted area by increased multiple cropping. The large international corporations manufacturing tractors have considerable competence and strong incentive to promote sales and to organize distribution and service facilities in developing countries, and the local climate of opinion often favours special encouragement for the introduction of tractors because they are seen as a symbol of modernity.

Although the agricultural sector is necessarily subject to a severe purchasing-power constraint until considerable structural transformation has taken place, a sub-sector of large farm units that accounts for most of the increase in commercial production is able to escape that constraint. (This means, of course, that the remaining farm units will continue to be subject to a purchasing-power constraint that is now more binding.) It is also easy for such farms to realize the economies of scale associated with the use of tractors. In addition, reliance on rapid expansion of output by a 'modern' sub-sector of large and 'progressive' farm enterprises bypasses the problems and costs associated with involving a large fraction of the farm population in the modernization process.

This type of bias toward 'reinforcing success' is often linked with the assumption that there is necessarily a sizable trade-off between growth and equity goals. William Nicholls, for example, asserts that Brazil 'must unfortunately face a hard choice between equity and productivity', and he argues that 'the present large holdings' must be relied upon to satisfy the growth of commercial demand (Nicholls, 1971). Moreover, there is frequently a failure to recognize the extent to which bimodal and unimodal strategies are mutually exclusive alternatives. Although a unimodal strategy has important economic and social advantages under the conditions that characterize a late-developing country, logical arguments and historical evidence will obviously not be the only factors determining a country's pattern of agricultural development. It is also necessary to take account of political factors.

Political constraints

It was suggested earlier that it is a gross oversimplification to speak of a situation in which policy makers 'opt for a unimodal or bimodal strategy'. In a remarkably perceptive essay, Colin Leys (1971, p. 137) notes that 'the process of "choice" rarely consists of an explicit "moment" at which some appropriate person or committee reviews the alternatives, weighs their pros and cons and consciously selects one of them. It is, generally, a continual process of options foregone, through the passage of time, and through the taking of other decisions which have the often unforeseen consequence of closing off possibilities in spheres not considered at all in the context of the decision.'

Of equal importance is the need to recognize that decisions and policies are shaped by all sorts of conflicting interests. If there is any overriding concern it is preoccupation with staying in power, not the comparatively abstract goal of development.

To the outsider the problem of poverty might appear to be a supreme challenge and a spur to sacrifice and selfless action. But there is a large element of truth in the observation by Leys (1971, p. 125) that the political and administrative élites in most developing countries 'are rarely eager for measures that would entail redistribution of wealth or any threat to their own status or prospects'. After all this tends to be true of élites in all countries, the principal difference in less-developed countries being the more limited force of 'institutions which can hold these tendencies in check and make the élite give the public value for its privileges'.

References

Falcon, W. P. 1973: Agricultural employment in less developed countries: general situation, research approaches, and policy palliatives, mimeo, Economic Staff Working Paper no. 113, April, Washington, DC: International Bank for Reconstruction and Development.

Johnston, B. F., and Kilby, P. 1975: *Agriculture and Structural Transformation*. New York: Oxford University Press.

Kocher, J. E. 1973: Rural development, income distribution, and fertility decline. Occasional Paper of the Population Council, New York.

Leys, C. 1971: Political perspectives, in D. Seers and L. Joy (eds), *Development in a Divided World*. Harmondsworth: Penguin.

Mill, J. S. 1870: *Principles of Political Economy*, 5th edn, book 2, New York.

Mueller, E. 1971: Agricultural change and fertility change: the case of Taiwan, mimeo, Ann Arbor: University of Michigan.

—— 1973: The impact of agricultural change on demographic development in the Third World, in *Demographic Growth and Development in the Third World*. Belgium: International Union for the Scientific Study of Population.

Musgrave, R. A. 1969: Cost-benefit analysis and the theory of public finance. *Journal of Economic Literature* 7 (3) (September), 804.

Nicholls, W. H. 1971: The Brazilian food supply: problems and prospects. *Economic Development and Cultural Change* 19 (3) (April), 387–8.

8 John Harriss,

'Does the "Depressor" Still Work? Agrarian Structure and Development in India'

Excerpts from: *Journal of Peasant Studies* 19 (2), 189–227 (First published as 'Does the "Depressor" Still Work? Agrarian Structure and Development in India: A Review of Evidence and Argument') (1992)

The aim of this article is to review recent evidence about the trends of change in the agrarian structure and agricultural development of India, having in view Daniel Thorner's argument of more than 30 years ago in his lectures on 'The Agrarian Prospect in India': '(The) complex of legal, economic, and social relations uniquely typical of the Indian countryside served to produce an effect which I should like to call that of a built-in "depressor"' (Thorner, 1956, p. 16). With this term he referred to agrarian production relations which made it paying for landlords to live by appropriating rents, usurious interest and speculative trading profits from the impoverished mass of the peasantry – and thereby limited the possibilities of productivity raising investment in technical charge. His was a penetrating statement. How well has it stood the tests of time and the accumulation of evidence? What is the state, generally, of our understanding of agrarian social differentiation in India?

Later, in the 1970s much was written, not only in South Asia, about the mode(s) of production and the directions of structural change in the predominantly rural economies of Africa and Asia, and in Latin America. This work was inspired, of course, by classical studies of capitalist transformation in Europe and Russia – especially by Lenin's *Development of Capitalism in Russia* – as well as by the contemporary literature of development and underdevelopment; and it was influenced by practical political concerns comparable with Lenin's and Mao's, when they tried to answer the question, 'Who are the friends of the revolutionary Proletariat in the countryside?' The specifically Indian version of the debate (reviewed by Harriss (1980) and by Thorner (1982)) shared in the general failings of neo-Marxist development theory which led to what has been described as 'the impasse' in development sociology in the 1980s (Booth, 1985). In concluding my own review of the Indian debate I wrote:

The author is grateful to Dr Barbara Harriss of Queen Elizabeth House Oxford; to Terry Byres and participants in his seminar at the School of Oriental & African Studies; and to Alice Thorner, Brigitte Silberstein and their colleagues at the Centre d'Etudes de L'Inde et de l'Asie du Sud, in Paris, for comments on various earlier versions of this article.

> It is a striking feature of the contributions to the debate that while they are concerned with history, they mostly include very little historical analysis. Different authors refer to historical facts, certainly, but rather to support or illustrate their theoretical arguments than actually to subject them to analysis. The debate has been influenced, though largely indirectly, by an Althusserian conception of mode of production, which entails the explanation of all social phenomena as the effects of structural determinations . . . historical, empirical analysis can yield no surprises, bring about no changes in our understandings, and gives no purpose to the study of history except the provision of illustration . . . In some recent work on the problems at issue in the Indian debate a strong reaction against this influence is apparent, and it is explicitly stated in Banaji's work (1977). Yet Banaji's theoretical approach contains its own teleology . . . Criticism of teleological and historicist conceptions of Marxism does *not* imply a rejection of the theorisation of problems in favour of simple empiricism, but the restoration of historical, empirical analysis to the place occupied by purely structural conceptions of causality. The Indian debate (however) has been marked by rationalistic abstraction and a lack of historical analysis, and by sweeping empirical generalisation (1980, pp. 73, 75).

Booth argued similarly that behind the impasse in the sociology of development as a whole lay

> . . . a metatheoretical commitment to demonstrating that the structuring and processes that we find in the less developed world are not only explicable but necessary under Capitalism. This general formula covers two variants: the type of necessity entailed by the Marxian insistence that the chief features of capitalist national economies and social formations can be 'read off' from the concept of the capitalist mode [of] production and its laws; and another, also inspired in Marx's theory, that involves a system teleology or functionalism (1985, p. 776).

It seems to me that this aptly applies to the Indian mode of production debate, which exhibits both variants of the general formula. It was couched in the terms that, 'If the mode of production is correctly specified, then its laws of development will be understood, and "correct" lines of political action determined'. It thus involved a teleological conception of social change, a realist view of the concept of social class, and subordinated any examination of history to this particular historiography. It gave rise, nonetheless, in the Indian context, to some contributions of lasting significance concerning the process of commercialisation and commoditisation of agriculture to which I shall return below. For the moment I wish to consider evidence about the kind of structural changes which are taking place in rural India.

Agricultural change and agrarian social structure

The 1980s began with some powerful statements of the proposition that the Indian rural economy displays strong tendencies of capitalist development. Byres, in an influential article (1981), drew on a distinction made earlier by C. H. Hanumantha Rao to argue that the new agricultural technology, though theoretically scale neutral, was not in practice 'resource neutral', with the result that because of the advantages accruing to richer cultivators:

> That the new technology has hastened the process of differentiation seems beyond doubt . . . It has served to consolidate the rich peasantry as a powerful, dominant class . . . (and brought about) . . . a process of proletarianisation of the peasantry, or depeasantisation: one that is complex, but as yet partial in its impact (1981, p. 52).

Byres pointed out that the process of change might aptly be described as one of *partial proletarianisation*, given that the evidence strongly suggests increasing dependence upon wage labour, and the loss by poor peasants of an increasing share of the *operated* area to rich peasants, though without their necessarily losing the ownership of the land.

This was a much more nuanced view than was presented by others at around this time – Omvedt, in a series of articles which appeared in *Frontier* (and also in Omvedt (1983)), or C. T. Kurien in an analysis of change in Tamil Nadu (1980). These writers depended heavily upon the simple observation from census data of the increasing proportion of 'Agriculture Labourers' in the rural workforce – ignoring '. . . the obvious warning: no one should use these data to analyse decade-by-decade changes in the Indian economy' (Centre for Monitoring the Indian Economy, 1985; cited by Omvedt (1988)). Apart from the problems caused by shifting definitions of occupational status from census round to census round (to which the Centre refers), there are also those arising from the under-counting of women as 'peasant cultivators', and from the hiding of increasingly complex patterns of occupational multiplicity in the category of 'agricultural labourer' (see Harriss (1989a) for an extended discussion of this point). Omvedt now concedes, '. . . it was primarily looking at such census data that led many, including myself for some time, to *imagine* a reality of a vastly increasing proletarianisation in agriculture' (1988, p. 18). The National Sample Survey, too, are not without problems, but they suggest only a modest increase in the incidence of agricultural labour households (those in which the majority of income is derived from wage labouring on others' lands).

The National Sample Survey data on land ownership clearly indicate that the incidence of landlessness *declined* between the mid-1950s and the early 1970s (very clearly in nine out of 15 major states – see Table 8.1, and Sanyal (1988)), and again between 1972 and 1982 in five states. In other states the incidence of landlessness in 1982 compared closely with what it had been in 1962 (according to the evidence of the 37th Round of the NSS). Only in Andhra Pradesh, Madhya Pradesh and in Maharashtra does it appear that there may have been an upward trend in landlessness. Decreasing landlessness has been associated with a curtailment of large holdings in all states, and thus with a modest decline in the inequality of land ownership. This reflects an increase in marginal and small holdings which has occurred (although perhaps not to the extent that might have been expected; see Table 8.2).

These sample survey findings are borne out in some micro-studies of changes in land ownership – by Attwood for a village in Maharashtra (1979); by Cain for

Table 8.1 Percentage of landless households and of households not operating land

State	Percentage of Landless Households Round				Percentage of Households Not Operating Land Round			Percentage of Households Owning But Not Operating Round		Percentage of Households Neither Owning Nor Operating Round	
	8th	*17th*	*16th*	*37th*	*8th*	*17th*	*26th*	*17th*	*26th*	*17th*	*26th*
(1)	*(2)*	*(3)*	*(4)*	*(5)*	*(6)*	*(7)*	*(8)*	*(9)*	*(10)*	*(11)*	*(12)*
Andhra Pradesh	–	6.64	6.95	11.93	–	37.95	36.05	32.03	29.68	5.92	6.37
Assam	41.57	27.77	24.99	–	19.49	36.22	28.39	15.27	13.76	20.95	14.63
Bihar	16.56	8.63	4.34	4.11	23.84	21.71	20.65	15.28	17.52	6.43	3.13
Gujarat	–	14.74	13.44	16.83	–	31.74	33.75	18.11	25.47	13.63	8.28
Jammu/Kashmir	17.31	10.93	0.95	6.85	14.74	11.09	6.64	5.30	6.07	5.79	0.57
Karnataka	–	18.64	12.46	13.70	–	24.11	29.77	10.51	20.28	13.60	0.49
Kerala	36.95	30.90	15.74	12.76	20.14	23.76	11.69	8.91	1.44	14.85	10.25
Madhya Pradesh	–	9.14	9.58	14.89	–	22.65	16.95	16.66	12.09	5.99	4.86
Maharashtra	–	16.03	15.83	21.24	–	26.29	30.97	12.41	21.09	13.88	9.88
Orissa	12.29	7.84	10.57	7.66	22.19	32.59	25.13	26.22	17.30	6.37	7.83
Punjab	36.86	12.33	7.14	6.41	38.92	39.09	58.61	30.51	52.90	8.58	5.71
Rajasthan	24.85	11.84	2.91	8.13	18.28	11.84	7.83	9.64	5.52	2.20	2.31
Tamil Nadu	–	24.20	17.01	19.13	–	44.91	41.95	23.03	27.86	21.88	14.09
Uttar Pradesh	9.36	2.78	4.55	4.85	23.47	20.76	24.26	18.49	20.42	2.27	3.84
West Bengal	20.54	12.56	9.78	17.21	24.30	33.88	30.94	24.21	23.09	9.67	7.85
All India	23.09	11.68	9.64	11.33	28.21	26.86	27.41	17.42	20.51	9.44	6.90

Source: Sanyal (1988); and *Sarvekshana*, XI, 2, Oct. 1987.

Table 8.2 Distribution of ownership land holdings

All-India	*marginal (up to 1.01 ha)*		*small (1.01–2.02 ha)*		*small–medium (2.03–4.04 ha)*		*medium (4.05–10.12 ha)*		*large (>10.12 ha)*	
	house-holds	area	house-holds	area	house-holds	area	house-holds	area	house-holds	area
1961–62	60.06	7.59	15.16	12.40	12.86	20.54	9.07	31.23	2.85	28.24
1971–72	62.62	9.76	15.49	14.68	11.94	21.92	7.83	30.73	2.12	22.91
1982	66.64	12.22	14.70	16.49	16.78	23.38	6.45	29.83	1.42	18.17

Source: 37th Round of NSS; in *Sarvekshana*, XI, 2, Oct. 1987.

villages in the semi-arid tropics of India (1981; Walker and Ryan, 1990, Ch. 6); in my own work in villages in Northern Tamil Nadu (1987); and in that of Athreya, Djurfeldt and Lindberg in the middle Kaveri valley in Tiruchi District, in Tamil Nadu (1990). These writers find a tendency for poorer landholders at inheritance to gain land and the richer to lose it. Their findings cannot be generalised, and Swaminathan in a study in the Cumbum Valley of Tamil Nadu (*Development and Change*, forthcoming), Bliss and Stern in Palanpur (in Western UP), Kathleen Gough in Thanjavur (1989), and I myself in studies in Birbhum District in West Bengal, have all found evidence of an increasing concentration of land ownership in comparable 'dynamic studies' at the micro-level.

The micro-studies really make the point that all-India, or even 'all-Tamil Nadu', or 'all-Maharashtra' generalisations are bound to be misleading. There are certain fallacies of aggregation. But it is at least possible to conclude that the available data do not support the view that the process of dispossession/depeasantisation and proletarianisation has been taking place universally. All-India data on land operation point in the same direction (presented by Omvedt (1988)), although in some states, as Byres noted, there has been a slight tendency for the proportion of households not operating any land to increase (see also Table 8.1). At the time of writing data from the 37th Round of the NSS, relating to land operation in 1982, are not available. The preliminary reports on the survey do show, however, that the proportions of households leasing out land, of those leasing in land, and the areas leased in and leased out have all declined appreciably since the early 1970s. What can be stated with some confidence is that, according to NSS data, 'self-employment' has declined in importance while casual daily paid wage employment has become more significant through the 1980s (see Table 8.3 from Minhas and Mazumdar (1987)).

Byres' argument that '. . . the most significant contribution (of the new technology) has been to throw into increased wage employment large numbers of poor peasants who continue to own some land' (1981, p. 53), might seem to have been borne out by the evidence and analysis which has become available in the 1980s. It is *not* clear however that the new technology is directly responsible for 'throwing' people into wage employment; and the negative connotation of this expression does not allow for findings such as those which Walker and Ryan report from the ICRISAT village studies, showing that participation in the labour market plays an important role in dampening fluctuations in income and consumption – fluctuations which frequently undermine the livelihoods of poor households in the long run (Walker and Ryan, 1990). And an interesting micro-study from Bangladesh by Bhaduri, Rahman and Arn (1986), shows how polarising tendencies are moderated because of the increasing availability of other sources of income. This makes it possible for small landholders to retain their property so that '. . . there is a marked tendency towards a stabilisation of land ownership even among the smaller size groups' (Bhaduri *et al.*, 1986); see also the debate on this article in *Journal of*

Table 8.3 Employment and casual labour 1972/1983. Percentage distribution of all workers according to usual occupation by category of employment separately for males and females in rural and urban India.

Category of employment	*Male*			*Female*		
	round			*round*		
	27	*32*	*38*	*27*	*32*	*38*
			RURAL			
Self-employment	65.90	62.77	60.40	64.48	62.10	62.21
Regular wage/ salaried work	12.06	10.57	10.77	4.08	2.84	3.10
Casual wage labour	22.04	26.66	28.83	31.44	35.06	34.69
			URBAN			
Self-employment	39.25	40.38	40.67	48.40	49.47	46.50
Regular wage/ salaried work	50.69	46.41	44.58	27.89	24.94	26.23
Casual wage labour	10.06	13.21	14.75	23.71	25.59	27.27

Source: *Sarvekshana*, Vol. 9, No. 4 (April 1986), p. S-112.

Rural Persons (M+F)

	27th		*32th*		*38th*	
	Casual workers in Labour Force %	*Incidence of unemployment %*	*Casual workers in Labour Force %*	*Incidence of unemployment %*	*Casual workers in Labour Force %*	*Incidence of unemployment %*
1 Andhra Pradesh	36.21	11.23	39.88	10.47	41.59	9.93
2 Assam	10.23	1.65	16.76	1.54	17.59	5.33
3 Bihar	26.84	10.02	47.06	7.96	34.31	8.41
4 Gujarat	23.76	5.44	31.86	9.20	34.77	5.46
5 Haryana	9.11	2.94	12.06	6.26	14.57	5.76
6 Himachal Pradesh	2.94	0.57	8.65	1.58	6.36	1.50
7 Karnataka	31.22	8.59	39.07	8.91	40.47	8.29
8 Madhya Pradesh	19.47	3.42	25.70	2.76	27.24	2.35
9 Kerala	40.02	23.50	33.85	25.79	35.78	26.69
10 Maharashtra	37.43	9.43	38.37	7.20	39.75	7.11
11 Orissa	30.50	10.19	30.50	10.69	35.71	9.25
12 Punjab	13.97	3.94	17.93	4.69	17.03	7.05
13 Rajasthan	5.18	3.26	11.02	2.69	11.69	4.10
14 Tamil Nadu	37.1	12.08	39.60	15.71	43.58	20.32
15 Uttar Pradesh	11.17	3.38	17.31	3.73	17.26	3.60
16 West Bengal	32.56	10.67	34.10	9.40	35.66	16.70
17 Delhi	8.56	3.18	9 58	10.46	3.53	13.59

Source: Minhas and Mazumdar (1987).

Peasant Studies, Vol. 14, No. 4, 1987. This work helps to make the important methodological and theoretical point that it is no longer enough, if ever it was, to attempt to study rural differentiation by examining agricultural production and landholding in isolation from other activities.

For India the National Sample Survey data referred to above show a significant and sustained tendency towards the diversification of rural occupations, and a crucial question on which there is as yet little evidence is that of whether this reflects the existence of growth linkages from agriculture and a shift into dynamic, productive activities (as in Mellor's optimistic scenario for a 'green revolution'-based, food and employment first strategy for development, presented in *The New Economics of Growth* (1976); see also Mellor and Johnston (1984)), or a shift to a residual sector with lower productivity than in agriculture. Vaidyanathan, in his R. C. Dutt Memorial Lecture of 1986 (published 1988) was inclined to a pessimistic view, partly because of the association between the incidence of non-agricultural employment and that of unemployment (see Table 8.4). Thus increased rural non-agricultural employment '. . . may *reflect* at least in part a tendency for the unemployed to crowd into casual labour outside agriculture' (Vaidyanathan, 1988, p. 34), although it would be possible to draw other inferences from the correlation. Vaidyanathan also reports that '. . . higher inequality of operational holdings seems to go with lower incidence of non-agricultural employment' (1986, p. A-143), perhaps bearing out the suggestion made by Rizwanul Islam (1986) that rural inequality will tend to limit the positive effects of agricultural growth on the non-agricultural economy. This may well explain the observation that 'The available evidence (on the impact of the new technology in agriculture on rural non-agricultural employment) does not suggest a very substantial increase of employment in these sectors' (Basant, 1987, p. 1360); see also Harriss and Harriss (1984).

The impact of the green revolution

What has been said already, about trends of agrarian change and their analysis, has inevitably involved some reference to the question of the impact of the introduction of new technology into agriculture, or 'the Green Revolution'. Yet this has been and remains such an important subject that it should be considered further. There is an extensive debate in which sharply polarised positions are to be found, with some insisting that modern varieties 'cannot do any significant good' because they have such an adverse effect upon the livelihoods of the poor.

Considerable effort has been expended on measuring the incidence of poverty in rural India and on analysing its determinants, and in this extensive literature there is a particular concern with the relations between agricultural growth and poverty. Ahluwalia's arguments in 1978, suggesting that there was a negative relation between poverty and agricultural performance in the period 1956–57 to 1973–74 met with a barrage of criticism. Yet now one of the critics writes that 'In retrospect the criticisms of Ahluwalia's method of establishing

Table 8.4 Non-agricultural employment and unemployment

	1971–73 to 1977–78 Change in			*1977–78 to 1983 Change[1] in*		
	Non-Agricultural Employment (Per Cent Points)	*Foodgrain Production (Per Cent)*	*Unemployment Rate (Per Cent Points)*	*Non-Agricultural Employment (Per Cent Points)*	*Foodgrain Production (Per Cent)*	*Unemployment Rate (Per Cent Points)*
	1	*2*	*3*	*4*	*5*	*6*
Andhra Pradesh	+ 1.8	+ 34	+ .3	+ 4.8	+28	+ .3
Assam	− 9.1	+ 2	− .2	+ 6.3	+11	+1.9
Bihar	− 1.2	+ 6	−1.4	+ 2.8	− 2	− .6
Gujarat	− 3.1	+ 75	+1.0	+ 4.5	+50	−1.1
Haryana	+ 5.4	+ 31	+3.5	+ 5.1	+29	+ .1
Jammu and Kashmir	+ 7.2	+ 11	−2.7	+10.2	+ 1	+9.8
Karnataka	− 1.3	+ 58	+0.3	+ .8	+ 1	−1.8
Kerala	− .8	− 6	+2.6	+ 1.7	− 4	−3.6
Madhya Pradesh	+ .9	+ 16	− .3	+ .2	+24	+ .1
Maharashtra	−10.8	+243	−2.2	+ 2.8	+ 5	+ .3
Orissa	− 1.8	+ 11	+ .2	+ 7.1	+23	+ .8
Punjab	+ 5.3	+ 35	.0	+ 5.3	+45	+1.0
Rajasthan	+ .6	+ 39	− .6	+ 4.2	+41	+1.6
Tamil Nadu	+ 6.6	+ 8	+6.7	+ 7.6	−20	+5.2
Uttar Pradesh	+ 1.1	+ 17	+1.0	+ 2.3	+38	− .4
West Bengal	+ .8	+ 32	− .2	+ 5.3	+ 4	+7.1

Note: 1 The correlation coefficients between change in non-agricultural employment on the one hand and the changes in foodgrain output and unemployment on the other are as follows:

	1971–77	*1977–83*
Foodgrain output	−.55	−.11
Unemployment	38	87

Sources: 1 Relates to population in the age group 15–59 years; 1972–73 estimates are from the 28th round of NSS; 1977–78 estimates are from the 32nd round of NSS; *Sarvekshana*.

2 Relate to population aged 5 years or more; estimates for 1977–78 from *Sarvekshana*, Jan.–April 1981 and for 1983, NSS Report No. 315.

Source: Vaidyanathan (1986).

the trend appear less convincing' and the same author now agrees with Ahluwalia's conclusions from his up-dated analysis – to 1977–78 (Ahluwalia, 1985) – that (no matter what the relationship with agricultural growth) '. . . there has been an underlying tendency for the incidence of rural poverty to decline' (Ghosh, 1989). But perhaps what is most important is that even Ahluwalia himself argues that reliance on growth of agricultural output alone will not bring about a large decrease in the incidence of poverty. Another recent analysis by Gaiha (1989) reaches the conclusion that the inverse relationship between agricultural growth and poverty is sometimes weak or absent, while the influence of price fluctuations is consistently strong and often decisive. Most specifically unanticipated inflation in consumer prices aggravates poverty – partly reflecting the strong relationship identified in other studies, between dependence upon casual wage labour and poverty. An extremely important but very little recognised fact about poverty in India is that poor people are *not* primarily 'small farmers' but those dependent upon irregular and unreliable wage incomes – and who are thus exposed to the hazard of sudden deterioration in their exchange entitlements in circumstances of inflation.

Gaiha's findings concerning the influence of food prices and their fluctuations on poverty is borne out by the findings of the ICRISAT village studies in three regions of the Indian semi-arid tropics (Walker and Ryan, 1990). In a panel of households studied between 1975/76 and 1983/84 poorer households experienced as much or more growth in real incomes per capita as the richer households and some improvements in household wealth. The researchers explain this finding as being the result of increased cereal productivity, mainly elsewhere, which has decreased prices and thus led to income growth for the poorer households (net buyers of food), together with the fact that government policies have tended to stabilise cereal prices so that there has been only very mild food price inflation. The other essential condition of improved livelihoods was the tightening of rural labour markets, to which I shall return.

Limited though I believe it is by its positivistic emphasis on measurement, the poverty literature has at last shown up the dependence of the poor upon wages, and thus that the impact of agricultural growth on employment and wages is of crucial importance (see, in addition to Gaiha (1989), Sundaram and Tendulkar (1988)). This perception firmly underpins the analysis of the impact of the green revolution by Lipton and Longhurst (1989), which is based on the South Asian literature. These authors are scathing in their criticism of the considerable literature which maintains that the modern varieties have led to impoverishment, showing that it is based on a partial, 'adding-up' approach, involving isolated attempts to measure effects on different groups of farmers, or labourers, or consumers, and rarely looking beyond first round effects. What is required, they argue, is a systemic view, taking account not only of the economy as a whole, over successive rounds of effects, but also of the relations of the economy with power structures.

They believe that though much of the work which condemns the modern variety (MV) revolution is unconvincing (for the reason just stated), there is still a problem. The biological characteristics of the MVs are such that they should be beneficial (and are indeed increasingly so, as a result of recent developments in agricultural research) to poor people as cultivators, labourers and consumers. Yet, in spite of the strong evidence that 'small farmers' have benefited from MVs, albeit after a lag in adoption; that agricultural employment has been increased; and that cereals prices have been reduced, it *still* seems that there has been little change in the incidence of poverty, even in MV-lead areas such as Punjab. What accounts, then, for 'the MV-poverty mystery'? Part of the explanation is that the focus in research on 'small farmers' and MVs was misplaced, given that the rural poor in South Asia are predominantly those dependent upon casual wage labour. And although MVs do increase labour absorption in agriculture (see also Basant, 1987) the sheer abundance of labour means that agricultural workers' real wages and incomes have risen little if at all, even where labour has not been displaced by mechanisation. At the same time there has been less of a shift into productive rural non-agricultural activities than might have been predicted (see above). And it may also be the case that employers are able to reduce wages in 'responsive wage deceleration' when food prices are reduced. Here Lipton and Longhurst rely on a paper by Papanek in which it is reported that in India 'over a two-year period nominal wages fully adjust to the price changes, resulting in constant real wages'. Thus potential gains to poor people as consumers from cheaper foodgrains are eroded.

These arguments are broadly supported in recent reviews of evidence on employment in Indian agriculture by Bhalla (1987) (see Table 8.5), and on agricultural wages by A. V. Jose (1988) (and see Table 8.6). Bhalla shows that 'What has happened in India is that for those crops and those states where the Green Revolution came early, the usual initial response was a sustained rise in labour use per hectare. This trend characteristically peaked in the mid-1970s or shortly afterwards and subsequent increases in yield were associated with declines in per hectare labour absorption' (1987, p. 540); (see also Vaidynathan (1986)). Jose, although he reports an increase in real wage rates for male workers between 1970–71 and 1984–85 in almost every state (and in all states for women, and to a greater extent), also notes two major exceptions – the MV-lead states of Punjab and Haryana. Over the longer run (from 1956–57) 'stagnation or decline in real wages . . . appears to have been the characteristic feature in a number of Indian states' and spurts in real wages are found characteristically to be of short duration (1988, p. A-53). Although agricultural output and wages are usually positively associated, the Punjab experience '. . . which after having recorded one of the highest rates of growth in agricultural output among Indian states shows stagnancy and even a marginal decline in real wages' (1988, p. A-56) suggests that there is probably '. . . an upper limit to any possible increase in wages that can come about in the agriculturally upbeat regions of India' (1988, p. A-57). This is because of both circular migration of agricultural labour into dynamic regions and technological change leading to labour displacement. Jose's conclusion is that '. . . agricultural wage increases in so far as they are currently taking place in some regions of India might turn out to be a short-lived phenomenon' (1988, p. A-57).

I referred above to the findings of the ICRISAT village studies that between 1975/76 and 1983/84 poorer households got a little better off, and mentioned that lower and stable cereal prices and tightening labour markets were the conditions of this improvement. The tightening of labour markets in these mainly 'non-green revolution' dryland villages was due not so much (if at all) to increases in demand for agricultural labour, but to a variety of factors operating differently in different villages, including the effects of the Maharashtra Employment Guarantee Scheme and of land allotments to previously landless

Table 8.5 Trend rates of growth in labour absorption by states, 1971–72 to 1983–84 for 'all crops' and related variables

States (ranked according to col. (2) growth rate)	*Trend rates of growth in*			*Growth rates in*		*Per cent change in employment 1977–78 over 1972–73 (NSS)*
	Labour absorption		*Gross cropped area under all crops*	*Production (49 crops)*	*Labour productivity*	
	Total	*Per hectare*				
(1)	*(2)*	*(3)*	*(4)*	*(5)*	*(6)*	*(7)*
1 Andhra Pradesh	3.663	2.779	0.571	3.31	1.78	12
2 Gujarat	3.574	3.191	0.371	3.92	2.38	14
3 Maharashtra	3.435	1.730	1.676	5.60	4.44	30
4 Karnataka	2.614	1.483	1.114	2.44	0.75	9
5 West Bengal	2.037	2.096	0.057	0.91	−0.59	10
6 Rajasthan	1.393	0.737	0.659	2.47	0.97	− 9
7 Orissa	1.331	0.795	0.531	2.28	1.15	4
8 Punjab	1.079	−0.887	2.002	3.92	2.63	− 2
9 Uttar Pradesh	0.598	−0.145	0.743	3.09	1.72	− 6
10 Haryana	0.230	−0.357	0.589	3.31	1.47	−11
11 Bihar	0.074	1.037	−0.954	0.49	−0.68	8
12 Madhya Pradesh	−0.014	−0.976	0.972	1.65	0.03	− 1
13 Tamil Nadu	−0.510	0.960	−1.457	1.12	0.26	−13

Source: Bhalla (1987).

Table 8.6 Real wages of agricultural labourers

A: INDEX NUMBERS OF REAL WAGES OF MALE AGRICULTURAL LABOURERS

(1970–71=100)

Year	States							
	Andhra	*Assam*	*Bihar*	*Gujarat*	*Haryana*	*Himachal*	*Karnataka*	*Kerala*
1970–71	100.00	100.00	100.00	100.00	100.00	100.00	100.00	100.00
1971–72	95.17	92.20	97.08	104.95	97.48	110.54	103.46	111.09
1973–74	82.89	90.10	90.85	79.64	79.22	88.21	90.81	99.37
1974–75	73.54	78.87	82.32	66.07	74.39	78.69	76.18	82.72
1975–76	87.57	89.65	114.47	91.71	81.08	92.73	94.51	94.44
1976–77	97.29	95.12	138.95	124.13	83.76	93.27	115.82	108.39
1977–78	101.27	98.09	119.57	114.22	91.86	85.78	127.87	111.55
1978–79	112.79	95.46	119.55	120.91	97.09	97.60	130.04	113.73
1979–80	111.82	90.39	111.62	112.16	92.84	94.61	121.78	119.75
1980–81	108.31	92.48	106.85	106.95	82.95	103.24	108.70	131.30
1982–83	122.25	103.10	117.52	119.76	98.21	102.73	107.97	135.69
1983–84	134.98	109.36	123.53	129.05	103.12	106.52	104.76	120.91
1984–85	145.09	120.06	145.90	151.60	100.05	104.91	103.56	131.57

B: INDEX NUMBERS OF REAL WAGES OF MALE AGRICULTURAL LABOURERS

(1970–71=100)

Year	States							
	MP	*MR*	*Orissa*	*Punjab*	*Rajasthan*	*Tamil Nadu*	*UP*	*WB*
1970–71	100.00	100.00	100.00	100.00	100.00	100.00	100.00	100.00
1971–72	99.29	96.50	99.49	98.29	103.51	98.71	100.38	104.07
1973–74	88.04	77.27	89.75	83.09	87.67	92.80	84.57	90.39
1974–75	75.02	64.31	73.67	78.82	71.35	74.48	74.86	79.41
1975–76	100.76	66.12	87.68	90.35	90.63	92.93	119.99	98.48
1976–77	116.25	74.28	121.36	99.71	118.27	91.52	126.70	102.06
1977–78	109.90	77.75	115.58	94.95	114.09	89.92	100.22	110.88
1978–79	115.40	88.19	119.63	97.02	112.23	100.69	109.76	112.05
1979–80	103.94	86.11	108.60	92.63	106.41	106.82	103.99	107.39
1980–81	99.32	78.07	105.12	84.98	105.25	104.44	91.54	103.15
1982–83	125.43	91.52	96.05	86.15	119.51	100.81	108.22	92.03
1983–84	137.77	108.40	120.54	92.53	122.75	98.49	123.33	98.21
1984–85	144.72	122.64	140.62	97.43	112.90	118.46	131.45	103.84

C: INDEX NUMBERS OF REAL WAGES OF FEMALE AGRICULTURAL LABOURERS

(1970–71=100)

Year	States							
	Andhra	*Assam*	*Bihar*	*Gujarat*	*Haryana*	*Himachal*	*Karnataka*	*Kerala*
1970–71	100.00	100.00	100.00	100.00	100.00	100.00	100.00	100.00
1971–72	94.18	95.14	95.30	106.09	99.65	97.81	104.67	126.21
1973–74	80.05	97.07	90.49	87.91	76.44	81.80	91.45	106.21
1974–75	72.72	78.25	85.44	73.77	73.00	75.42	83.64	91.13
1975–76	94.94	92.09	128.57	106.61	82.94	91.56	104.30	102.37
1976–77	101.44	103.78	135.26	154.94	101.47	92.49	141.58	120.97
1977–78	104.98	105.68	145.50	145.57	98.56	89.41	161.96	128.34
1978–79	119.47	101.18	136.81	151.21	96.37	91.80	154.08	137.17
1979–80	120.21	95.91	130.39	138.53	109.38	88.32	145.71	137.33
1980–81	111.92	96.14	117.34	125.42	107.84	90.71	120.13	148.36
1982–83	124.70	104.85	141.09	128.65	140.87	99.98	120.17	155.33
1983–84	136.82	112.70	148.20	145.27	137.20	109.02	119.22	136.58
1984–85	146.33	125.43	165.48	155.81	129.94	110.92	122.91	157.70

D: INDEX NUMBERS OF REAL WAGES OF FEMALE AGRICULTURAL LABOURERS

(1970–71=100)

Year	States							
	MP	MR	Orissa	Punjab	Rajasthan	Tamil Nadu	UP	WB
1970–71	100.00	100.00	100.00	100.00	100.00	100.00	100.00	100.00
1971–72	97.63	92.06	98.85	125.42	100.82	99.64	94.77	117.41
1973–74	79.78	76.85	95.63	95.78	72.13	93.71	85.92	106.16
1974–75	74.60	65.72	81.60	87.18	71.21	71.20	78.01	99.78
1975–76	104.42	67.54	118.36	92.63	119.01	95.05	105.12	119.09
1976–77	145.16	81.07	166.18	0.00	159.83	97.95	125.58	137.09
1977–78	136.62	88.91	156.34	0.00	135.31	113.36	95.11	156.11
1978–79	140.69	98.56	157.01	132.88	145.87	119.66	107.09	144.72
1979–80	123.84	95.10	125.71	132.41	135.64	144.49	106.65	141.42
1980–81	119.41	84.90	124.50	120.50	113.18	110.91	90.74	135.66
1982–83	144.43	95.84	109.91	132.48	129.10	120.36	125.92	132.78
1983–84	156.87	114.10	136.85	137.29	139.62	110.62	139.52	142.32
1984–85	168.25	121.58	148.14	125.51	133.85	121.25	145.12	140.08

Source: Jose (1988).

households, as well as (in one village more than others) the expansion of off-farm employment and of temporary/circular migration for work in towns. My findings in my restudies in northern Tamil Nadu in 1983–84 were much the same: 'It seems that the diversification of occupations locally and increased migration for work in larger, distant urban centres have tightened the agricultural labour market' (Harriss, 1985, 1991). V. K. Ramachandran, too, in the different circumstances of the Cumbum Valley (Madurai District) in Tamil Nadu, found that between 1977 and 1986, in spite of a decline in employment in agriculture, especially for women, 'there was an *increase* in the real wage-rates paid for cash-paid, daily-rated operations' (1990, p. 244). The reasons for this change are not explored. Ramachandran does note elsewhere, however, 'a diversification of the occupation structure of the village population' (1990, p. 256).

The ICRISAT researchers conclude that in rainfall unassured villages in the semi-arid tropics '. . . changes in off-farm employment will likely play a larger role in conditioning wage levels and contractual relations than opportunities for agricultural employment within the village' (Walker and Ryan, 1990, p. 150). The same observation seems to me to apply to the groundwater irrigated, 'green revolution' area of northern Tamil Nadu and to be generalisable. It underlines the vital importance of rural non-agricultural activity, and confirms that a major concern in development policy in India *must* be with the sluggish growth of such activity.

Agrarian change and rural labour

Evidence has accumulated, therefore – from both large-scale surveys and from detailed local studies that, contrary to the beliefs and expectations of many, the process of polarisation has not, generally, brought about 'depeasantisation'. On the contrary, though generalisations are doubtful, it is at least as likely that there has been a tendency for landlessness to decline together with some proliferation of marginal holdings. It is also true that dependence on self-employment has declined in favour of dependence on casual wage labour so that it is possible that small property is reproduced significantly through the participation of land-holders in agricultural and non-agricultural employment. As Ben White has argued, the challenge in analysis of agrarian social differentiation is to incorporate the opposing but co-existing tendencies of polarisation and proletarianisation, and of the tenacity (at the same time) of small/poor peasant households (White, 1989).

In India as in the South-East Asian regions to which White refers, the extent and nature of rural non-agriculture is often of crucial importance. Levels of livelihoods almost certainly depend significantly upon the availability of productive non-agricultural employment (local rural, distant-rural, or non-rural in location), for it seems likely that it is this, together with the effects of state interventions – notably those influencing employment – as much as changes in agriculture itself, which accounts for the tightening of rural labour markets which, it seems, may well have occurred fairly generally in the recent past; and thus for increasing real wages and modest improvements in livelihoods, notably those outside areas of high 'green revolution'.

In this context, together with the cost/prices squeeze on cultivation occasioned by the steep rise in the value of modern inputs per unit of output in circumstances in which the relative price of agricultural products has declined (Mundle, 1990), and a broad setting in which the 'stickiness downwards' of wage rates is characteristic of rural labour markets, there are clear indications of change in labour contracts and in some cases of the segmentation of the labour market, evidently in the interests of control over the labour process.

It is worth dwelling for a moment upon the 'stickiness downwards' of rural wage rates. Drèze and Mukherjee in their review *Labour Contracts in Rural India: Theories and Evidence* (1987) conclude that it is the rigidity of daily money wages, in a context in which alternative contracts are available, and especially their *rigidity downwards* which is most striking and most difficult to explain in terms of any of the conventional theories of labour market functioning and wage determination (given that the resistance to wage cuts clearly comes from labourers rather than employers – so that the efficiency wage thesis is ruled out). As Drèze and Mukherjee also hint, it seems to me that the 'stickiness downwards' of standard daily money wages depends upon the kind of mutuality which exists amongst labourers in a local village labour market. As I, and, I believe, others have observed, rather than endure the shame of rate-busting locally, labourers who are desperately in need of scarce employment may go outside to try to find it. And outside their own villages they will perhaps feel able to work for a lower rate. Such mutuality is associated with another feature of rural labour markets over much of India – the fact that they are fragmented, or parcellised with, *inter alia*, different wage rates obtaining for the same tasks even in villages which are very close to each other (see Kapadia (1990) for a striking example from Tamil Nadu).

Ashok Rudra has argued that neither the analytical tools of neoclassical economics, nor those of Marxist theory is adequate to explaining these characteristics of rural labour markets. His own explanation advances the view that 'Even when labourers are not organised in unions they have a sense of community and an understanding of collective self-interest which is an integral part of the ethos of village society' (Rudra, 1984, p. 1265); see also Bardhan and Rudra (1986). Exactly. Unfortunately, as the case described by Kapadia (1990) shows, mutuality may be confined to, and is in a sense defined by, the boundaries of a small neighbourhood and caste or kin group so that labourers within even a small area may enter into competition with each other.

And one of the responses of employers to the actual or possible tightening of rural labour markets, and the mutuality of labour, has been deliberately to segment the market, as in the case of the plain of South Gujarat, vividly described by Jan Breman. His monograph, *Of Peasants, Migrants and Paupers* (1985) shows how

> Over the last two decades agriculture has become a dynamic industry (on the south Gujarat plain) which in addition has created a number of new economic activities in the countryside itself. Together these ensure great pressure of work nearly the whole year round. But it is precisely in Surat District, where intensification and diversification of production has been most rapid, that this development has led to growing unemployment amongst the local landless . . . (1985, p. 343).

Alongside this unemployment of local labour there exists substantial seasonal migration of labour from outside. The reason for such an apparently odd situation is that outside labourers are more amenable to control: 'In contrast to the Halpatis the seasonal workers cannot retreat to a milieu of their own order to make a stand against the farmers' dominance'. The rural labour market is thus characterised by the deliberate exclusion of local labour and is segmented in the interests of establishing control over labour, with the result that an area of accelerated economic growth has seen the progressive impoverishment of a large group of local labourers.

From Tamil Nadu there are several reports of changes in the recent past in the mode of employment of agricultural labour which also seem to have the effect of segmenting the labour market, though with what results is not so clear as in the case described by Breman. Athreya, Djurfeldt and Lindberg report that in the irrigated tracts of Tiruchi District '. . . intensive cultivation has brought about a system of gang labour wherein a gang is employed on a collective piece-rate to perform a certain task, like harvesting a field, and where the gang leader receives the payment and distributes it in equal proportions among the gang members' (1990, p. 307); see my similar observations in North Arcot District, reported in Harriss, (1985); and those of Kapadia (1990). Athreya *et al.* argue that 'This makes for a segmentation of the rural proletariat (elsewhere they compare it with what Gillian Hart refers to as "exclusionary arrangements" (Hart, 1986)), dividing the casual labourers into an "elite" of young and strong workers – who are members of the gangs, who are earning comparatively high wages – and a fringe of workers, mostly consisting of elderly people and children, who can only be sure of getting employment during the peak periods' (1990, p. 307). The politics of these production relations need further analysis.

In other cases the response of employers to the tightening of labour markets has been to attempt to re-forge 'attached labour relationships'. My re-studies of northern Tamil Nadu, in 1983–84 showed that the numbers of regular farm servants, or '*padials*', had increased during the period of green revolution and a tightening labour market. Brass insists that such labour contracts should be described as 'bonded labour', and he provides detailed reports of deliberate 'deproletarianisation' or unfreeing of labour though the creation of debt bondage in the green revolution heartland of Haryana. He argues of this:

> '. . . in Haryana unfreedom constitutes a central component in the struggle between landholders and agricultural workers. *In a context where alternative and better paid employment is available* (my emphasis), and labourers are as a result capable of sustaining their unwillingness to enter attachment at any price, landholders have not only to make labour power available but also to make it available cheaply. Both these objectives are realised through the operation of the debt bondage mechanism, which constricts, or eliminates the free movement of labour in the market' (1990, p. 55).

In circumstances like those of northern Tamil Nadu where, historically, the supply of labour has been more abundant, but where there are still pressures associated with control over labour in the circumstances of the costs-prices squeeze and a tightening labour market, I believe that landholders have not had to resort to the same extent to debt bondage as a mechanism for attaching labour but have been able to operate through the institutions and ideology of patronage. But the effect is comparable: the re-creation of a supposedly 'traditional' institution in order to extend control over the agricultural labour process.

In circumstances in which labour is relatively abundant and not, as in Haryana, in circumstances in which alternative and better paid employment is available, then labour attachment may exist but reflect very different dynamics. Ramachandran reports from the Cumbum Valley that 'For many workers . . . non-wage employment in the form of labour service is a means of booking a place for wage employment when it is available, of gaining tenure with an employer . . . (or) . . . In G. A. Cohen's insightful formulation "the worker here agrees to be a bit of a serf in order to qualify for proletarian status"' (Ramachandran, 1990, p. 254). As this author remarks, at the conclusion of his book '. . . unfreedom and the system of labour service have not withered away or been consigned to the past; on the contrary, they play an important role in the daily lives of agricultural labourers and in production relations' (1990, p. 262).

The failure of MVs to make more of a dent in the problems of rural poverty is thus perhaps not such a mystery after all. It has to do with what Washbrook, in a review of the recent historiography of early nineteenth century India describes as 'the logic of South Asia's own process of "indigenous" capitalist development – in which accumulation was sustained through coercion and the decline in the share of the social product accorded to labour rather than by putting valuable capital at risk by investment' (1988, p. 90). This in turn has to do with demography – the supply of labour – and with structures of social control. And, as Lipton and Longhurst argue, the character of the 'MV revolution' is such as to have reinforced existing local power structures, with the effects on real wages and livelihoods that have been described.

. . . .

The political economy of South Asian agriculture: do institutions matter?

Daniel Thorner's idea of 'the built-in depressor' in Indian agriculture, articulated a theory about the agrarian problem in India which appeared in the writings of the nationalist historians on the one hand, and influenced some aspects of colonial agrarian policy on the other. It appears again in the report of the *Task Force on Agrarian Relations* of 1973: 'local power structures are insurmountable hurdles in the path of the spread of modern technology'. And it is of course a theory about the institutional constraints on agricultural development which has been refined and elaborated by contemporary economists – notably in the work of Bhaduri and Bharadwaj concerning the impact of commercialisation upon Indian rural economy and society (see especially Bhaduri (1983) and Bharadwaj (1985)).

They argue that the process of commercialisation as it proceeds in circumstances in which there is already sharp inequality in the ownership of productive assets (so that individuals don't enter the market on equal terms), forces the involvement in markets of producers who lack net marketable surplus. This 'compulsive involvement' in the market (Bharadwaj), or 'forced commerce' (Bhaduri) is on terms which are extremely adverse so that what is produced is substantially appropriated by dominant owners of means of production or merchant capitalists. It may be the case that these dominant economic agents link transactions across markets (land and labour; land, money and product markets; labour and money markets), and are able by this means to drive up the rate of surplus appropriation – although the extent of such interlinkage is contested empirically (Rudra and Bardhan, 1983).

And, in so far as it does exist, it is explained by others as a technically efficient solution to problems of resource allocation in circumstances in which markets are incomplete or imperfect, in which information is imperfect or costly and in which the costs associated with all transactions are consequently high. But for Bhaduri and Bharadwaj 'compulsive involvement' is the cause of the 'underdevelopment' of the mass of rural producers – in the sense that they are unable to improve their productivity and incomes – and also the base of an efficient mode of surplus appropriation. The counter-arguments are quite neatly summarised alongside the presentation of a lot of basic empirical material by Inderjit Singh (1988). His argument is that an institution such as share-cropping is not a cause of agrarian underdevelopment, acting as a barrier to technological innovation, but a reflection of it – so that tenancy reform is either irrelevant (because technical and infrastructural development is, anyway, rendering share-cropping unimportant – as does *seem* to be the case in most of South Asia, according to available evidence) or may be counterproductive because of leading to greater inefficiency in resource allocation. Here I shall not attempt to offer a judgement on the theoretical merits of the opposing cases, but rather put forward the implications of recent substantive (descriptive – analytical) research.

An argument about the impact of commercialisation, and the effect of a 'built-in depressor' appears in the analyses of the agrarian economies of North Arcot District in Tamil Nadu and of Birbhum District in West Bengal by Barbara Harriss and myself (B. Harriss, 1981, 1983; J. Harriss, 1982a, 1983) as well as in work in Bangladesh by Wood (1981) and by Westergard (1985). I shall summarise our arguments about North Arcot as a paradigmatic case.

(1) North Arcot has an agrarian structure characterised by high incidence of landlessness (38 per cent of rural households); and altogether about 80% of rural households control insufficient means of production to be able autonomously to produce their own livelihood requirements. A small class of surplus appropriators – rich peasants and a few small landlords – is able to dominate labour, money and product markets as well as land, now as, according to Washbrook's analysis (e.g. 1978), they were able to in the later nineteenth century. (The agrarian structure is closely comparable with that recently described by Athreya, Djurfeldt and Lindberg from a region further south in Tamil Nadu. In a careful analysis they established that only four per cent of rural households in 'dry' villages were surplus appropriators and show that 'The majority of the middle peasantry is pushed below the level of autonomous production and depends upon non-farm sources of income for their reproduction' (1990, p. 231).

(2) There is consequently a high level of compulsive involvement in markets. Most are compelled to sell their paddy or groundnuts at harvest, and then to draw loans for subsistence and for the renewal of production. Such compulsive involvement was increased in the late 1960s and early 1970s by the introduction of green revolution technology. Small producers' accounting is in terms of total product of grain rather than profit. Increasing the number of bags of paddy from the field made HYVs attractive to such producers and they would borrow to purchase the necessary seeds and other inputs. It is easy to see how this could, in many cases, lead to increased debt and further 'compulsive involvement'.

(3) For surplus appropriators and those with money, money-lending, trade and non-agricultural production were more profitable than investment in agricultural production. Such monied capitalists thus profited from the compulsive involvement in the market of the mass of producers. In more abstract terms merchant capital was dominant, and acted as a 'built-in depressor'. The mass of producers lacked resources for investment in production and the surplus appropriators (rich peasants and landlords) had little incentive to invest substantially in production. The advent of the 'green revolution' and higher land productivity had given the system a little kick by 1973/74, and brought about some increase in production, but it did not appear at all likely to be a sustainable effect, because it was simply not profitable enough for the surplus appropriators to be encouraged to change the system of production relations.

In fact, follow-up research in 1983/84 showed that we were wrong and that 'the built-in depressor' did not operate in the way in which we supposed. Agricultural production had continued to grow, exceptionally for Tamil Nadu, at more than three per cent per annum through the 1970s and early 1980s; cultivation of HYVs had increased from less than 20 per cent of the paddy acreage to more than 90 per cent; investments in ground water irrigation and in fertilisers had increased from what were already comparatively high levels; and there were widely distributed although very modest increases in real incomes. What had happened to offset the depressor? Basically what happened involved a number of conjunctural factors, and state interventions lying beyond the reach of the local power structures which are emphasised in the notion of 'the depressor'.

The conjunctural factors were (i) the chance that the variety IR20 was evidently well-suited to local agro-climatic conditions, proving higher yielding than local varieties even in moisture-stressed and low fertility conditions; (ii) the equalisation of prices between local and HYVs which occurred in the mid 1970s; (iii) the expansion of non-agricultural employment locally and in more distant towns and cities, not linked to local agriculture, but playing a significant part in the tightening of the local labour market. The state interventions were (i) those which continued the expansion of supply of cheap formal sector credit; legislated against usurious private moneylending and sought to regulate agricultural marketing; (ii) those which subsidised the supply of electric power – and thus continued the expansion of ground water irrigation – and, for a time at least, maintained the fertiliser/paddy price ratio; (iii) land reform legislation which, though not effective in bringing about much redistribution of land, has limited the accumulation of land in large holdings and provided allotments for some landless people, thus improving their credit worthiness even if not supplying them with a livelihood holding; (iv) welfare interventions, notably the Noon Meals Scheme for children, which has had *some* impact on living standards, and also contributed to the generation of off-farm employment. This has tightened labour markets and with other interventions referred to and conjunctural factors helped to reduce compulsive involvement.

In short, in the words of Athreya, Djurfeldt and Lindberg – used to describe their almost identical findings from Tiruchi District – the point is that '. . . relations of production in *themselves* do not explain agrarian change. Several state interventions here proved strategic . . . and analysis (points to) the fundamental role of politics for economic change' (1990, p. 14).

. . . .

Conclusion

In the earlier part of this review I argued that the crucial changes which have occurred in the agrarian economy of India in the recent past are the increasing numbers of and area under marginal and small holdings; the increasing importance of casual wage labour and the relative decline in significance of 'self-employment'; and the tightening of rural labour markets, for which there is some macro-level, all-India evidence, as well as a good deal of evidence from micro-level research. This 'tightening', which has apparently occurred rather generally, and has resulted in some increases in real wages, has been brought about as much by changes in non-agriculture as in agriculture itself. These trends have been influenced by state interventions, including notably, those intended to bring about agrarian reform, public employment programmes, and the supply of subsidised credit. Such intervention in turn reflects the 'mode of domination' which obtains in India.

It is held by a range of scholars and activists that the Indian state is characterised by a compromise of power between the bourgeoisie and the rich peasantry, and perhaps public sector professionals. And for all the possible imperfections of Indian parliamentary democracy, regimes, which do not necessarily represent these dominant class interests directly, do need to mobilise and maintain political support through elections. Whereas at one time it may well have been the case that political power depended on command over 'vote banks' controlled by local power brokers (in the system eloquently described by F. G. Bailey (1963)) there is strong evidence that this is no longer so, and that regimes have sought to manage their legitimacy through the pursuit of populist politics which have yielded the kinds of results to which I have referred. The recently published research of Marguerite Robinson, carried out in part of Andhra Pradesh over 15 years, and covering altogether a period of 25 years of local politics (Robinson, 1988) demonstrates this persuasively.

Robinson shows that

> The concept of the vote bank is appropriate for votes as they were cast from 1957 to 1972. (Within the region) a single dynamic ensured both that the vote banks of local leaders B and C went to candidate A, and that B and C were able to deliver their respective vote banks as promised . . . (but) . . . The situation was altered during the mid-1970s: both the economic position and the political power of the Mallannapalle

landlords changed significantly. While still powerful, they were shown to be vulnerable; as a result the poor have begun to believe that their misery may not be inevitable (1988, p. 16).

Robinson's argument is that political change and economic change are interrelated. In the 1970s state interventions driven mainly by central government – agrarian reform and debt legislation, legislation to abolish bonded labour, and the increased supply of cheap credit – for all their weaknesses and limitations, have increased the 'countervailing power' of labourers and poor peasants and weakened the landlords. Thus she reports:

> What landlords fear most is a population without fear. As Lakshma said about labour (under the working conditions he provided): 'Without fear people will not work'. He was right. Credit for productive activities among the rural poor, despite all the difficulties of implementation, has played an essential role in this regard' (1988, p. 261).

In the same period there have been major changes in the way in which the political system works at state and regional levels – changes brought about above all, in my view, by Mrs Gandhi's dismantling of the Congress 'Dominant Party System' in the early 1970s (see, for example, Manor, 1983). Robinson reports that, although 'the pattern of local politics which underlay the elections in Narsipur taluk between 1957 and 1972 depended upon the delivery of vote banks to candidates by village leaders . . . the 1977 Parliamentary elections were structured differently' (1988, p. 248):

> The components themselves were undergoing change (in varying degrees): landowners were both losing land and competing for labour; the moneylenders had to compete with government credit sources; the Harijans no longer supplied force on demand. In addition, the new *sarpanch*, an agricultural labourer of middle caste rank, represented for the first time in the political arena the 'other' two-thirds of the taluk's population: those who are not landowners, Komatis, or Harijans. Under these circumstances, the village vote bank collapsed and people voted – or did not vote – as individuals (1988, p. 248).

Thereafter voters watched for performance from their candidates and 'they can be expected to change their votes if not satisfied' (1988, p. 265). Thus economic changes and political changes have tended to reinforce each other. Once the old system of managing political power was broken politicians came under increased pressure to deliver, even unto the poor. . . .

I am not claiming that India has a democratic, socialist regime. Rather am I suggesting that the way in which political management has been achieved in India in the recent past has introduced a systemic inclination to populist policies – with the kinds of consequences at the macro-economic level which are finally becoming apparent. There is sufficient evidence from different parts of the country, in addition to the inferences which we may draw from macro-economic trends, to show that the pattern of politics described by Robinson obtains quite widely, even though – as is clearly shown in the accounts given by

Brass and by Breman of the exercise of power over labour – the dominant rural class has surely not been weakened everywhere to the extent that it has in Medak District. Jodha's re-studies of villages in Rajasthan yield findings similar to Robinson's concerning the improvement in the political position of 'the poor' (Jodha, 1989); so do my own from northern Tamil Nadu (Harriss, 1982a, 1985); so do Cain's researches in the ICRISAT villages in the semi-arid region of peninsular India (1981).

In conclusion, therefore, debates over agrarian structure in India have tended to ignore the fact – obvious though it is once it is pointed out – that agrarian processes must be understood in the context of larger political economic forces. Local power structures are themselves influenced by circumstances such as demographic change, price levels and legislative interventions which are beyond their reach, and which are significant in determining productivity and patterns of growth and change (Herring, 1984). It is also true of course that the state is influenced by local processes, but it can never be simply an 'aspect' or reflection of them. The nature of the state and the relations between the state and local society exercise a powerful influence upon agrarian structure and processes of agrarian social differentiation.

In the case of Tamil Nadu, the Dravidian parties which have been in power since 1967 – save for now two periods of President's Rule – have managed political support and maintained their legitimacy by the pursuit of populist politics, which have encouraged the adoption of costly welfare programmes like the Noon Meals Scheme, and made them susceptible to demands for the continued subsidisation of rural electricity and rural credit – even at the cost of accumulation within the state. (I think it is not far-fetched to link the declining status of Tamil Nadu as an industrial centre to the opportunity costs of populist rural welfarism in terms of industrial power generation foregone; or indeed the poor agricultural growth performance of the state as a whole to the preference for welfare subsidy rather than improvement in irrigation or dryland agriculture.) Political management does not depend upon local big men or faction leaders but upon ideological and material appeal 'to the people'. It has thus encouraged the mobilisation of the broad-based demands from rural people which are championed by Farmers' Movements elsewhere; and has helped to blunt though not eliminate polarising tendencies in rural society. The structure of agrarian social relations would be different in the context of a different sort of regime with another mode of political management – as is the case, for example, in Bangladesh where, according to Cain's analysis (1981) the political system allows, even encourages, the continuing accumulation/polarisation of land ownership by local bosses.

As Gillian Hart has argued 'The neglect of power and politics results in an almost exclusive focus on commercialisation and technology as the main sources of rural change and portrays agrarian change as a unilinear process leading to a determinate outcome' (1989, p. 31) – when we should rather take a dialectical view of change.

Selected References

Ahluwalia, M. 1978: Rural Poverty and Agricultural Performance in India. *Journal of Development Studies* 14 (3), 298–323.

——, 1985: Rural Poverty, Agricultural Production and Prices: a reexamination, in Mellor J. and Desai, G. (eds), *Agricultural Change and Rural Poverty*. Baltimore, MD and London: Johns Hopkins UP.

Athreya, V., Djurfeldt, G. and Lindberg, S. 1990: *Barriers Broken: Production Relations and Agrarian Change in Tamil Nadu*. Delhi, Newbury Park and London: Sage Publications.

Attwood, D. 1979: Why Some of the Poor Get Richer: Economic Change and Mobility in Rural Western India. *Current Anthropology* 20, 495–516.

Banaji, J. 1977: Capitalist Domination and the Small Peasantry: Deccan districts in the late Nineteenth Century. *Economic and Political Weekly* 12 (33–34), 1375–404.

Bardhan, P. and Rudra, A. 1986: Labour Mobility and the Boundaries of the Village Moral Economy. *Journal of Peasant Studies* 13, 90–115.

Basant, R. 1987: *Agricultural Technology and Employment in India: A Survey of Recent Research. Economic and Political Weekly* 1297–1308, 1348–64.

Bhaduri, A. 1983: *The Economic Structure of Backward Agriculture*. London: Academic Press.

——, 1986: Forced commerce and agrarian growth. *World Development* 14.

——, *et al.* 1986: Persistence and Polarisation: A Study in the dynamics of agrarian contradiction. *Journal of Peasant Studies* 13, 82–9.

Bhalla, S. 1987: Trends in Employment in Indian Agriculture, Land and Asset Distribution. *Indian Journal of Agricultural Economics* 42 (4), 537–60.

Bharadwaj, K. 1985: A View on Commercialisation in Indian Agriculture and the Development of Capitalism. *Journal of Peasant Studies* 12 (4), 7–25.

Booth, D. 1985: Marxism and Development Sociology: interpreting the impasse. *World Development* 13, 761–88.

Bouton, M. 1985: *Agrarian Radicalism in South India*. Princeton, NJ: University Press.

Boyce, J. 1987: *Agrarian Impasse in Bengal: Institutional Constraints to Technological Change*. Oxford: OUP.

Brass, T. 1990: Class Struggle and the Deproletarianisation of Agricultural Labour in Haryana (India). *Journal of Peasant Studies* 18 (1), 36–67.

Breman, J. 1985: *Of Peasants, Migrants and Paupers: rural labour circulation and capitalist production in west India*. New Delhi: OUP.

Byres, T. J. 1981: The New Technology, Class Formation, and Class Action in the Indian Countryside. *Journal of Peasant Studies* 8 (4), 405–54.

Cain, M. 1981: Risk and Insurance: Perspectives on Fertility and Agrarian Change in India and Bangladesh. *Population and Development Review* 7 (3), 435–74.

Drèze, J. and Mukherjee, A. 1987: Labour Contracts in Rural India: Theories and Evidence, Discussion Paper No. 7, Development Research Programme, Suntory Toyota International Centre for Economics and Related Disciplines, London School of Economics.

Frank, A. G. and Fuentes, M. 1987: Nine Theses on Social Movements. *Economic and Political Weekly* 22, 1503–1510.

Gaiha, R. 1989: Poverty, Agricultural Production and Prices in Rural India – A Reformulation. *Cambridge Journal of Economics* 13, 333–52.

Ghosh, A. 1989: Rural Poverty and Relative Prices in India. *Cambridge Journal of Economics* 13, 307–31.
Gough, K. 1989: *Rural Change in Southeast India 1950s to 1980s*. New Delhi: OUP.
Harriss, B. 1981: *Transitional Trade and Rural Development*. Delhi: Vikas.
——, 1983: Paddy and Rice Marketing in a Bengal District. *Cressida Transactions* 2, 77ff.
—— and Harriss, J. 1984: 'Generative' and 'Parasitic' Urbanism? Some Observations from the Recent history of a South Indian Market Town. *Journal of Development Studies* 20 (3), 82–101.
Harriss, J. 1980: Contemporary Marxist Analysis of the Agrarian Question in India. Madras Institute of Development Studies, Discussion Paper.
——, 1982a: *Capitalism and Peasant Farming: Agrarian Structure and Ideology in Northern Tamil Nadu*. Bombay: OUP.
——, 1982b: Character of an Urban Economy, 'Small-Scale' Production and Labour Markets in Combatore. *Economic and Political Weekly* 17 (23 and 24), 945–54 and 993–1002.
——, 1983: Making out on Limited Resources, what happened to semi-feudalism in West Bengal. CRESSIDA Transactions 2, 16–76.
——, 1985: What Happened to the Green Revolution in South India? Economic Trends, Household Mobility and the Politics of an 'Awkward Class', University of East Anglia, School of Development Studies, Discussion Paper No 175.
——, 1987: Capitalism and Peasant Production: the Green Revolution in India, in T. Shanin (ed.), *Peasants and Peasant Societies* (2nd edn), Oxford: Blackwell.
——, 1989a: Knowing About Rural Economic Change: Problems Arising from a Comparison of The Results of 'Macro' and 'Micro' Research in Tamil Nadu, in P. K. Bardhan (ed.), *Conversations between Economists and Anthropologists*. Delhi: OUP.
——, 1989b: Indian Industrialisation and the State, in H. Alavi and J. Harriss (eds), *Sociology of Developing Societies: South Asia*. London: Macmillan.
——, 1991: The Green Revolution in North Arcot: Economic Trends, Household Mobility, and the Politics of an 'Awkward Class, in P. Hazell and C. Ramasamy (eds.), *The Green Revolution Reconsidered: The Impact of High-Yielding Rice Varieties in South India*, Baltimore, MD and London: Johns Hopkins UP.
Hart, G. 1986: Exclusionary Labour Arrangements: Interpreting Evidence and Employment Trends in Rural Java. *Journal of Development Studies* 22 (4), 681–96.
——, 1989: Introductory essay in G. Hart *et al.* (eds).
——, Turton, A. and White, B. 1989: *Agrarian Transformations: The State and Local Processes in Southeast Asia*. Berkeley, CA: University of California Press.
Hartmann, B. and Boyce, J. 1983: *A Quiet Violence: View from a Bangladesh Village*. London: Zed Books.
Herring, R. 1984: Economic Consequences of Local Power Configurations in Rural South Asia, in M. Desai *et al.* (eds), *Agrarian Power and Agricultural Productivity in South Asia*. Delhi: OUP.
Islam, R. 1986: Non-Farm Employment in Rural Asia: Issues and Evidence, in R. T. Shand (ed.) *Off-farm Employment in the Development of Rural Asia*. Canberra: ANU.
Jodha, N. S. 1989: Social Science Research on Rural Change: Some Gaps, in P. K. Bardhan (ed.). *Conversations between Economists and Anthropologists*. Delhi: OUP.

Jose, A. V. 1988: Agricultural Wages in India. *Economic and Political Weekly*. Review of Agriculture, June, A-46–A-58.

Kapadia, K. 1990: Gender, Caste and Class in Rural South India, Ph.D. thesis. University of London.

Kurien, C. T. 1980: Dynamics of Rural Transformation: A Case Study of Tamil Nadu. *Economic and Political Weekly*, Annual Number, 365–90.

Lieten, G. K. 1990: Depeasantisation Discontinued: Land Reforms in West Bengal. *Economic and Political Weekly* 25 (40), 2265–71.

Lindberg, S. 1990: Civil Society against the State: Farmers' Agitations and New Social Movements in India, paper to the Seventh Annual Conference of the Nordic Association for Southeast Asian Studies, and the XII World Congress of Sociology, Madrid.

Lipton, M. with Longhurst, R. 1989: *New Seeds and Poor People*. London: Unwin Hyman.

Manor, J. 1983: The Electoral Process amid Awakening and Decay: Reflections on the Indian General Elections of 1980, in P. Lyon and J. Manor (eds), *Transfer and Transformation: Political Institutions in the New Commonwealth*. Leicester: Leicester UP.

Mellor, J. 1976: *The New Economics of Growth*. Ithaca, NY: Cornell UP.

——, and Johston, B. F. 1984: The World Food Equation: Interrelations among Development, Employment and Food Consumption. *Journal of Economic Literature* 22, 531–74.

Minhas, B. S. and Mazumdar, G. 1987: Unemployment and Casual Labour in India: An Analysis of Recent NSS data. *Indian Journal of Industrial Relations* 22, (3), 237–53.

Mundle, S. 1990: Food, Finance and Foreign Trade: The Limits of High Growth in India, seminar paper, presented at the University of East Anglia.

Moore, M. 1985: *The State and Peasant Politics in Sri Lanka*. Cambridge: CUP.

Omvedt, G. 1983: Capitalist Agriculture and Rural Classes in India. *Bulletin of Concerned Asian Scholars* 15 (3), 30–54.

——, 1988: The 'New Peasant Movement' in India. *Bulletin of Concerned Asian Scholars* 20 (2), 14–23.

Patnaik, U, 1987: *Peasant Class Differentiation: A Study in Method with Reference to Haryana*. Delhi: OUP.

Ramachandran, V. K. 1990: *Wage Labour and Unfreedom in Agriculture: An Indian Case Study*. Oxford: OUP.

Robinson, M. 1988: *Local Politics: The Law of the Fishes*. Delhi: OUP.

Rudolph, L. I. and Rudolph, S. H. 1984: Determinants and Varieties of Agrarian Mobilisation, in M. Desai *et al.* (eds), *Agrarian Power and Agricultural Productivity in South Asia*. Delhi: OUP.

—— and ——, 1987: *In Pursuit of Lakshmi: The Political Economy of the Indian State*. IL Chicago: Chicago UP.

Rudra, A. 1984: Local Power and Farm Level Decision-making, in M. Desai *et al.* (eds) *Agrarian Power and Agricultural Productivity in South Asia*. Delhi: OUP.

—— and Bardhan, P. 1983: *Agrarian Relations in West Bengal: Results of Two Surveys*. Bombay: Somaiya Publications.

Sanyal, S. K. 1988: Trends in Landholding and Poverty in Rural India, in T. N. Srinivasan and P. K. Bardhan, *Rural Poverty in South Asia*. New York: Columbia UP.

Singh, I. 1988: *Small Farmers in South Asia; Tenancy in South Asia; Land and Labour in South Asia*. Discussion Papers 31–33. Washington, DC: The World Bank.

Sundaram, K. and Tendulkar, S. D. 1988: Towards an Explanation of Interregional Variations in Poverty and Unemployment in India, in T. N. Srinivasan and P. K. Bardhan, *Rural Poverty in South Asia*. New York: Columbia UP.

Thorner, A. 1982: Semi-feudalism or Capitalism? Contemporary Debate on Classes and Modes of Production in India. *Economic and Political Weekly* 17, 49, 50 and 51, 1961–68, 1993–99, 2061–65.

Thorner, B. 1956: *The Agrarian Prospect in India* (Second Edition 1976). New Delhi: Allied Publishers.

Vaidyanathan, A. 1986: Labour Use in Rural India: A Study of Spatial and Temporal Variation. *Economic and Political Weekly*. Review of Agriculture, Dec., pp. A-130–A-146.

——, 1988: *India's Agricultural Development in a Regional Perspective*. Madras: Sangam Books.

Walker, T. S. and Ryan, J. G. 1990: *Village and Household Economies in India's Semi-Arid Tropics*, Baltimore, MD and London: Johns Hopkins UP.

Washbrook, D. 1978: Economic Development and Social Stratification in Rural Madras: The 'Dry Region', 1878–1929, in C. Dewey and A. Hopkins (eds), *The Imperial Impact: Studies in the Economic History of Africa and India*, London: Athlone Press.

——, 1988: Progress and Problems: South Asian Economic and Social History c. 1720–1860. *Modern Asian Studies* 22 (1), 57–96.

Westergard, K. 1985: *State and Rural Society in Bangladesh: A Study in Relationships*. London: Curzon Press.

White, B. 1989: Problems in the Empirical Analysis of Agrarian Differentiation, in G. Hart *et al.* (eds) 15–30).

Wood, G. 1981: Rural Class Formation in Bangladesh 1940–1980. *Bulletin of Concerned Asian Scholars* 13 (4), 2–17.

9 Robert H. Bates,

'Governments and Agricultural Markets in Africa'

Excerpts from: R. Bates (ed.), *Toward a Political Economy of Development*, chapter 10. Berkeley: University of California Press (1988)

Governments in Africa intervene in agricultural markets in characteristic ways: they tend to lower the prices offered for agricultural commodities, and they tend to increase the prices that farmers must pay for the goods they buy for consumption. And although African governments do subsidize the prices farmers pay for the goods they use in farming, the benefits of these subsidies are appropriated by the rich few: the small minority of large-scale farmers.

Other patterns, too, are characteristic of government market intervention. Insofar as African governments seek increased farm production, their policies are project-based rather than price-based. Insofar as they employ prices to strengthen production incentives, they tend to encourage production by lowering the prices of inputs (that is, by lowering costs) rather than by increasing the prices of products (that is, by increasing revenues). A last characteristic is that governments intervene in ways that promote economic inefficiency: they alter market prices, reduce market competition, and invest in poorly conceived agricultural projects. In all of these actions, it should be stressed, the conduct of African governments resembles the conduct of governments in other parts of the developing world.

One purpose of this paper is to describe more fully these patterns of government intervention. A second is to examine a variety of explanations for this behavior.

The regulation of commodity markets

It is useful to distinguish between two kinds of agricultural commodities: food crops, many of which can be directly consumed on the farm, and cash crops, few of which are directly consumable and which are instead marketed as a source of cash income. Many cash crops are in fact exported; they provide not only cash incomes for farm families but also foreign exchange for the national economies of Africa.

Export crops

An important feature of the African economies is the nature of the marketing systems employed for the purchase and export of cash crops. The crops are grown by private farm families, but they are then sold through official, state-controlled marketing channels. At the local level, these channels may take the form of licensed agents or registered private buyers; they may also take the form of cooperative societies or farmers' associations. But the regulated nature of the marketing system is clearly revealed in the fact that these primary purchasing agencies can in most cases only sell to one purchaser: a state-owned body, commonly known as the marketing board.

Background

The origins of these boards are diverse. In some cases, particularly in the former settler territories, they were formed by farmers themselves. At the time of the Great Depression, commercial farmers banded together in efforts to stabilize the markets for cash crops; in effect, with the support of the colonial states they dominated, they sought to create producer-run cartels. More commonly, the origins of the marketing boards lay in an alternative source of cartel formation: in the efforts of the purchasers and exporters of cash crops to dominate the market and to force lower prices on farmers.[1]

In either case, World War II led to the institutionalization of the regulation of export markets. During the war, Britain sought to procure agricultural commodities and raw materials from her colonial dependencies. Some materials, such as food for troops in North Africa, were needed for the war effort; others were needed to generate foreign exchange for the purchase of armaments from North America; and the purchase of still other goods was required to provide prosperity for the colonial areas and thereby to lessen the likelihood of political instability at a time when British armed forces were already spread perilously thin. To secure the regularized purchase of raw materials, the British government created a ministry of supply. The ministry signed bulk-purchasing agreements with the colonial governments in each of the African territories. To administer the terms of these agreements, the colonial authorities created official marketing agencies. In those territories in which large-scale producers had already begun to operate market-stabilizing schemes, the producer associations running these schemes were recruited to staff and administer the state marketing boards. In the territories where purchasers' cartels held a predominance of market power, the state procurement schemes in effect conferred legal standing on the merchant-based cartels; the cartels became the instruments for securing raw materials.[2]

In either case, upon independence many African governments found themselves the inheritors of bureaucracies that held an official monopoly over the purchase and export of commodities in the most valuable sector of their domestic economies. These new states possessed extremely powerful instruments of market intervention. They could purchase export crops at an administratively set, low domestic price; they could then market these crops at the prevailing world price; and they could accumulate the revenues generated by the difference between the domestic price at which the goods were purchased and the world price at which they were sold.

Government taxation

Initially, the revenues accumulated by the marketing boards were to be used for the benefit of the farmers, in the form of price assistance funds. At times of

[1] P. T. Bauer, *West Africa Trade* (London: Routledge and Kegan Paul, 1964); William O. Jones, 'Agricultural Trade within Tropical Africa: Historical Background,' in Robert H. Bates and Michael F. Lofchie, eds, *Agricultural Development in Africa: Issues of Public Policy* (New York: Praeger, 1980).

[2] Charlotte Leubuscher, *Bulk Buying from the Colonies: A Study of the Bulk Purchase of Colonial Commodities by the United Kingdom Government* (London: Oxford University Press, 1956); Elspeth Huxley, *No Easy Way: A History of the Kenya Farmers Association and Unga Limited* (Nairobi: private printing, 1957), pp. 137ff.; and Bauer, *West Africa Trade*.

low international prices, these funds were to be employed to support domestic prices and so to shelter the farmers from the vagaries of the world market. For example, 70 percent of the western Nigerian marketing board's revenues were to be retained for such purposes. But commitments to employ the funds for the benefit of the farmers proved short-lived. They were overborne by ambitions to implement development programs and by political pressures on governments from nonagricultural sectors of the economy.

The Cotton Price Assistance Fund, for example, was accumulated by the Lint Marketing Board in Uganda. In the 1950s it was employed to stabilize prices, but thereafter it was increasingly used for other purposes. In the pre-independence period, for example, it was used to secure revenues for the building of the Owen's Falls Dam; although the fund purchased shares in the Uganda Electricity Board, the agency responsible for the dam, it has received no dividends from these shares (and they have declined in value). In the 1960s, the fund 'lent' 100 million Uganda shillings to the government for investment in the capital budget, interest free. Still later, it was employed to capitalize the Cooperative Development Bank by contributing twelve million shillings interest free, repayable over thirty-five years.[3]

In West Africa, too, the revenues of the marketing boards were increasingly diverted to uses other than the stabilization of farmers' incomes. In Nigeria, for example, funds were first loaned to the regional governments; later, they were given to these governments in the form of grants; later still, the legislation governing the use of these revenues was altered such that the boards became instruments of direct taxation.[4] We have already noted that the statutes governing the marketing boards in western Nigeria reserved 70 percent of the trading surpluses for price stabilization; an additional 7.5 percent was to be employed for agricultural research, and the remaining 22.5 percent for general development purposes. But Helleiner notes that, following self-government,

> the Western Region's 1955–1960 development plan announced . . . the abandonment of the '70–22.5–7.5' formula for distribution of the Western Board's right to contribute to development, and provided for 20 million in loans and grants to come from the Board for the use of the Regional Government during the plan. . . . [The Board] was now obviously intended to run a trading surplus to finance the regional Government's program. The Western Region Marketing Board had by now become . . . a fiscal arm of the Western Nigerian Government.[5]

This transition was followed as well in Ghana, where 'the government decided to remove . . . legal restrictions on its access to the funds of the Board.'[6]

[3] Uganda, Treasury Department, 'Statement of Cotton Price Assistance Fund at 31st October 1977,' 11 November 1977 (typescript); David Walker and Cyril Ehrlich, 'Stabilization and Development Policy in Uganda: An Appraisal,' *Kyklos* 12 (1959): 341–53.

[4] H. M. A. Onitiri and Dupe Olatunbosun, *The Marketing Board System* (Ibadan: Nigerian Institute of Social and Economic Research, 1974).

[5] Gerald K. Helleiner, *Peasant Agriculture, Government, and Economic Growth in Nigeria* (Homewood, Ill.: Richard D. Irwin, 1960), pp. 170–1.

[6] Bjorn Beckman, *Organizing the Farmers: Cocoa Politics and National Development in Ghana* (New York: Holmes and Meier, 1976), p. 199.

The movement from an instrument of price stabilization largely for the benefit of farmers to an instrument of taxation with the diversion of revenues to nonfarm sectors can be seen as well in changes in the price-stabilizing policies employed by the marketing boards. Investigations clearly suggest that what was stabilized was not the domestic price paid to farmers but, rather, the difference between the domestic and the world price, that is, the off-take from the farmer which was appropriated by the government.[7]

Food crops

African governments also intervene in the market for food crops. And, once again, they tend to do so in ways that lower the prices of agricultural commodities.

One way African governments attempt to secure food cheaply is by constructing bureaucracies to purchase food crops at government-mandated prices. A recent study by the United States Department of Agriculture examined the marketing systems for food crops in Africa and discovered a high incidence of government market intervention. In the case of three of the food crops studied, in over 50 percent of the countries in which the crop was grown the government had imposed a system of producer price controls, and in over 20 percent the government maintained an official monopsony for the purchase of that food crop (Table 9.1); in these instances, the government was by law the sole buyer of the crop.

Table 9.1 Patterns of Market Intervention for Food Crops in Africa

	Countries in which crop is grown	*Countries with producer price controls*		*Countries with legal monopoly*	
	Number	*Number*	*%*	*Number*	*%*
Rice	26	25	96	11	42
Wheat	12	8	67	4	33
Millet and sorghum	38	9	24	7	18
Maize	35	24	69	9	26
Roots and tubers	33	6	18	1	3

Source: United States Department of Agriculture, *Food Problems and Prospects in SubSaharan Africa* (Washington, D.C.: USDA, 1980), p. 173.

Regulation of food markets entails policing the purchase and movement of food stocks and controlling the storage, processing, and retail marketing of food. An illustration is offered by the maize industry of Kenya; according to subsection 1 of section 15 of the Maize Marketing Act, 'All maize grown in

[7] See Bates, *Markets and States in Tropical Africa* (Berkeley: University of California, 1981).

Kenya shall, subject to the provision of this Act, be purchased by and sold to the Board, and shall, without prejudice to the Board's liability for the price payable in accordance with section 18 of this Act, rest in the Board as soon as it has been harvested.'

According to the Maize Marketing (Movement of Maize and Maize Products) Order, all movements of maize require a movement permit, which is only valid for twenty-four hours and which must be obtained from the Maize and Produce Board. The sole exceptions are the movement of maize or maize products within the boundaries of the farm; the movement of not more than two bags (180 kg) accompanied by the owner; and the movement of not more than ten bags within the boundaries of a district, accompanied by the owner and intended for consumption by the owner or his family.[8]

The controls over the market for food crops increase the costs of marketing. In part, this is because the government-imposed barriers to entry confer excess profits on the public agents who operate in the market. Their nature and magnitude are perhaps most vividly illustrated by the bribes they extract from farmers and traders.[9] A second major consequence of the regulated maize market is that many consumers pay higher prices and many producers receive lower prices than would be the case were maize to be moved more easily between places and over time.

More directly relevant to the concerns of this paper, however, is the impact of food marketing controls on food prices. For insight into this subject we can turn to Doris Jansen Dodge's study of NAMBoard, the food marketing bureaucracy in Zambia. Over the years studied by Dodge (1966–1967 to 1974–1975) NAMBoard depressed the price of maize by as much as 85 percent; that is, in the absence of government controls over maize movements, the farmers could have gotten up to 85 percent higher prices for their maize than they were able to secure under the market controls imposed by NAMBoard. Gerrard extends Dodge's finding for Zambia to Kenya, Tanzania, and Malawi; Dodge herself extends them to eight other African countries.[10] The result is a weakening of incentives to produce food.

Projects

In order to keep food prices low, governments take additional measures. In particular, they attempt to increase food supplies. This can be done either by

[8] Guenter Schmidt, 'Maize and Beans in Kenya: The Interaction and Effectiveness of the Informal and Formal Marketing Systems,' Institute for Development Studies, University of Nairobi, Occasional Paper No. 31, 1979.

[9] *Ibid.*, p. 68.

[10] See Doris J. Janzen, 'Agricultural Pricing Policy in Sub-Saharan Africa of the 1970s,' unpublished paper, 1980 (mimeo.); Christopher David Gerrard, 'Economic Development, Government Controlled Markets, and External Trade in Food Grains: The Case of Four Countries in East Africa,' Ph.D. dissertation, University of Minnesota, 1981; and Doris Jansen Dodge, *Agricultural Policy and Performance in Zambia* (Berkeley: Institute of International Studies, 1977).

importing food or by investing in food production projects. Foreign exchange, however, is scarce; especially since the rise of petroleum prices, the cost of imports is high. So as to conserve foreign exchange, then, African governments attempt to become self-sufficient in food. To keep prices low, they invest in projects that will yield increased food production.

In some cases, governments turn public institutions into food production units: youth-league farms and prison farms provide illustrative cases. In other instances, they attempt to provide factors of production. In Africa, water is commonly scarce and governments invest heavily in river-basin development schemes and irrigation projects. Capital equipment is also scarce; by purchasing and operating farm machinery, governments attempt to promote farm production. Some governments invest in projects to provide particular crops: rice in Kenya, for example, or wheat in Tanzania. In other instances, governments divert large portions of their capital budgets to the financing of food production schemes. Western Nigeria, for example, spent over 50 percent of the Ministry of Agriculture's capital budget on state farms over the period of the 1962–1968 development program.[11]

Nonbureaucratic forms of intervention

Thus far I have emphasized direct forms of government intervention. But there is an equally important, less direct form of intervention: the overvaluation of the domestic currency.

Most governments in Africa maintain an overvalued currency.[12] Foreign money therefore exchanges for fewer units of local currency. A result is to lower the prices received by the exporters of cash crops. For a given sum earned abroad, the exporters of cash crops receive fewer units of the domestic currency. In part, overvaluation inflicts losses on governments; deriving a portion of their revenues from taxes levied by the marketing boards, the governments command less domestic purchasing power as a result of overvaluation. But because their instruments of taxation are monopolistic agencies, African governments are able to transfer much of the burden of overvaluation: they pass it on to farmers, in the form of lower prices.

In addition to lowering the earnings of export agriculture, overvaluation lowers the prices paid for foreign imports. This is, of course, part of the rationale for a policy of overvaluation: it cheapens the costs of importing plant, machinery, and other capital equipment needed to build an industrial sector.

[11] Frances Hill, 'Experiments with a Public Sector Peasantry,' *African Studies Review* 20 (1977): 25–41; and Werner Roider, *Farm Settlements for Socio-Economic Development: The Western Nigerian Case* (Munich: Weltforum, 1971).

[12] International Bank for Reconstruction and Development, *Accelerated Development in Sub-Saharan Africa: An Agenda for Action* (Washington, D.C.: IBRD, 1981); and Franz Pick, *Pick's Currency Yearbook, 1976–1977* (New York: Pick Publishing, 1978).

But items other than plant and equipment can be imported, and among these other commodities is food. As a consequence of overvaluation, African food producers face higher levels of competition from foreign foodstuffs. And in search of low-price food, African governments do little to protect their domestic food markets from foreign products – products whose prices have artificially been lowered as a consequence of public policies.

Industrial goods

In the markets for the crops they produce, African farmers face a variety of government policies that serve to lower farm prices. In the markets for the goods that they consume, however, they face a highly contrasting situation: they confront prices for consumers that are supported by government policy.

In promoting industrial development, African governments adopt commercial policies that shelter local industries from foreign competition. To some degree they impose tariff barriers between the local and the international markets. To an even greater extent, they employ quantitative restrictions. Quotas, import licenses, and permits to acquire and use foreign exchange are all employed to conserve foreign exchange, on the one hand, while, on the other, protecting the domestic market for local industries. In connection with the maintenance of overvalued currencies, these trade barriers create incentives for investors to import capital equipment and to manufacture domestically goods that formerly had been imported from abroad.[13]

Not only do government policies shelter industries from low-cost foreign competition, they shelter them from domestic competition as well. In part, protection from domestic competition is a by-product of protection from foreign competition. The policy of allocating licenses to import in conformity with historic market shares provides an example of such a measure. The limitation of competition results from other policies as well. In exchange for commitments to invest, governments guarantee periods of freedom from competition. Moreover, governments tend to favor larger projects; seeking infusions of scarce capital, they tend to back the proposals that promise the largest capital investments. With the small markets typical of most African nations, the result is that investors create plants whose output represents a very large fraction of the domestic market; a small number of firms thus come to dominate the market. Finally, particularly where state enterprises are concerned, governments sometimes confer virtual monopoly rights upon particular enterprises.

[13] J. Dirck Stryker, 'Ghana Agriculture,' paper for the West African Regional Project, 1975 (mimeo.); Scott R. Pearson, Gerald C. Nelson, and J. Dirck Stryker, 'Incentives and Comparative Advantage in Ghanaian Industry and Agriculture,' paper for the West African Regional Project, 1976 (mimeo.); International Bank for Reconstruction and Development, *Kenya: Into the Second Decade* (Washington, D.C.: IBRD, 1975); and International Bank for Reconstruction and Development, *Ivory Coast: The Challenge of Success* (Washington, D.C.: IBRD, 1978).

The consequence of all these measures is to shelter industries from domestic competition.

One result is that inefficient firms survive. Estimates of the use of industrial capacity range as low as one-fifth of the single-shift capacity of installed plant.[14] Another consequence is that prices rise. Protected from foreign competition and operating in noncompetitive market settings, firms are able to charge prices that enable them to survive despite operating at very high levels of cost.

Farm inputs

By depressing the prices offered farmers for the goods they sell, government policies lower the revenues of farmers. By raising the prices that consumers – including farmers – must pay, governments reduce the real value of farm revenues still further. As a consequence of these interventions by governments, then, African farmers are taxed. Oddly enough, while taxing farmers in the market for products, governments subsidize them in the market for farm outputs.

Attempts to lower input prices take various forms. Governments provide subsidies for seeds and fertilizers, the level of the latter running from 50 percent, in Kenya, to 80 percent, in Nigeria. They provide tractor-hire services at subsidized rates – up to 50 percent of the real costs in Ghana in the mid-1970s.[15] They provide loans at subsidized rates of interest for the purchase and rental of inputs. And they provide highly favorable tax treatment for major investors in commercial farming ventures.[16] Moreover, through their power over property rights African governments have released land and water to commercial farmers at costs that lie below the value they would generate in alternative uses. The diversion of land to large-scale farmers and of water to private tenants on government irrigation schemes, without paying compensation to those who had employed these resources in subsistence farming, pastoral production, fishing or other ventures, represents the conferring of a subsidy on the commercial farmer – and one that is paid at the expense of the small-scale, traditional producer. This process has been documented in northern Ghana,[17]

[14] Ghana, *Report of the Commission of Enquiry into the Local Purchasing of Cocoa* (Accra: Government Printer, 1967); and Tony Killick, *Development Economics in Action: A Study of Economic Policies in Ghana* (New York: St. Martin's Press, 1978), p. 171.

[15] Stryker, 'Ghana Agriculture'; C. K. Kline, D. A. G. Green, Roy L. Donahue, and B. A. Stout, *Industrialization in an Open Economy: Nigeria 1945–1966* (Cambridge, England: Cambridge University Press).

[16] See, for example, David Onaburekhale Ekhomu, 'National Food Policies and Bureaucracies in Nigeria: Legitimization, Implementation, and Evaluation,' paper presented at the African Studies Association Convention, Baltimore, Maryland, 1978 (mimeo.).

[17] *West Africa*, 3 April 1978.

Nigeria,[18] Kenya,[19] Ethiopia,[20] and Senegal.[21] It was, of course, common in settler Africa as well.

In the case of land and water use, then, a major effect of government intervention in the market for inputs is to augment the fortunes of large-scale farmers at the expense of small-scale farmers. To some degree, this is true of programs in support of chemical and mechanized inputs as well. And even where there is no direct redistribution, it is clear that government programs that seek to increase food production by reducing the costs of farming reach only a small segment of the farming population: the large farmers. In part, this is by plan. The programs are aimed at the 'progressive farmers' who will 'make best use of them.' Because the large farmers have the same social background as those who staff the public services, the public servants feel they can work most congenially and productively with these people.[22] Moreover, to favor the large farmer is politically productive. I will elaborate this argument below.

Discussion

Governments intervene in the market for products in an effort to lower prices. They adopt policies which tend to raise the price of the goods farmers buy. And while they attempt to lower the costs of farm inputs, the benefits of this policy are reaped only by a small minority of the richer farmers. Agricultural policies in Africa thus tend to be adverse to the interests of most producers.

Studies in other areas suggest that this configuration of pricing decisions is common in the developing nations.[23] Indeed, it is argued by some that the

[18] Ekhomu, 'National Food Policies'; Janet Girdner and Victor Oloransula, 'National Food Policies and Organizations in Ghana,' paper presented to the annual meeting of the American Political Science Association, New York, 1978.

[19] Apollo I. Njonjo, 'The Africanization of the "White Highlands": A Study in Agrarian Class Struggles in Kenya, 1950–1974,' Ph.D. dissertation, Princeton University, 1977.

[20] John Cohen and Dov Weintraub, *Land and Peasants in Imperial Ethiopia: The Social Background to Revolution* (Assen: Van Gorcum, 1975).

[21] Donal B. Cruise O'Brien, *The Mourides of Senegal: The Political and Economic Organization of an Islamic Brotherhood* (Oxford: Clarendon Press, 1971).

[22] See David M. Leonard, *Reaching the Peasant Farmer: Organization Theory and Practice in Kenya* (Chicago: University of Chicago Press, 1977); and H. U. E. Van Velzen, 'Staff, Kulaks and Peasants,' in Lionel Cliffe and John Saul, eds, *Socialism in Africa*, vol. 2 (Dar es Salaam: East African Publishing House, 1973).

[23] Raj Krishna, 'Agricultural Price Policy and Economic Development,' in M. Southworth and Bruce F. Johnston, eds, *Agricultural Development and Economic Growth* (Ithaca: Cornell University Press); United States General Accounting Office, *Disincentives to Agricultural Production in Developing Countries* (Washington, D.C.: Government Printer, 1975); Carl Gotsch and Gilbert Brow, 'Prices, Taxes and Subsidies in Pakistan Agriculture, 1960–1976,' World Bank Staff Working Paper no. 387 (Washington, D.C.: World Bank, 1980); Keith Griffin, *The Green Revolution: An Economic Analysis* (Geneva: United Nations Research Institute, 1972); Michael Lipton, *Why Poor People Stay Poor: Urban Bias in World Development* (Cambridge, Mass.: Harvard University Press, 1977).

principal problems bedeviling agriculture in the developing areas originate from bad public policies. In the words of Theodore Schultz, given the right incentives, farmers in the developing world would 'turn sand into gold.'[24] 'Distortions' introduced into agricultural markets by governments, he contends, furnish the most important reasons for their failure to do so.[25] While Schultz's position is perhaps an extreme one, it nonetheless underscores the importance of understanding why Third World governments select this characteristic pattern of agricultural policies. In the remaining sections, I will advance several explanations for their choices.

Governments as agents of the public interest

The first approach derives from development economics. Public policy represents a choice by government made out of a regard for what is socially best. The overriding public interest of poor societies is in rapid economic growth. And the policy choices of Third World governments represent their commitment to rapid development, a commitment that implies supplanting agriculture with industry.

In common with most political scientists, I remain skeptical of such a benevolent theory of government. It is therefore unsettling to have to admit that, *confining attention to export crops*, the implications of this approach are consistent with many of the facts.

All the governments of Africa seek industrial development. Most seek to create the social and economic infrastructure necessary for industrial growth and many are committed to the completion of major industrial and manufacturing projects. To fulfill their plans, governments need revenues; they also need foreign exchange. In most of the African nations, agriculture represents the single largest sector in the domestic economy; and in many it represents the principal source of foreign exchange. It is therefore natural that in seeking to fulfill these objectives for their societies, the governments of Africa should intervene in markets in an effort to set prices in a way that transfers resources from agriculture to the 'industrializing' sectors of the economy: the state itself and the urban industrial and manufacturing firms.

An explanation based on the development objectives of African regimes is thus consistent with the choices made in the markets for export goods. It is also consistent with other well-known facts. The policy choices that have been made are, for example, in keeping with the prescriptions propounded in leading development theories. According to these theories, to secure higher levels of per capita income, nations should move from the production of primary

[24] Theodore W. Schultz, *Transforming Traditional Agriculture* (New York: Arno Press, 1976), p. 5.

[25] Theodore W. Schultz, ed., *Distortions of Agricultural Incentives* (Bloomington: Indiana University Press, 1978).

products to the production of manufactured goods. Savings take place out of the profits of industry and not out of the earning of farmers. Resources should therefore be levied from agriculture and channeled into industrial development. And agriculture in the developing areas, it is held, can surrender revenues without a significant decline in production. These were, and remain today, critical assertions in development economics. Many policymakers in Africa were trained by development specialists; and important advocates of these arguments have served as consultants to the development ministries of the new African states. It is therefore credible to account for the policy choices made by African governments – ones that systematically bias the structure of prices against agriculture and in favor of industry – as choices made in accordance with prescriptions of how best to secure the welfare of people in poor societies.

Such an approach ultimately proves unsatisfactory, however, for it fails to generate explanatory power, and where it does offer explanations, they are often wrong. Although the social welfare-maximizing interpretation of government could not be rejected on the basis of the actions exhibited by governments concerning export crops, the deficiencies of this approach become apparent when it is applied to government policies concerning food crops.

To secure social objectives, governments can choose among a wide variety of policy instruments; and knowledge of the public objectives of a program fails to give insight into why a particular policy instrument is chosen. For example, an important objective of African governments is to increase food supplies. To secure more supplies, governments could offer higher prices for food or they could invest the same amount of resources in food production projects. There is every reason to believe that the former is the more efficient way of securing the objective. But governments in Africa systematically prefer project-based policies to price-based policies.

To strengthen the incentives for food production, African governments can increase the price of farm products or subsidize the costs of farm implements. Either would result in higher profits for producers. But governments systematically choose the latter policy.

To increase output, African governments finance food production programs. But given the level of resources devoted to these programs, they often create too many projects; the programs then fail because resources have been spread too thin. Such behavior is nonsensical in terms of the social objectives of the program.

To take a last example: in the face of shortages, governments can either allow prices to rise or maintain lower prices while imposing quotas. In a variety of markets of significance to agricultural producers, African governments exhibit a systematic preference for the use of quotas – a preference that cannot readily be accounted for in terms of their development objectives.

A major problem with an approach that tries to explain agricultural policies in terms of the social objectives of governments, then, is that the social objectives that underlie a policy program fail to account for the particular form the policies assume. The approach thus yields little in the way of explanatory power. A second major difficulty is that when explanations of governmental behavior are made in terms of the objectives of public policy, they often prove false.

This problem is disclosed by the self-defeating nature of many government policies. To secure cheaper food, for example, governments lower prices to producers; but this only creates shortages which lead to *higher* food prices. To increase resources with which to finance programs of development, governments increase agricultural taxes; but this leads to declines in production and to shortfalls in public finances and foreign exchange. And to secure rapid development, governments seek to transfer resources from agriculture to industry; but this set of policies has instead led to reduced rates of growth and to economic stagnation.

The policy instruments chosen to secure social objectives are thus often inconsistent with the attainment of these objectives. And yet the choices of governments are clearly stable; despite undermining their own goals, governments continue to employ these policy instruments. Some kind of explanation is required, but one based on factors other than the social objectives of governments.

Governments as agents of private interests

An alternative approach would not view governments as agencies that maximize the social welfare. Rather, it would view them as agencies that serve private interests. And, rather than interpreting government policies as choices made out of a regard for the public interest, it would instead view them as decisions made in order to accommodate the demands of organized private interests. This approach would view public policy as the outcome of political pressures exerted by groups that seek satisfaction of their private interests from political action.

Particularly in the area of food price policy, this approach has much to recommend it. Put bluntly, food policy in Africa appears to represent a form of political settlement, one designed to bring about peaceful relations between African governments and their urban constituents. And it is a settlement in which the costs tend to be borne by the mass of the unorganized: the small-scale farmers.

The urban origins of African food policies are perhaps most clearly seen in Nigeria. If one searches out the historical origins of government food policy in Nigeria, one is drawn to the recommendations of a series of government commissions – the Udoji Commission, the Adebo Commission, and the Anti-Inflation Task Forces, for example – which were impaneled to investigate sources of labor unrest and to resolve major labor stoppages.[26] The fundamental

[26] See Nigeria, Federal Ministry of Information, *Second and Final Report of the Wages and Salaries Review Commission, 1970–71* (Lagos: Ministry of Information, 1971); Nigeria, *Public Service Review Commission* (Lagos: Government Printer, 1974); Nigeria, Federal Ministry of Information, *First Report of the Anti-Inflation Task Force* (Lagos: Government Printer, 1978).

issue driving urban unrest, they noted, was concern with the real value of urban incomes and the erosion of purchasing power because of inflation. While recommending higher wages, these commissions also noted that pay increases represented only a short-run solution; in the words of the Adebo Commission, 'It was clear to us that, unless certain recommended steps were taken and actively pursued, a pay award would have little or no meaning.' 'Hence,' in the words of the commission, 'our extraordinary preoccupation with the causes of the cost of living situation.'[27] As part of its efforts to confront the causes of the rising cost of living, the commission went on to recommend a number of basic measures, among them proposals 'to improve the food supply situation.'[28] The origins of many elements of Nigeria's agricultural program lie in the recommendations of these reports.

Urban consumers in Africa constitute a vigilant and potent pressure group demanding low-priced food. Because they are poor, much of their income goes for food; some studies suggest that urban consumers in Africa spend between 50 and 60 percent of their incomes on food.[29] Since changes in the price of food have a major impact on the economic well-being of urban dwellers in Africa, they pay close attention to the issue of food prices.

Urban consumers are potent because they are geographically concentrated and strategically located. Because of their geographic concentration, they can be organized quickly; and because they control transport, communications, and other public services, they can impose deprivations on others. They are therefore influential. Urban unrest frequently heralds a change of government in Africa, and the cost and availability of food supplies are major factors promoting urban unrest.

It should be noted that it is not only the workers who care about food prices. It is also the employers. Employers care about food prices because food is a wage good; with higher food prices, wages rise and, all else being equal, profits fall. Governments care about food prices not only because they are employers in their own right but also because as owners of industries and promoters of industrial development programs they seek to protect industrial profits. Indicative of the significance of these interests is that the unit that sets agricultural prices often resides not in the Ministry of Agriculture but in the Ministry of Commerce or of Finance.

When urban unrest begins among food consumers, the political discontent often spreads rapidly to upper echelons of the polity: to those whose incomes come from profits, not wages, and to those in charge of major bureaucracies. Political regimes that are unable to supply low-cost food are seen as dangerously incompetent and as failing to protect the interests of key elements of the social order. At times of high prices, influential elites are likely to ally with the

[27] Nigeria, *Public Service Review Commission*, p. 10.

[28] *Ibid.*, p. 93.

[29] Hiromitsu Kaneda and Bruce F. Johnston, 'Urban Food Expenditure Patterns in Tropical Africa,' *Food Research Institute Studies* 2 (1961): 229–75.

urban masses, to shift their political loyalties and replace those in power. Thus it was that protests over food shortages and rising prices formed a critical prelude to the coups and coup attempts in Ghana, Liberia, Kenya, and Guinea.

It is ironic but true that among those governments most committed to low-cost food are the 'radical' governments in Africa. Despite their stress on economic equality, they impose lower prices on the commodity from which the poorest of the poor, the peasant farmers, derive their income. A major reason for their behavior is that they are deeply committed to rapid industrialization; moreover, they are deeply committed to higher real wages for urban workers and have deep institutional ties to organized labor.

We can thus understand the demand for low-cost food. Its origins lie in the urban areas. It is supported by governments, both out of political necessity and, on the part of more radical ones, out of ideological preference. Food is a major staple and higher prices for such staples threaten the real value of wages *and* profits.

. . . .

Governments as agencies that seek to retain power

The interest-group model thus accounts for major elements of the food policies maintained by African governments. It explains the political pressures for low food prices and thus helps to explain why, when governments want more food, they prefer to secure it by building more projects rather than offering higher prices. By the same token, it helps to account for the governments' preference for production subsidies rather than higher food prices as incentives for food production.

Nonetheless, an interest-group explanation too is incomplete. Its primary virtue is that it helps to account for the essentially draconian pricing policies adopted by African governments. Its primary limitation is that it fails to explain how governments get away with such policies. How, in nations where the majority of the population are farmers and the majority of the resources are held in agriculture, are governments able to succeed in implementing policies that violate the interests of most farmers? In search of answers to this question, a third approach is needed, one that looks at agricultural programs as part of a repertoire of devices employed by African governments in their efforts to secure political control over their rural populations and thus to remain in power.

Organizing a rural constituency

We have already seen that adopting policies in support of higher prices for agricultural commodities would be politically costly to African governments. It is also important to note that such a stance would generate few political benefits. From a political point of view, conferring higher prices offers few attractions for politicians, for the benefits of the measure could be enjoyed by rural opponents and supporters alike. The benefits could not be restricted to the faithful or withheld from the politically disloyal. Pricing policies therefore cannot be employed by politicians to organize political followings.

Project-based policies suffer less from this liability. Officials can exercise discretion in locating projects; they can also exercise discretion in staffing them. Such discretion allows them to bestow benefits selectively on those whose political support they desire. Politicians are therefore more likely to be attracted to project-based policies as a measure of rural development.

The relative political utility of projects explains several otherwise puzzling features of government agricultural investments. One is the tendency to construct too many projects, given the budgetary resources available. A reason for this proliferation is that governments often wish to ensure that officials in each administrative district or electoral constituency have access to resources with which to secure a political backing.[30] Another tendency is to hire too large a staff or a staff that is technically untrained, thus undercutting the viability of the projects. A reason for this is that jobs on projects – and jobs in many of the bureaucracies involved with agricultural programs, for that matter – represent political plums, given by those in charge of the programs to their political followers. State farms in Ghana were staffed by the youth brigade of the ruling Convention People's Party, and the cooperative societies in Zambia were formed and operated by the local and constituency-level units of the governing party, to offer just two examples of the link between staffing and political organization.

Again and again, from an economic point of view, agricultural projects fail; they often fail to generate earnings that cover their costs or, when they do so, they often fail to generate a rate of return comparable to that obtainable through alternative uses of government funds. Nonetheless, public agencies revive and reimplement such projects. A major reason is that public officials are frequently less concerned with using public resources in a way that is economically efficient than they are with using them in a way that is politically expedient. If a project fails to generate an adequate return on the public investment but nonetheless is privately rewarding for those who build it, provision it, staff it, or hold tenancies in it, then political officials may support it, for the project will serve as a source of rewards for their followers and as an instrument for building a rural political organization.

. . . .

[30] See Bates, *Rural Responses*; Jerome C. Wells, *Agricultural Policy and Economic Growth in Nigeria, 1962–1968* (Ibadan: Oxford University Press for the Nigerian Institute of Social Science and Economic Research, 1974); Alfred John Dadson, 'Socialized Agriculture in Ghana, 1962–1965,' Ph.D. dissertation, Harvard University, 1970.

Conclusion

Governments in Africa, like governments elsewhere in the developing world, intervene in agricultural markets in ways that violate the interests of most farmers. They tend to adopt low-price policies for farm products; they tend to increase the prices farmers must pay for the goods they consume. And while they subsidize the prices of goods that farmers use in production, the benefits of these subsidies are appropriated by the richer few. In addition, the farm policies of African governments are characterized by a stress on projects rather than prices; when price policies are used, by a preference for lowering farm costs rather than increasing farm revenues; and by widespread economic inefficiency.

I have examined several political explanations for this configuration of agricultural policies. I conclude by commenting on their durability.

The pattern of price interventions, I have argued, represents the terms of a political pact among organized political interests, the costs of which are transferred to unorganized interests who are excluded from the price-setting coalition. Members of the pact are labor, industry, and government; small-scale farmers constitute its victims; and large-scale farmers stand as passive allies, politically neutralized through subsidy programs.

No member of the winning coalition possesses an incentive to alter its political demands unilaterally. Organized labor, for example will not unilaterally alter its demand for cheap food. Nor will industry call for reforms that raise food prices, and thus wages, unless other members of the coalition make credible commitments to offsetting concessions. In the short term, then, the coalition and the price structure that supports it appear stable.

Over the longer run, however, the structure of the payoffs achieved by the coalition changes. Farmers adjust; in response to pricing policies, they produce less. The result in food markets is lower supplies at higher prices. The result in export markets is fewer exports and less foreign exchange. The costs which once were externalized upon the unorganized agrarian sector are now internalized, through the operation of markets, onto the dominant coalition. The farmers have transferred the costs of the political settlement to the intended beneficiaries. And as these costs mount, the pact among them becomes less stable.

As the payoffs from this basis for governance in Africa erode, opportunities arise for the introduction of new pricing policies. And as the costs of the present policies are disproportionately borne by one of the more influential of the coalition partners, the governments themselves, the likelihood of policy changes is enhanced. To support low food prices, governments must provide additional supplies, either by subsidizing local production or by financing imports from abroad. But, throughout Africa, states are undergoing a fiscal crisis; they lack both revenues and foreign exchange. One consequence is that governments are less willing or able to bear the costs of current agricultural policies. Another is the reallocation of political power. At moments of fiscal crisis, finance ministers and directors of the central banks gain greater influence over public policy. Moreover, they find allies among foreign donors and international creditors, who pressure governments to make adjustments that will lessen their burden of debt. In league with international agencies, these figures have assumed greater influence over public policy.

The set of public policies described in this paper have thus formed the basis for a political pact among organized interests. But they have set in motion economic forces which erode their economic and political value. Moreover, the fiscal crisis in contemporary Africa has restructured power relations within African governments and has brought new players into the policymaking process. The result is that the commitment to these policies may not be stable and they may in fact be subject to change.

10 Robert Chambers, 'Rural Poverty Unobserved: The Six Biases'

Excerpt from: R. Chambers, *Rural Development: Putting the Last First*, pp. 13–23. Harlow: Longman (1983)

Many biases impede outsiders' contact with rural poverty in general, and with the deepest poverty in particular. These apply not only to rural development tourists, but also to rural researchers and local-level staff who live and work in rural areas. Six sets of biases stand out:

(i) Spatial biases: urban, tarmac and roadside

Most learning about rural conditions is mediated by vehicles. Starting and ending in urban centres, visits follow networks of roads. With rural development tourism, the hazards of dirt roads, the comfort of the visitor, the location of places to visit and places for spending the night, and shortages of both time and fuel dictate a preference for tarmac roads and for travel close to urban centres. The result is overlapping urban, tarmac and roadside biases.

Urban bias concentrates rural visits near towns and especially near capital cities and large administrative centres. But the regional distribution of the poorest rural people often shows a concentration in remoter areas – north-eastern Brazil, Zambia away from the line of rail, lower Ukambani in Kenya, the Tribal Districts of Central India, the hills of Nepal. In much of the developing world, some of the poorest people are being driven from those densely populated areas better served with communications and are being forced, in order to survive, to colonise less accessible areas, especially the savannahs and forests. Hard to reach from the urban centres, they remain largely unseen.

Tarmac and roadside biases also direct attention towards those who are less poor and away from those who are poorer. Visible development follows main roads. Factories, offices, shops and official markets all tend to be at the sides of main roads. Even agricultural development has a roadside bias: in Tamil Nadu agricultural demonstrations of new seeds and fertilisers have often been sited beside main roads; and on irrigation systems, roads follow canals so that the farms seen are those of the topenders who receive more water and not those of the tailenders who receive less or none. Services along roadsides are also better. An improved tarmac or all-weather surface can bring buses, electricity, telephone, piped water supply, and better access to markets, health facilities and schools. Services near main roads are better staffed and equipped: Edward Henevald found that two schools near a main highway in Sumatra had more than their quota of teachers, while a school one kilometre off the road had less than its quota.

When roads are built, land values rise and those who are wealthier and more influential often move in if they can. In Liberia, new rural roads were followed by speculators rushing to acquire deeds and to buy or to displace local farmers (Cobb *et al.*, 1980, pp. 12–16). For part of Western Kenya, Joseph Ssennyonga had described a similar tendency for the wealthier and more influential to buy up roadside plots, creating an 'elite roadside ecology' (1976, p. 9). So the poorer people shift away out of sight. The visitor then sees those who are better-off and their houses, gardens, and services, and not those who are poorer and theirs. Ribbon development along roadsides gives a false impression in many countries. The better the road, the nearer the urban centre, and the heavier the traffic, so the more pronounced is the roadside development and the more likely visitors are to see it and be misled.

Nor does spatial bias apply only to main roads. Within villages, the poorer people may be hidden from the main streets and the places where people meet. M. P. Moore and G. Wickremesinghe, reporting on a study of three villages in the Low Country of Sri Lanka, have this to say about 'hidden poverty':

> In retrospect at least, one of the most obvious aspects of poverty in the study villages is the extent to which it is concealed from view . . . the proportion of 'poor' households . . . varies from 14 per cent in Wattegama to 41 per cent in Weligalagoda. Yet one could drive along all the motorable roads in the villages and scarcely see a single 'poor' house. Here, as in most of rural Sri Lanka, wealthier households use their social and economic power to obtain roadside homestead sites. Not only do these confer easier access to such tangible services as buses, electricity connections or hawkers, but they provide such intangible benefits as better information and gossip from passers-by. Equally, the roadside dweller has a potential site for opening a small shop, especially if located near the all-important road junctions, which provide the focus of commercial and social life in almost all rural areas. *To even see the houses of the poor one often has to leave the road. Many visitors, including public officers, appear not to do so very often*. (1980, p. 59; emphasis added)

The same can be said of Harijan colonies in or near villages in South India, and of Basarwa (Bushmen) in or outside the villages of the Kalahari. Peripheral residence is almost universal with the rural poor.

It is not just the movements of officials that are guided by these spatial biases of rural development tourism. Social science researchers are far from immune. There are honourable exceptions, but urban and tarmac biases are sometimes evident in choices of villages to study. Of all specialists, social anthropologists are perhaps the least susceptible, but even they sometimes succumb: as they have grown, Bangalore and Bangkok have each swallowed up a social anthropologist's village. Again, when Indian institutions were urged to adopt villages, two research and training organisations in Bangalore, unknown to each other, included the same village: it can scarcely be a coincidence that it was close to the main Bangalore–Mysore road, a decent but convenient distance from Bangalore itself. Within villages, too, the central, more prosperous, core is likely to attract researchers.

Moore, again describing three villages in Sri Lanka, writes:

> Apart from the roadside issue, the core can exercise a great pull on the outsider who decides to do a few days' or a week's fieldwork. Apart from the facilities and the sense of being at the strategic hub of local affairs, it can claim a sense of history and tradition, to which sociologists especially appear vulnerable. (1981, p. 48)

He considers that sociologists writing on Sri Lanka have mostly focussed on core areas and completely ignored the peripheries. One may speculate about how generally the location of good informants and of facilities at the cores of villages prevent perception by social scientists of the peripheral poor.

Urban bias is further accentuated by fuel shortages and costs. When fuel costs rise dramatically, as they have done in recent years, the effect is especially marked in those poor countries which are without oil and also short of foreign exchange. The recurrent budgets of government departments are cut. Staff are difficult to shed, so the cuts fall disproportionately on other items. Transport votes are a favourite. Rural visits, research and projects shrink back from more distant, often poorer areas to those which are closer, more prosperous, and cheaper to visit.

In Zambia, the travel votes of the Ministry of Agriculture and Water Development could buy in 1980 only one fifth of the petrol they could buy in 1973 (ILO, 1981, p. 74) and senior agricultural extension staff were virtually office-bound. In Bangladesh, similarly, district agricultural officers have been severely restricted in their use of vehicles. In India, cuts have occurred in transport allocations for staff responsible for supervising canal irrigation: the likely effects include less supervision leading to less water reaching the already deprived areas and less staff awareness of what is happening there. Every rise in oil prices impoverishes the remoter, poorer people by tilting the urban-rural terms of trade against them, and at the same time reduces the chances of that deprivation being known. Visits, attention and projects are concentrated more and more on the more accessible and more favoured areas near towns.

(ii) Project bias

Rural development tourism and rural research have a project bias. Those concerned with rural development and with rural research become linked to networks of urban-rural contacts. They are then pointed to those rural places where it is known that something is being done – where money is being spent, staff are stationed, a project is in hand. Ministries, departments, district staff, and voluntary agencies all pay special attention to projects and channel visitors towards them. Contact and learning are then with tiny atypical islands of activity which attract repeated and mutually reinforcing attention.

Project bias is most marked with the showpiece: the nicely groomed pet project or model village, specially staffed and supported, with well briefed members who know what to say and which is sited a reasonable but not excessive distance from the urban headquarters. Governments in capital cities need such projects for foreign visitors; district and subdistrict staff need them too, for visits by their senior officers. Such projects provide a quick and simple reflex to solve the problem of what to do with visitors or senior staff on inspection. Once again, they direct attention away from the poorer people.

The better known cases concern those rural development projects which have attracted international attention. Any roll of honour would include the Anand Dairy Cooperatives in India; the Chilalo Agricultural Development Unit in Ethiopia; the Comilla Project in Bangladesh; the Gezira Scheme in Sudan; the Intensive Agricultural Districts Programme (IADP) in India; Lilongwe in Malawi; the Muda Irrigation Project in Malaysia; the Mwea Irrigation Settlement in Kenya; and some *ujamaa* villages in Tanzania. These have been much visited and much studied. Students seeking doctorates have read about them and then sought to do their fieldwork on them.

Research generates more research; and investment by donors draws research after it and funds it. In India, the IADP, a programme designed to increase production sharply in a few districts which were well endowed with water, exercised a powerful attraction to research compared to the rest of India. An analysis (Harriss, 1977, pp. 30–4) of rural social science research published in the Bombay *Economic and Political Weekly* showed an astonishing concentration in IADP districts, and an almost total neglect of the very poor areas of central India. In a different way, the Comilla Project may also have misled, since Comilla District has the lowest proportion of landless of any district in Bangladesh. Research on *ujamaa* in Tanzania in the clusters of villages (the Ruvuma Development Association, Mbambara, and Upper Kitete) which were among the very few in the whole country with substantial communal agricultural production, sustained the myth that such production was widespread. Research, reports and publications have given all these atypical projects high profiles, and these in turn have generated more interest, more visitors, and yet more research, reports and publications.

Fame forces project managers into public relations. More and more of their time has to be spent showing visitors around. Inundated by the celebrated, the curious, and the crass – prime ministers, graduate students, women's clubs, farmers' groups, aid missions, evaluation teams, school parties, committees and directors of this and that – managers set up public relations units and develop a public relations style. Visitors then get the treatment. A fluent guide follows a standard route and a standard routine. The same people are met, the same buildings entered, the same books signed, the same polite praise inscribed in the book against the visitors' names. Questions are drowned in statistics; doubts inhibited by handouts. Inquisitive visitors depart loaded with research papers, technical evaluations, and annual reports which they will probably never read. They leave with a sense of guilt at the unworthy scepticism which promoted their probing questions, with memories of some of those who are better-off in the special project, and impressed by the charisma of the exceptional leader or manager who has created it. They write their journey reports, evaluations and articles on the basis of these impressions.

For their part, the project staff have reinforced through repetition the beliefs which sustain their morale; and their projects take off into self-sustaining myth. But in the myth is the seed of tragedy, as projects are driven down this path which leads, step-by-step to self-deception, pride, defensiveness, and ultimately debunking.

(iii) Person biases

The persons with whom rural development tourists, local-level officials, and rural researchers have contact, and from whom they obtain impressions and information, are biased against poorer people.

(a) .*Elite bias* 'Elite' is used here to describe those rural people who are less poor and more influential. They typically include progressive farmers, village leaders, headmen, traders, religious leaders, teachers, and paraprofessionals. They are the main sources of information for rural development tourists, for local-level officials, and even for rural researchers. They are the most fluent informants. It is they who receive and speak to the visitors; they who articulate 'the village's' interests and wishes; their concerns which emerge as 'the village's' priorities for development. It is they who entertain visitors, generously providing the expected beast or beverage. It is they who receive the lion's share of attention, advice and services from agricultural extension staff. It is they who show visitors the progressive practices in their fields. It is they too, who, at least at first, monopolise the time and attention of the visitor.

Conversely, the poor do not speak up. With those of higher status, they may even decline to sit down. Weak, powerless and isolated, they are often reluctant to push themselves forward. In Paul Devitt's words:

> The poor are often inconspicuous, inarticulate and unorganised. Their voices may not be heard at public meetings in communities where it is customary for only the big men to put their views. It is rare to find a body or institution that adequately represents the poor in a certain community or area. Outsiders and government

> officials invariably find it more profitable and congenial to converse with local influentials than with the uncommunicative poor. (1977, p. 23)

The poor are a residual, the last in the line, the most difficult to find, and the hardest to learn from: 'Unless paupers and poverty are deliberately and persistently sought, they tend to remain effectively screened from outside inquirers' (*ibid*., p. 24).

(b) *Male bias* Most local-level government staff, researchers and other rural visitors are men. Most rural people with whom they establish contact are men. Female farmers are neglected by male agricultural extension workers. In most societies women have inferior status and are subordinate to men. There are variations and exceptions, but quite often women are shy of speaking to male visitors. And yet poor rural women are a poor and deprived class within a class. They often work very long hours, and they are usually paid less than men. Rural single women, female heads of households, and widows include many of the most wretched and unseen people in the world.

(c) *User and adopter biases* Where visits are concerned with facilities or innovations, the users of services and the adopters of new practices are more likely to be seen than are non-users and non-adopters. This bias applies to visitors who have a professional interest in, say, education, health or agriculture, to local-level officials, and to researchers. They tend to visit buildings and places where activity is concentrated, easily visible, and hence easy to study. Children in school are more likely to be seen and questioned than children who are not in school; those who use the health clinic more than those who are too sick, too poor, or too distant to use it; those who come to market because they have goods to sell or money with which to buy, more than those who stay at home because they have neither; members of the cooperative more than those who are too poor or powerless to join it; those who have adopted new agricultural, health or family planning practices more than those who have not.

(d) *Active, present and living biases* Those who are active are more visible than those who are not. Fit, happy, children gather round the Jeep or Land Rover, not those who are apathetic, weak and miserable. Dead children are rarely seen. The sick lie in their huts. Inactive old people are often out of sight; a social anthropologist has recorded how he spent some time camping outside a village in Uganda before he realised that old people were starving (Turnbull, 1973, p. 102). Those who are absent or dead cannot be met, but those who have migrated and those who have died include many of the most deprived. Much of the worst poverty is hidden by its removal.

(iv) Dry season biases

Most of the poor rural people in the world live in areas of marked wet–dry tropical seasons. For the majority whose livelihoods depend on cultivation the most difficult time of the year is usually the wet season, especially before the first harvest. Food is short, food prices are high, work is hard, and infections are prevalent. Malnutrition, morbidity and mortality all rise, while body weights decline. The poorer people, women and children are particularly vulnerable. Birth weights drop and infant mortality rises. Child care is inadequate. Desperate people get indebted. This is both the hungry season and the sick season. It is also the season of poverty ratchet effects, that is, of irreversible downward movements into poverty through the sale or mortgaging of assets, the time when poor people are most likely to become poorer.

The wet season is also the unseen season. Rural visits by the urban-based have their own seasonality.

Nutritionists take care to plan
to do their surveys when they can
be sure the weather's fine and dry,
the harvest in, food intake high.

Then students seeking Ph.D.s
believe that everyone agrees
that rains don't do for rural study
– suits get wet and shoes get muddy

And bureaucrats, that urban type,
wait prudently till crops are ripe,
before they venture to the field
to put their question: 'What's the yield?'

For monsoonal Asia, which has its major crop towards the end of the calendar year, it is also relevant that:

The international experts' flights
have other seasons; winter nights
In London, Washington and Rome
are what drive them, in flocks, from home

since they then descend on India and other countries north of the equator in January and February at precisely the time of least poverty and when marriages and celebrations are to be seen and heard.

Some opposite tendencies, however, deserve to be noted:

And northern academics too
are seasonal in their global view
For they are found in third world nations
mainly during long vacations.

North of the equator this means visits at the bad time of the monsoon in much of Asia and of the rains of West Africa. There are also professionals like agriculturalists and epidemiologists whose work demands rural travel during the rains, for that is when crops grow and bugs and bacteria breed.

But the disincentives and difficulties are strong. The rains are a bad time for rural travel because of the inconveniences or worse of floods, mud, landslides, broken bridges; and getting stuck, damaging vehicles, losing time, and enduring discomfort. In some places roads are officially closed. In the South Sudan there is a period of about two months after the onset of the rains when roads are impassable but when there is not yet enough water in the rivers for travel by boat. Many rural areas, especially those which are remote and poor, are quite simply inaccessible by vehicle during the rains. The worst times of the year for the poorer people are thus those that are the least perceived by urban-based outsiders.

Once the rains are over such visitors can however travel more freely. It is in the dry season, when disease is diminishing, the harvest in, food stocks adequate, body weights rising, ceremonies in full swing, and people at their least deprived, that there is most contact between urban-based professionals and the rural poor. Not just rural development tourism, but rural appraisal generally is susceptible to a dry season bias. A manual for assessing rural needs warns of an experience when 'Once, the jeeps needed for transporting the interviewers were recalled for a month *during the few precious months of the dry season*' (Ashe, 1979, p. 26; my emphasis). Whole institutes concentrate their field research in the dry seasons; the rains are for data analysis and writing up with a good roof over one's head. Concern to avoid inconveniencing respondents when they are busy and exhausted with agricultural activities provides a neat justification, both practical and moral, for avoiding research during the rains. Many factors thus conspire to ensure that the poorest people are most seen at precisely those times when they are least deprived; and least seen when things are at their worst.

(v) Diplomatic biases: politeness and timidity

Urban-based visitors are often deterred by combinations of politeness and timidity from approaching, meeting, and listening to and learning from the poorer people. Poverty in any country can be a subject of indifference or shame, something to be shut out, something polluting, something, in the psychological sense, to be repressed. If honestly confronted, it can also be profoundly disturbing. Those who make contact with it may offend those who are influential. The notables who generously offer hospitality to the visitor may not welcome or may be thought not to welcome, searching questions about the poorer people. Senior officials visiting junior officials may not wish to examine or expose failures of programmes intended to benefit the poor. Politeness and prudence variously inhibit the awkward question, the walk into the poorer quarter of the village, the discussion with the working women, the interviews with Harijans. Courtesy and cowardice combine to keep tourists and the poorest apart.

(vi) Professional biases

Finally, professional training, values and interests present problems. Sometimes they focus attention on the less poor: agricultural extension staff trained to advise on cash crops or to prepare farm plans are drawn to the more 'progressive' farmers; historians, sociologists and administrators, especially when short of time, can best satisfy their interests and curiosity through informants among the better-educated or less poor; those engaged in family welfare and family planning work find that bases for the adoption of any new practices can most readily be established with better-off, better-educated families. But sometimes, in addition, professional training, values and interests do focus attention directly on the poor. This is especially so in the fields of nutrition and health, where those wishing to examine and to work with pathological conditions will tend to be drawn to those who are poorer.

More generally, specialisation, for all its advantages, makes it hard for observers to understand the linkages of deprivation. Rural deprivation is a web in which poverty (lack of assets, inadequate stocks and flows of food and income), physical weakness and sickness, isolation, vulnerability to contingencies, and powerlessness all mesh and interlock. But professionals are trained to look for and see much less. They are programmed by their education and experience to examine what shows up in a bright but slender beam which blinds them to what lies outside it.

Knowing what they want to know, and short of time to find it out, professionals in rural areas become even more narrowly single-minded. They do their own thing and only their own thing. They look for and find what fits their ideas. There is neither inclination nor time for the open-ended question or for other ways of perceiving people, events and things. 'He that seeketh, findeth.' Visiting the same village, a hydrologist enquires about the water table, a soils scientist examines soil fertility, an agronomist investigates yields, an economist asks about wages and prices, a sociologist looks into patron-client relations, an administrator examines the tax collection record, a doctor investigates hygiene and health, a nutritionist studies diets, and a family planner tries to find out about attitudes to numbers of children. Some of these visiting professionals may be sensitive to the integrated nature of deprivation, but none is likely to fit all the pieces together, nor to be aware of all the negative factors affecting poorer people.

Specialisation prevents the case study which sees life from the point of view of the rural poor themselves; but where such case studies are written their broader spread helps understanding and points to interventions which specialists miss. In contrast, narrow professionalism of whatever persuasion leads to diagnoses and prescriptions which underestimate deprivation by recognising and confronting only a part of the problem.

Selected references

Ashe, J. 1979: *Assessing rural needs: A manual for practitioners*. Mount Rainier, MD: Volunteers in Technical Assistance.

Cobb, R., Hunter, R., Vandervoort, C., Bledsoe, C. and McCluskey, R. 1980: *Impact of rural roads in Liberia*. Washington, DC: Agency for International Development.

Devitt, P. 1977: Notes on poverty-orientated rural development. *ODI Occasional Paper* 2, 20–41.

Harriss, J. 1977: Bias in perception of agrarian change in India. In Farmer, B. H. (ed.), *Green revolution?* London: Macmillan.

ILO 1981: *Zambia: Basic needs in an economy under pressure*. Geneva: ILO.

Moore, M. 1981: Beyond the tarmac road: a guide for rural poverty watchers. *IDS Bulletin* 12 (4), 47–9.

Moore, M. and Wickremesinghe, G. 1980: *Agriculture and society in the Low Country (Sri Lanka)*. Colombo: Agrarian Research and Training Institute.

Ssennyonga, J. 1976: The cultural dimensions of demographic trends. *Populi* 3 (2), 2–11.

Turnbull, C. 1973: *The mountain people*. London: Picador.

SECTION THREE

HOW POOR PEOPLE SURVIVE: THE WEAPONS OF THE WEAK

Editor's Introduction

Until about the mid-1970s rural development studies were dominated by an academic and policy-related agenda which prioritized state actions in support of institutional change and the diffusion of new agricultural technologies. This is not to say that a micro-sociological concern for farmer decision making was undeveloped (see Schultz, 1964; Lipton, 1968), or that anthropologists, in particular, were not advancing our understanding of the local contexts for agrarian change and rural development; they clearly were. It is to suggest that a macro-approach to rural development studies was not then as closely interwoven with a micro-approach as it would be in the late 1970s and 1980s. A first wave of rural development studies found little room for the actions of individual men or women, farming households, or social classes in their accounts of the processes and problems of rural development. Men and (sometimes) women were credited with responding to government projects in the countryside, or to price signals, but they were less often seen as the authors of their own rural (or rural *and* urban) livelihoods. This was as true of accounts emanating from *dependencia* perspectives and Marxism as it was of accounts associated with the modernization paradigm.

The balance between the macro and the micro began to shift quite noticeably in the mid-1970s and through the 1980s and early 1990s. In development economics the counter-revolution was beginning to catch hold, and many of its proponents were keen to place individual economic agents at the centre of their accounts of development. Mainstream development economics was condemned for denying the wants and capacities of rational economic actors, and for imposing on them the *dirigiste* and rent-seeking ambitions of a modernizing state. Meantime, in sociology, political science and geography there was a general reaction against the so-called structuralist turn, or the tendency to present men and women as little more than the bearers of some underlying social relations of class or status. Pierre Bourdieu and Tony Giddens were at the forefront of a rethinking of the structure–action dichotomy

(Bourdieu, 1971; Giddens, 1976, 1977). The theory of structuration associated with Giddens pointed to an 'intersection between knowledgeable and capable human agents and the wider social systems and structures in which they are implicated' (Gregory, 1994, p. 600). Structurationists accept that men and women, and poor men and women in particular, are not free to make their lives as they would wish, but they insist that in most cases they are able to shape the structures of production, exchange, power, governance, patriarchy and care in which they find themselves. The space for action is accordingly opened up – or recognized! – and with it a sensitivity to the politics of the local: to relations within the household and between people and their immediate environments, as well as to the high politics of the capital city, the so-called nation-state, or the international community.

In development studies this movement away from a guiding structuralism, or structural functionalism, coincided with a growing sensitivity to the skills and voices of the people who were presented as the beneficiaries of an externally-generated 'development'. Robert Chambers (**Reading 10**) was one of those taking a lead in heeding what soon would be called local knowledges (Hobart, 1993), or indigenous ecological or agricultural systems. Other leads came from an actor-oriented rural sociology (Long, 1977; Long and van der Ploeg, 1994), and from feminists who prised open the household to explore the different resources and life-chances enjoyed by Third World men and women, as well as the young and the elderly. Still another lead came from thematic enquiries into the relationships between development and population, the environment and migration. In all of these cases a powerful body of work emerged which highlighted the *isolation paradox* that links economic and social actions at a household level to wider processes of structural transformation in a developing society.

In the case of demography the new orthodoxy of the 1970s and 1980s held that 'poor people are not poor because they have large families. Quite the contrary: they have large families because they are poor' (Mamdani, 1972, p. 14). That is to say, rates of population growth that might in some way be undesirable at a national or regional level, emerge, paradoxically, from the rational decisions of isolated households struggling to survive in an often harsh social and natural environment. Couples are held to want children as economic assets (and boys especially in parts of South and East Asia, where up to 100 million women are 'missing': Sen, 1991; see also Miller, 1982; Das Gupta, 1987), and as a means of insurance against the risks of being old in societies lacking in state welfare systems. It follows that government attempts to curb national population growth rates cannot rely on campaigns to combat 'ignorance' about family planning, or which dole out modern forms of contraception. Poor households have to be persuaded that large numbers of children are not required to keep their families afloat, or they have to be provided with other forms of insurance against endemic risk. Women, too, need to be empowered to take command of their own bodies. In India, birth rates are lowest in Kerala, where female literacy rates are high and where women participate more equally in the economic and political life of the State than is common elsewhere in the country (Sen, 1994).

Recent evidence on demographic trends in the developing world suggests that population growth rates *are* slowing down, and quite rapidly so in parts of Asia and Latin America. Neo-Malthusianism continues to rear its head periodically at international conferences on the 'population problem', but it enjoys little support in the social scientific community. Some recent evidence also suggests that the demand theory of fertility outlined above might be taking an unduly pessimistic view of the role played by state extension services and family planning clinics in dampening down high local rates of fertility (Cleland, in Cleland and Thomas, 1993). In Bangladesh, total fertility rates seem to be dropping quite sharply, notwithstanding a failure to change local structures of risk significantly (World Bank, 1993). But even though new data are accumulating on this and related topics, the value of an approach to population studies in terms of a household risk-avoidance model is now widely accepted. In Africa such a model is associated with John Caldwell and his co-workers (Caldwell, 1977; Caldwell *et al.*, 1978). In South Asia, the early and dominantly qualitative work of Mamdani in the Punjab has for many years been complemented by the rigorous empirical work carried out by Mead Cain in Char Gopalpur, Bangladesh and in several villages in India.

Reading 11 finds **Mead Cain** drawing on his studies in Char Gopalpur to highlight the consequences of reproductive failure for the elderly of rural South Asia. Cain argues that rural poverty increases adult mortality risks for both men and women, and 'increases the probability of reproductive failure'. The poor are thus driven to have large numbers of children partly to ensure the survival of a son to guard against 'a couple's vulnerability to economic crises'. Cain also shows that 'a young widow with no mature son on whom to depend faces the prospect of further rapid economic decline and very high mortality risk'. Although this mortality risk is heightened by the patriarchal nature of rural Bangladeshi society, it remains the case that 'women have little choice but to depend on males for their survival'.

Cain's work is also instructive in two other respects. It is so, first, because it clarifies the nature and range of what used to be called household survival strategies, at least in regard to family formation in rural South Asia. (The fact that these strategies are usually gendered, and might involve adaptive mechanisms as much as a disposition to survive, indicates why the notion of a household survival strategy is less often referred to in the 1990s than it was in the 1980s.) In addition, Cain's work is crucially dependent on a mass of local survey work and ethnographic information that is perhaps not as robust as he would like or as some development economists might demand. But this is not intended as a criticism of Cain. The point is that our knowledge of the household and intra-household adaptive mechanisms adopted in rural areas of the Third World necessarily depends on data drawn from small sample populations at particular points in time and space. It is partly for this reason that Cain himself was drawn into debate throughout the 1980s with Carol and Murray Vlassoff, both of whom challenged Cain's account of 'old age security and the utility of children in South Asia'. Writing of a village in Maharashtra, India, the Vlassoffs found – or seemed to find – that children were perceived as being of little material value to their ageing parents (M. Vlassoff and C. Vlassoff, 1980; see also Vlassoff, 1990). Notwithstanding certain problems in their analysis (Datta and Nugent, 1984), it is surely possible that both Cain and the Vlassoffs are telling us plausible stories about different places in rural South Asia on the basis of the available evidence. Put another way, there is much about this and other topics that we still do not know.

The isolation paradox apparent in the relationship between population and development is also apparent in the relationship between the environment and development, and for much the same reasons. In **Reading 12 Bina Agarwal** presents a disturbing account of the nature of the fuelwood problem in the rural Third World. Like Cain, she is attentive to the gender dimension of a paradox that sees many women cutting down trees for fuelwood in their immediate vicinity and then having to walk ever greater distances to collect fuelwood at a later date. There are now many studies which show that rural women are well aware of the consequences they face from a local deforestation (see also **Reading 25**), just as the West has been made aware of the global climatic changes that can be induced by large-scale deforestation (World Bank, 1992). Many women have also demonstrated a willingness to join in forest protection and management schemes with local forest departments or non-governmental organizations. But deforestation seems set to continue, and for reasons that are quite understandable at a household level. In the absence of alternative sources of fuel, and where many forests are reserved by the state for the production of major timber by commercial contractors, people in rural areas of the Third World are driven to mine their village forests or to encroach on state forest lands. In some cases a struggle over forest access and the forest usufruct ensues, as in the Chipko struggles of the Uttarakhand, India (Guha, 1989) or in Amazonia, where Chico Mendes paid with his life (Hecht and Cockburn, 1989). Agarwal's paper sets out the basic issues that lie behind this growing environmental crisis. She also suggests some tentative solutions to the Third World's fuelwood problem. Some of these solutions bear on questions of common or joint property management more generally (see also Singh, 1994; Wade, 1988).

In many parts of the rural Third World forest environments are not available to be exploited as part of a survival strategy or as a means of coping with poverty. For an increasing number of poor men, women and children the only option open to them is to migrate to fields and cities elsewhere in search of paid work. For such people, entitlements to food come primarily from cash wages earned from daily paid labour. Despite many global images to the contrary, most parts of the Third World have long been in constant flux, with personal mobility being a *sine qua non* of household survival or advancement. This mobility in turn has prompted academics to rethink the role and nature of migration in the Third World. In the 1960s and 1970s the positive model of permanent rural to urban migration associated with W. Arthur Lewis was challenged both theoretically and empirically. The empirical challenge suggested that circular migration was more common than permanent rural to urban migration outside Latin America, and that much migration was intra-rural. The theoretical challenge came from John Harris and Michael Todaro, who suggested that rural migrants might stream to cities in excess of a 'warranted rate' because of high expected inequalities in living standards between the countryside and the city (Harris and Todaro, 1970).

In the 1980s these and other studies were given added weight and an important new twist. Several authors began to suggest that migration was not just prompted by individual economic agents maximizing their likely incomes, but was sometimes organized by labour contractors seeking to create local gluts of labour as a way of driving down real wages. The work of **Jan Breman** deserves a special mention in this regard. **Reading 13** sets out Breman's theoretical perspective on rural labour circulation in western India *and* makes the reader aware of what life is like for poor people trapped in 'involuntary' migration. Breman also shows how difficult it is for rural labourers to organize effectively against their bosses when they are deliberately disorganized on the basis of area of origin, dialect and community. Breman's story is far removed from the optimistic accounts of migration as self-improvement that prevailed in the 1950s, 1960s and 1970s.

The Breman Reading is taken from a book published in 1985, a year which also saw the publication of a seminal text on the weapons of the weak by James Scott (Scott, 1985). Scott's book is also set in a Green Revolution area of Asia, but the village of Sedaka that is foregrounded in his study is located in Malaysia, not India. The importance of Scott's work lies mainly in his observation that poor people in rural areas of the Third World are intent not only on surviving poverty, but also on challenging those they hold responsible for their plight (however carefully or 'invisibly'). Scott's work on Sedaka continues his earlier studies of the moral economy of the peasantry (Scott, 1976, 1987). He contends that peasants are most likely to resist those who would extract surplus labour from them when the local moral economy breaks down; that is, when the rural elite engages only in labour exploitation and not in the paternalistic duties expected of a reputable employer. Scott's work is also directed against those who have looked to the peasantry as a revolutionary vanguard in the Third World. Scott maintains both that 'peasant wars of the twentieth century' (Wolf, 1971) are few and far between, and yet that peasant resistance is everywhere around us on a daily basis. The absence of revolution should not be read as evidence of peasant conservatism or acquiescence. Far from it: 'Most forms of [peasant] struggle stop well short of outright collective defiance. . . . I have in mind the ordinary weapons of relatively powerless groups: foot dragging, dissimulation, desertion, false compliance, pilfering, feigned ignorance, slander, arson, sabotage, and so on. These Brechtian – or Schweikian – forms of class struggle have certain features in common. They require little or no coordination or planning; they make use of implicit understandings and informal networks; they often represent a form of individual self-help; they typically avoid any direct, symbolic confrontation with authority' (Scott, 1985, p. xvi).

Scott's book has been criticized on several grounds – for not attending to gender issues (Hart, 1991), for neglecting class (Brass, 1991), for not positioning the author squarely in his ethnographic reports – but none of this gainsays the view that it is a classic text of the 1980s. The phrase 'the weapons of the weak' slipped quickly into the vocabularies of peasant and development studies, and a number of authors built upon the conceptual framework that Scott set out. **Reading 14** by **Judith Carney and Michael Watts** is a good example of a piece of work that is indebted to Scott, but which moves beyond Scott in its sensitivity to gender issues and 'the manufacture of symbolic and material dissent' in central Gambia. Carney and Watts also signal a debt to Michael Burawoy's work on production politics and the manufacturing of consent (here dissent) in industrial North America (Burawoy, 1985).

The Carney and Watts Reading is taken from a much longer paper which pays attention to the history of 'rice politics' in Gambia between 1949 and 1986, and the changing nature of Mandinka land rights and household structures. In the Reading published here we are offered a finely observed account of various 'struggles over work, struggles over meaning' that have taken shape in the wake of new forms of contract rice farming introduced to central Gambia by the government and foreign donors in the late 1970s. The authors take care to colour in the ambiguous and sometimes contradictory nature of these struggles, and the centrality of gender conflicts in what is often written up as 'peasant politics'. Carney and Watts also offer a cautionary note in respect of Scott's account of the weapons of the weak. While accepting that a subaltern society like the Mandinka 'are engaged in a dialectical struggle between active and passive, acceptance and resistance', they also warn that: 'Much of what passes as resistance is more akin to adaptation: it is all too easy to romanticise the prosaic and everyday, and to create a fantasy land of class struggle'.

The final Reading in this section, **Reading 15** by **Cathy Schneider**, neatly complements the article by Carney and Watts. Schneider too is concerned with the politics of protest, but her point of focus is a set of urban popular movements in Pinochet's Chile. Again on the basis of painstaking fieldwork, Schneider is able to show how various local attempts to take command of urban spaces were more or less bound up with the activities of the Chilean Communist Party and with 'organized politics'. Schneider also shows how the chronology of these urban movements had more to do with the life-cycle characteristics of a protest movement than with some 'objective' and controlling economic forces in the country at large. In this fashion, she confirms that everyday patterns and processes of 'development', or democratization, or even anti-development, are made by the everyday actions of men and women responding to, and challenging, certain deeper, structural or external factors beyond their immediate control. The 'micro' and the 'macro' have to be held together in development studies if the subject is not to fall into an anaemic formalism or a romantic localism.

Guide to further reading

General

Agarwal, B. 1990: Social security and the family in rural India: coping with conditionality and climate. *Journal of Peasant Studies* 17, 341–412.

Beck, T. 1992: Survival strategies and survival amongst the poorest in a West Bengal village. In Wilber, C. and Jameson, K. (eds), *The political economy of development and underdevelopment*. New York: McGraw Hill, 478–97.

Becker, G. 1981: *Treatise on the family*. Cambridge, MA: Harvard UP.

Bourdieu, P. 1971: *Outline of the theory of practice*. Cambridge: CUP.

Chambers, R., Longhurst, R. and Pacey, A. (eds) 1981: *Seasonal dimensions to rural poverty*. London: Pinter.

Dasgupta, P. 1993: *An inquiry into well-being and destitution*. Oxford: OUP.

Dreze, J. and Sen, A. K. 1989: *Hunger and public action*. Oxford: OUP.

Evans, A. 1991: Gender issues in household rural economics. *IDS Bulletin* 22, 51–9.

Folbre, N. 1986: Cleaning house: new perspectives on households and economic development. *Journal of Development Economics* 22, 5–40.

Giddens, A. 1976: *New rules of sociological method*. London: Hutchinson.

Giddens, A. 1977: *Studies in social and political theory*. London: Hutchinson.

Gregory, D. 1994: Structuration theory. In Johnston, R. *et al.* (eds), *The dictionary of human geography*. Oxford: Blackwell, 600–3.

Hobart, M. 1993: *An anthropological critique of development: The growth of ignorance*. London: Routledge.

Kabeer, N. 1994: *Reversed realities: Gender hierarchies in development thought* (Chapter 5). London: Verso.

Lipton, M. 1968: Game against nature: theories of peasant decision-making. *Two talks given on the BBC Third Programme*; published in Harriss, J. (ed.), *Rural development*. London: Hutchinson, 1982.
Long, N. 1977: *An introduction to the sociology of rural development*. London: Tavistock.
Long, N. and van der Ploeg, J. 1994: Heterogeneity, actor and structure: towards a reconstitution of the concept of structure. In Booth, D. (ed.), *Rethinking social development*. Harlow: Longman, 62–89.
Longhurst, R. 1988: Cash crops, household food security and nutrition. *IDS Bulletin* 19, 28–36.
Ravallion, M. 1989: Income effects on undernutrition. *Economic Development and Cultural Change* 38, 489–516.
Roberts, P. 1991: Anthropological perspectives on the household. *IDS Bulletin* 11, 60–4.
Schultz, T. 1964: *Transforming traditional agriculture*. New Haven: Yale UP.
Wade, R. 1988: *Village republics*. Cambridge: CUP.
World Bank 1990: *World development report, 1990*. Oxford: OUP/World Bank.
World Bank 1992: *World development report, 1992*: Oxford: OUP/World Bank.

Population, fertility and risk

Banister, J. 1987: *China's changing population*. Stanford: Stanford UP.
Bongaarts, J., Frank, O. and Lesthaeghe, R. 1984: The proximate determinants of fertility in sub-Saharan Africa. *Population and Development Review* 10, 511–37.
Cain, M. 1982: Perspectives on family and fertility in developing countries. *Population Studies* 36, 159–75.
Cain, M. and Vlassoff, M. (in debate) 1991: On widows, sons and old-age security in Maharashtra, India. *Population Studies* 45, 519–35.
Caldwell, J. 1977: The economic rationality of high fertility: an investigation with Nigerian survey data. *Population Studies* 31, 15–27.
Caldwell, J., Reddy, P. and Caldwell, P. 1978: *The causes of demographic change*. Delhi: OUP.
Clay, D. and Johnson, N. 1992: Size of farm or size of family: which comes first? *Population Studies* 46, 491–505.
Cleland, J. and Thomas, N. (in debate) 1993: Equity, security and fertility. *Population Studies* 47, 345–59.
Das Gupta, M. 1987: Selective discrimination against female children in rural Punjab, India. *Population and Development Review* 13, 77–100.
Datta, S. and Nugent, J. 1984: Are old age security and utility of children in India really unimportant? *Population Studies* 38, 507–9.
Dyson, T. and Moore, M. 1983: On kinship structure, female autonomy and demographic behaviour in India. *Population and Development Review* 9, 32–60.
Ehrlich, P. and Ehrlich, A. 1990: *The population explosion*. New York: Simon and Schuster.
Enke, S. 1971: Economic consequences of rapid population growth. *Economic Journal* 81, 800–11.

Greer, G. 1984: *Sex and destiny: The politics of human fertility*. London: Secker and Warburg.
Guz, D. and Hobcraft, J. 1991: Breastfeeding and fertility: a comparative analysis. *Population Studies* 45, 91–108.
Harriss, B. and Watson, E. 1987: The sex ratio in South Asia. In Momsen, J. and Townsend, J. (eds), *The geography of gender in the Third World*. London: Hutchinson.
Kirk, M. 1984: The return to Malthus? The global demographic future. *Futures* 16, 124–38.
Mamdani, M. 1972: *The myth of population control*. New York: Monthly Review Press.
McNamara, R. 1984: Time bomb or myth: the population problem. *Foreign Affairs* 62, 1107–31.
McNicoll, G. and Cain, M. (eds) 1990: *Rural development and population: Institutions and policy*. Oxford: OUP/Population Council.
Meadows, D. H., Meadows, D. L., Randers, J. and Behrens, W. 1972: *The limits to growth*. London: Pan.
Miller, B. 1982: *The endangered sex: Neglect of female children in North India*. Ithaca: Cornell UP.
Nieuwenhuys, O. 1994: *Children's lifeworlds: Gender, welfare and labour in the developing world*. London: Routledge.
Palloni, A. 1981: Mortality in Latin America: emerging patterns. *Population and Development Review* 7, 623–49.
Raju, S. 1991: Gender and deprivation. *Economic and Political Weekly*, December, 2827–39.
Rodgers, G. and Standing, G. (eds) 1981: *Child work, poverty and underdevelopment*. Geneva/ILO.
Sen, A. K. 1991: More than 100 million women are missing. *New York Review of Books*, 20 December.
Sen, A. K. 1994: Population: delusion and reality. *New York Review of Books*, 22 September.
Simon, J. 1981: *The ultimate resource*. Princeton: Princeton UP.
Vlassoff, C. 1990: The value of sons in an Indian village: how widows see it. *Population Studies* 44, 5–20.
Vlassoff, M. and Vlassoff, C. 1980: Old age security and the utility of children in rural India. *Population Studies* 34, 45–59.
White, B. 1975: The economic importance of children in a Javanese village. In Nag, M. (ed.), *Population and social organization*. The Hague: Mouton, 127–46.
World Bank 1984: *World development report, 1984*. Oxford: OUP/World Bank.
World Bank 1986: *Population growth and policies in sub-Saharan Africa*. Washington, DC: World Bank.
World Bank 1993: *Success in a challenging environment: Fertility decline in Bangladesh*. Washington, DC: World Bank/Population Reference Bureau.

Environment and migration

Adams, W. 1990: *Green development*. London: Routledge.
Agarwal, B. 1986: *Cold hearths and barren slopes: The woodfuel crisis in the Third World*. London: Zed.

Albert, B. 1992: Indian lands, environmental policy and military geopolitics in the development of the Brazilian Amazon: the case of the Yanomami. *Development and Change* 23, 35–70.

Blaikie, P. 1985: *The political economy of soil erosion*. Harlow: Longman.

Breman, J. 1985: *Of peasants, migrants and paupers: Rural labour circulation and capitalist production in West India*. Delhi: OUP.

Breman, J. 1990: 'Even the dogs are better off': the ongoing battle between capital and labour in the cane fields of Gujerat. *Journal of Peasant Studies* 17, 546–608.

Bunker, S. 1985: *Underdeveloping the Amazon: Extraction, unequal exchange and the failure of the modern state*. Urbana: University of Illinois Press.

Croll, E. and Parkin, D. (eds) 1992: *Bush base and forest farm: Culture, environment and development*. London: Routledge.

Crush, J. and James, W. 1991: Depopulating the compounds: migrant labour and mine housing in South Africa. *World Development* 19, 301–16.

Eckholm, E. 1982: *Down to Earth*. London: Pluto.

Ekins, P. 1992: *A new world order: Grassroots movements for global change*. London: Routledge.

Guha, R. 1989: *The unquiet woods: Ecological change and peasant resistance in the Himalaya*. Delhi: OUP.

Hardin, G. 1968: The tragedy of the commons. *Science* 162, 1243–8.

Harris, J. and Todaro, M. 1970: Migration, unemployment and development: a two-sector analysis. *American Economic Review* 60, 126–42.

Hecht, S. and Cockburn, A. 1989: *The fate of the forest: Developers, destroyers and defenders of the Amazon*. London: Verso.

Hill, P. 1963: *The migrant cocoa-farmers of southern Ghana: A study in rural capitalism*. Cambridge: CUP.

Kamarck, A. 1976: *The tropics and economic development: A provocative enquiry into the poverty of nations*. Baltimore: Johns Hopkins UP/World Bank.

Lipton, M. 1980: Migration from rural areas of poor countries: a review. *World Development* 8, 651–77.

Mabogunje, A. 1990: Agrarian responses to outmigration in sub-Saharan Africa. In McNicoll, G. and Cain, M. (eds), *Rural development and population*. Oxford: OUP/Population Council, 324–42.

Molnar, A. 1988: *Women and social forestry issues*. World Bank: Asia Technical Department.

Moore, H. and Vaughan, M. 1994: *Cutting down trees: Gender, nutrition and agricultural change in the northern province of Zambia, 1890–1990*. London: James Currey.

Nesmith, C. 1991: Gender, trees and fuel: social forestry in West Bengal. *Human Organization* 50, 337–48.

Pearce, D. (ed.) 1991: *Blueprint 2: The greening of the global economy*. London: Earthscan.

Radcliffe, S. 1990: Ethnicity, patriarchy and incorporation into the nation: female migrants as domestic servants in Peru. *Society and Space* 8, 379–93.

Redclift, M. 1984: *Development and the environmental crisis*. London: Methuen.

Schmink, M. 1982: Land conflicts in Amazonia. *American Ethnologist* 9, 341–57.

Shiva, V. 1989: *Staying alive: Women, ecology and development*. London: Zed.

Shrestha, N. 1985: The political economy of economic underdevelopment and external migration in Nepal. *Political Geography Quarterly* 4, 289–306.
Singh, K. 1994: *Managing common pool resources*. Delhi: OUP.
Smil, V. 1984: *The bad Earth: Environmental degradation in China*. London: Zed.
Sontheimer, S. (ed.) 1991: *Women and the environment: A reader*. London: Earthscan.
Todaro, M. 1969: A model of labour migration and urban unemployment in less developed countries. *American Economic Review* 59, 138–48.
Wolpe, H. 1972: Capitalism and cheap labour-power: from segregation to apartheid. *Economy and Society* 1, 425–6.
World Bank 1989: *Sub-Saharan Africa: From crisis to sustainable agriculture*. Washington, DC: World Bank.

The weapons of the weak

Brass, T. 1991: Moral economists, subalterns, new social movements, and the (re-)emergence of a (post-)modernized (middle) peasant. *Journal of Peasant Studies* 18, 173–205.
Burawoy, M. 1985: *The politics of production*. London: Verso.
Comaroff, J. 1985: *Body of power, spirit of resistance: The culture and history of a South African people*. Chicago: University of Chicago Press.
de Certeau, M. 1984: *The practice of everyday life*. Berkeley: University of California Press.
Eckstein, S. (ed.) 1989: *Power and popular protest: Latin American social movements*. Berkeley: University of California Press.
Escobar, A. and Alvarez, S. (eds) 1992: *The making of social movements in Latin America*. Boulder: Westview.
Fox, J. 1992: Democratic rural development: leadership accountability in regional peasant organizations. *Development and Change* 23, 1–36.
Guha, R. and Spivak, G. C. (eds) 1988: *Selected subaltern studies*. Oxford: OUP.
Hart, G. 1991: Engendering everyday resistance; politics, gender and class formation in rural Malaysia. *Journal of Peasant Studies* 19, 93–121.
Mainwaring, S. 1987: Urban popular movements, identity and democratization in Brazil. *Comparative Political Studies* 20, 131–59.
Scheper-Hughes, N. 1992: *Death without weeping: The violence of everyday life in Brazil*. Berkeley: University of California Press.
Scott, J. 1976: *The moral economy of the peasant*. New Haven: Yale UP.
Scott, J. 1985: *Weapons of the weak: Everyday forms of peasant resistance*. New Haven: Yale UP.
Scott, J. 1987: Resistance without protest and without organization. *Comparative Studies in Society and History* 29, 434–57.
Slater, D. 1991: New social movements and old political questions. *International Journal of Political Economy* 21, 32–65.
Wolf, E. 1971: *Peasant wars of the twentieth century*. London: Faber.

11 Mead Cain,

'The Consequences of Reproductive Failure: Dependence, Mobility, and Mortality among the Elderly of Rural South Asia'

Reprinted in full (bar the Conclusion) from: *Population Studies* 40, 375–88 (1986)

In an earlier paper I described the pattern of family formation and dissolution in rural Bangladesh, the demographic sources of variation in the course of the life cycle, and their implications for the economic mobility of households.[1] Given the high-risk environment of Bangladesh, I suggested three ways in which economic mobility could be influenced by the course of the household life cycle, all of which revolved around the timely production of sons. The first is through the cumulative effect of child labour, net of the costs that they engender. While children of either sex begin work at young ages, the division of labour in Bangladesh is such that boys are far more productive than girls. Thus the producer/consumer ratio depends as much on the sex composition of the family as on the ages of the children, and households with a higher proportion of sons are more favourably positioned with respect to mobility prospects than those with a higher proportion of daughters. The second link to mobility is through sons as insurance against risk. Here the cumulative product of child labour is less important than the presence of sons during crises that threaten a household with loss of property: a debilitating illness that strikes the household head, for example, or a natural disaster that destroys the crop. A son, in such circumstances, may help to avert property loss through distress sales by filling in for a sick father or permitting the diversification of income sources. The third link, a special case of the second, is distinguished because the crisis or potential crisis is occasioned by the death of the patriarch and thus the dissolution of the household. In the absence of a mature son survivors are at substantial risk of economic decline during the transitional period because of the insecurity of property rights and the appalling vulnerability of women in this society.

It was further argued that these linkages between demographic processes and economic mobility contribute to economic polarization in rural areas. Evidence was presented to suggest that fewer children of the poor survive to maturity and that those who do survive leave their parents' households at an earlier age. Other evidence shows that fertility rates are lower among the poor than among the wealthy. Moreover, as is shown by the incidence of widowhood, mortality is higher among poor adult males than among the more wealthy. Consequently, the poor should face a higher risk of property loss and

[1] Mead Cain, 'The household life cycle and economic mobility in rural Bangladesh', *Population and Development Review* **4**, 3 (1978), pp. 421–438.

economic decline during intermittent periods of economic crisis and at the time of household dissolution. The poor are further disadvantaged because they accrue proportionally less child labour, the labour of their children is less productive, and they exert less control over their children.

The combination of a harsh physical and social environment and extreme economic dependence creates special hazards for women in Bangladesh. These were explored in another paper.[2] Under ideal demographic circumstances, a woman's progression through life is marked by the successive transfer of her dependency from one category of male to another: first father, then husband and, finally, son. As a dependent member of a male-headed household, a woman's fortunes are determined by the forces that govern household mobility. This in itself is an unenviable position, because a woman has little control over decisions that most directly affect household welfare. Thus, an intelligent woman who marries a fool will usually suffer the fool's fate. Similarly, loss caused by poor judgment or bad luck of a father or son is likely to be shared by the daughter or widowed mother. Things can get worse quickly, however, under less than ideal demographic circumstances: that is, if a daughter is orphaned, a wife is divorced or widowed at a young age, or if an older woman has no surviving son on whom to depend. As was illustrated by several case studies, in these circumstances women are at risk of abrupt declines in economic welfare.

In a subsequent paper this analysis was extended by examining the empirical pattern of economic mobility in detail and introducing data from several locations in rural India for comparative purposes.[3] Focusing on the gain or loss of land assets by households between inheritance and the time of the survey, the analysis showed that overall mobility has been much greater in the Bangladesh sample, that the pattern of mobility in Bangladesh has resulted in increasing inequality in the distribution of land while inequality has decreased in the Indian samples, and that those who began as small landowners in Bangladesh experienced a higher incidence of loss and a lower incidence of gain than large owners, while the opposite is true for India. Among the several factors contributing to the contrasting experience of the two settings are differences in the dependence and vulnerability of women. Despite similarities in kinship organization (patriarchal authority structure, patrilineal descent and inheritance, and patrilocal residence), women in the Indian localities are less dependent upon men and less vulnerable in the absence of a male on whom to depend than women in rural Bangladesh. This was suggested by a comparison of the experience of women widowed at an early age or with no surviving mature son. Of 14 such women in the Bangladesh sample, ten were left with some land at the time of their husband's death and four were landless. Of the

[2] Mead Cain, S. R. Khanam and S. Nahar, 'Class, patriarchy, and women's work in Bangladesh', *Population and Development Review* **5**, 3 (1979), pp. 405–438.

[3] Mead Cain, 'Risk and insurance: perspectives on fertility and agrarian change in India and Bangladesh', *Population and Development Review* **7**, 3 (1981) pp. 435–474.

ten who initially had land, all subsequently sold or otherwise lost at least some of their land by the time they were interviewed, and four had lost all their land through distress sales. Of the four who were initially landless, one became a beggar, and all were destitute. In the Indian villages, of twelve women widowed under similar circumstances three were landless at the time of their husband's death and nine were left with some land. Only one of these nine subsequently lost or sold land. The three landless widows have managed to subsist on their own, through wage employment and operating small businesses.

In the present paper we extend this earlier work. The focus is on the elderly, their means of support and quality of life, in selected areas of rural Bangladesh and India. Of particular interest are the material consequences of reproductive failure, defined as failure to produce a son who survives and is able and willing to assume responsibility for parents who are no longer able to care for themselves. Emphasis is given to the fate of elderly women; however, because of the dependent status of women in the region this necessarily entails a consideration of the fate of elderly men also. Women's economic dependence links their fortunes to those of their husbands, for better or worse. Because wives are typically five to ten years younger than their husbands, however, the probability of being widowed is high for a woman, and in this their dependence sets them apart: the consequences of reproductive failure for them are potentially much more severe. Data on the living arrangements of the elderly and on other aspects of their material well-being are drawn from fieldwork conducted in a village in Mymensingh District, Bangladesh (Char Gopalpur) from 1976 to 1978, several villages in Maharashtra and Andhra Pradesh, India in 1980, and two villages in Raisen District, Madhya Pradesh, India in 1983.[4]

Living arrangements of the elderly

The evolution of the typical household is similar in both Bangladesh and the relevant areas of India. As noted in the introduction, residence patterns in both settings are patrilocal. When a daughter marries, she moves to her husband's home. When a son marries, he and his wife normally remain, for a period, as members of his parents' household. As other sons mature and marry, he may eventually establish a separate household, adjacent to his parents' dwelling. In some cases, two or more married sons remain in a single large household together with their parents, until their parents die. More often, however,

[4] Descriptions of study designs, sample characteristics, and the setting of Char Gopalpur, the two villages in Andhra Pradesh, and four villages in Maharashtra are given in Cain, *loc. cit.* in note 3. The research in Madhya Pradesh, was conducted in November 1983 in collaboration with the Economics Programme of ICRISAT, and focused on samples of 40 households from each of two villages in Raisen District. As in the case of the six other Indian villages, the samples were stratified by size of land-holding.

elderly parents live with one son, usually their youngest, with other sons maintaining separate, adjacent households.

Where parents are fully dependent, all sons usually contribute to their support, even if they do not all live in the same household. This arrangement may take the form of a larger allocation of family land to the son with whom the parent is actually living. Where land is still cultivated jointly by all sons, this will be reflected in the distribution of agricultural produce. Alternatively, each son might cultivate land independently and transfer a portion of his yield to his parents. Less frequently, an elderly parent circulates from son to son, taking meals for a period from each in turn.

Table 11.1 shows the distribution of elderly people according to living arrangement for the combined Indian sample (320 households) and Char Gopalpur, Bangladesh (343 households). The elderly are defined as those whose reported age is 60 years or older. The classification of living arrangements in Table 11.1 refers to whether or not a mature son lives with, or adjacent to, the elderly person. Those in the first category live in the same household with one or more married sons. They may have additional sons living in adjacent households. The second category includes those who live with one or more unmarried sons aged 15 years or older. They, too, may have one or more married sons who live in separate, adjacent households. The third category includes elderly people who have neither a married nor a mature unmarried son living with them, but who do have one or more married sons living in adjacent households. People in these first three categories can be said to have achieved reproductive success: they have mature sons living with them, or in close proximity, upon whom they can depend.

Table 11.1 Percentage distribution of persons aged 60 or older according to living arrangement in several rural areas in India and Bangladesh

Living arrangement	*India*	*Bangladesh*
With married son(s)	65	62
With unmarried mature son(s)*	11	16
No mature son:		
(1) Married son(s) adjacent	4	13
(2) Son(s) not adjacent†	7	6
With married daughter	6	2
Other‡	6	1
(*N*)	(114)	(94)

* Mature sons are defined as those aged 15 or older.
† Includes elderly people living alone, with a spouse, with sons less than 15 years old, and/or with unmarried daughters.
‡ Other arrangements: for example, living with a brother's or nephew's family.

The remaining three categories in Table 11.1 contain elderly people who have experienced reproductive failure – they do not live with, or in close proximity to, a mature son – and whose living arrangements are, thus, abnormal. The first (no mature son and no son adjacent) contains elderly people who live alone, with sons who are less than 15 years old, and with unmarried daughters of any age. The next category contains those who live with a married daughter and her husband. Finally there is a residual category that includes persons who live with such other adult relatives as a brother, nephew or uncle.

The majority of elderly people in both the Indian (65 per cent) and Bangladesh (62 per cent) samples live with one or more married sons. Overall, there is a marked correspondence between the distributions for India and Bangladesh, as one would expect, given similar systems of kinship and household formation. Altogether, 81 per cent of the elderly in the Indian sample and 91 per cent of the elderly in Char Gopalpur live either with, or adjacent to, a mature son.

There are differences between the two areas, however. One is that a higher proportion of elderly parents in Bangladesh have sons living adjacent to, rather than with them. This is reflected by a higher proportion of persons recorded in both the category 'no mature sons and married son(s) adjacent' (13 per cent for Bangladesh and 4 per cent for India) and the category 'with unmarried mature son(s)', many of whom also have married sons living in adjacent households. While this difference might be interpreted as a reflection of weaker ties between parents and children in Bangladesh than in India, it is in fact more a reflection of differences in settlement and housing materials. In rural Bangladesh the settlement pattern is dispersed, while in the relevant areas of India the pattern is nuclear, with houses and huts clustered together. Thus the land for housing is relatively cheaper in Bangladesh than in India. Moreover, building materials in rural Bangladesh are a good deal simpler and cheaper than in the areas of rural India that comprise our sample. Therefore, it is considerably less costly to form a new household in rural Bangladesh.

The significance of a household boundary separating elderly parents from sons varies a great deal between families. In a few cases it signals serious disaffection between father and son. Often it reflects a simple preference for privacy and physical independence on the part of the elderly or the young, while intergenerational bonds of affection and obligation remain strong. In other cases, specific strains between the father and daughter-in-law or the mother and daughter-in-law precipitate the creation of an independent household, while relations between father or mother and son remain strong. It is important to note also that such boundaries are not immutable. One is more likely to find elderly couples living in separate households than single widowed parents. When one spouse dies, the remaining parent is usually absorbed into a son's household.

More significant in Table 11.1 are the differences in the proportion of elderly people living either with a married daughter or in other arrangements. In both India and Bangladesh these are second-best alternatives to the cultural norm of living with sons, and they are chosen only when the preferred arrangement is not possible: in almost all cases, that is, when elderly people have no surviving son. As the table suggests, however, these second-best arrangements are more viable in the Indian than in the Bangladesh setting. The final two categories account for 12 per cent of the elderly in the Indian sample, but for only three per cent in Char Gopalpur.

In the absence of one's own son, depending on a daughter – which, in effect, means the daughter's husband – is precarious because in neither setting is there a strong, socially recognized obligation for a man to support his wife's parents. When he agrees to do so, it may be with considerable reluctance, and it is through generosity rather than a sense of obligation that he accepts the responsibility. The quality of care, therefore, is likely to be inferior; and regardless of the adequacy of material support, an elderly person in this position is likely to experience emotional strain and unhappiness. A more satisfactory option, for those with property, is more akin to adoption. In this case, a daughter's marriage is arranged with an explicit understanding that her husband will move to her parents' residence and assume the duties of a natural son. But this is an implicit contract that requires property to negotiate and sustain. In the Bangladesh sample only two people aged 60 or older were living with a married daughter, both of them in the home of their son-in-law. In the Indian sample, of seven elderly people living with married daughters, only one lived in his own home.

Adopting a son is perceived to be another option for those with no son of their own, at least in the Indian villages. In fact, however, there are few such cases in our samples, and only one in which the adopted son is old enough to care for the elderly. In this case, a relatively wealthy widower in one of the Madhya Pradesh villages adopted one of his sister's sons and now resides with him, his wife, and their children. Among those aged 60 or older in the Indian sample, there is one other case of adoption, but the boy is only 13 years old. The Bangladesh sample contains only one instance of adoption. In this case, too, the boy was less than 15 years old at the time of our study and, in addition, the parents had natural sons of their own. In practice, the pool of children who might be candidates for adoption is small, being comprised of the offspring of brothers or sisters. In addition, of course, the brother or sister must be willing to part with a son, a condition that is by no means automatically met. Adoption is thus a highly constrained option. It is the more so in Bangladesh, where the prevailing Islamic doctrine prohibits full adoption.[5]

These areas of India and Bangladesh also differ in the reliability of brothers as sources of support in old age, in the absence of sons, married daughters, and adoption. An earlier analysis, drawing on data from many of the same samples, indicated that there was greater economic interdependence and cohesiveness between brothers in India than in Bangladesh.[6] In the Indian villages, for

[5] Muhammad Abdul-Rauf, *The Islamic View of Women and the Family* (New York: Robert Speller and Sons, 1977), pp. 88–90.

[6] Mead Cain, 'Perspectives on family and fertility in developing countries', *Populations Studies* **36**, 2 (1982), pp. 159–175.

example, the incidence and average period of joint agricultural production among brothers whose father had died was much greater than in Char Gopalpur. Furthermore, in Char Gopalpur a large proportion of partners in land transactions – the great majority of which were distress sales – were brothers or other close patrikin, suggesting that extended kin function poorly as a mutual support group or insurance co-operative in that setting. The distinction between India and Bangladesh in the closeness of brothers appears to apply also to support in old age. Among the elderly in Table 11.1 there are, in fact, two cases of old men, one a widower and one never-married, living with brothers in the Indian sample, and none in Char Gopalpur. In interviews with adult males of all ages in the two Madhya Pradesh villages the help of brothers was regularly mentioned, along with adoption and assistance from a daughter and her husband, as a potential source of old-age support in the event of childlessness or no surviving son. This was not a common expectation in Char Gopalpur.

These options exhaust the institutional or semi-institutional solutions, in the absence of sons, to the welfare needs of the elderly in these settings. In each case, the three alternatives are less readily available and satisfactory in Bangladesh than in the Indian villages. Neither rural India nor Bangladesh maintains public or community-based institutions to provide for the needs of the elderly. The remaining varieties of living arrangements in Table 11.1 (not yet mentioned in the discussion), all of which are in the final residual category, include, for India: two widowed co-wives who live together; a widow living with her step-grandson; a widow living with her widowed daughter, daughter's son, and his wife; and a widower living with his sister's daughter and her husband. There is only one person in this residual category for Bangladesh, a widower living with his father's sister's grandson. These are exceptional arrangements and in themselves have little significance; however, their prevalence in the Indian villages, and rarity in Char Gopalpur, indicate the systemic tolerance of such arrangements in the former setting and relative intolerance in the latter.

Material consequences of reproductive failure

The distributions in Table 11.1 suggest that one consequence of reproductive failure is an early death. This is clearer in the case of Bangladesh. The proportion of people in a population who, given prevailing reproductive, mortality and marriage regimes can expect to have no surviving son at age 60 and over, can be predicted with some precision. For populations with demographic parameters similar to those found in these parts of rural India and Bangladesh, this figure should be about 17 or 18 per cent.[7] Our sample from India comes close: the proportion of elderly people in Table 11.1 with no

[7] This result, derived from a simulation model developed by John Bongaarts, assumes an age at first marriage (female) of 18, and an expectation of life at birth of 50 for females and 47 for males.

surviving son is 16 per cent. For Bangladesh, however, the proportion with no surviving son is only three per cent. Assuming, as we can, that the difference is real rather than an artefact of small sample size, the only possible explanation for this deficit is that a disproportionate number of people in Bangladesh, who are either childless or without living sons, die before their 60th birthday. More will be said about this pattern of differential mortality later; for the present we continue the description of living arrangements, focusing on variations by sex and economic status.

In Table 11.2 we present distributions of elderly people aged 60 or over according to living arrangements, broken down by sex and economic status for India and Bangladesh. As regards differences by sex, we recall that among once-married spouses the husband in these societies is, on average, between five and ten years older than the wife. For any given age, therefore, women will be that much further along than men in the development of their families. This is reflected in Table 11.2, which shows for both Indian and Bangladesh samples a larger proportion of women than men living with married sons and a smaller proportion living with sons less than 15 years old (the fourth category). In addition, a considerably higher proportion of women aged 60 or older are currently unmarried than men of this age. In our sample, 73 per cent of women in the Indian villages are not married, compared to 37 per cent of men, while in Char Gopalpur the figures are 67 and 11 per cent, respectively, for women and men. These sex differences in marital status reflect both the age difference

Table 11.2 Living arrangements of persons aged 60 or older by sex and economic status, India and Bangladesh (percentages)

	India		*Bangladesh*	
Living arrangement	*Males*	*Females*	*Males*	*Females*
With married son(s)	60	71	56	73
With unmarried mature son(s)	13	10	16	15
No mature son:				
(1) Married son(s) adjacent	6	2	15	9
(2) Son(s) not adjacent	10	4	10	0
With married daughter	5	8	2	3
Other	6	6	2	0
(*N*)	(62)	(52)	(61)	(33)
	India		*Bangladesh*	
	Poor	*Less poor*	*Poor*	*Less poor*
With married son(s)	62	67	42	74
With unmarried mature son(s)	8	14	25	10
No mature son:				
(1) Married son(s) adjacent	2	6	14	12
(2) Son(s) not adjacent	12	3	14	2
With married daughter	10	3	3	2
Other	4	8	3	0
(*N*)	(48)	(66)	(36)	(58)

between spouses and different rates of remarriage; the latter are considerably higher for men than for women, and considerably higher among Bangladeshi than among Indian men. As noted above, a single surviving parent is less likely than a surviving couple to maintain a separate household.

Differences by economic status are more striking than those by sex. Economic status here is measured by ownership of arable land. For Bangladesh the poor include the landless and those owning less than one and a half acres. Similarly, in the Indian sample, the poor are either landless or owners of small amounts of land. In both samples the rate of reproductive failure is higher among the poor than the less poor. In Bangladesh, 96 per cent of the relatively wealthy elderly live either with, or adjacent to, a mature son, compared to 81 per cent of the poor. In India the figures are 87 and 73 per cent, respectively, for the relatively wealthy and the poor. As noted in the introduction, the poor are less likely to achieve reproductive success because their fertility is slightly lower and the mortality of their children higher. The poor are also disadvantaged because they lack the control over sons that comes with property ownership. In Table 11.2 the consequences of this disadvantage are more evident for Bangladesh, where only 42 per cent of the poor live with married sons as compared to 74 per cent of the relatively wealthy.

There are further clues regarding the material consequences of reproductive failure and how these differ between Bangladesh and India to be found in Table 11.2. The impact of reproductive failure on the mortality of parents should be less severe for the relatively wealthy than for the poor. Therefore, if there were no economic mobility over the course of the life cycle – that is, if there were no transition between our categories of 'poor' and 'less poor' – *ceteris paribus*, a higher proportion of elderly people with no surviving son would be expected among the relatively wealthy than among the poor. Furthermore, this proportion should approach the predicted level of 17 or 18 per cent of those aged 60 or older. In fact, however, the reverse seems to be the case: the difference (deficit) between the observed and the predicted proportion of people with no surviving son is substantial among the relatively wealthy, and greater than among the poor. Thus in Bangladesh two per cent of the relatively wealthy have no living son, compared to six per cent of the poor, while in India the figures are 14 and 18 per cent, respectively. This suggests, more strongly in the case of Bangladesh than India, that reproductive failure is associated with consequences for mobility, so that those who start off in the 'less poor' category and then experience reproductive failure suffer economic decline at a disproportionate rate. The figures in Table 11.2 also suggest that women are quite likely to bear the brunt of the consequences of reproductive failure on mortality. Again, this is clearer in the case of Bangladesh. A simple comparison of the number of men and women in this age group yields a sex ratio of 1.8 (men to women) for Char Gopalpur and 1.2 for the Indian sample.

Table 11.3 shows persons aged 50–59 in addition to those aged 60 and older and the samples are further disaggregated by sex and economic status. First, we consider the ratios of men to women in Table 11.3 by age group and economic status:

	India		Bangladesh	
Age	Poor	Less poor	Poor	Less poor
50–59	1.2	1.1	1.4	1.1
60+	1.5	1.0	2.6	1.5

This pattern of sex ratios suggests that poor elderly women are at greatest risk of excess mortality. In the Indian sample a substantial deficit of women relative to men is found only among the poor aged 60 or older. In Char Gopalpur, a deficit of women is also found in the younger age group of the poor, and among those who are less poor and aged 60 or older.

A comparison of persons aged 50–59 in Table 11.3 with those aged 60 or older permits a fuller description of how living arrangements evolve. I focus initially on males. The critical category in Table 11.3 is that containing males who have no mature son living with them and no son living adjacent when they are in their fifties. Men in this category are already in an unenviable position relative to others in their cohort who have a married son living with or close to them, and their prospects for the future are not good. By this age their physical powers have probably begun to decline and their susceptibility to illness has increased; thus their need for a son on whom to depend has become greater.

Table 11.3 Living arrangements of the elderly by sex, class and age, India and Bangladesh (percentages)

	India							
	Females				Males			
	Poor		Less poor		Poor		Less poor	
Living arrangement	*50–59*	*60+*	*50–59*	*60+*	*50–59*	*60+*	*50–59*	*60+*
With married son(s)	56	63	55	76	38	62	63	58
With unmarried mature son(s)	24	5	24	12	29	10	24	15
No mature son:								
(1) Married son(s) adjacent	6	0	11	3	2	3	5	9
(2) Son(s) not adjacent	9	11	5	0	29	14	7	6
With married daughter	6	16	3	3	0	7	0	3
Other	0	5	3	6	2	3	0	0
(*N*)	(34)	(19)	(38)	(33)	(42)	(29)	(41)	(33)
	Bangladesh							
With married son(s)	50	60	74	78	0	35	45	71
With unmarried mature son(s)	15	30	0	9	39	23	40	11
No mature son:								
(1) Married son(s) adjacent	10	10	21	9	7	15	5	14
(2) Son(s) not adjacent	15	0	0	0	54	19	10	3
With married daughter	10	0	5	4	0	4	0	0
Other	0	0	0	0	0	4	0	0
(*N*)	(20)	(10)	(19)	(23)	(28)	(26)	(20)	(35)

They find themselves in this position because they have either been in childless marital unions, or in unions that produced no sons, or their sons have died, or were born late in their marriage. (An additional possibility is that people in this category have mature sons who live elsewhere. I found no such case among men aged 50–59; however, among those aged 60 or older, there was one instance in Char Gopalpur and two in the Indian villages.)

As men in this group age, they face several possibilities. Those with no surviving son must look for alternative arrangements, or else continue in their current situation – living alone or with unmarried daughters. As we saw earlier, very few men in Bangladesh succeed in finding an alternative. (In India the chances are better.) Those who have sons less than 15 years old will, depending on the age of their eldest son, remain in the same category for a considerable period or 'graduate' to one of the more desirable living arrangements. In the Bangladesh sample of men aged 50–59 with no mature son and no son living adjacent, six have no son and 11 have one or more sons less than 15 years old. Of the latter, the age of the eldest son ranges from two to twelve. Those without sons may be considered the most vulnerable; however, with the prospect of increasing debility, those with young sons are not in a much better position and those whose sons are less than ten years old may, in fact, be worse off because of the added burden of dependency that the sons represent. Another outcome for men in this category is, therefore, an early death. The survival prospects for their wives and young children are also not so good.

The inference drawn from Table 11.2 regarding the consequences of reproductive failure on mobility seems to be substantiated by a comparison of the distributions of poor and less poor men in Table 11.3, particularly for the age group 50–59. The probabilities of childlessness and the composition by sex and birth order are roughly the same for those who are better off as the poor. The survival chances of children of wealthier parents should be better than those of the poor, but not to the extent implied by the distributions in Table 11.3 – for Bangladesh, a fivefold difference between the poor (54 per cent) and the less poor (10 per cent) in the proportion with no mature son and no son adjacent. A similar, although smaller, difference is found in the Indian sample (29 and 7 per cent for the poor and less poor, respectively). The numbers on which the percentages in the cells of Table 11.3 are based are rather small and, thus, allow only tentative conclusions; however, these differences are consistent with the hypothesis that reproductive failure greatly increases the probability of economic loss. The evidence is stronger for Bangladesh because the difference in distributions by economic status is more pronounced and the proportion of poor men in the relevant category for Bangladesh (54) is almost twice that of India (29).[8]

[8] Also, in the case of India, the possibility that sampling variability may be partly responsible for the inter-class difference is suggested by an apparent inconsistency between the younger and older cohorts of the less poor group: that is, the proportion in the last three categories of living arrangement in the younger cohort is smaller (7) than in the older cohort (18), whereas it should be the reverse.

To explore the mobility hypothesis further, I attempted to reconstruct the economic histories of each of the individuals in the Bangladesh sample who live neither with, nor adjacent to, a mature son. This category contains 26 people aged 50 or older, three of them women and 23 men. Two of the women are married to men in the same category, and thus there are 24 unique histories. The mobility profiles are shown in Table 11.4. The information available for constructing these profiles was not equally good or complete for all cases. In the course of fieldwork in Char Gopalpur a sample of approximately one-third of all households was singled out for intensive study. For individuals from this sample there are detailed records of gains and losses of land assets from the time of inheritance to the date of our survey. For those not in this sample it was possible, in most cases, to determine whether they had experienced net gain or loss. For example, in a number of cases persons not included in the intensively studied sample had a brother who was included. For these individuals determining the amount of inheritance was straightforward. Current information on land ownership was gathered for all households; therefore, net change in assets could be estimated with precision. In other cases it was only possible to establish that some loss had occurred, without being able to pin down the exact

Table 11.4 Economic mobility profiles of persons aged 50 or older who live neither with, nor adjacent to, a mature son, Char Gopalpur, Bangladesh

Case	*Arable land inherited**	*Land owned in 1976*	*Net change*	*Comment*
1	?	0	–	
2	20	1	–	
3	?	0	–	
4	16	8	–	
5	11	4	–	
6	?	0	–	
7	6	0	–	
8	12	0	–	Dead by 1980 at age 56
9	18	0	–	
10	1	0	–	
11	28	0	–	
12	114	86	–	
13	?	0	–	Widow
14	0	0	0	Dead by 1980 at age 55
15	0	0	0	Dead by 1980 at age 60
16	?	14	?	Leases all land to others
17	?	0	0 or –	
18	?	0	0 or –	
19	?	0	0 or –	
20	?	0	0 or –	
21	?	1	0 or –	
22	?	0.5	0 or –	
23	?	32	+	
24	17	49	+	Sold some land in recent years

* Units of land are tenths of an acre.

amount. There is, in fact, only one case (number 16 in Table 11.4) about which we can say nothing regarding the net change in assets between inheritance and the time of our study.

Twenty-one of the 23 individuals in Table 11.4 for whom we have information, or 91 per cent, have either lost land assets or registered no change. The majority of them have lost assets, and of those whose position did not change, almost all were landless or practically landless to begin with. Furthermore, the majority of those who lost assets had lost everything and, by 1976, were landless. Recalling that our category 'less poor' includes households with one and a half acres or more, it can be seen that four cases (nos. 2, 4, 9 and 11) shifted from the 'less poor' to the 'poor' category during the period between inheritance and 1976. Others for whom we do not have information on amount of land inherited may also have made this transition, but it is not possible to be certain. What is very clear from Table 11.4, however, is that the experience of this sub-group elderly people with respect to economic mobility is extraordinarily bad. Compare their record with that of a representative sample of 114 households from the same village, of which 39 per cent gained assets between inheritance and the date of our study, 45 per cent lost, and 16 per cent registered no change.[9]

The frequency of land transactions in the Indian villages was much lower than in Char Gopalpur, and the proportion of all sales that could be classified as distress sales was also much smaller.[10] Only 18 per cent of all households in samples from three of the Indian villages experienced a decline in their land asset position between inheritance and the time of our survey. The great majority of those who lost land came from the largest-size ownership group, and they disposed of land for reasons that were demonstrably not related to reproductive failure. This is not to say that the material consequences of reproductive failure are negligible in these Indian settings; however, they are not reflected in changes in land owned, and are certainly less severe than in Bangladesh.

During a visit to Char Gopalpur in 1980 I learned of the death of the three individuals noted in Table 11.4.[11] All three were landless at the time of their death and two had never owned land. Their deaths cannot be directly attributed to the absence of mature sons to depend on during the periods of illness immediately preceding their deaths, but the lack of someone to depend on must have affected their chances of survival. Life is harsh for all villagers, and mortality risks are high for all regardless of reproductive histories. Life is harsher and mortality risks are higher for the relatively poor: work, when available, is physically demanding; diet, at the best of times, is poor; shelter, clothing and medical care are inadequate; drinking water is contaminated and

[9] Cain, *loc. cit.* in note 3, table 5, p. 446.

[10] *Ibid.*

[11] Others in Table 4 may have died by 1980. I visited only one-third of the village households.

the environment is otherwise unhealthy. Seen through Western eyes, practically all adults look older than their years. The added difficulty for those in Table 11.4 without land is that if they fall sick and are unable to work, there is no one in their households who can replace them as the family provider. Their children are too young, and their wives, no matter how willing, are excluded from most kinds of wage work and other forms of economic activity due to the strictness of purdah, the prevailing division of labour and extreme labour-market segregation. Therefore, illness brings a loss of income which in turn may exacerbate the illness and hasten death.

There is only one woman in Table 11.4, a widow aged 52 in 1976, with nine daughters, four of whom were still unmarried and living at home. The woman lives a marginal existence, earning some income by processing paddy, selling rice, and working in others' households. She receives occasional assistance from a son-in-law and from neighbours. Her husband left some land, which has since been sold. The reason why there are so few women in this category, either divorced or widowed, is quite simply that it is an economically untenable living arrangement – they cannot (and do not) survive. The widows of the three men in Table 11.4 who died must find another adult male on whom to depend or they, too, will probably die. Alternatives are scarce in Bangladesh; thus, as indicated by the distribution of elderly women in Tables 11.2 and 11.3 and the very high sex ratios at older ages, a great many women in this position die before their 50th birthday.

All but one of the men in Table 11.4 were married at the time of our study, and it is important to remember that their wives share their economic losses. The only difference between men and their wives is that wives experience the loss when they are younger. As regards economic mobility, the consequences of reproductive failure for widows with property are worse than for men. Judging from the experience of widows in Char Gopalpur, as reported in the introduction, property loss for these women is a certainty, and their transition to poverty can be very abrupt.

A more immediate potential consequence of reproductive failure, also with adverse economic implications for women, is divorce. In Char Gopalpur, a childless marriage will often be terminated by divorce. (Less commonly, a man will take additional wives.) Regardless of which partner is infertile or who initiates the divorce, a woman's 'value' as a potential spouse is severely reduced as a consequence, and if she remarries it will most probably be to someone who is less desirable (e.g. poorer or older) than her first spouse. Furthermore, this process delays the start of childbearing (assuming that she is fertile) and thus lessens her chances of achieving reproductive success. Divorce may also be initiated by a man if a union produces daughters but no sons. A sterile man can thus 'damage' several women. One man in Table 11.4 has been married six times, for example, and another five times. A number of others have been married twice. There is again a contrast between this situation and that in the Indian villages, where repeated divorce and remarriage in such circumstances is tolerated less, and thus occurs less frequently.

Summary

The material consequences of reproductive failure in Bangladesh are summarized in Figure 11.1. The broken lines in the figure relate to both men and women; the solid lines indicate processes that affect women alone. Poverty increases adult mortality risks for both men and women. It also increases the probability of reproductive failure. The poor are slightly less fertile than the wealthy, and their children face higher mortality. They are thus less likely to produce a son who survives. However, the primary determinant of reproductive failure or success is not economic status, but chance. Those who are unlucky enough not to produce a surviving son, regardless of their initial economic position, face a high probability of loss. Absence of a mature son at later ages increases a couple's vulnerability to economic crises. Through impoverishment, reproductive failure induces heavier mortality among ageing parents. The material consequences of childlessness or of failure to bear and raise a son are more severe for women than for men. Women have little choice but to depend on males for their survival. Infertility may trigger divorce, an event that automatically lowers the value of a woman in the marriage market and delays the process of family building. For unions that remain intact, reproductive failure, through impoverishment, increases the probability of widowhood. A young widow with no mature son on whom to depend faces the prospect of further rapid economic decline and very high mortality risk.

The vicious cycle portrayed in Figure 11.1 is less pronounced in the areas of rural India that we have considered. Women there are less vulnerable than in rural Bangladesh, and thus the adverse consequences of divorce and widowhood, although present, are less severe. In general, the environment of risk in the Indian settings is more benign, and better means of adjusting to risk have evolved. Therefore, reproductive failure does not so automatically lead to

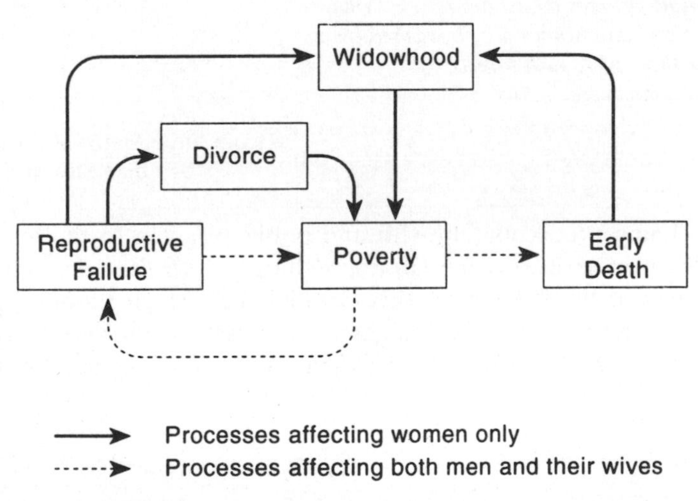

Figure 11.1 The consequences of reproductive failure.

the distress sale of assets and economic decline. For the elderly with no mature sons there exist alternative living arrangements which, although less desirable than the cultural ideal, are at least economically viable. The apparent contrast between Bangladesh and India may in part be a function of the criterion of economic mobility that we have used – gain or loss of land assets. More sensitive criteria and analysis are likely to diminish the contrast. Nevertheless, in view of the evidence on mortality and mobility differentials, it seems clear that the material consequences of reproductive failure are worse in rural Bangladesh than in rural India.

12 Bina Agarwal,
'The Woodfuel Problem'

Excerpts from: B. Agarwal, *Cold Hearths and Barren Slopes*, chapter 1. London: Zed Books Ltd (1986)

Remember those forests of oak and rhododendron,
fir and spruce,
those trees of pine and deodar
that have vanished? . . .

The trees near the streams have been felled,
the rivers have run dry;
the wild fruit, the herbs are gone,
the berries, the wild vegetables have disappeared. . . .

With the felling of trees landslides have started;
our fields, barns and homes are all washed away.
Where once there were lush forests
there is now sparseness. . . .

– from three Garhwali folk songs
by Ghan Shyam Shailani

The overwhelming preoccupation with and publicity given to the world oil crisis over the past several years, contrasts sharply with the relatively little recognition given to another energy crisis – a quiet one – which pre-dates that of oil. This is the crisis of woodfuel[1] shortages facing a vast proportion of people in the Third World, for whom it is the principal, and often, the sole source of inanimate energy.[2]

[1] The term 'woodfuel' refers throughout to wood-based fuel (essentially firewood and charcoal).

[2] This includes all energy sources other than human and animal.

The shortages are manifest in the longer hours spent, especially by women and children, for gathering fuel and fodder, in families going hungry because they do not have enough firewood to cook the food, in the rising prices of firewood in cities, in the shift to cattle dung and other substitute fuels for domestic use, in the growing confrontations between forest communities and forestry officials, and so on.

As forests in the Third World are beginning to disappear (attributable in large measure to historical and ongoing malpractices, and State policies) more and more people are being forced to depend on fewer and fewer tree resources for fuel. Anecdotal evidence on people's growing desperation is often dramatic. One observer notes how with the gradual denudation of the Sahelian countryside people have started cutting down saplings (Floor, 1977). Another mentions seeing people strip the bark off the trees that line the roads in Peshawar, Pakistan (Eckholm, 1976, p. 6). A third makes the following comment about the hills of Nepal: 'Nowhere can there be seen a tree or bush unscarred by axes, knives and browsing domestic animals. The imprint of people searching for fuel and fodder is to be seen everywhere' (Hughart, 1979, p. 28). Outside Ougadougou, in Burkina Faso, the land is noted to have been stripped of trees for 45 miles in all directions (French, 1978, p. 1), and the same is observed to be true for Niamey (capital of Niger) around which there is now a virtual desert in a 70 kilometer radius (Spears, 1978, p. 3).

In the Himalayan foothills of Nepal a journey to gather firewood and fodder took an hour or two a generation ago – today it is observed to take a whole day (Eckholm, 1976, p. 7). In Bara (Sudan) where ten years ago fuelwood was said to have been available after a 15 to 30 minute walk from the village, women now have to walk for at least one to two hours (Digerness, 1977, p. 15). In some villages of Gujarat, in India, where the surrounding forests have been completely denuded, women spend long hours collecting weeds and shrubs and digging out the roots of trees (Nagbrahman and Sambrani, 1983, pp. 36, 37). Inadequacies of fuel are driving people in several regions to shift to foods that are less fuel-consuming but of lower nutritional value, or to miss some meals altogether and go hungry. As one author puts it: 'None of the principal foodcrops of the tropics is palatable unless it has been cooked first. Lack of fuel can be as much a cause of malnutrition as the lack of food' (Poulsen, 1978, p. 13).

Yet the 'discovery' of a woodfuel crisis by policy makers in the Third World countries is not merely recent but also incidental – incidental on the one hand to a general assessment by these countries of their energy 'options' and on the other to the connection made, often unjustifiably, between deforestation (with its unignorable, adverse ecological consequences) and the gathering of wood for fuel.

The discovery, in turn, has brought in its wake the offering of a spate of 'solutions' in the form of improved wood-burning devices (especially domestic cooking stoves), a variety of tree-planting schemes, and improved wood-conversion hardware (especially charcoal kilns). These three have broadly been termed here as 'woodfuel innovations'.[3] Yet most of the attempts to promote these innovations have so far been ineffective.

The major part of the present study is devoted to an examination of the problems faced in the diffusion of these innovations and to an identification of the factors underlying the failure of many such promotional efforts. It is shown here that the nature of the woodfuel crisis, the causes underlying it, as well as the possibilities of alleviating it are linked to a complex set of social, economic and political factors. And even if the attempted solutions are confined to schemes for promoting hardware and reforestation, a range of problems, stemming from these factors, are likely to be encountered in the implementation of such schemes. These are problems which cannot be solved in the laboratory, and which highlight the need for a radical departure from the typical 'top-down' approach to innovation diffusion, and for some basic institutional and structural changes.

The case of woodfuel innovations is considered here within the general context of the diffusion of innovations in rural areas, and is seen as being in many ways similar to, and in many others distinct from that of other rural innovations. Given the overall dearth of case study literature relating specifically to the diffusion of woodfuel innovations, an examination of the experience of other innovations is not only useful, but often crucial to an understanding of the problems relating to woodfuel innovations.

However, the search for an effective solution to the woodfuel crisis, it is argued here, cannot be confined only to the issue of using available wood supplies more efficiently (hardware), or of increasing the supplies of wood available (forestry). It must also take account of the need to alter the existing distribution between uses and users, both of wood and of non-wood-related energy sources.

. . . .

1. The importance of wood as a source of fuel

The 1973 oil price hike initiated by the Organisation of Petroleum Exporting Countries (OPEC) provoked a world-wide alarm about an impending 'energy' crisis. The message was treated as ominous, especially by the oil-importing Third World countries, since it meant that the foreign exchange costs of such imports would become increasingly prohibitive. Added to this was the fear that even at the going price adequate supplies would not long be available. The same was seen to hold true for other fossil fuels.[4]

[3] The term 'innovation', here and subsequently, has been used in a broad sense to include both objects and practices perceived as being new by an individual or group, even if it has previously been in existence or in use elsewhere. Hence, improved stoves and kilns and tree-planting schemes could all be termed innovatious in this sense.

[4] Coal, oil and natural gas.

The realisation that there are physical limitations to the world availability of such fuels (quite apart from the financial costs) has led to a search for alternatives among renewable energy sources,[5] not merely to substitute for existing requirements but also to provide for the future. And one of the most important sources of renewable energy presently in use in large parts of the Third World is woodfuel. This includes wood burnt directly as a fuel (here termed 'firewood' or 'fuelwood') as well as fuels derived from wood, such as charcoal, producer gas, water gas, methanol and gasoline. Among these derived fuels, charcoal is the most important in the present context.

Unfortunately, no precise macro-estimates exist of the use of woodfuel in different countries. Most studies rely on the figures published by the United Nation's Food and Agricultural Organisation in its *Annual Yearbooks of Forestry Products*. These figures relate to the annual production of fuelwood plus charcoal (for domestic and other uses) which is typically assumed to equal annual woodfuel consumption. For some countries the figures given are either based on FAO's own estimates or obtained from unofficial sources. For others, they are official figures supplied by the countries and would usually cover only recorded removals from forests. Since the fuelwood consumed domestically is often gathered in the form of twigs and branches from forest floors or from trees located on open roads and in fields – removals which go largely unrecorded – the official figures of production would tend to underestimate actual consumption. This underestimation may be quite substantial in some regions, as was noted by Openshaw (1971) for Tanzania, when he compared the figures for recorded production of fuelwood and charcoal in 1970 with the consumption figures arrived at on the basis of survey data. A similar observation is made by Earl (1975, p. 58) for the 1970 FAO figures for Nepal. Further, although FAO gives country-specific figures for imports and exports of fuelwood plus charcoal, the world aggregates do not balance: the 1982 world export figure was only 74 per cent of the world import figure (FAO, 1984). Hence the FAO woodfuel-related statistics need to be treated with caution. Nevertheless, these figures are still generally considered the best available at the macro-level, and in the absence of better information they continue to be useful in providing broad pointers and giving an overall picture.

On the basis of the FAO figures, it is estimated that currently two-thirds of all inanimate energy in Africa, one-third in Asia, and one-fifth in Latin America comes from fuelwood and charcoal. Table 12.1, which gives data for 1982, indicates that in many countries such as Benin, Chad, Ethiopia, Nepal, Rwanda, Uganda and Upper Volta, firewood and charcoal together provide close to 90 per cent or more of the total inanimate energy used, and for the majority of other countries in Africa, and many in Asia and Central America, the figure is well over 50 per cent.

Again, while there are no accurate macro-estimates of the proportion of

[5] These cover solar, hydro-electric, wind, ocean-thermal, tidal, biomass (including wood) and nuclear.

Table 12.1 Commercial and woodfuel energy consumption in the Third World (1982)

Countries	*Commercial energy per capita kg. CE**	*Woodfuel† consumption per capita*		*Woodfuel as a per cent of total energy (commercial+woodfuel)*
		Cu.m	*kg. CE*	
Africa				
Angola	126.0	0.96	319.0	71.7
Benin	39.7	1.05	348.4	89.8
Central African Republic	41.7	1.06	353.0	89.4
Chad	21.5	1.62	541.3	96.2
Ethiopia	31.0	0.83	276.2	89.9
Ghana	122.3	0.60	198.8	61.9
Ivory Coast	287.0	0.78	260.4	47.6
Kenya	102.6	1.48	492.3	82.8
Madagascar	66.8	0.59	197.4	74.7
Malawi	50.0	0.90	301.3	85.8
Mali	29.6	0.58	194.2	86.8
Niger	47.3	0.58	191.7	80.2
Nigeria	185.6	0.84	278.7	60.0
Rwanda	21.3	1.11	370.9	94.6
Senegal	199.7	0.55	183.1	47.8
Sudan	84.0	1.75	583.7	87.4
Uganda	25.3	1.77	590.2	95.9
Upper Volta	31.4	1.04	346.7	91.7
Zaire	68.9	0.91	303.5	81.5
Zambia	389.2	0.87	289.6	42.7
Zimbabwe	636.1	0.96	321.4	33.6
Asia/Pacific				
Afghanistan	50.1	0.30	98.9	66.4
Bangladesh	49.8	0.32	108.5	68.5
China	581.3	0.16	51.7	8.2
India	200.6	0.29	96.8	32.5
Indonesia	234.9	0.74	245.1	51.1
Korea (Republic of)	1439.0	0.18	58.8	3.9
Malaysia	985.8	0.50	165.2	14.4
Nepal	11.2	0.90	298.6	96.4
Pakistan	225.8	0.21	68.7	23.3
Philippines	329.6	0.55	182.4	35.6
Sri Lanka	121.9	0.49	163.8	57.3
Thailand	364.4	0.73	243.5	40.0
Papua New Guinea	293.2	1.78	594.4	67.0
South and Central America				
Brazil	704.8	1.25	417.5	37.2
Chile	908.9	0.50	167.9	15.6
Colombia	897.2	0.50	168.2	15.8
Cuba	1440.4	0.29	95.4	6.2
Guatemala	217.5	0.84	279.3	56.2
Honduras	233.2	1.03	342.1	59.5

Notes: (1) * CE = coal equivalent.
† Woodfuel consumption figures relate to fuelwood+charcoal.

(2) The total energy estimates do not include other traditional sources of energy such as cattle dung, crop residues, etc. because of the absence of micro-estimates for these sources.
(3) Per capita consumption has been computed by dividing:
  (a) available figures of total commercial energy in kg. CE, and total woodfuel energy consumption in cubic metres; and
  (b) computed figures of total woodfuel energy consumption in kg. CE; by available estimates of populations in mid-1982.
(4) Conversion factor used for woodfuel: 1 cubic metre = 0.33 metric tons CE. (This is taken from *Energy Statistics Yearbook 1982*; p. xxvi.)

Sources: (1) Data on total commercial energy taken from United Nations (1984): *Energy Statistics Yearbook 1982*, Dept. of International Economic and Social Affairs. Statistical Office, New York.
(2) Data on total woodfuel consumption taken from FAO (1984): *1982 Yearbook on Forestry Products*, FAO Forestry Statistics Series, Rome. (Note: Production is assumed as equal to consumption in this data source.)
(3) Data on population taken from World Bank (1984): *World Development Report, 1984*, Washington D.C.

woodfuel burnt directly as wood (i.e. firewood) and that burnt as charcoal etc., broad assessments suggest that firewood is by far the more important. From FAO estimates, the world production of charcoal in 1982 comes to only 6.6 per cent of the world woodfuel production in that year. A range of macro-studies indicate that the use of charcoal is predominantly concentrated in the urban areas. In rural areas, wood is usually burnt directly as a fuel in most households.

The consumption of woodfuel in per capita terms varies greatly between the Third World countries as seen from Table 12.1. Leaving aside countries such as China, S. Korea and Cuba where woodfuel energy constitutes less than 10 per cent of the total energy consumed, and basing our comparison on the others, we note from the table that the lower end of the range is occupied largely by the countries of South Asia, with per capita consumption in most being less than 0.5 m^3, and being especially low in Bangladesh, India and Pakistan. The upper end of the range is occupied largely by the countries in Africa, many of which have a per capita consumption above 1.0 m^3, and some such as Chad, Sudan and Uganda, consume over 1.5 m^3. In between fall the majority of countries in Africa, S.E. Asia and South and Central America, with consumption ranging between 0.5 and 1.0 m^3. One of the important factors influencing these variations in consumption among countries where woodfuel is an important energy source, is the relative availability of wood for fuel in different ecological regions. It can be seen from Table 12.2 that the per capita availability of fuelwood varies from levels several times above estimated needs in the high forest areas of the world, to much below needs in the arid and sub-arid zones.

Table 12.2 Fuelwood* use by ecological regions (1980) (in m^3 per capita per year)

	Fuelwood*	
	Needs	Availability
Africa (South of Sahara)		
Arid and sub-arid areas	0.5	0.05 to 0.01 (AS)
Mountainous areas	1.4 to 1.9	0.5 to 0.7 (AS)
Savanna areas (a)	1.0 to 1.5	0.8 to 0.9 (D)
(b)		1.8 to 2.1 (PD)
High forest areas (a)	1.2 to 1.7	1.8 to 2.0 (PD)
(b)		5.0 to 10.0 (S)
Asia (Far East)		
Mountainous areas	1.3 to 1.8	0.2 to 0.3 (AS)
Indo-gangetic plains (S. Asia)	0.2 to 0.7	0.15 to 0.25 (D)
Low land areas in S.E. Asia (Plains and islands)	0.3 to 0.9	0.2 to 0.3 (D)
High forest areas	0.9 to 1.3	1.0 to 6.0 (S)
Latin America		
Andean plateau	0.95 to 1.6	0.2 to 0.4 (AS)
Arid areas	0.6 to 0.9	0.1 to 0.3 (AS)
Semi-arid areas	0.7 to 1.2	0.6 to 1.0 (D)
Sub-tropical and temperate areas	0.5 to 1.2	1.9 to 2.3 (PD)
Abundant forest areas	0.5 to 1.2	2.5 to 10.0 (S)

Note: * Wood for charcoal is included here. Hence, strictly in terms of the definition used in this book, these figures would relate to 'woodfuel', not just to 'fuelwood'.
AS: Acute Scarcity; D: Deficit; PD: Prospective Deficit; S: Satisfactory.
Source: FAO (1981): *Map of the Fuelwood Situation in Developing Countries – Explanatory Note*, FAO, Rome.

. . . .

2. Firewood collection: from where and by whom?

In most rural areas, firewood has been and still largely tends to be non-monetized so that people usually have to depend on what they can themselves gather. By one estimate, for rural India, only 12.7 per cent of total firewood consumed is purchased, the rest being obtained either from one's own land or from the land of others (Government of India, 1982, p. 6). In Nepal, in some rural areas, (e.g. Pangua village) all firewood needs are met by self-collection, nothing is purchased (Bajracharya, 1983). The same is true for some villages in Peru (Skar, 1982). In fact, the self-collection of firewood still appears to be the commonly observed pattern in most of the rural Third World.

Rural households with land can gather firewood from trees located on their own land, supplemented by crop residues, etc. The landless, however, have to depend on wood from common land, or where allowed to do so, gather it from other people's land by say contributing labour in return. As incomes decline, the dependency for fuel collection from sources other than one's own increases: Table 12.3 based on survey data illustrates this for south India.

Table 12.3 Fuel collection from own sources by income class in rural south India (1979–80)

Income class Rs./annum	*Per cent of total households collecting fuel from own sources*		
	Firewood	*Dung cake*	*Vegetable wastes*
Less than 3000	33.8	69.1	43.8
3001–8000	70.2	92.6	77.9
8001 and above	83.8	93.0	98.1
All classes	60.9	84.5	69.9

Source: ITES (1981, p. 176): *Rural Energy Consumption in Southern India*, Institute of Techno-Economic Studies, Madras.

Inequalities in land ownership patterns also lead to high inequalities in fuel availability between households. In Bangladesh, Briscoe (1979, p. 628) noted, for example, that in the village he surveyed 80 per cent of all fruit and firewood trees were owned by 16 per cent of the families, who also owned 55 per cent of the cropped land and 46 per cent of the cattle. Islam (1980, p. 73) again, on the basis of a survey of Bangladeshi villages, noted that 52.4 per cent of the trees were owned by 11.4 per cent of the households. A similar pattern is found in village studies in India: a six-village study in Karnataka highlights the close positive relationship between average landholding size and the quantity of firewood consumed in the household (ASTRA, 1981). In other words, access to land affects not only the household's access to food but also to the fuel used to cook it.

The collection of firewood in most parts of the Third World is done primarily (and sometimes exclusively) by women and children, with men usually (but not always) providing some supplementary labour. The actual time taken for collection varies in different regions according to the availability of tree resources, but in most cases it is a strenuous and time-consuming task. Table 12.4 brings together some of the existing studies which indicate the time taken and distances travelled in rural households for firewood collection. In a significant number of cases the time is 3–4 hours per day or more. In some areas, as in the Sahel, women have to walk up to 10 km for this purpose; in Gambia it takes from midday to nightfall to gather an evening's supply, while in parts of India women spend five hours per day on an average, travelling 5 km or more over difficult mountainous terrain. Fleuret and Fleuret (1978, p. 318) note in the context of Tanzania: 'Every aspect of fires and fuels is the work of women in Kwemzitu and no other task is considered to be as tiring or as demanding or to have so little to show for itself'. In some parts of East Africa, the collection of

Country	Region	Year of data	Firewood collection*		Data source
			Time taken	Distance travelled	
Asia					
Nepal	Tinan (hills)	1978	3 hr/day	n.a.	Stone (1982)
	Pangua (hills)	late 1970s	4–5 hr/bundle	n.a.	Bajracharya (1983)
	n.a.	n.a.	0.62 hr/day	n.a.	Acharya and Bennett (1981)
India	Chamoli (hills)				
	(a) Dwing	1982	5 hr/day†	over 5 km	Swaminathan (1984)
	(b) Pakhi		4 hr/day	over 3 km	
	Gujarat (plains)				
	(a) Forested		Once every 4 days	n.a.	Nagbrahman and Sambrani (1983)
	(b) Depleted	1980	Once every 2 days	4–5 km	
	(c) Severely depleted		4–5 hr/day	n.a.	
	Madhya Pradesh (plains)	1980	1–2 times/week	5 km	Chand and Bezboruah (1980)
	Kumaon Hills	1982	3 days/week	5–7 km	Folger and Dewan (1983)
	Karnataka (plains)	n.a.	1 hr/day	5.4 km/trip	Batliwala (1983)
	Garhwal (hills)	n.a.	5 hr/day	10 km	Agarwal (1983)
Bangladesh	Chargopal	1977	0.4 hr/day	n.a.	Cain, *et al.* (1979)
Indonesia	Java	1972–73	0.3 hr/day	n.a.	White (1976)
Africa					
Sahel	n.a.	c.1977	3 hr/day	10 km	Floor (1977)
	n.a.	1981	3–4 hr/day	n.a.	Ki-Zerbo (1981)
Niger	n.a.	c.1977	4 hr/day	n.a.	Ernst (1977)
Upper Volta	n.a.	n.a.	4½ hr/day	n.a.	Ernst (1977)
Sudan	Bara	1966–67	0.33 hr/day	n.a.	Digerness (1977)
		1976–77	1–2 hr/day	n.a.	
Tanzania	(hills)	1975–76	1.6 hr/day	n.a.	Fleuret and Fleuret (1978)
Kenya	n.a.	n.a.	3–3½ hr/day	n.a.	Earthscan (1983)
Ghana	n.a.	n.a.	4–5 trips/week	2½–7 miles	DEVRES (1980)
Latin America					
Peru	(a) Pincos (highlands)	1981	1.33 hr/day	n.a.	Skar (1982)
	(b) Matapuan (highlands)	1981	1.67 hr/day	n.a.	

Notes: * Firewood is noted to be collected principally by women and children in all the studies listed, with the exception of Java where the labour put in is primarily by men.
† Average computed from information given in the study.
n.a. = Information not available.

firewood is considered appropriate only for married women, although in most countries, children especially young girls, help their mothers (DEVRES, 1980, p. 21).

While the collection of firewood is primarily for self-consumption, yet increasingly as other sources of livelihood get eroded, selling firewood for an income is becoming common in some areas. The sellers are usually rural women belonging to the poorer households, who take the wood to nearby towns. In the Bara (Sudan), firewood sales represent the sole source of income in many cases (Digerness, 1977). This is also true in parts of India, as for example in certain areas of Bihar, where women of poor tribal households travel 8–10 km in search of firewood (which is usually procured illegally), then catch a truck or train to the nearby town (Ranchi), spend the night at the station, and return with a meagre earning of Rs.5.50, on an average, for a headload of about 20 kg of wood (Bhaduri and Surin, 1980). 'Headloading' as it is called has become common here after a drought left many households destitute, some years ago. By one estimate 2–3 million rural people in India are headloaders and are spread across several States (Agarwal and Deshingkar, 1983). Wood is also sometimes converted into charcoal and sold by the rural poor for a livelihood in several Third World countries (DEVRES, 1980).

3. Implication of shortages

The availability of wood, even though it constitutes a *renewable* resource, is limited, and becoming increasingly short in supply. Large tracts of land which were earlier thickly wooded, today lie barren. By one estimate there will be virtually no forests left in the Third World countries in 60 years time, in the absence of serious measures to counteract this; and the present rate of reforestation is assessed by him to be less than 10 per cent of that necessary to supply the minimum needs of these countries by the year 2000 (Spears, 1978, pp. ii, 15).

Deforestation, in turn, can have a range of devastating consequences both for the society and for the individual. At the social level these relate to the effects of soil erosion, flooding, climatic maleffects, the spread of deserts, the drying of previously perennial streams, an increased frequency of landslides in the hill areas, the rapid siltation of rivers and reservoirs, and so on (Eckholm, 1976; Digerness, 1977). As a result, there is also an adverse effect on agricultural production which has implications both for society as a whole and for those dependent on agriculture for their livelihood. By one estimate, 40 per cent of the farmers of the Third World live in valley lands and depend heavily for their irrigation water on the 'sponge effect' of forests in surrounding catchment areas (Myres, 1978, p. 951). The disappearance of forests produces a tendency for rain water to be released in floods during the wet-season, followed by drought in other seasons. In the forested zones of Indonesia, Malaysia and the Philippines, the green revolution is assessed to be losing its momentum because farmers can no longer find regular supplies of irrigation water for multiple rice farming (see Myres, 1978, p. 951).

For the individual, in addition to the indirect implications of ecological destruction caused by deforestation (including the effects on farm production) are the direct ones, relating especially to the decreasing supplies of wood for fuel. Some of these noted or likely direct consequences are enumerated below.

To begin with there is a substantial increase in the time and energy spent (especially by women and children) to gather firewood. In parts of Sudan, over the past decade or so, the time taken to collect firewood has increased over four-fold (Digerness, 1977, p. 15). . . . In India, in parts of Bihar where 7–8 years ago women of poor rural households could get enough wood for self-consumption and sale within a distance of 1.5 to 2 km, they now have to trek 8–10 km every day. Similarly, in some villages of Gujarat (India) the women, even after spending several hours searching, do not get enough for their needs, and have to depend increasingly on the roots of trees and on weeds and shrubs which do not provide continuous heat, thus increasing their cooking time as well (Nagbrahman and Sambrani, 1983, p. 36, 37). At times even straw is used. Given the already heavy working day of most rural women, any further increase in their time spent on firewood collection and cooking becomes an overwhelming burden. Also, Hoskins (1983) notes how in parts of Africa, because of the extra time needed to collect firewood, daughters are now taken out of school to help their mothers. For the mothers themselves there might be a resultant economic cost in terms of employment or other income-earning opportunities foregone.

A related aspect is the greater time now spent by women in cooking because they have to adapt their cooking methods to economize on firewood. In ecological zones where wood has generally been scarce, conventional cooking practices are already such as to conserve wood at the cost of greater time and effort spent on preparing the meal. For example, in parts of the African Sahel, women light only the ends of logs and branches, placed like the spokes of a wheel, and cook with the heat generated from the ends of the spokes.

Where adequate fuel is not obtainable despite the extra time and effort spent in firewood collection, there have been noticeable changes in consumption patterns. In some areas families have had to reduce the number of meals cooked, as in Bangladesh and the African Sahel: in the latter region there has been a shift from cooking two meals to one meal a day. In other cases there is a shift to less nutritious foods: in Guatemala, to conserve firewood, families are shifting away from beans (which were a part of the staple diet) because they take too long to cook. In the Sahel, the diet of millet is rarely supplemented by meat, especially because a large quantum of firewood is needed in its preparation; also due to fuel shortages, families in this region are noted to be shifting from millet to rice because rice takes less time to cook. Again, in eastern Upper Volta, attempts by the government to introduce the cultivation of soyabeans is being resisted by the women because of the longer cooking time and greater quantity of fuel that soyabeans require relative to the traditional cowpeas.

Necessity is also driving people in some areas to shift to food which can be eaten raw, or to eat partially cooked food (which could be toxic), or to eat cold leftovers (with the danger of food rotting in a tropical climate), etc. All this increases the vulnerability to ill-health and infection. In some regions, adverse nutritional consequences are also found to result from the trade-off between time spent in gathering fuel and that spent in cooking. In Peru, for instance, in areas where firewood is scarce and its collection requires women to travel far from home, taking several hours, much less time can be spent by them in cooking: this is noted to adversely affect the quality of the family's diet.

Increased deforestation also adversely affects the availability of forest produce other than firewood, which is gathered by the local people for their own needs and often for sale as well. This includes not only fruit and fodder, but also herbs and plants for local medicines, turpentine and resin, certain kinds of flowers, such as *mahua* in India, which provide oil, liquor and cattle feed, the seeds and pods of various trees such as sal and *neem* which provide oil, the twigs of the *neem* which are used to brush teeth in India, the leaves of the sal which are used as disposable plates, the leaves of the *tendu* from which *biri* (a type of Indian cigarette) is rolled, various types of fibre and floss, tans and dyes, and so on. By one estimate some 30 million people in India (mostly tribals and forest dwellers) depend on such minor forest produce for some part of their livelihood (Kulkarni, 1983, p. 194); and they are getting increasingly marginalised.

Additionally, deforestation has been noted to lower the water table, making it much more difficult and time consuming for women to obtain water, especially during the dry season. This is noted for instance in Senegal (Hoskins, 1983), in India (Bahuguna, 1984), and in Nepal (Dogra, 1984). As a woman of the Uttarkhand hills in India puts it:

> 'When we were young, we used to go to the forest early in the morning without eating anything. There we would eat plenty of berries and wild fruits . . . drink the cold sweet (water) of the *Banj* (oak) roots . . . In a short while we would gather all the fodder and firewood we needed, rest under the shade of some huge tree and then go home. Now, with the going of the trees, everything else has gone too.' (in Bahuguna, 1984, p. 132).

In some areas, such shortages are found to be linked crucially to life and death questions. In the above mentioned region of India, for example, a woman grass-roots activist has noted several cases during the past three years of young women committing suicide because of the growing hardship of their lives with ecological deterioration. Their inability to obtain adequate quantities of water, fodder and fuel leads to scoldings by their mothers-in-law (in whose youth forests were plentiful), and soil erosion has made it much more difficult for the women to produce enough grain for subsistence in a region of high male out-migration. In one year seven such cases of suicide were observed – four of them in a single village where shortages are especially acute (Bahuguna, 1984).

Another manifestation of shortages is the significant rise in the price of firewood along with its increasing commercialisation. While this price rise is observed mainly in the urban areas, with prices in many Third World cities having more than doubled over the past decade, it also adversely affects those among the rural population who have to depend on firewood purchase. These are mostly the poor households with few personal resources of land or cattle to provide fuel. Also, as firewood gets commercialised, it will become increasingly difficult for such households to gather it free or at minimal cost.

Further, in general, there is noted to be a shift from firewood to other fuels such as cattle dung, maize stalks, etc. The use of cattle dung, in particular, has a high opportunity cost in terms of the loss of agricultural output that could have been produced had the dung been used as manure instead. By one rough estimate, for every ton of cattle dung burnt there is a loss of 50 kg of foodgrains, and given the further estimate that in the Third World about 400 million tons net weight of cattle dung is burnt every year, this could amount to a loss of 20 million tons of grain annually (Spears, 1978, pp. 4, 5).

In short, the signs of an existing woodfuel crisis are undoubtedly apparent in many areas, and those of an impending one in many others.

. . . .

4. Causes of the crisis

In considering the causes of the woodfuel crisis a complex set of factors is encountered. Impinging on the issue are three closely interrelated aspects: one, of the absolute availability of wood for all uses (which depends on the tree resources of the country, barring imports); two, of the availability of wood for fuel (which depends on the distribution of available wood supplies between different uses); and three, of the availability of woodfuel to the poor (which depends on the distribution of available woodfuel supplies between people). As elaborated below, woodfuel shortages being faced by the poor today emerge from the particular uses to which forest land and wood resources have been put over the years by specific classes of people, and cannot be traced in any straightforward way to the gathering of firewood by the poor for their domestic use.

. . . .

Woodfuel shortages being experienced today are closely related to the form and degree of forest exploitation that has taken place over several decades *in the past* (as a result of specific State policies, private profit consideration, demographic pressures, etc.). This exploitation has taken a variety of forms, such as the clearing of forest land for agriculture, plantations, pastures and other uses; the commercial logging of trees for timber either for building purposes or for exports; the use of wood as an industrial raw material and as a fuel in small-scale and cottage industries, and so on, and is not necessarily associated with the use of wood as a domestic fuel. Even today, the deforestation that is occurring cannot be attributed straightforwardly to the gathering of wood by poor rural households for their domestic use, since a good deal is likely to be as twigs, etc. It is noteworthy that the use of wood for almost all purposes other than as a domestic fuel in the rural areas, requires the cutting down of trees.

Even to the extent that in some areas trees are being cut or the barks of trees being stripped off by the poor to obtain fuel, it must be seen as a *symptom* of the crisis – a reaction of the people who are the severest hit by it but who cannot be held responsible for having *caused* the crisis. As one landless woodcutter in the Sudan put it: 'We take trees belonging to other people. We cut them when they are too young. We never pay royalty . . . We are in a miserable state after our animals starved to death during the drought. We must live for something. What else can we do?' (Digerness, 1977, p. 12). What indeed! With no land and no employment, cutting trees for fuel and producing charcoal for sale is their only means of survival in this context.

This last observation brings us to the major aspect relating to the experience of woodfuel shortages as a 'crisis', namely the maldistribution of material resources between different classes and social groups within a country, which affects both the absolute availability of wood and its relative distribution between uses and users. For example, the use to which land is put, and the effect this has on deforestation, is related to the unequal distribution in the ownership of and access to land among people. Thus, on the one hand, large tracts of forests on both private and public land are often cut down for the private gain of a few, and on the other hand, an increasing number are forced to survive on diminishing tracts of 'common' forest, which they are then blamed for destroying. As one recent study in the villages of Rajasthan and Madhya Pradesh (India) noted, common property resources are crucial for the survival needs of those with little or no land: 86 per cent of agricultural labourers and small farmers in the surveyed Rajasthan villages and 98 per cent in the Madhya Pradesh villages depend on common land for fuel while none of the large farmers in either region do so. As more and more of such land is being brought under private ownership (with much of it going to the larger farmers) the poorest are forced to depend on shrinking areas of common land (Jodha, 1983, pp. 8, 9).

Here the issue is not one of 'weaning tribal people from their age-old practice of shifting cultivation' by 'careful demonstration' (of improved methods) and 'persuasive measures' as some observers emphasize, but rather of radical agrarian reform for bringing about a more egalitarian pattern of land ownership and control, and of ensuring alternative, viable means of livelihood for the poor.

Again, the persistence of the illegal cutting down of government-owned forests by local timber merchants who make large profits on public resources (as observed in parts of south Asia) is closely linked to the economic and political power that these men command. This enables them not only to control the local villagers but also to ensure the allegiance of the local government officers. The landless, in contrast, are liable to face persecution if caught collecting even small amounts of firewood without a permit (BRAC, 1980; Swaminathan, 1982).

In other words, the maldistribution of material resources, by determining who has access to available wood supplies, also affects the *relative* availability of woodfuel to different sections of the population, and hence the relative cost of the shortages to different people.

Further, the severity of the woodfuel crisis is accentuated by the high and rising prices of alternative fuels such as kerosene. This, in turn, is closely related to the dynamics of supply–demand that operate between the oil-exporting countries and the developed oil-importing ones. The unequal access of non-oil-rich Third World countries *vis-à-vis* the rest of the world constitutes the international dimension of a country's internal crisis.

Selected references

Acharya, M. and Bennett, L. 1981: *The rural women of Nepal: An aggregate analysis and summary of eight village studies*. CEDA, Tribhuvan University, Nepal.

Agarwal, A. 1983: Cooking energy systems – problems and opportunities. *Discussion Paper*. Delhi: Centre for Science and Environment.

Agarwal, A. and Deshingkar, P. 1983: Headloaders: hunger for firewood – 1. *Report Number 118*. Delhi: Centre for Science and Environment.

ASTRA 1981: *Rural energy consumption patterns: A field study*. Bangalore: Indian Institute of Science.

Bahuguna, S. 1984: Women's non-violent power in the Chipko movement. In Kishwar, M. and Vanita, R. (eds), *In search of answers: Indian women's voices from Manushi*. London: Zed Books.

Bajracharya, D. 1983: Fuel, food and forest: dilemmas in a Nepali village. *Working Paper 83–1*. Honolulu: East-West Center.

Batliwala, S. 1983: Women and cooking energy. *Economic and Political Weekly* 18, 24–31 December.

Bhaduri, T. and Surin, V. 1980: Community forestry and women headloaders. In *Seminar report: Community forestry and people's participation*. Ranchi: Consortium for Community Forestry.

BRAC 1980: *The net: Power structure in ten villages*. Dhaka: Bangladesh Rural Advancement Committee.

Briscoe, J. 1979: Energy use and social structure in a Bangladesh village. *Population and Development Review* 5 (4).

Cain, M., Khanam, S. R. and Nahar, S. 1979: Class, patriarchy, and the structure of women's work in rural Bangladesh. *Working Paper 43*. New York: Population Council.

Chand, M. and Bezboruah, R. 1980: Employment opportunities for women in forestry. In *Community forestry and people's participation – seminar report*. Ranchi: Consortium for Social Forestry.

DEVRES 1980: *The socio-economic context of fuelwood use in small rural communities*. AID Evaluation Special Study No. 1, USAID, August.

Digerness, T. 1977: Wood for fuel: the energy situation in Bara, the Sudan. *Mimeo*: Department of Geography, University of Bergen.

Dogra, B. 1984: *Forests and people*. Bharat Dogra, A-2/184 Janakpuri, New Delhi.

Earl, D. 1975: *Forest energy and economic development*. Oxford: Clarendon Press.

Earthscan 1983: Earthscan Press Briefing Document 37. (*mimeo*) London: International Institute for Environment and Development.

Eckholm, E. 1976: The other energy crisis. *World Watch Paper 1*. World Watch Institute, USA.

Ernst, E. 1977: Fuel consumption among rural families in Upper Volta, West Africa. (*mimeo*) Ouayadougou: Peace Corps.

FAO 1984: *1982 Yearbook on forestry products*. Rome: FAO.

Fleuret, P. and Fleuret, A. 1978: Fuelwood use in a peasant community: a Tanzanian case study. *Journal of Developing Areas* 12.

Floor, W. 1977: The energy sector of the Sahelian countries. *Mimeo*: Ministry of Foreign Affairs, The Netherlands.

Folger, B. and Dewan, M. 1983: Kumaon hills reclamation: end of year site visit. (*mimeo*) Delhi: OXFAM America.

French, D. 1978: Renewable energy for Africa: needs, opportunities, issues. *Paper submitted to USAID*, Washington, DC, AFR/DR/SDP, July.

Government of India 1982: *Report of the Fuelwood Study Committee*. New Delhi: Planning Commission.

Hoskins, M. 1983: Rural women, forest outputs and forestry projects. *Discussion Paper MISC/83/3*. Rome: FAO.

Hughart, D. 1979: Prospects for traditional and non-conventional energy sources in developing countries. *World Bank Staff Working Paper* 346.

Islam, M. 1980: *Study of the problems and prospects of biogas technology as a mechanism for rural development: Study of a pilot area of Bangladesh*. Report to the International Development Research Center, Ottawa.

Jodha, N. 1983: Market forces and the erosion of common property resources. *Paper presented at the International Workshop on Agricultural Markets in the Semi-Arid Tropics, ICRISAT*, October.

Ki-Zerbo, J. 1981: Women and the energy crisis in the Sahel. *Unasylva* 33 (11).

Kulkarni, S. 1983: Towards a social forestry policy. *Economic and Political Weekly* 18 (6).

Myres, N. 1978: Forests for people. *New Scientist* 21/28 December.

Nagbrahman, D. and Sambrani, S. 1983: Women's drudgery in firewood collection. *Economic and Political Weekly* 1–8 January.

Openshaw, K. 1971: *Present consumption and future requirements of wood in Tanzania*. Rome: FAO.

Poulsen, G. 1978: *Man and trees in tropical Africa*. Ottawa: International Development Research Center.

Skar, S. 1982: Fuel availability, nutrition and women's work in highland Peru. *World Employment Programme Research Paper 10/23*. Geneva: ILO.

Spears, J. 1978: Wood as an energy source: the situation in the developing world. *Paper presented to the 103rd Annual Meeting of the American Forestry Association*, October.

Stone, L. 1982: Women in natural resources: perspectives from Nepal. In Stock, M., Force J. and Ehrenreich, D. (eds), *Women in natural resources: An international perspective*. Moscow, USA: University of Idaho Press.

Swaminathan, S. 1982: Environment: tree versus man. *India International Centre Quarterly* 9, 3–17.
Swaminathan, M. 1984: Eight hours a day for fuel collection. *Manushi* March–April.
White, B. 1976: Population, involution and employment in rural Java. In Hansen, G. (ed.), *Agricultural development in Indonesia*. Ithaca: Cornell UP, 17–31.

13 Jan Breman,

'The Creation of a Labour Surplus in Surat District, Gujarat, India'

Excerpt from: J. Breman, *Of Peasants, Migrants and Paupers: Rural Labour Circulation and Capitalist Production in West India*, pp. 327–38. Delhi: OUP (1985)

The extensive labour mobility which has come to exist in Surat District as the result of the transformation to a modern capitalist agriculture, leads us very easily to the notion that rural society there is in a state of flux, and that a part of the population is – temporarily or permanently – drifting around the countryside. Earlier I myself evoked the image of a stream of migrants which like a locust swarm descends on the plain every year, yet this does not mean that labour circulation takes place in an undirected, uncontrolled manner. Structural changes in the home area compel land-hungry peasants to migrate, but they do not wander aimlessly about. Their coming to the plain is systematically encouraged, and is moreover often well planned.

This is clearly apparent from the thorough manner in which sugar factories in the vicinity of Bardoli must set to work in order to ensure the services of tens of thousands of cane cutters each year. Mass seasonal migration has become an important aspect of the functioning of the regional market for agricultural labour, but this also applies in principle to the coming and going of gangs of harvest workers from the tribal hinterland, who report for work with the farmers in the plain at the peak periods of the agrarian cycle. Their presence suggests a much greater degree of spontaneity and casualness compared with the tight organization with which labour is imported for cutting sugarcane. The impression of a lack of planning is linked with the character of this pattern of migration, of which the following elements are essential:

the separation into a great number of small, self-constituted groups (gangs) which are not firmly integrated and which also vary in size;
the lack of standardised and uniform labour conditions;
the short duration of stay, and therewith the need to move on at frequent intervals;
the fleeting, impersonal contact with a multitude of employers.

Are these features characteristic of a structureless labour market? Fisher reaches this conclusion in his analysis of the deployment of Mexican migrant labourers in the harvesting of agricultural products in California (1951, pp. 463–91). In contrast I want to argue that the recruitment and employment of migrants is in fact structured and is moreover consistent with the functioning of the labour market. I am alluding here especially to the particularistic orientation and fragmented character of this circulation for work – two aspects which typify, to a great extent, the organization of the overall labour regime.

The whole migration process has a decidedly particularistic colouring. Regarding the questions posed earlier – who goes away, in what manner, to what destination, for how long, etc. – can only be answered by taking personal ties as a starting point; these exert a strongly regulating influence. What at first sight appears to be a kind of tidal movement – a capricious pattern of wholesale influx, a haphazard swarming over the villages in the plain to be followed in time by a gradual dispersal – is in reality a movement of labour which is prestructured by highly personalized control mechanisms, that bridge the distance between home area and destination. A key figure in this is the jobber who, in providing the link between supply and demand, is on the one hand bound to the employers with whom he maintains regular contacts and via whom he enlarges his territory, while on the other hand striving to secure labourers in numbers which match his current needs. The *mukadam* does not operate as an independent agent in a free labour market, but depends instead on a limited number of employers who have subordinated him to their needs by extending credit. In his turn the jobber maintains a clientele in his own home territory, on whom he likewise imposes the obligation to supply the required labour power at a given moment, by provision of advances. For the composition of his gang he makes use of primary loyalties – selection on the basis of kinship or neighbourhood (*falia*) – which enable him to make demands on his clients. The farmers of the plain also make use of this kind of personal control mechanism, for instance by recruiting additional labour from the hinterland through migrants already employed as farm servants.

The periodic migration of small and more loosely structured groups consisting of members of one or more families, as well as of neighbouring households, usually takes place on the basis of already existing contacts. The preference is for a familiar route – one or more villages which several of the migrants have already visited on a previous trip, and where tentative arrangements have been made with farmers (and perhaps confirmed through written or verbal communication) for a subsequent visit. Naturally seasonal workers sometimes leave their homes without having a definite aim, beyond that of trying their luck in trekking around at random . . . Nevertheless it is my opinion that such a lack of structuring applies only to a small proportion of the total stream of migrants. Whenever the movement of labour over great distances lacks a formal framework, other, more personal channels for regulating this traffic take its place. Mechanisms of a particularistic nature ensure a very effective linkage between supply and demand in the hinterland and in the plain respectively. For the rest, these arrangements based on primordial ties are also recognizable in the organization of cane cutting by the sugar factories around Bardoli. The *mukadam* is the connecting link here, between factory management and the work force attracted anew every year for the duration of the harvesting campaign. The informal setting within which the *mukadam* operates merges with a more formal sphere of management in which he functions as the entry point (Breman, 1979, p. 80).

Isolation is another basic trait of labour circulation. The migrant's mobility has mainly to do with the changing environment in which he finds himself, and much less with the type of work he performs. The majority of those going to the plain remain involved in agricultural work and see no chance of penetrating into another sphere of labour. At best, seasonal workers may remain behind with farmers who take them on as farm servants. Sometimes formal barriers are raised which prevent a longer duration of stay, or transfer to another segment of the labour market (see also Fisher, 1951, pp. 484–5). Thus the *mukadams* who lead the cane cutters are contractually obliged to return home with their gangs after the campaign is over. The temporary nature of their stay in the plain, and their confinement to agricultural labour alone, automatically follows from the isolation within which seasonal migrants operate. Even unskilled employment outside agriculture is mostly unattainable for them; in my opinion this is more because they lack contacts than because they are unfit for such labour. Their temporary residence in the plain is organized in such a way that they remain screened off from other employment opportunities. Conversely, non-agrarian seasonal migrants are bound to a large extent in the same way to the specific work for which they were recruited. Thus canal diggers, road workers, brickfield coolies, etc. are confined to mutually exclusive work sites, passing their time in the plain segregated from local labourers and from each other. The group character of the migration and the fragmentation of the labour market tend to reinforce each other.

Confinement to a separate arena is not characteristic of the life led by seasonal migrants only, since the same phenomenon is found among those who have established themselves more permanently in the plain. A large proportion of the labour in the paper mills, for instance, is imported from other places, workers were brought here by those who were foremen and fellow villagers at the same time. Although working here for years they remain strangers who live in the shadow of the factory and who make themselves invisible, as it were, to the local population. From these enclaves links are maintained with the home area, forming a recruitment channel by means of which (it goes without saying) vacant work places are filled. The trouble it costs to gain entry to another sphere of employment is a result of the inclination to keep the already occupied work places free from interlopers.

Those migrants who do not come to the plain as members of groups appear to be more flexible and are not confined to contacts existing previously. Departure from home is often enforced by a situation of acute need arising, for instance, from a natural catastrophe or a crisis in the family household. This is the type of migrant encountered in improvised camps or shelters set apart on the borders of Bardoli town and in various places along the main road to Surat. Although they have not entirely lost contact with the home milieu, nevertheless the way back is cut off for most of them. Conversely they do not succeed – even after a sojourn of many years – in capturing a recognizeably permanent place for themselves in the labour structure of the plain. One day they work on the land, the next day in construction work, the day after that as navvies; thus this category corresponds most closely to the image of a free and mobile rural proletariat. The continual changing of the work situation is not based on personal choice but is, rather, a necessity imposed from without, and indicates the greatest possible insecurity of livelihood.

On the other hand this does not mean that the migrants who – for the length of their residence in the plain – remain confined to one and the same type of activity (through all kinds of binding mechanisms of a particularistic nature) lead a more secure existence. The fragmented way in which most migrants are recruited and employed prevents them from making a common stand and hinders the spontaneous transition to better-paid work. This fragmentation is, as already stated, a distinguishing feature of the whole labour regime. The barriers thrown up within and between the agrarian and non-agrarian reservoirs of employment in the south Gujarat countryside, result in the enclosure of labour in separate compartments, as far as possible.

Having drawn attention to the sharply compartmentalized structure, I would like to stress further the relative autonomy of the urban labour order compared with the rural. In contrast to the suggestion, in much of the literature, of a comparatively simple town-country duality, further mitigated by the mobility of labour, I am inclined to emphasize the internal heterogeneity of each of these networks as much as the lack of a direct linkage between them. My interpretation runs parallel to that of Bardhan:

> . . . the process of migration from rural to urban areas is structured connecting specific groups of rural migrants (from specific regions and for specific groups and communities) to particular types of urban work. The mechanism of such structuring is a combination of factors: specific needs (like the need for seasonal jobs) and skills of the migrants, personalized or community based systems of contacts and information, and the use of gang leaders and other intermediaries for labour recruitment in industry and civil construction works (Bardhan, 1977, p. A 46).

I have particular doubts about the validity of the theory which says that displacement of labour follows a phased pattern, with rural migrants passing through and remaining briefly in one or more intermediate work situations, en route for the big city. This model of step migration presumes personal choice and freedom of movement, an assumption which conflicts with the manner in which labour relations are structured in south Gujarat. It is already clear from my previous argument that migrants arriving in the rural areas of Surat District are indeed seldom encountered in the city. The limited data which I collected on the phenomenon of urbanization lead me to the opposite conclusion that those who have found employment in the cities have come directly from their villages. By way of illustration, witness the following diary fragment noted in Udhna, the industrial headquarters of Surat.

> *15 April.* On a building site I encounter a group of 32 labourers, nearly all from Khandesh. They are usually split up in work units usually consisting of man and wife, with or without working children. Next come a number of single people, women as well as men. Only a few are older than the late twenties or early thirties.
>
> The group has been assembled by a *mukadam*, who received his instructions from a contractor. Today he has sent for 12 people from his native place in Jalgaon. The rest were recruited in the *bazars* of Udhna and Surat, either personally or with the help of some trusted labourers in the gang who regularly work for him. Those who have already spent a long time in Surat – a period of two to three years is not exceptional – also came straight here from their villages, in the same way as newcomers today. For the landless among them there is really no way back. They enter a life of circulation within the urban economy without a fixed residence. As long as work is in progress they have the use of a temporary shelter made from building materials or the waste from these. After a job is finished the group breaks up and the camp disappears. The labourers hang around the *bazar* and other meeting places in the hope of a new assignment. Do they use this time to make a return trip to their home villages? Contact is not entirely lost, but those who stay longer in the city and have their wives and children with them, seldom seem to return. The answer given by one informant was repeated by others in various tones: 'if there's work there's no time – if there's no work, there's no money'.
>
> The *mukadam* is a key figure in the urban market for unskilled labour. He does not operate independently, but maintains a permanent relationship with one employer, or sometimes with several. The jobber at this building site has already worked for years for the same contractor. When there is no work his boss still gives him an allowance by way of compensation and advance wages on services still to be rendered. The work begun today is paid by piece rate. The men receive Rs 6, the women – for somewhat lighter work – Rs 4 per brass. The mukadam's remuneration consists of a commission of Re 1 a day on every labourer working under him, while in addition he passes on to his workers a lower rate per brass than the contractor gives him. When a husband and wife work together they earn no more than Rs 10 daily, Rs 70 per week. This evening the contractor has come, to provide an advance that will enable the labourers to buy food for the coming week. He provides a total of Rs 600 which the jobber divides out among the couples of labourers in portions of Rs 30–40. Those who are known as hardworking and trustworthy receive a higher amount, while newcomers have to be content with less. Once a week the settlement of accounts takes place, but the jobber does not pay out everything to the labourers, so that they are tied to him for the duration of the job.

It should be clear from this example that the structure and dynamics of the urban market for unskilled labour do not differ fundamentally from the rural market, which is not to say that the two are linked with each other in a direct way. The organization of work is directed towards maintaining labour in separate compartments, not only between city and countryside but also within these. What appears at first sight to be a loose, amorphous structure is in fact a systematic breaking up of the lower echelons of the whole labour regime, with the result that migrants are also mobile along a great number of separate channels (Joshi and Joshi, 1975, ch. V).

Further study of the trek to the city falls outside the scope of the present study. I content myself here with the conclusion that there is no reason for accepting the idea that those circulating in the countryside view the city as their ultimate destination. There is a great diversity of migration routes and the subdivision of these into a rural and an urban component does not, in itself, do justice to the much more complex reality. Particularism and fragmentation are the regulatory mechanisms for the supply of labour to the south Gujarat plain. The effect is an enormous extension of supply, far more than is required inside and outside agriculture. In the previous section I have rejected standard explanations for labour migration which assume local shortage of manpower either in a quantitive or in a qualitative sense. On the basis of my outline of the structure of the labour market, I can now be clearer about this. Labour from elsewhere is rendered mobile not in order to satisfy a shortage in the area of destination but in order to create a surplus. The advent of migrants enables farmers and other employers in Bardoli and its neighbourhood to make less use of the local landless. Why this preference for outsiders?

The first motive to be considered is of a strictly material nature. There is no difficulty in finding acceptance for the idea that migrants are cheaper than local labourers. When the going wage rate amounted to Rs 3, the outsiders were prepared to accept only Rs 2, as I have repeatedly been able to confirm. The underpayment of outsiders is greater in all kinds of tasks; for instance, the wives of migrant labourers who now and then do household work receive only food for this, with no extra money. The farmers themselves maintain that they operate a uniform rate, and instruct even their temporary labourers to deny, if questioned, any difference in wage payment. I gained the impression that this difference in payment diminished or even disappeared entirely with increasing pressure of work, yet the migrants still remain cheaper because, for the same daily amount, they work two to three hours longer than the local landless. Besides this, the strangers are much easier to cheat, being forced to accept their ultimate dismissal with even less than the sum agreed upon.

Apart from this direct profit – a lower price for the same effort or the same price for a better output – there is yet another economic advantage for the farmers in employing outsiders. The massive supply of migrant labourers results in cost cutting because of its depressive effect on local wages, and thus the bargaining position of the local landless is weakened. Does this mean that the extra labour supply is a deliberate attempt to keep the wage level low for agricultural labour, by the creation of a surplus? In his analysis of the migration of Mexican harvest workers to California, Fisher rejects this proposition, which according to him is much too simple. 'The lowering of wage rates where it occurs is as much an effect as it is an objective of oversupply . . . the demand of farmers for more labor than an outside observer would think they require is a rational response to the peculiar circumstances of the harvest, and is largely independent of the effect that the supply of labour may have upon its price' (1951, p. 471). This conclusion is also applicable to the situation I have been discussing. Migrant labourers are cheaper, but this is an incidental advantage and not the principal reason why they are preferred.

A second motive is the farmers' far greater control over migrants, who are in a strange environment and are not familiar with local usages against which they might be able to judge the treatment meted out to them. I have already indicated that the migrants worked much longer hours than the local landless. The demands imposed on their time and working strength are not specific or subject to fixed agreements. Those who have been allocated accommodation in or close to the farmer's house, are continually available for all kinds of tasks, from early morning until late evening. It is no wonder that the farmers praise their diligence, conscientiousness, uprightness and docility. This high degree of versatility and pliability makes the migrants greatly valued as labourers. At the same time this is an implicit judgement upon the shortcomings of the local landless, for they are perceived as lacking in all these desirable qualities. In contrast to the Halpatis, the seasonal workers cannot retreat to a milieu of their own in order to make a stand against the farmers' dominance. To be sure this resistance among the local landless is mainly of a passive nature – a counter point culture in a situation of oppression which finds expression in escape, sabotage, and the like, more than in organized action. Underlying this is a common stand, a community of interests of which farmers as much as agricultural labourers are becoming conscious to an increasing degree, and which is also becoming clearly articulated. The migrants are excluded from this, the isolation which characterizes their life in the plain preventing them from forming a common front against their employers. They are split up into small groups and the particularistic links accompanying these, keep them separated from the local landless and from each other. In this respect also, the fragmentation of the labour force is clearly to the advantage of the farmers.

Finally, a third motive is that contacts with migrant labourers in the plain need to be only fleeting and of short duration. Farmers can limit the relationship with the strangers to the actual sphere of work alone, much more easily than they can in the case of the local landless. The relationship concerns – literally – a work arrangement. i.e. a commitment embracing nothing more than the performance of labour and the payment for it. The reduction of labour to a commodity means that it is deprived of all social significance. The farmer is not interested in the agricultural labourer's personal qualities and prefers to ignore his background and the burden of dependent family members. It is obvious that the employment of outsiders suits to perfection this desire for a contractual relationship. While in the plain, the migrant labourers maintain no social life which can create claims conflicting with the demands made by the farmers on them. The kind of circumstances which sometimes keep Halpatis from their daily work (to the great discontent of their bosses) such as sickness in the home, a family celebration, religious ceremonies, etc. do not affect the migrants.

The local landless still expect a wage that enables them to survive, and are prepared to work in exchange for this. When there happens to be nothing for them to do, they depend on the kind of relief which will enable them to keep their households intact. As I demonstrated earlier, the farmers are increasingly less willing to respect this claim to survival, and the advent of seasonal workers provides them with an alternative. After all, these migrants are paid on the basis of actual performance and expect nothing more. It becomes obvious that it is the home area which bears the cost of their reproduction.

While reporting on his research into rural labour relations in eastern Germany at the end of the nineteenth century, Weber drew attention to the importance of this factor:

> The migrant worker, torn from his family and usual environment, is regarded as simple labour power both by the landlords and by himself. The barracks of the migrant workers are the money-economy equivalents of the slave barracks of antiquity. The estate owner saves on workers' housing, since accommodation for the migrants costs little or nothing. He has also no need to allocate plots of land, but above all he is not regulated by laws governing conditions of work and pay. Thus while the seasonal wage rates are higher, taken over the year the employer lays out no more, usually less, than he used to for a resident worker throughout the year. (Weber, 1979, p. 193)

Besides being compartmentalized the migrants in the plain are also reduced to a commodity. The farmers making use of their services do not see them as individuals and they also remain hidden as a collective entity in the social environment at large. This underlines their isolation.

References

Bardhan, K. 1977: Rural employment, wages and labour markets in India: a survey of research. *Economic and Political Weekly* XII, A34–A64, 1062–74, 1101–18.

Breman, J. 1979: Seasonal migration and co-operative capitalism: the crushing of cane and of labour by the sugar factories of Bardoli, south Gujarat. *Economic and Political Weekly* XIV, 1317–60.

Fisher, J. 1951: The harvest labour market in California. *Quarterly Journal of Economics* LXV, 463–91.

Joshi, H. and Joshi, V. 1975: *Surplus labour and the city: A study of Bombay*. Bombay: OUP.

Weber, M. 1979: *Economy and society*. Berkeley: University of California Press.

14 Judith Carney and Michael Watts, 'Manufacturing Dissent: Work, Gender and the Politics of Meaning in a Peasant Society'

Excerpts from: *Africa* 60 (2), 207–40 (1990)

[A]s men and women transform raw material into useful things, they also reproduce particular social relations as well as an experience of those relations. [A]longside the organization of work . . . there are apparatuses of production which regulate production relations. . . . [Yet] as we slaved away on our machines . . . we manufactured not only parts of diesel engines, not only relations of co-operation and domination, but also consent to those activities and relations. (Michael Burawoy, *The Politics of Production*)

In studying such [radical] transformations, it is always necessary to distinguish between the material transformation of the economic conditions of production . . . and the legal, political, religious, artistic or philosophic – in short, the ideological forms in which men [*sic*] become conscious of this conflict and fight it out. (Karl Marx, *A Contribution to the Critique of Political Economy*)

Introduction: the manufacturing of dissent

This article addresses the changing nature of farm work in a peasant society in The Gambia, West Africa. The practice of farm labour has been transformed in the most palpable way by the advent of radically new technical and social relations of production associated with mechanised double-cropping of irrigated rice. Technical change, agricultural intensification and a new labour process are, however, all built upon the bedrock of household production, since peasant growers are socially integrated into the new scheme as contract farmers, specifically as contracted sharecroppers. Family labour continues to be the dominant social form in which labour power is mobilised, but under conditions directly determined and shaped by the contractors, namely project management. Irrigated double-cropping of rice production is particularly labour-demanding and makes expanded claims on customary structures of domestic labour recruitment. These new economic practices subject the culturally dominant representations of work, labour obligations and property rights – the constituents of custom and tradition – to the test of social practice. In our examination of Mandinka rice growers we suggest, following T. J. Clark, that 'society is a battlefield of representations on which the limits and coherence of any given set are being fought for and regularly spoilt' (Clark, 1984, p. 6). The introduction of a new production regime has converted rural Mandinka society into a contested social terrain; the primary struggle is a contest over gender and the conjugal contract in which property, or more accurately constellations of property rights, is at stake. By seeing economic life as, among other things, a realm of representations, we argue that the struggles over meaning and the manufacture of symbolic and material dissent in central Gambia – a proliferation of intra-household conflicts, juridical battles over divorce in the local courts, renegotiations of the conjugal contract – are the idioms of what Burawoy (1985) calls production politics.

. . . .

Of contracts, gender and production politics

> Family enterprise is a battleground over patriarchy where property is immediately at stake. (Harriet Friedmann, 1986, p. 192)

A series of poor agricultural seasons, domestic cheap food policies and a high fertility regime all have contributed to the continued growth in Gambian rice imports over the past twenty years. The conspicuous failure of the small-scale, community-based projects to mitigate this trend contributed to the new food security programme launched by the Gambian government during the late 1970s. To ensure marketable rice surpluses, the government and foreign donors have pioneered a distinctive strategy, a form of contract production, which removes both community and household jurisdiction over irrigated land. Household participation in irrigated rice projects is now tied to meeting two criteria designed to secure compliance with production targets: (i) double-cropping; and (ii) repayment in full of the production inputs. The Jahaly-Pacharr project, which began operation in 1984, is the first scheme implemented with these production objectives. The contract specifies production routines and service provision in return for a share of the grower's harvest but it also contains the threat of eviction. Jahaly-Pacharr was developed on lease land, which grants to the state the right to withdraw plots from non-compliant or unproductive farmers. The project poses, in other words, a threat to the family property system which has hitherto underpinned the social organisation of household productive activities.

The Jahaly-Pacharr irrigated rice project is located in central Gambia near Sapu, 280 kilometres east of the nation's capital, Banjul. Completed in late 1987, the project involves over 2000 rural households from 70 villages. Nearly 1500 hectares are developed for rice production, of which 560 are pump-irrigated, some 700 tidal-irrigated and 200 hectares for improved rain-fed cultivation. The project involves four major ethnic groups: the Mandinka, The Gambia's ethnic majority and primary rice cultivators, as well as Wolof, Serrahuli and Fula. Jahaly-Pacharr is a large-scale irrigation scheme with centralised water delivery and management. Project officials set the rhythm for household productive activities by establishing the cultivation calendar for cropping, weeding, transplanting and harvesting as well as specifying the application dates and types of seed varieties, fertilisers and pesticides used. Smallholders are informed of these decisions through the project's extension agents who provide technical assistance. The project's paddy is marketed through the government's co-operatives.

In so far as growers are contracted as households, the Jahaly-Pacharr project was predicated on the availability and intensification of family labour. From its inception, however, female family labour has been widely viewed as *the* precondition for meeting production objectives:

> . . . women are better than men as far as transplanting is concerned and they are also better than men as far as working in the water . . . so quite frankly we expect a lot of labour from women, more so than from men.[1]

Female rice farmers were also the focus of donor interest in the project.[2] Responding to mounting international concern over the adverse consequences for development interventions for women and in particular the failure of previous Gambian irrigation schemes to reach female cultivators, the donors viewed the Jahaly-Pacharr project as an opportunity to implement equity objectives:

> In assisting the Government [of The Gambia] to reach its goal of rice self-sufficiency and to improve the lot of the rural poor, the project makes a special reference to women, who traditionally have been the major rice growers under arduous swamp conditions.[3]

The donors' intention was not only integration into the project but to make women the primary beneficiaries by granting them direct access to developed land. However, the reallocation of land rights and the new production routine conflicted directly with the donors' equity objectives.

The retention of plot usufruct in Jahaly-Pacharr depends on compliance with the stipulated cultivation practices, cropping schedules, production targets and loan repayment schedules. The production strategy strikes to the heart of family relations because it imposes new and demanding claims on household labour; skilled female family labour in particular is critical to fulfilling production targets. Growers are not only under considerable pressure to perform work tasks according to a regimented work schedule but the tasks themselves

[1] The Jahaly-Pacharr project manager quoted in BBC's Global Report, 'The Lost Harvest', 1983, produced by S. Hobson. According to von Braun and Webb (1989 p. 519) women provided 39 per cent of total labour on the pump-irrigated plots, identical to that of men.

[2] The Jahaly-Pacharr project is funded with a US$16.5 million loan on easy lending terms through the International Fund for Agricultural Development (IFAD), the African Development Fund (ADF), the World Food Program and the governments of the Netherlands and West Germany.

[3] International Fund for Agricultural Development (IFAD), 1981.

are long (extending through the dry season) and arduous. The labour intensity is reflected in the fact that under fully water-controlled conditions, rice production requires 60 per cent more labour per unit of land than traditional swamp rices despite the deployment of machine-powered ploughing services (von Braun and Johm, 1987; von Braun and Webb, 1987). With the onset of project rice production, men's wet season work in agriculture has actually decreased (reflecting a shift of labour time from upland crops to rice) whereas women's total agricultural work has increased. Women naturally experience mechanised rice production as radically new claims on their bodies and the imposition of enormously demanding work routines to produce a commodity which they see as 'women's sweat'.

Project implementation has, however, placed Mandinka women in a particularly vulnerable position which transcends the onerous demands of farm labour. As the only Gambian ethnic group in which women have no individual crop rights on the uplands, the Mandinka women have found that the absorption of pre-existing individual and household fields into the scheme meant that their entitlement to their crop rights would henceforth depend on access to the project's plots. Unless their property rights were protected, Mandinka women had no alternative income-earning plots in the area and no protection from household claims to their labour. Women were particularly sensitive to the loss of a fundamental right they had customarily enjoyed within the Mandinka farming system. Their position was forcefully presented by one Mandinka woman:

> It seems this project is just like the Chinese one when we suffered before. We aren't going to put up with that again. . . . I have this to say to you men. We were asleep then. But now we are awake.[4]

Indeed, the importance of property control in activating claims to female labour was vividly portrayed in the project's first land distribution. The original female rice tillers were to receive irrigated plots but the first distribution revealed how fraught with conflict such compensatory allocation would be in practice. Less than one third of the plots were registered in women's names because men opposed women's ownership on the grounds that divorce would alienate land from household control. The failure of land distribution to conform to donor objectives prompted IFAD to intervene directly. In early 1984 a mission sent to The Gambia succeeded in re-registering the project's plots in women's names and IFAD was accordingly praised for implementing

[4] Mariama Koita, a Mandinka woman in one project village, quoted in BBC's 'Lost Harvest'. Her reference is to the development approach pioneered by the Taiwanese Mission and followed in subsequent small-scale irrigated rice projects. By targeting male household heads for technological diffusion, the bilateral missions provided a legitimation for male control of pump-irrigated perimeters. Women suffered doubly: (i) they were excluded from the benefits derived from technical improvements in rice production; and (ii) they lost many of their individual tidal rice plots which were absorbed by irrigation projects. See Carney (1988, pp. 334–49) and Dey (1980, 1982).

gender equity in its rural development project. In practice, however, the land registration obfuscated the profound transformations in household property control.

IFAD was able to re-register the plots in women's names because male heads of household had already achieved *de facto* control. Project management had intervened to overcome male resistance to donor wishes by concurring with men that the household, and not the individual woman, irrespective of prior claims had *final* control over crop disposition on the irrigated plots. The Land Committees invoked tradition and historical precedent in classifying the new developed area as communal land but the overriding concern of the project was productivity. In so far as the project's production routine is predicated on the intensification of family labour, government officials were unwilling to champion principles that might reduce its availability. As a labour category, *maruo* [household production] alone provided the institutional base within the household to make claims on family labour. The management's decision paved the way for men to declare the project's plot *maruo*. In sum, women were not compensated for the loss of their swamp fields with irrigated plots, and experienced new demands on their labour for communal rice cultivation. Whereas individual fields in the traditional swamps are cultivated during the same wet-season months as irrigated fields on the project, the new communal labour obligations were at the expense of women's customary *kamanyango* [individual] production.

The importance of land taxonomies and systems of classification in establishing control over land, and simultaneously over dependent labour, was immediately evident. During the first cropping season, all the pumped plots were considered *maruo*. Women provided most of the agricultural labour, but household heads typically refused to grant them the customary *kamanyango* crop rights. During the second wet-season cropping period when it was clear that women would not receive any crop rights on the pumped-irrigated plots, they pressed for official acknowledgement to designate the project's tidal and rainfed areas as individual fields. Despite the management's earlier support of women's individual crop rights on project land, it subsequently withdrew any sort of official sanction. Similarly, IFAD refused to intercede at this critical juncture to safeguard women's rights in the project. Consequently, women were forced to negotiate for *kamanyango* crop rights through the same patriarchal village and household social structures that sought to deny them access to irrigated land.

The differential effects of project development on household members were clearly discernible during the first year. Women provided a substantial share of the field labour, but the household head maintained his control over the product of their labour. The consequences of what amounted to a concentration of land rights was especially vivid. The Gambian government raised the producer price of paddy, which transformed rice from a subsistence to a cash crop and ensured that high-yielding irrigated rice production generated very substantial profits. While part of the communal rice production is devoted to subsistence consumption, the household head markets and sells the plot's surplus rice and pockets the revenues from paddy sales.[5] The naming of the project's plots as household fields thus enabled the household head to make claims to women's unpaid labour when in practice the plot functions in part as his individual field capable of generating investable surpluses for personal accumulation.

It is clear that technological innovation and a restructuring of the labour process had fundamental consequences for the micro-politics of the Mandinka household. Male household heads apparently expanded their hegemony through a manipulation and centralisation of crop and labour rights. Dependent women, conversely, faced attenuated land claims or actual loss of individual rice fields, and new communal labour obligations directly in conflict with their capacity to cultivate traditional rice (assuming they maintained some access to lowland fields); men conversely appropriated large investable surpluses in the guise of communal food production. As von Braun and Webb (1989, p. 525) observed, 'women not only end up growing crops with technologies that result in lower new returns to their labor time but they also exhibit lower average labor productivity levels than men in the *same* crop'. This recomposition of labour and the reclassification of property rights naturally manufactured considerable stresses within the social architecture of the Mandinka household. Many of the new claims, new rights and new obligations were contested on a variety of material, symbolic and ideological terrains. Emotions and material interest arose, then, from the same social – in this case domestic – matrix which is deeply embedded in 'property relations, working processes and the structure of domination' (Medick and Sabean, 1984, p. 4).

Struggles over work, struggles over meaning

> The issue is rather one of on-going struggle over meaning. The struggle is formed in the context of the social relations of individuals, groups, classes and cultures which at the same time are constituted by the struggle. Reciprocity, dependency and resistance . . . are not 'structurally given' . . . they come into being only in the struggle for meaning. (Medick, 1987, p. 98)

By the end of the project's first cropping season, women received no *kamanyango* rights in the project. Nor had they been remunerated in cash or kind for their labour on the pumped plots. The importance of their labour to the retention of plot usufruct, however, provided the basis for women's struggle to improve their position within the household. At the onset of the project's second year, women demanded from their husbands remuneration

[5] Von Braun and Webb (1989, 529), note that the market surplus of rice declined as male control of irrigated rice increased. While rice is consumed domestically, the yields on irrigated plots are *four times* greater than traditional swamp rice and accordingly generate substantial surpluses. Von Braun and Webb do not refer to the sale of communal rice and who accumulates the income derived from the rapid commoditisation of project rice.

for their work on the pump-irrigated plots. In doing so they contested the basis of the conjugal contract – namely the right of their husbands to lay claim on their labour power – and thereby effected from within the household a challenge to patriarchal family relations.

Women's success in bargaining for labour compensation has been conditioned by three primary factors: (i) ethnicity; (ii) the economic status of their household; and (iii) conjugal relations. The determinant factor in women's bargaining position was whether or not they retained access to individual fields. Project development did not change the land and crop rights of Wolof and Serrahuli women who retained upland groundnut *kamanyango*. They maintain the right, like junior males, to withdraw their labour from household demands during the rainy season – a consideration which has enhanced their intra-household bargaining position. As a result, in households with adequate project land to generate marketable surpluses, these women typically receive a portion of the plot's paddy as labour remuneration.[6]

The pattern is more complex and contradictory among the Mandinka. Most Mandinka women lost their *kamanyango* rice fields with project development and without alternative land access, they were vulnerable to their husband's demands to provide labour for rice production. Project expectations for Mandinka women to perform most of the on-farm operations were, of course, not without historical and cultural precedent. Women are skilled rice growers and since the nineteenth century the Mandinka have responded to economic commoditisation by intensifying the spatial and gender basis of production – specifically, by increasing food cropping in the tidal swamps and female labour in rice production. The implementation of contract cropping in Jahaly-Pacharr required an intensification of family labour to extend labour recruitment over twelve months and under routinised and regimented work conditions. The struggles that ensued in Mandinka households were over who would work under what conditions, and specifically the degree to which dependent female labour would bear new work burdens. Mandinka women's weakened negotiating position with the loss of traditional swamp land had one implication of enormous consequence: even though the Mandinka received more land than other ethnic groups – due to their dominance in pre-project rice cultivation – women's capacity to negotiate labour remuneration was tightly circumscribed.

By the end of the third cropping year, Mandinka women responded to the changes that had occurred in household property and family relations in three basic ways:

[6] This refers principally to Serrahuli women who grew rice as a *maruo* crop and groundnuts as *kamanyango* prior to project development. Based on their previous rice cultivation in the Jahaly and Pacharr swamps, Serrahuli and Mandinka villages have received more project land than Wolof and Fula villages. Von Braun and Webb (1989) refer to the growth of upland cultivation by women as individual rice production declines, but their data are not disaggregated by ethnicity. There is, in fact, no evidence that Mandinka women are significantly engaging in upland groundnut cultivation as a compensation for loss of lowland rice swamps.

1. Women provided labour as required on pump-irrigated plots and received compensation through usufruct on tidal-irrigated and rainfed plots.
2. Women provided labour on the pumped plots for which they were compensated in paddy, receiving a fixed share of the pumped plot's yield for their labour.
3. Women did not provide labour on the household *maruo*. The household head could not, or refused to, offer labour remuneration in paddy on the pumped plot or through access to irrigated areas on the project.

In the first case, the households involved were typically resource-sufficient, controlling adequate project land for both subsistence and cash needs. Given that nearly seventy villages are involved in the Jahaly-Pacharr project and the fact that tidal land was still being developed in January 1987 at the time of the survey, it is difficult to estimate the proportion of Mandinka households in the first category. In one village intensively surveyed (Pacharr) which received land in both the pumped and non-pumped areas, women have received *kamanyango* rights to tidal-irrigated plots. However, their economic opportunities in the project have been tempered by four considerations. First, most plots are typically cultivated by several women; each one-hectare plot may be shared by as many as four females. Second, in contrast to the pumped plots, the majority of tidal, and all the rainfed, areas are cropped only during the rainy season, which limits their income-earning capacity. Third, these plots are substantially less productive than the pump-irrigated areas. Finally, a woman's *kamanyango* crop rights are only protected to the extent that she remains in the household. Upon divorce, a woman cannot guarantee continued control of the field. Despite such considerations, *kamanyango* usufruct to tidal and rainfed plots do provide Mandinka women some form of labour compensation and, thus, cash-earning possibilities in the project.

The second response to Mandinka women's 'wage' demands occurred in households blessed with ample pump-irrigated but little tidal or rainfed land. Women have sometimes successfully negotiated elementary sharecropping relations, an adaptation which rewards them for their overall labour productivity. In exchange for their unrestricted labour on the pumped fields, women are granted a fixed proportion (share) of the plot's yield. The external (grower–management) sharecropping relation is, in other words, *internalised* within the conjugal contract. The percentage of households involved in sharecropping arrangements in one Mandinka village intensively surveyed (Wellingara) was roughly 20 per cent (Carney, 1987). The share women received varied but averaged about 10 per cent of the total paddy output, with a value that ranged between 150 to 500 dalasis per year (US$20–70).[7] Most of

[7] According to von Braun and Webb (1989, p. 523) an unspecified sample of such arrangements produced an average compensation of 2 kg of paddy per woman's labour day; this is equivalent by value to the average labour productivity of women in their *least* remunerative cash crop (cotton).

Wellingara's households, however, fell within the third category in which women were unsuccessful in renegotiating and bargaining labour compensation. In a few cases this was due to the household's economic position: roughly 15 per cent of Wellingara's households received no project land. In these cases, the plot is shared with other households and functions primarily for subsistence rather than income generation. Such circumstances limit women's ability to compensate for their *kamanyango* loss and have led them to pursue other economic strategies, notably hiring themselves as wage labourers, cultivating market gardens, and/or engaging in petty trade. The impact of changing intra-household resource control on Mandinka women is most clearly visible in the majority of Wellingara households with enough land to provide for subsistence and cash needs, but in which household heads refused to remunerate women for work on communal rice fields. As a consequence women have withdrawn their labour from the rice fields, an explosive tactic with far-reaching impact on the social organisation of household production.

Many Mandinka households find great difficulty in meeting the stringent cropping production schedules. This has contributed to relatively low yields. In the project's first year, average dry season pump-irrigated rice yields were 7.5 tons per hectare but in the 1985 and 1986 dry seasons, yields have averaged only 5.7 tons (Table 14.1). The yield declines due to labour withdrawal do not threaten the Jahaly-Pacharr project in a serious way, but they have seriously dampened project enthusiasm and explain the management's belief that Mandinka growers are poor and uncommitted farmers.

Table 14.1 Seasonal yields on *maruo* plots

Season	*Year*	*Yield/hectare (tons)*
Dry	1984	5.0
Wet	1984	6.5
Dry	1985	5.7
Wet	1985	4.0*
Dry	1986	5.7
Wet	1986	4.8†

Source: Carney, 1987, p. 54.
* This decline reflects the lowered production of the tidal and rainfed plots which began being developed in the project's second year.
† Estimated.

Another consequence of female labour withdrawal has been an intensification of male labour in rice cultivation. But given the high rate of polygamous marriages in the project area, men cannot compensate for the loss of skilled female labour through an intensification of their own. Thus, there has been an increase in labour hire by these households, particularly in the rainy season when men also plant groundnuts. Even in the project's first year when women provided unpaid labour on the plots, only 25 per cent of Wellingara households were able to complete transplanting without wage labour. By the project's third year, when the effects of female labour withdrawal were felt, more than half the village's households were hiring labour for harvesting and threshing (Table 14.2). For the 1986 cropping year, households spent on average US$30–85 for hired labour, a figure which closely approximates the value of paddy that women receive in sharecropping arrangements. Female labour withdrawal has radically affected the capacity of the household to intensify its labour and hence to meet production schedules.

Table 14.2 Wellingara: compounds hiring labour on *Maruo (%)*

Activity	*1984**	*1986†*
Transplanting	67	66
Harvesting	42	61
Threshing	21	63

Source: Carney: fieldwork data 1984, 1987.
* Sample: 25 households (of 40).
† Sample: 38 households (of 40).

Finally, most of the wage labourers in Mandinka villages are women who have been proletarianised by the project. Ironically, the creation of a local labour market is in response to a smallholder project, a development which has simultaneously contributed to a transformation of women's reciprocal labour networks. Traditionally, village women's age grades (*kafos*) provide gang labour for rice transplanting. Payment for *kafo* labour was typically deployed for individuals in financial distress and for life-cycle ceremonies such as marriage. These reciprocal networks also functioned however as an auxiliary labour reserve for sick or pregnant women. Nowhere are the project's effects on the labour process more visible than in the reconfiguration and individualisation of women's *kafos*. Dispossessed of rice land, women currently work for hire on the project's plots and the *kafos* now recruit labour gangs for critical rice operations. They provide an organisational framework in which women pool their labour for hire in transplanting, weeding, harvesting or threshing, but in contrast to the pre-project period, payment is no longer retained for the mutual need but is divided up among the individual members as a *de facto* wage. By forming work groups or 'companies', women are able to bid-up the rural wage; a group of twenty women usually charge 80 dalasis for transplanting one irrigated plot and can typically accomplish two plots a day. Each individual wage worker earns 8 dalasis rather than 5, the prevailing daily wage rate. For women who no longer have access to land, their economic needs are increasingly met through wage labour on the project's rice plots, a development which has decisively individualised their reciprocal labour networks.

The complex and contradictory ways in which intra-household dynamics are shaped by a new technology reaffirms the need to see work and patriarchal politics as cross-cut by bargaining, negotiation and, on occasion, conflict. Such a perspective stands in sharp contrast to von Braun and Webb's analysis of the Jahaly-Pacharr project which sees the household as a 'mini-state' in which family members are taxed through communal work and allocative inefficiencies abound. Their analysis poses a rather unproblematic centralisation of authority and a capitulation by junior members and wives to paternal authority. Conversely, our analysis is rooted in specific ethnic conditions and focuses on the spaces in which women can negotiate and the tactics employed to poach upon property, conjugality and notions of familial obligation. The question of what makes work worth doing is complex but the Mandinka case suggests multiple resolutions to struggles over work conditions. At the very least, these negotiations starkly reveal how difficult it is to intensify production and simultaneously to produce a disciplined domestic labour force. How, or whether, these struggles are resolved represents a local contestation of state power and to this extent can dialectically shape larger structures of which peasants are subordinate parts.

The Mandinka domestic struggles over work, property rights and resources are clearly material processes, but they are also simultaneously cultural and symbolic. The two moral axes of Mandinka domestic life—*badingya* and *fadingya*—have been in some fundamental sense realigned. *Badingya* represents co-operation, obligation and harmony, and *fadingya* connotes ambition, selfishness, conflict and, at the limit, a domestic or community anomie. The oscillation between these poles naturally carries a strong normative but also interpretative current; that is to say, what is normal, just or simply 'traditional' is established through discursive practice. The Jahaly-Pacharr project has animated this discursive realm; it has manufactured a sort of *symbolic discontent*. This is not surprising because every field is the site of a struggle over what Bourdieu calls the definition of legitimate principles of the division of the field:

> In the symbolic struggle over the production of common sense, or more precisely, *legitimate naming* . . . agents engage the symbolic capital they have acquired in previous struggles . . . [that is] all the power they possess over instituted taxonomies. (Bourdieu, 1985, pp. 731–2, our emphasis)

It is largely through language that local practices are instituted, but meanings are never fixed and outside forces can often determine what is to be endowed with meaning (Mintz, 1985, p. xxix). The politics of classification and the power to name go some way towards explaining why the debate over the naming of irrigated plots is, in Mandinka society, such a charged issue. While it is within the language of everyday life that meaning is mobilised in the defence of domination, the outcome of struggles such as those initiated over customary rights in Mandinka farming households is far from overdetermined. This is not simply to suggest that the wage relation may have 'liberating' consequences for women, but more generally that culture contributes towards the creative, tense, and uncertain social reproduction of diverse kinds of relationship, that 'cultural reproduction always carries with it the possibility of producing . . . alternative outcomes'.[8] The discontent manufactured by the Jahaly-Pacharr project suggests subtle changes in gender practices and consciousness and the beginnings of a transgression which threatens the discipline of everyday domination. The politicisation of women's gang labour groups, the active presence of women on Land Allocation Committees, victories for women in land disputes resolved in local courts, female deputations to government ministers and the like, all suggest that, to use the words of Mariama Koita (a woman rice farmer from Kerewan Samba Sira), '[women] were asleep then. Now we are awake.'

Conclusion: the inflection of resistance

> Everyday life invents itself by poaching in countless ways on the property of others, . . . [T]he goal . . . is to bring to light the clandestine forms taken by dispersed, tactical and makeshift creativity of groups or individuals already caught in nets of 'discipline'. (Michel de Certeau, 1984, pp. xii–xiv)

The mediating role of domestic social structure is a critical ingredient in Scott's recipe for peasant resistance but we have added a slightly different flavour to what he calls the polyp-like qualities of peasant politics. The labour implications of a technological innovation such as rice contracting in Jahaly-Pacharr have acted as a major perturbation to the internal structure of Mandinka households. As sites of competing interests, households are always characterised by friction and negotiation; contract production, in its demand for a disciplined familial labour force, has converted the domestic arena into a terrain of explicit confrontation. To understand fully this contest, however, it is necessary to prise open the black box of the household, and examine the circuits of economic, symbolic and cultural capital which lie inside. These internal circuits shape, and are shaped by, state interventions and the ways in which capital endeavours to take hold of peasant agriculture (Kautsky, 1906). As Berry notes in her discussion of land rights in Africa:

> State intervention to redefine conditions of access to land tends to *provoke new struggles over meaning and power which are as likely to inhibit growth* . . . as to release the productive energies of small-scale enterprise, market competition or capitalist accumulation. (Berry, 1988, p. 70)

The case of Jahaly-Pacharr reveals that the transformation of land rights in conjunction with a radically new labour process has given rise to contradictory, and deeply gendered, developments with respect to domestic access to, control over and definition of resources. The struggles over meaning suggest that households experience capitulation or centralisation less in an unproblematic fashion but more through complex bargaining, renegotiation and multiple

8 Willis, 1981, p. 172. We have also been deeply influenced in this regard by Taussig, 1987.

resolutions. In addition, it is entirely unclear whether these contradictions, displacements and conflicts are compatible with the agricultural performance anticipated by project management.

The Mandinka example can be complemented by other cases derived from other ethnic groups within the project area and in other regions of The Gambia. In the Serrahuli village of Allunhare in Upper River Division, a quite different household structure in which dependent women customarily have access to their own individual upland groundnut farms—and greater domestic bargaining power more generally—the impact of irrigation on female income and autonomy is dramatic (Watts, forthcoming). Conversely, highly skewed landholding between lineages and households has produced complex sharecropping arrangements between land-rich and land-poor households. In Kenya, contract farming of labour-intensive tobacco introduced by the British Tobacco Company has had the effect of converting segmented Teso households into centralised, unitary domestic units—a process with obvious parallels to the Mandinka case – which has radically undermined existing conjugal relations with serious consequences for domestic conflict and dissent.

To emphasise that it is not uncommon for the politics of production in peasant societies to be inflected inward and to be deeply engendered is not to isolate such behaviour from the seamless web of normal resistance, a subterranean class-consciousness, as Scott outlines it. But this raises two important problems: the first is to understand why at this juncture the domestic arena appears so central to the conflicts over access to land and labour among the Mandinka, and the second is to see the connections between domestic and broader (i.e. extra-household) resource struggles. To the first question, a possible answer resides in the fusing of two theoretical realms: the labour process and the family enterprise. The unique quality of peasant forms of production in which labour and property are united raises the possibility, realised under contract farming, of collapsing production and family politics into one sphere, the domestic. As Harriet Friedmann so aptly puts it, 'the family enterprise is a battleground over patriarchy where property is immediately at stake' (Friedmann, 1986, p. 191). The new labour process at Jahaly-Pacharr immediately placed those property and gender relations at stake; indeed it raised the stakes. The project was immediately and decisively politicised at the point of production. In view of the history of resource conflicts around rice lands after the Second World War, domestic struggles necessarily were visible and potentially explosive.

The second question on linkages is more complex. On the one hand, all of the struggles should not be construed as narrowly gender-based or intra-familial, and they did indeed spill out of the household altogether. Jahaly women petitioned the management for access to tidal fields, organised a deputation to the Minister of Co-operatives; they have taken some land claims into the local courts and, not least, explicitly articulated collective interest through their group labour by bargaining up their daily wage rates. Tensions between growers and management are commonplace, and both men and women have occasionally attempted to divert project fertiliser to other farms, to pilfer seed, and to under-report harvests. In at least one village, the profitability of irrigated rice production has also intensified long-standing ethnic and communal tensions between Mandinka and Fula communities.

Some of the obvious means by which domestic politics can be bridged to social or class politics are less in evidence but this may reflect the immaturity of the project. Unlike other contract schemes, for example, local producers have not organised into grower associations or used either local co-operatives or political parties as mediators between the management and growers (Watts, 1990). In part this diffidence may reflect the fact that in The Gambia, as in much of sub-Saharan Africa, there is no landlord class, and while the project has appropriated the land on lease, there is enormous ambiguity among growers as regards what this means in practice. The project believes that peasants have no proprietary claims to the land and hence they can be evicted from the project in the event of low productivity or contract contravention. Conversely, farmers uniformly believe that irrigated plots are owned by the household, a belief reinforced by the legacy of earlier irrigation schemes which proved incapable of centralising land rights. In the first two years of operation (1985–87), no peasants were evicted and hence the point remains moot.

But this is a terrain on which domestic and class politics might meet, since the cases in which the management has exercised some discipline and sanction, warning some peasants for contract abrogation (late weeding, inadequate seedbed preparation), have almost uniformly involved male Mandinka farmers put into difficulties by the withdrawal of household labour by their wives and unable to meet the strictures of the new work routine. Indeed, the project management typically refers to male Mandinka growers as 'lazy', 'poor rice farmers' and largely uncommitted, exhibiting a cultural preference for leisure over hard work. Some institutions explicitly connect the domestic and macro-arenas, the most obvious being Islam. Appropriate familial behaviour is often legitimated in the vocabulary of Muslim faith, just as growers' expectations from Gambian project management are articulated in a Muslim idiom of largesse and justice. All this would suggest that until the project dramatically transforms local socio-economic differentiation, as it surely must in the medium-term given the profitability of rice farming, and until it acts upon its purported capacity to remove peasants as tenants from the land (which must raise the possibility of sharp conflicts between growers and management over dispossession), the politics beyond the domestic field will remain somewhat muted.

To recognise the primary quality of domestic struggle is not only to assert that gender conflicts are part of normal resistance but also to highlight the contradictory qualities of the enormous field that Scott has circumscribed. The repertories of these local forms of contention are Janus-faced and we must be sensitive to Genovese's claim (1974) that slave resistance was resistance *within* the confines of slavery. Much of what passes as resistance is more akin to adaptation: it is all too easy to romanticise the prosaic and everyday, and to create a fantasy land of class struggle. Nonetheless the Mandinka case reveals also how subaltern societies are engaged in a dialectical struggle between active and passive, acceptance and resistance, coercion and consent. Jahaly-Pacharr introduced new work conditions for producers, but as Mandinka men and women transformed raw material into commodities for sale they did not, as Burawoy (1985) suggests, simply reproduce particular social relations and an experience of those relations. The new labour process was experienced *differentially* by gender, and this manufactured a specific sort of dissent rather than consent. While it would be facile to suggest that domestic renegotiations constitute the radical transformation implied by Marx in his *Critique of Political Economy*, we have argued that the spaces opened up by prosaic conflicts in the property system offer the possibility of significant advance and social change.

Acknowledgements

Various aspects of the research on which this paper is based were conducted under the auspices of a University of Michigan project to study the Gambia River Basin (1983–85). Critical follow-up support was received from the Institute for Development Anthropology and the Ford Foundation. The authors wish to acknowledge the assistance and support of Dr John Sutter, Modi Sanneh, Oussainou Baldeh, Julia Morris and Chris Elias. The authors are grateful to the following individuals who commented upon earlier drafts of the manuscript: Pauline Peters, Ivan Karp, Jane Guyer, Gill Hart, Michael Burawoy, Richard Roberts, Donald Moore and Joel Samoff.

Selected references

Berry, S. 1988: Concentration without privatization: agrarian consequences of changing patterns of rural land use controls in Africa, in S. Reyna (ed.). *Land and Society in Contemporary Africa*. Hanover: University Press of New England, 53–75.

—— 1989: Social institutions and access to resources. *Africa* 59 (1), 41–55.

Bourdieu, P. 1985: The social space and the genesis of groups. *Theory and Society* 14 (6), 723–44.

Braun, J. von, and Johm, K. 1987: Tradeoffs in the rapid expansion of smallholder rice production in The Gambia. Paper prepared for the Conference on Cereal Consumption in West Africa, Dakar.

Braun, J. von and Webb, P. 1987: Effects of new agricultural technology and commercialization on women farmers in a West African setting. Washington: IFPRI.

—— 1989: The impacts of a new crop technology on the agricultural division of labour in a West African setting. *Economic Development and Cultural Change* 37 (3), 513–34.

Burawoy, M. 1985: *The Politics of Production*. London: Verso.

Carney, J. 1987: Contract farming in irrigated rice production. Working Paper no. 9. Binghamton, NY: Institute of Development Anthropology.

—— 1988: Struggles over crop rights within contract farming households in a Gambian irrigated rice project. *Journal of Peasant Studies* 15 (3), 334–49.
Certeau, M. de. 1984: *The Practice of Everyday Life*. Berkeley: University of California Press.
Clark, T. J. 1984: *The Painting of Modern Life*. New York: Knopf.
Dey, J. 1980: Women and Rice in The Gambia: the impact of irrigated rice development on the farming system, PhD thesis, Department of Sociology, University of Reading.
—— 1982: Development planning in The Gambia: the gap between planners' and farmers' perceptions, expectations and objectives. *World Development*, 10 (5), 377–96.
Friedmann, H. 1986: Patriarchy and property. *Sociologia Ruralis* 26 (2), 186–93.
Genovese, E. 1974: *Roll, Jordan, Roll*. New York: Random House.
International Fund for Agricultural Development (IFAD) 1981. *Revised Appraisal Report, Jahaly and Pacharr Smallholders Project*. The Gambia: Ministry of Agriculture, IFAD and ADF.
—— /ADF. 1984: Supervision Mission Memo Jahaly Pacharr Project, The Gambia. Project mimeo.
Kautsky, K. 1906: *La Question Agraire*. Paris, Maspero.
Medick, H. 1987: Missionaries in the row boat. *Comparative Studies in Society and History* 29 (1), 76–98.
—— and Sabean, D. 1984: Introduction, in H. Medick and D. Sabean (eds.), *Interest and Emotion*, 1–18. Cambridge: CUP.
Mintz, S. 1985: *Sweetness and Power*. New York: Harper.
Taussig, M. 1987: History as commodity. Unpublished manuscript. Bogotá.
Watts, M. 1990: Peasants under contract, in H. Bernstein and M. Mackintosh (eds.), *The Food Question*. London: Earthscan.
—— Forthcoming. *Manufacturing Discontent: production politics in a peasant society*. London: Heinemann.
Willis, P. 1981: *Learning to Labor*. New York: Columbia UP.

15 Cathy Schneider,

'Radical Opposition Parties and Squatters Movements in Pinochet's Chile'

Reprinted from: *Latin American Perspectives* 18 (68), 92–112. (This version reprinted in full from A. Escobar and S. Alvarez (eds), *The Making of Social Movements in Latin America: Identity, Strategy, and Democracy*, pp. 260–75, 1992. Boulder: Westview) (1991)

On September 11, 1973, the Chilean armed forces attacked the presidential palace of democratically elected Marxist President Salvador Allende. Within hours, the palace was in flames, the president dead, and leading members of the government imprisoned or in hiding. Before the flames were extinguished, tanks and helicopters assaulted Santiago's impoverished *poblaciones* ('urban slums and shantytowns'), forcing tens of thousands of Chileans from their homes.

In the months following the coup, over 100,000 civilians were detained, and most of them were brutally tortured. Thousands were summarily executed or simply 'disappeared.' Every traditional feature of Chilean society came under attack – congress, political parties, labor unions, neighborhood organizations, even local parishes.

For ten years, the only national institution capable of defying the regime was the Catholic church, which through its gradual absorption into the struggle for human rights, moved into direct confrontation with the regime. At the local level, the church acted as a safety umbrella, protecting activists and victims alike. In the shantytowns, small human rights and economic subsistence organizations began to appear. In factories, underground political networks were reestablished, and in universities, students defied regulations prohibiting political discussion and organized demonstrations in opposition to the regime. Yet, these struggles all shared a common denominator – they were localized, ephemeral, and easily repressed. As Alfred Stepan noted, 'In Chile, eight years of authoritarian rule passed without significant movement out of the initial authoritarian situation: civil society remained debilitated in the face of state strength' (Stepan, 1985, p. 322). The military appeared invincible. In May 1980, Augusto Pinochet held a plebiscite on a new constitution institutionalizing his rule and won overwhelmingly.

Suddenly, on May 11, 1983, a storm of protest swept through Santiago's streets. The 1982 economic crisis had created a widening fissure between the military and its supporters, opening the doors to a growing movement of opposition. And, as if a spell had been broken, unarmed students, workers, and shantytown residents flooded the streets, demanding an immediate end to military rule. At its helm were the same poblaciones that had been the target of military repression for almost a decade.

In the western zone of Santiago, residents erected burning barricades, drummed pots and pans, and organized marches. In Santiago's southern zone – poblaciones La Victoria, La Legua, El Piñar, Guanaco, Germán Riesco, and Villa Sur in the *comuna* ('district') San Miguel, and the poblaciones San Gregorio, Nuevo San Gregorio, Joáo Goulat, Yungay, and La Bandera in the comuna La Granja (CETRA/CEAL, 1983) – residents sprayed shantytown walls with political slogans, led mass marches, and cut electricity to large portions of the city by throwing metal objects at electric cables. When the armed forces attacked these seemingly defenseless communities, residents responded by digging trenches, erecting burning barricades, and pelting military tanks with rocks. The deluge of protests in the same shantytowns that had experienced severe military repression between 1973 and 1983 challenged the military's claim to have reshaped political loyalties in Chile. 'In 1983, with the protests,' Genaro Arriagada would later reflect, 'Chile rediscovered the part of its reality that, in the delirium of the economic miracle, it had forgotten. From the start of those mass demonstrations, names like La Pincoya, La José María Caro, La Victoria regained their Chilean citizenship and a place among the concerns of Chileans' (cited in Timerman, 1987, p. 70).

For three years, the protests raged. On July 2 and 3, 1986, they reached their peak with a massive nationwide strike. The military responded with a new wave of repression. Ten people were killed, raising the number of protest-related deaths to over 400. In the población Los Nogales, two teenagers (one a U.S. resident who had been in Chile less than a month) were arrested by a military patrol, doused with gasoline, and set on fire. In the población La Victoria, a thirteen-year-old girl was shot dead on her way to buy bread.

By August 1986, exhaustion had set in, and shantytown residents returned to the safety of their homes. Those who continued to resist the dictatorship through direct confrontation had become isolated. By 1987, only a small minority of *pobladores* ('residents of poblaciones') were either organized or active. As one organizer noted, 'In 1984 I went underground. I wanted to fight the dictatorship directly. I was not alone in this. By 1987 there was not a single public Communist working in my población.'

The Christian Democratic and 'renovated' sectors of the Socialist party abandoned the protests. In 1987, they formed a coalition for 'free elections,' and in the following year, the coalition became the basis for the '*Concertación* [Coalition] for the No,' a broad front aimed at defeating Pinochet in the upcoming, constitutionally mandated plebiscite. Pinochet lost the October 11, 1988, plebiscite on his presidency, and in the democratic elections held in December 1989, his hand-selected candidate was soundly defeated. On March 11, 1990, a democratically elected civilian government took office.

This story of protest and repression raises critical questions about the nature of political life in authoritarian Chile. Why did impoverished and almost defenseless shantytown residents risk arrest, torture, and even death to fight a regime that they seemed to have so little chance of defeating? Why did protests center in some shantytowns but not others? Why did they suddenly decline in 1986? And lastly, to what extent and in what way did the shantytown struggle contribute to the return to democratic rule in 1990?

Most of the literature on the Chilean protest movement fails to address these questions. Writers either treat protests as a spontaneous response to economic grievance, real or imagined, or they see them as the manipulation of a skilled political elite. Such approaches focus on protests only as they appear from outside or above. They omit history and historical continuities, leave aside the political context that determines how individuals organize around grievances, blur the distinctions between successful and unsuccessful centers of protest, and, most broadly, fail to flesh out the nexus between political and civil society.

Political scientists and sociologists such as Genaro Arriagada (1988) and Eduardo Valenzuela (1984), for example, focused on the deprivation, social dissolution, and anger produced by the 1982 economic crisis to explain the eruption of protest violence in Santiago's shantytowns. 'During the protest years the poblaciones were comprised of a mass of unorganized individuals and a few isolated, weak and unfinanced organizations of several thousand residents' (Arriagada, 1988, p. 61). Yet, this explanation runs counter to a major set of facts. In 1983, there was no association between the level of economic depression and the intensity and scope of protest action. Those poblaciones hit hardest by the crisis, in relative or absolute terms, responded weakly to the call for protests. And when protests did emerge in these areas, they were short-lived for the protesters were unable to withstand the repression.

Other sociologists, including Tilman Evers (1985), Fernando Ignacio Leiva and James Petras (1986), and Teresa Valdés (1987), focused on the construction of autonomous neighborhood organizations and the formation of a new social actor, the poblador movement, to explain the eruption of protests. But the new social movement interpretation fails to explain the peculiar configuration of protest in authoritarian Chile. Had the 1983 protests been the consequence of a new social movement in the shantytowns, the distribution of such protests would have been evenly spread, rather than concentrated in those poblaciones that had previously been the center of left-wing political activity. As sociologists such as Alain Touraine and Eugenio Tironi observed, those who joined soup kitchens or economic cooperatives did not necessarily participate in the protest or so-called poblador movement. As Tironi observed:

> The network of social organizations that they [the pobladores] managed to construct . . . resulted in the reactivation of the political militants within the poblaciones. . . . The so-called *poblador* movement has been completely confused with the political militants that we have identified above (1988, p. 74).

However, the 1983–1986 Chilean protest movement, if not the harbinger of a new social actor or a spontaneous response to the 1982 economic crisis, was more than the product of isolated political militants operating out of shantytowns. The capacity of these urban neighborhoods to mobilize mass political resistance, despite ten years of severe military repression, lay in the political heritage of decades of work in the popular culture and in the formation of a skilled generation of grass-roots militants. As Roger Burbach noted,

> Grassroots militants who had been active years before the coup played key leadership roles in all of these organizing activities. As one party leader noted 'we as parties were able to do little, but the political militants at the local level rebuilt the social movement during the darkest moments of the regime.' The ideological and political consciousness that broad sectors of the population had acquired over the years prepared them for political work at the community level and for organizing around local needs, even when their ties to the political leadership had been broken (1988, p. 5).

The Chilean protest movement was like its Spanish equivalent in the 1940s and 1970s, 'dependent on the underground survival of the parties of the Left. Those parties provided the strategies and the leaders, and it was the capacity of these parties to survive that kept the . . . resistance alive in the long and difficult period . . . and that later rekindled the struggle' (Maravall, 1978, p. 66). And it was in the same left-wing neighborhood that had been most active in the decades before the coup that resistance struggles began and were strongest.

The high degree of politicization, organization, and solidarity in such neighborhoods was a direct consequence of the historical connection between these areas and the Chilean Communist party, as well as the party's consistent emphasis on the creation of solidarity communities with skilled grass-roots leaders. Much like the left-wing enclaves of the U.S. civil rights movement described by Gitlin, 'these enclaves of elders and subterranean channels, rivulets, deep-running springs [nurtured] unconventional wisdom, moods and mystiques. With left wing politics in a state of collapse most of these opposition spaces were cultural – ways of living, thinking, and fighting oneself free of the . . . consensus' (1987, p. 28).

Those who became politicized as a result of contact with the Communist party or Communist neighborhoods – even those who became militants of other political parties or never joined a political party at all – shared a political conception: They identified their problems in structural terms and sought solutions through collective action. Contact with the Communist party and life in traditional Communist neighborhoods overcame 'the fundamental attribution error – the tendency of people to explain their situation as a function of individual rather than situational factors' (McAdam 1982, p. 50). As one Christian Democrat living in a Communist población explained, 'I used to be ashamed of my poverty, I saw it as personal failure. A communist neighborhood organizer explained to me that I needn't be ashamed. That we all shared the same problems.'

Residents of Communist-influenced neighborhoods absorbed the lesson that individuals can make a difference, that they can, in collective organization with others, take control over their lives. Those who lived in such communities continued to resist military rule despite the high personal cost because they believed that (1) their particular organizational skills were critical to movement success, (2) collective solidarity action was capable of defeating even the most powerful military regime, and (3) their own intense collective identity was such that passive acceptance of the regime was incompatible with the personal and individual sense of self.

These communities were critical in reestablishing political activity after the coup and the destruction of the traditional political parties. Christian Democratic poblaciones were weakened by the ambivalence of their party toward the military government, and the traditional Socialist party was paralyzed by internal division and infighting. The traditional right-wing parties had dissolved themselves in support of the military regime, and the Communist party as a formal institution was also decimated, with most of its leadership either dead or in exile. Yet, that party's traditional emphasis on the creation of coherent grass-roots communities and skilled grass-roots leaders had left a legacy that was not so easily destroyed.

After the coup, grass-roots activists in poblaciones that had traditionally depended on elite support or state resources found it impossible to maintain a defense against military repression. They either ceased political work or attempted to keep a semblance of political organization by working under the protective auspices of the Catholic church. Others moved to the traditional left-wing, Communist-influenced poblaciones, where a history of grass-roots political activity had created a supportive environment and set apart potential informants.

In these traditional communities, resistance activity often began within weeks of the coup. In Granadilla, for instance, the first political act of resistance was the building of a collective farm or *huerta*. This farm became the basis for a communal kitchen, which, in turn, inspired a series of women's workshops (producing clothes and toys) and youth groups (mainly cultural). In 1977, the leaders of the popularly elected neighborhood council convinced the mayor to fund the entire gamut of popular organizations through the Minimum Employment Program (PEM) and the Public Works Program for Heads of Household (POJH). Thus, in 1978, a Communist neighborhood council was administering the minimum employment program for the Pinochet government.

By 1979, political activity in the poblaciones accelerated. Activists began to branch out to develop organizations in the less traditional activist poblaciones, but they avoided working in the poblaciones they lived in for fear of informants. They also avoided the larger mass organizations set up by Pinochet – such as sports clubs or mothers' centers – for the same reason. Instead, they began to create parallel organizations, such as cultural centers and human rights groups. These groups developed a reputation in the area as a front for political activists and further isolated the activists from the nonpolitical members of the community.

When the 1982 economic crisis sparked a political crisis, alienating former supporters of the military regime, the poblaciones exploded. As the protests reached their peak, the less organized neighborhoods and political parties were swept into activity. The Christian Democratic party and the Renovated Socialists became critical to the movement and forged a powerful Democratic Alliance coalition. But those poblaciones without a tradition of political activity remained dependent on the political skills of their national leadership and found it increasingly difficulty to withstand the escalating military repression.

The traditional Communist-influenced poblaciones, on the other hand, maintained high levels of resistance throughout the protest cycle. These neighborhoods had been 'the small motor that later turned the larger motor of the mass movement' (Gitlin, 1987, p. 26). Even after the protests had ceased and the majority of Chileans had returned to their homes, the traditional Communist poblaciones maintained a high level of political mobilization, their continued activity and increasing militancy eventually isolating them from the population at large. When the military was finally defeated in 1988, these poblaciones, along with the Communist party, found themselves at the margins of political life. The following interviews stress the differences that existed, during the 1983–1986 cycle of protest, between traditional Communist poblaciones and those with less of an activist history.

Interviews

Low-mobilization poblaciones

Villa Wolf. Villa Wolf is typical of poblaciones that failed to mount significant protest after 1983. It is both very poor and very small (it has barely 1000 inhabitants, and the unemployment rate exceeds 65 percent). Despite its size, however, Villa Wolf is poorly organized. The only neighborhood organizations it has been able to sustain are the soup kitchens, and these are maintained with great difficulty. 'The problem here,' explained the head of the soup kitchen who is a member of the Christian Left,

> is lack of resources. . . . The community rejects the soup kitchens, they don't help with the construction. The other villas within La Pincoya receive more help from the community. *In comparison with other villas, Villa Wolf has very little sense of community, little solidarity and almost no social mobilization or protests.* The soup kitchens have tried to function as a source of information for women and as a primary health organization, but it is the same women who act in several capacities.

'Villa Wolf was originally established by the Christian Democrats,' observed another resident. 'And since the Christian Democrats supported the coup, many of the leaders of the poblacion are still with Pinochet.' According to the residents, the community lacks a collective identity, and the political parties that try to organize here are bitterly divided among themselves. The lack of community networks leaves Villa Wolf vulnerable to infiltration by government-linked groups, such as the *Unión Democrática Independiente* (UDI). The government then uses the information provided by such groups to persecute and arrest community activists and to frighten the other residents into passivity. 'The majority of protests,' argued the Left Christian organizer in Villa Wolf, 'have taken place in those sectors such as Pablo Neruda, which resulted from land occupations organized by the Communists.'

Sporadically combative poblaciones

The success of any day of national protest depended in large part on the level of protest activity in the sporadically combative poblaciones. Because these poblaciones were characterized by political infighting and factionalism, they were unable to maintain a constant level of resistance. They were pulled into the fray of battle, however, when changes in the external opportunity structure made protests costly.

Sara Gajardo. Located in the western comuna Cerro Navia, Sara Gajardo was originally founded as a land grant under *Operación Sitio*. In 1967, residents from a nearby land occupation were relocated to Sara Gajardo, but none of the political parties established deep roots in the community. There is a great deal of delinquency and drug addiction in Sara Gajardo, especially among the youth. On national protest days, many of these young people explode in rage, throwing stones and epithets at soldiers who are often backed up by tanks and helicopters. But the older members of this community are largely passive, even those who participated in the Popular Unity.

'The depth of participation within the población is of extreme importance,' insisted the twenty-one-year-old director of an art workshop. 'If there is a history of struggle and organization the regime can decapitate the organizations but the organizations rise again.' But another community leader explained that 'while the young people here have managed to recreate the neighborhood associations eliminated by the military, there is very little participation from the adults in the community. Consequently we have to learn everything on our own. We are given little direction.' The local priest concurred:

> There is a basic problem at the grass roots with the political parties. There is a tendency for the political parties to try to take control. The Communist Party is always trying to get in and control things. They get orders from outside and don't always consider the real needs of pobladores. There is a lot of disillusionment. People want to create their own future, not just leave things in the hands of a central committee. . . . The parties are not always honest and the people see them as utilitarian. They want socialism to transform the society radically but the parties don't seem suitable for that task.

The problem is not simply lack of direction: The lack of political solidarity raises the cost of political participation. 'It's more costly to organize here than in La Herminda,' contended a young activist, 'because the population is afraid of participating.' 'Disunity is the major problem,' a twenty-year-old political organizer suggested. 'People are afraid to get involved. There are many *soplones* [informants].' Another political organizer (a non-Communist) stated:

> The risk is not only in the streets, they attack us from helicopters. Although the repression is focused against La Herminda, Sara Gajardo has a higher rate of fatalities. La Herminda is better defended since it receives more back up from the community. There are more political militants in La Herminda, more Communist activists.

Villa O'Higgens. Villa O'Higgens is an extremely poor poblacion with sporadic levels of protest activity. On the day I conducted this interview, the military had occupied Villa O'Higgens in anticipation of the September 4 general strike. Armed soldiers patrolled the streets, making it impossible to cross from one house to the next without help from neighborhood sentries.

In one of the homes, I was greeted by a five-year-old boy with a toy gun. 'This is to stop the soldiers from breaking in,' he told me. But there were no protests in Villa O'Higgens on September 4. The high rate of fatalities on July 2 and 3 and the confusion surrounding the strike plans for September had convinced the vast majority that the risk was simply too great.

Villa O'Higgens, like many of the other sporadically mobilized poblaciones, is divided between land-grant sectors set up under Operación Sitio in 1968 and land-occupation sectors led by the Movimiento Izquierda Revolucionaria (MIR) in 1969. But the popular organizations died shortly after housing needs were settled.

Villa O'Higgens is still poorly organized, and the political parties are weak and divided. 'The people lack political consciousness,' complained a human rights worker. 'They are seduced by whatever organization provides economic assistance.'

'Even the church,' explained another human rights worker, 'while predisposed to help is not active, because there are not political parties demanding help as there are in La Victoria.' (The priest in Villa O'Higgens and the priest in La Victoria, for example, arrived from France together and shared political convictions. Their level of activism since their arrival in Chile, however, has reflected the level of activism in their respective communities.)

The weakness of the political parties means that protests in the población are tentative and easily discouraged. As a member of a cultural group explained:

> There were many protests here in 1983, but they were mostly spontaneous and lacked the organization to maintain themselves or defend the community. This was in marked contrast with those poblaciones where the defense was greater due to the strength of the political parties.

The level of political mobilization in Villa O'Higgens is, as in other sporadically mobilized poblaciones, dependent on changes external to the población itself. For instance, a human rights worker explained that the protests in Villa O'Higgens in 1985 reached a low due to lack of confidence, but in 1986,

> confidence increased due to what appeared to be the organization of the political parties, and the clarity of their new strategy. The July 2 general strike was especially strong because of a commando unit for the general strike made up of leaders of all political parties within the población. The organization and publicity began early, and there was support from many other sectors such as the mobilization of doctors and other professionals. But the costs were high. In Villa O'Higgens alone we sustained 14 bullet injuries, 5 injuries from bird shot, 34 arrests and one death.'

There is a lot of repression here now,' she continued, 'because this community is of strategic importance and there is little organization here to defend it.' (As we were talking the military arrived and began shooting at the homes and church even though no protests were taking place. The shooting continued without interruption for four hours.)

The focus of resistance: the combative poblaciones

The poblaciones that emerged as the central core of the resistance movement in 1983 were those formed by the Communist party. Political activity was not new to these areas. In fact, the character of these poblaciones had been forged by earlier waves of political activity. In such neighborhoods, political activists lived and organized in the same población and maintained an organic relationship with the residents. And they did more than organize protest. They held public events to strengthen community ties and revive cultural traditions, and they created new forms of grass-roots democracy.

La Victoria. La Victoria is a small población, no larger than three square kilometers in area and populated by 32,000 inhabitants. It was one of the earliest illegal land occupations in Chile. 'Most of us were *allegados* [homeless or living in the homes of other people],' explained one of the original participants of the *toma* ('illegal land seizure'). 'The government had no program to provide for the homeless,' argued Maria, a representative to the neighborhood council from 1960 to 1964.

> We began organizing for the land occupation six months early. We met clandestinely in different houses . . . On the day of the land occupation we were surrounded by police. There were 15 deaths from cold and disease. We faced fear by organizing. The solidarity was incredible . . . We just stayed on the land until the government agreed to sell us the rights. We formed a *Comando de Población* [Community Defense Organization] with committees on every block. Only after three years of struggle and organization did we win water rights in 1963. The Communists led the struggle. We created a school of adobe for the children with 14 rooms. Everyone in the población contributed something. Later we began to organize and fight for a health care clinic . . . It was only through struggle that we won the eight-hour working day. Everything we have, we won through struggle.

'Even the solidarity and activism of our local priests was won through struggle,' contended María. 'The first priest in the población was not progressive.' It was the community that demanded a different kind of church. 'The first church in this area was only interested in the rich. The priest tried to "buy" and divide the community, like a salesman offering bread. We chased him out in 1963. There was bread for all or there was bread for none . . . . We always worked for the good of all.' From then on, she explained, La Victoria always had priests who worked with the community, 'fighting with us, never against us.' 'We held elections with secret ballots,' contended another original participant. 'The Communist party won. The first president of the community was a Communist, and he remained president until 1968 when he died. We always held elections by block, and the Communists always won,' María insisted.

In 1973, the military attacked the población. La Victoria, along with four or five other poblaciones, constituted the main force of the resistance movement. Despite a lack of preparation or central leadership, residents confronted tanks alone. A journalist who lived in La Victoria during the coup explained that 'there were no armaments, no medical equipment for the injured, anyone who helped a resister was also killed.' La Victoria defended itself with pure commitment and solidarity through weeks of military assault, long after the Popular Unity government was destroyed. 'At the leadership level the political parties were in disarray,' explained the journalist, 'but at the base, in La Victoria, there were no divisions, we all worked together.'

By 1974, most of the leaders were in concentration camps or dead. Those who were still alive withdrew from politics or fled the población to organize clandestinely. It took ten years to rebuild community organization: 'We began with nothing,' they explained. In 1977, they reestablished the first cultural and human rights groups, and within two years, they had functioning popular economic subsistence organizations, *bolsas de cesantes* ('organizations of the unemployed'), *talleres laboral y cultural* ('technical and cultural workshops'), and so forth. These organizations allowed the pobladores to regroup, to share grievances, and to express dissatisfaction. Furthermore, it was difficult for the government to repress these groups because they served a function: They allowed people to survive during the implementation of the economic model. But they also allowed the atomized residents of the población to reorganize.

'We are more organized now,' one of the members of the senior citizens' club told me, 'than in 1983, in spite of the repression. We have an organization for every block, like Nicaragua.' La Victoria boasts 30 social organizations in addition to the block organizations. Each block elects its own delegates, and the block delegates and social organizations together elect the president of the *Comando Nacional* of La Victoria.

La Herminda de Victoria. Like many of the combative poblaciones, La Herminda de Victoria arose from an illegal land occupation organized by the Communist party's Committee of the Homeless, and it has the distinction of being the first land occupation in Pudahuel (what was the Barrantas and is now Cerro Navia). It was also the first población under authoritarian rule to gain the official recognition of a democratically elected neighborhood council.

Like the occupation of La Victoria, the occupation of this neighborhood in 1967 was met with extreme police repression. The población was roped in by police forces and virtually under siege for nine months. During the siege, a young girl named Herminda took ill and, unable to reach the hospital, died: The población was named in her honor. Finally, an agreement was reached with the government. The occupants were transported to a vacant area, to be named La Herminda de Victoria, and were sold rights to the land. The residents were given 45 hectares for 1464 families and asked to sign a contract requiring that they pay on a sliding scale. In three years, the new occupants would be provided with the deeds to their homes. Corporación de Vivienda (CORVI), the government housing authority, would provide water and sanitary services.

But in 1973, the población was hit very hard; most of the political leaders were killed and all the organizations were decapitated. In addition, the government simply ignored the contract that gave residents of La Herminda de Victoria deeds to their homes. In 1979, Juan Araya, the ex-president of the Committee of the Homeless, and the *Metropolitana de Pobladores* (the población coordinating organization linked to the Communist party) began a series of meetings in La Herminda. The Church-linked *Agrupación Vecinal* (Organization of Neighbors, AVEC) also participated. The idea was to reclaim not only housing rights but also rights in general and to learn the provisions of the law.

These meetings became the basis of the Committee of Pobladores, which led the struggle to recover the rights granted in the original housing contract. 'We pointed to a clause in the contract that guaranteed that all pre-1976 agreements be honored,' asserted a member of the committee:

> The government responded with threats. The residents of the población began to gather to defend us. They held meetings with lawyers. The first week 40 attended, the second week 200 attended, the third week 500 attended. Then there were no more meetings. We gathered in front of the official neighborhood council to denounce them. They responded with extreme repression, attempting to eliminate the Committee of Pobladores. The secret police threatened the leaders of the committee with arrest or even death.
>
> We had, however, our ideas very clear, and we simply continued. The official press launched a campaign against us. The secret police began an investigation and followed up with threats. The carabineros surrounded the church and the Christian community. But the residents of the community responded en masse and continued to meet every Sunday. If the police or secret police tried to single out one of us for punishment, the rest of the población responded, and made it impossible for the police to arrest or 'disappear' only one.

Finally, in April 1980, the government gave in, and 95 percent of La Herminda de Victoria's residents were given deeds to their homes. After the official neighborhood council resigned in embarrassment, the municipality was obliged to call for free elections. 'Ninety-five percent of the población voted for us,' exclaimed Pablo, an ex-member of the democratic council. Through community participation and solidarity, the población was able to resist the regime's repression and reclaim its housing rights.

The community did not cease political activity once housing claims had been satisfied. They demanded recognition of their democratically chosen neighborhood council, as well, and became the first community in Santiago to reestablish democratic self-government. As Pablo explained:

> Pedro was voted president. He and I were the only two in the junta [neighborhood council] that were not Communists. We were elected in June 1980. On December 25, 1980, we held a Christmas festival to celebrate our victory. 2500 children participated. All of the pobladores contributed something.
>
> On the 16th of March we celebrated our anniversary. Everyone participated. We invited people to dance in the streets, to join in parades.

But the political leaders were unable to maintain the relationship with the community that had made their victory possible. The government took advantage of this weakness, pitting the council against political militants by giving the former the responsibility of keeping order on protest days. The willingness of the council members to assume this role and act against sectors of their own community permitted the military to recapture control of the población. As the former council member explained:

> Our problems began because we were obligated to use the office of the official neighborhood council. In this way we became dependent on the municipality. We lost contact with the pobladores. We were required to be in charge of the workers on the minimum employment program. We were put in charge of maintaining order during the protests, criticizing acts of vandalism. This badly split the población. Finally, a group of former directors of the Communist party argued that we had completed our mission, that it was time to call elections for a new neighborhood council.

'The new council,' explained a young political leader, 'lacked knowledge of the laws, and a clear project or analysis of future consequences.' Pablo contended that 'they began to divide among themselves.' He stated:

> Finally they had to resign and the government appointed its own neighborhood council, and all that we had done was lost. Other communities learned from our mistakes. In such communities the democratic councils are functioning well because they are clear on what they want to accomplish. Here the Communists simply wanted to dominate the council.

Yungay. Yungay, originally Villa Lenin, is a small población of 14,000 inhabitants, but it is one of the most combative. Unlike the residents of La Victoria and Herminda, the people of Yungay never partook in an illegal land occupation. Rather, their strong sense of solidarity, belief in the power of collective action, and confidence in themselves grew out of their relationship with the Communist party. Although Yungay was settled legally during the government of Allende, it was the Communist party that organized the población and led the struggle for housing, water, schools, and electricity. More importantly, Communists in Yungay lived in Yungay and stressed the needs of the población over the needs of the national party.

The maturity of the political parties and their deep roots in the población allowed them to work with the church toward a common goal. In poblaciones without a long tradition of organization, the local priest observed, both the political parties and the church tend to direct the población from the outside, resulting in dogmatism and rigidity on both sides. But in Yungay, 'there is more maturity in both the political parties and in the Christian community, there is more participation from the base in both. Thus, there is more cross membership, more linkages between members, and more people that participate in both organizations.'

It was this responsiveness to the community that allowed grass-roots activists to reconstruct democracy at the local level. These individuals began reorganizing in the población in 1974, when they established the *Central Juvenil Latina* (Center for Latin American Youth). Later, they created a sports club and, later still, a soup kitchen under the auspices of the local parish. By 1978, they had re-created two cells of the Communist Party, and in 1980, they created a Committee on Human Rights and a neighborhood cooperative (*comprando juntos*). They also created a women's center and a health center and began the process of democratizing the neighborhood council.

They began meeting with lawyers in 1979 to seek legal authorization for a neighborhood council, and by 1984, they had grouped representatives of every block in the población. The block representatives then chose twelve delegates to form the new democratic neighborhood council. The president of the neighborhood council explained:

> The entire población participated in the selection. We then challenged the legitimacy of the government's neighborhood council. We embarked on a plan of action. As we began to produce ideas, we awakened the población. We began a momentum, a rhythm. Demanded that the municipality provide a building and a telephone. We told them that we would take them to court if they didn't comply. They submitted. We made a public declaration and a contract requiring them to install a telephone in our building. We now have a telephone, which is open to the community from 8:00 a.m. to 12:00 midnight.

The democratic neighborhood council used the money collected from telephone calls to buy the community an ambulance (they charged 20 pesos a call, the normal price for a telephone call in Chile). In March, Yungay was the only población in the area with a telephone and in December, it became the only población with its own ambulance.

'The neighbors began to have more confidence in us,' the president of the neighborhood council observed. 'They became more involved in the process, they worked more. With unity and democracy, all of the sectors in the población got involved.' Next, the neighborhood council opened a library, the first library in the area. They called it Pablo Neruda Library, after the famous Communist poet. 'All of this was done on our own,' the president proudly acknowledged.

> We received no help from anyone outside the población, neither economic nor material. Next we began to create sidewalks. We never had sidewalks. We constructed 5,000 meters of sidewalks. The neighbors laid the cement. The municipality provided the materials. They say they gave them to us, but in reality they robbed us. The materials were bought with the money they took from the community. They have to provide these things.

But the achievements of Yungay were not simply material. With each successful struggle, the residents grew bolder. They developed more confidence in themselves and a broader conception of their rights as citizens. As the president of the new democratic neighborhood council observed, 'We used to think that we had to give something to receive something from the municipality. Now we demand that which we deserve, what corresponds to us as citizens of this country. We demand what is legally ours. We demand our rights.'

Community solidarity enabled the neighborhood council to challenge the legitimacy of the government's appointed representatives. And it allowed the community to reassert its democratic will. The president of the neighborhood council explained:

> When we demand these things the other neighborhood council does not say anything, for fear that the mayor will simply fire them. We, on the other hand, have the

población on our side. He can't fire us, he didn't hire us. When he tries to throw us out, the entire población says NO. Unity rises from democratic foundations.

Sometimes the neighborhood council members use the law to apply pressure, but usually they rely on the people themselves – the community – to do this. Every week they call a meeting in every sector. 'We are always in meetings,' insisted the president of the council, adding, 'we work through grass-roots democracy.' Another member of the neighborhood council emphatically concurred:

> Here the political parties work from the base, not like the United States where the parties run things from the top. Here we have the capacity to organize and resist. They can repress and kill us as individuals, but the organizations survive and the resistance rises again. Here we have a history of combativeness, dating from the struggles in the nitrate mines at the turn of the century.

'The people support the neighborhood council with their own funds, every one contributes something even when they don't have enough to eat,' a member of the high school support group told me. 'The mayor has created his own neighborhood council, with all the government money, but we ignore it.'

Thus, in Yungay, a history of radical militancy and political solidarity, along with the continued presence and integration of grass-roots political militants, has allowed the población not only to maintain a high level of militant resistance to the regime but also to reestablish democracy at the local level.

Conclusions

In all of the less combative poblaciones, the political parties and party militants were viewed with suspicion. Residents sought individual solutions to their problems or accepted their fate. They distrusted political activists, whom they saw as outsiders and opportunists callously exploiting the suffering of the pobladores. The political organizers, lacking support, responded, in turn, with a rigidity and dogmatism uncharacteristic of those in the more combative poblaciones. This tension between the political activists and the pobladores paralyzed the población. Residents in these communities often expressed deep frustration and anger but lacked the direction and organization to act on these frustrations. Even the more radical residents found themselves restricted by the weakness of political organizations. And residents of Sara Gajardo complained that the repression aimed against La Herminda de La Victoria resulted in more fatalities in their own población. Only when a nonpartisan coalition from the center, such as the *Asamblea de Civilidad*, led the call for protests were the pobladores in these communities moblized.

The only poblaciones that were highly mobilized throughout the protest cycle were those that had originally been created by the Chilean Communist party. The Communist party and resistance movements survived in such shantytowns because the party's work in the popular culture before 1973 had created a skilled generation of grass-roots militants, capable of maintaining community support. But individual militants did not determine the extent or success of protest action. They were often killed or arrested and thus replaced by other youths who found themselves heirs to the same tradition. What was important was the extent to which community solidarity and a shared political vision had made all the residents potential militants. Even the priests in such shantytowns took a more active role because residents demanded political commitment from their clergy.

By 1986, the protest cycle was drawing to a close. The Christian Democratic and Socialist parties moved to a negotiating stance, concentrating their efforts on the 1988 plebiscite. The Communist party became less crucial to the movement, especially after the *Frente Patriótico Manuel Rodriguez* (a guerrilla group linked to the party) attempted to assassinate General Pinochet on September 7, 1986. Indeed, the Communist party's national strategy from 1983 onward weakened grass-roots organizing efforts in several respects. Although the party's focus prior to 1983 was on rebuilding social organizations and bringing together people who shared grievances, after 1983, it stressed an explicitly insurrectional strategy. In trying to control the grass-roots organizations, the party alienated the non-Communist members. Furthermore, the party also began to pull its leaders out of mass organizations to employ them in clandestine military operations, thus isolating itself from its mass base. Finally, the Communist party's armed strategy and unwillingness to consider alternative strategies (most notably its reluctance to participate in the 1988 plebiscite) cost it support even in some of the traditional Communist strongholds.

The Communist party's isolation coincided with the final phase of the protest cycle – the government's concession to the moderates, which led to the plebiscite in 1988 and the government's defeat. As a consequence, the more radical movement activists were relegated to the margins of political life, and the more moderate activists were reabsorbed into normal political channels.

The Chilean protests followed a pattern similar to successful cycles of protest in Western democracies (Tarrow, 1989). They began in traditionally active grass-roots communities, and as the protest movement gathered momentum, it absorbed new adherents, often people with no previous political experience. A political crisis provided a new political opportunity, energizing the movement and extending its scope. And as the protest cycle reached its peak, organizers found themselves engaged in increasingly costly and violent confrontations with the state. In response, the more moderate political leaders and activists moved to a negotiating stance, and the state was able to defuse the movement by granting significant concessions to the more moderate movement organizers. Thus, the políticos were reabsorbed into normal political channels, having won their most essential demand, while the more committed organizers were stripped of their mass support.

What becomes then, of the grass-roots activists and popular organizations now that political parties have returned to the forefront? Will they be rendered obsolete by the return to 'politics as usual'? The answer appears to be yes.

Although the democratic opening created by the 1988 plebiscite and election campaigns originally encouraged a flourishing of grass-roots political activity, this activity lasted only as long as the campaigns themselves. Since the inauguration of the democratic government in March 1990, many grass-roots militants have returned to their homes, and participation in social organizations has dramatically declined.

Still, the vitality of the new democracy will continue to depend on the ability of the parties to channel the energies of grass-roots militants into a deepening of the democratic process. If grass-roots organizations are demobilized and their militants excluded from political participation, the consequential alienation of this sector will weaken the democratic forces. To some extent, this process has already begun, leaving bitterness and disillusionment in its wake.

If, on the other hand, the decentralization of political power and the redemocratization of local and municipal governments guarantee a political space for popular participation, the new democracy may be fortified by its grass-roots support. The same militants that allowed political parties and resistance to survive authoritarian rule will then ensure the commitment of political parties to the needs of pobladores.

References

Arriagada, G. 1988: *Pinochet: The politics of power*. Boston: Unwin Hyman.

Burbach, R. 1988: *Chile: A requiem for the left*? Berkeley: Center for South American Studies.

CETRA/CEAL 1983: Tercera y cuarta protestas. *Páginas Sindicales* 6 (57).

Evers, T. 1985: Identity: the hidden side of new social movements in Latin America. In Slater, D. (ed.), *New social movements and the state in Latin America*. Dordrecht: CEDLA, 100–24.

Gitlin, T. 1987: *The sixties: Years of hope, days of rage*. New York: Bantam Books.

Leiva, F. and Petras, J. 1986: Chile's poor in the struggle for democracy. *Latin American Perspectives* 13, 5–25.

Maravall, J. 1978: *Dictatorship and political dissent*. London: Tavistock.

McAdam, D. 1982: *Political process and the development of black insurgency, 1930–1970*. Chicago: University of Chicago Press.

Stepan, D. 1985: State power and the strength of civil society in the Southern Cone of Latin America. In Rueschemeyer, D., Skocpol, T. and Evans, P. (eds), *Bringing the state back in*. Cambridge: CUP, 131–47.

Tarrow, S. 1989: *Democracy and disorder*. New York: OUP.

Timerman, J. 1987: *Chile: Death in the south*. New York: Alfred Knopf.

Tironi, E. 1988: Pobladores e integración social. In Tironi, E. (ed.), *Proposiciones: Marginalidad, movimientos sociales y democracia*. Santiago: SUR, 111–25.

Valdés, T. 1987: El movimiento de pobladores, 1973–1985: la recomposición de las solidaridades sociales, In Borja, J., Valdés, T., Pozo, H. and Morales, E. (eds), *Decentralizacion del estado: Movimiento social y gestion local*. Santiago: FLASCO, 211–24.

Valenzuela, E. 1984: *La rebelión de los jovenes*. Santiago: SUR.

SECTION FOUR

URBANIZATION AND INDUSTRIALIZATION

Editor's Introduction

In the 1950s it was widely assumed that developing countries were bound for an urban–industrial future. Urbanization and manufacturing industry were the most potent indicators of modernity. In the case of urbanization, however, the optimism of the 1950s soon gave way to fears that cities in the developing world were parasitic and not generative – a distinction outlined by Hoselitz in 1957 (Hoselitz, 1957). Hoselitz believed that most Third World cities would serve a generative economic and cultural role for the areas around them. Some primate cities would be parasitic – would exert an unfavourable impact on local economic growth – but they would be in a minority of developing world cities. Others disagreed. In the 1960s a view took hold that cities growing up in advance of industrialization were bound to be parasitic and, functionally, only semi-urban. Such cities were all too common in Latin America, where migrants were deserting the countryside to live as best they could in the informal sector of countless fake cities. Slums and shanties were the new pathologies associated with this unhappy 'over-urbanization' (Berry, 1961) and some cities duly became coded as sites of depravity, where local cultures of poverty echoed the dysfunctionality of rapid urban growth for the economy as a whole. The rapid urbanization of parts of the Third World became a problem for policy-makers, and not a sign of progress as it had been just a few years earlier.

Such rapid urbanization can partly be explained with reference to the isolation paradox we met in Section Three. Migrants to the city often left the countryside unaware that countless other individuals and families were also bound for the bright lights. Competition for urban jobs proved more intense than many expected. Migrants needed special favours or kinship links to break into industries closely controlled by trade unions (Holmstroem, 1984). Yet urbanization is clearly also stimulated by the belief that life *will* be better for poor people in the city; this, after all, is the message of the Harris–Todaro model of migration. It is the expectations of potential migrants that count. This in turn led some commentators to speak of a persistent urban bias in the development plans of many Third World countries.

Michael Lipton is most often associated with the idea of urban bias, and **Reading 16** finds him discussing the complicated links between migration, urbanization and urban bias. The Lipton Reading is taken from his widely cited book, *Why Poor People Stay Poor: A Study of Urban Bias in World Development* (Lipton, 1977). In that book Lipton maintains that urban bias is enforced by price twists which undervalue rural outputs relative to urban outputs, by education systems which encourage the young and able to leave the countryside as part of a rural skill drain, by tax policies which discriminate against the countryside, and by investment policies which fail to recognize that scarce capital is better employed in rural areas than in urban areas. Lipton argues that urban-biased policies are perpetrated by governments in tow to the narrow interests of an urban class; indeed, *Why Poor People Stay Poor* begins with the striking claim that: 'The most important class conflict in the poor countries of the world today is not between labour and capital. Nor is it between foreign and national interests. It is between the rural classes and the urban classes. The rural sector contains most of the poverty, and most of the low-cost sources of potential advance; but the urban sector contains most of the articulateness, organisation and power. So the urban classes have been able to "win" most of the rounds of the struggle with the countryside; but in so doing they have made the development process needlessly slow and unfair' (Lipton, 1977, p. 13).

Lipton has returned to his urban bias thesis on several occasions since 1977, subtly refining his arguments in the face of more and less well directed criticisms of his work (see the Guide to Further Reading). In **Reading 16**, Lipton sketches out two of the key arguments that make up a wider perspective on urban bias. In the first part of the Reading some familiar ground on urban bias is covered, some of which resonates with the work of Robert Bates and others (**Reading 9**). Lipton notes that migrants moving from villages to towns in the Third World do so in the belief that they will be better off as a result, and that, very often, they *will* be better off because of government spending policies that favour urban areas. The second part of the Reading is less often cited and concerns what Lipton calls 'The Facts of Pseudo-Urbanisation'. Lipton takes issue with the view that we are witnessing a mass urbanization in most of Africa and South Asia. He believes that much of the apparent increase in urbanization in the Third World is fuelled by circular, not permanent, migration to urban areas, and by the inclusion of working villages in the legal boundaries of expanding urban administrations. Lipton concludes that the 'urban problem' in the Third World is less pressing than many have assumed. He also insists that such problems as exist need to be tackled by means of investment in the rural source-areas of city-bound migration.

Lipton is careful not to extend his comments on pseudo-urbanization to Latin America, where it is generally accepted that industrialization and urbanization have proceeded much more quickly than in South Asia and Africa. According to the World Bank, about 75 percent of Latin America's population lived in urban settlements in 1992, compared to 26 percent and 27 percent in India and China respectively, and just 12 percent in Malawi and Uganda (World Bank, 1994, Table 31). Some of the most important research projects on Third World urbanization and urban problems have also been focused on Latin America, including those led by **Alan Gilbert** and his co-workers (Gilbert, 1983; Gilbert and Ward, 1985; Gilbert and Varley, 1991).

In **Reading 17** Gilbert begins by noting how much our knowledge of Latin America's slums and shanties has improved since the mid-1960s. The work of Abrams, Mangin and Turner convinced many planners and donor agencies, including, crucially, the World Bank, that shanty towns are as much 'slums of hope' as 'slums of despair' (Abrams, 1964; Mangin, 1967; Turner, 1967, 1968, 1976). On the basis of his work in Mexico and Peru, Turner concluded that most shanty town residents are not new migrants to the city, but are people who have accumulated sufficient belongings to escape the more precarious existence of a street or slum dweller in the inner-city. The primary existential need of new arrivals in the city is the 'opportunity' to be close to their place of work. Turner's 'bridgeheaders' need to be able to sell their wares or services to city-centre customers, even if this means living on the street or in rented slum accommodation. In time, though, many migrants acquire funds and belongings and then search for new accommodation away from the city centre. The first need of these bridgeheaders-turned-consolidators is security of tenure; their desire is for a place from which they and their possessions cannot easily be evicted. In Latin America new shanty towns tend to emerge as groups of 'consolidators' invade (or purchase) vacant and poorly-served lands on the periphery of cities, often with the support of political parties. The politics of clientalism and patronage loom large in this process of centre–periphery urban filtering. Once established in a peripheral settlement, or shanty town, residents look to politicians to recognize their *de facto* tenancies, and to provide them with the security necessary to encourage them to upgrade their housing stock. Many government housing departments and donor agencies now recognize the need to provide these 'spontaneous' settlements with basic services like water, electricity and transportation. In some cases land plots are provided for families to build their own houses: this describes a sites and services approach to the upgrading of a peripheral urban settlement or colony. Self-help is the guiding philosophy.

All of this represents an about-turn from the negative stereotyping of slums and shanties that prevailed until the mid-1970s; a time when governments were encouraged to bulldoze 'illegal housing' and make life difficult for street sellers. During the Emergency in India (1975–77) Sanjay Gandhi became infamous for ordering the demolition of squatter settlements in parts of New and Old Delhi. (This unhappy episode is brilliantly satirized in Salman Rushdie's epic novel, *Midnight's Children*: Rushdie, 1982.) Such a turnaround also illustrates how new ideas can inspire new policies, particularly when these policies can be sold to the authorities as a less expensive alternative to an existing policy – in this case, building public sector housing complexes.

Since the 1970s work on the housing needs of the urban poor has continued apace, but several new points of focus have emerged. Increasing attention is being paid to the role of female single-headed households in peripheral settlements (Chant, 1985), and to the rates at which real land prices might be rising in different parts of Third World cities, if at all (Jones and Ward, 1994). Work is also continuing on those people in Latin America who have not yet joined the ranks of 'owner-occupiers' in peripheral city locations, and who must rent rooms in difficult conditions in inner-city (and other) areas. It is precisely these people who feature in the Gilbert Reading reprinted here. Gilbert also contends that it is not always meaningful to distinguish between inner-city tenants and edge-of-city owner-occupiers, at least not in Bogota, Colombia or Puebla, Mexico. Indeed, 'what appears to be happening is that new rental accommodation is being created in the consolidating shanty towns. As owners manage to improve their homes and extend the accommodation, they let rooms to other families. . . . It could well be that the payments of tenants are one of the main sources of funds through which self-help accommodation is extended'. Such payments can make possible the addition of a more solid door to a shanty town house, or new guttering, or a second storey of accommodation.

Gilbert's work highlights yet again the need for rigorous and patient data collection if we are to improve our understanding of the cross-cutting actions that build up the fabric of urban settlements in the Third World. Although Gilbert endorses some aspects of the Turner model of intra-urban filtering, he also insists that the possibility of moving from an inner-city location to a shanty town is constrained in many cities by physical circumstances and by persistent poverty or unemployment. This is even more the case in parts of Africa and Asia, where the slums of hope thesis is not always appropriate.

A version of the slums of hope thesis has also been advanced by students of the so-called formal and informal sectors in Third World cities. In the 1970s and 1980s a degree of dissatisfaction set in with urban models and policies that favoured manufacturing industries and the formal sector, and which neglected the skills and contributions of the many people in service industries and the informal sector. At first, this dissatisfaction was voiced in studies which referred to the Third World city as a 'shared space' (Santos, 1979). It was rightly pointed out that the formal and informal sectors of Third World cities cannot be understood in isolation from one another, but are mutually informing and constitutive. To take but one example, the 'modern' sector in most Third World cities cannot function without the very flexible forms of transportation that are conventionally associated with the informal sector. Some writers have even detected an articulation of capitalist and pre-capitalist modes of production within Third World cities, with formal sector bosses and politicians actively seeking the reproduction of an informal economy that supplies labour and commodities to the formal sector at lower prices than would otherwise prevail (Bromley and Birkbeck, 1988).

More recent work on the informal economy has tended to play up its intrinsic worth, regardless of what might be happening in the formal sector. The informal sector has been lauded as a major source of employment. It has also been commended for its high levels of efficiency, which some say compare favourably with labour and management attitudes in the sheltered formal enterprises of the urban economy. This view finds favour with neoliberals, for obvious reasons. Perhaps the most startling recent account of life and conditions in the informal economy, however, has come from Hernando de Soto and his colleagues at the Institute for Liberty and Democracy in Lima. Certainly, de Soto's account of *The Other Path: The Invisible Revolution in the Third World* has reached a wider audience than most texts on development can ever hope to. In the United States it was greeted with the sort of interest usually reserved for books dealing with America's decline and the rise and fall of great powers!

Notwithstanding that de Soto's text has been widely touted on the Right, it is clear that *The Other Path* is written for de Soto's friends and colleagues on the 'Left' as much as for anyone else. But de Soto is rightly sceptical of these binary political divides. His perspective is best described as urban populist, insofar as he commends the skills and activities of the urban poor as the best means by which the urban poor might be empowered. De Soto takes issue with all those, on the Left and Right, who are minded to see poor people as a problem and who would like to take actions on their behalf (or claim as much). De Soto rejects the idea that the state is, or could easily be, a disinterested social and economic actor that is willing to put to rights the wrongs that condemn so many people to a life of poverty. De Soto's view of the state has something in common with the views of Anne Krueger or Robert Bates (even allowing for important differences of approach in the work of these two authors). State functionaries are charged with having an agenda of their own. They exact rent or a surplus from those they are meant to act on behalf of, and treat such people, at best, as clients and occasional recipients of patronage. More bleakly still, de Soto paints a picture of a Peru in the 1980s where 'legality is a privilege available only to those with political and economic power'.

This last quotation comes from the pen of Peru's most famous modern novelist, and erstwhile Presidential candidate, **Mario Vargas Llosa**. Vargas Llosa's Foreword to de Soto's account of *The Other Path* is published here as **Reading 18**, both because it is a pithy rendition of a much longer argument and because it pinpoints the involvement of a member of the Third World's cultural intelligentsia in everyday politics and scholarly work. One only has to think of Soyinka in Nigeria and Achebe in Kenya to see that this involvement is not at all uncommon. Vargas Llosa also writes in a generous prose style that drives right to the heart of the Kafkaesque logics of corruption and bureaucratic mismanagement that de Soto's team uncovered in urban Peru. As an economic liberal himself, Vargas Llosa puts his own gloss on de Soto's findings. He argues that the poor in Peru need to be set free to produce the wealth they are demonstrably capable of creating. The poor also need to be protected by a legal system that works dispassionately and without favour. As Vargas Llosa puts it: 'in order to produce wealth, it is necessary that the state's actions not obstruct the actions of its citizens, who, after all, know better than anyone else what they want and what they have to do'. Failure to follow this Other Path to development – failure to set free the invisible revolution that is already bubbling away in the informal/illegal economy – will likely drive people to a political extremism of the Right or Left; quite possibly, in the case of Peru, to the Shining Path of the Maoist Sendero Luminoso. That way, Vargas Llosa suggests, lies chaos and underdevelopment.

Not everyone will agree with the prospectus for economic liberalism that Vargas Llosa puts his name to. Vargas Llosa maintains that the privatization of state productive activities 'does not mean that the state will wither away and die', but the role for state actions that he foresees is more limited than many students of development would feel comfortable with. The counter-revolution has clearly challenged the confident accounts of the developmental state that flourished before 1980, but most scholars still believe that the state has important functions to discharge in terms of 'nation-building' and in providing a measure of public support for private investment. Even the new emphasis on human capital formation has not dented the view that developing countries suffer from low levels of capital formation, and that such countries should make good this 'gap' by recourse to foreign aid, private foreign investment, or government schemes to mobilize domestic savings. To this extent, at least, the legacy of the Harrod–Domar growth model lives on. The Harrod–Domar model suggests that the gross national product (GNP) of a developing country will grow at a rate inversely proportional to the domestic capital–output ratio and directly proportional to the national savings rate. If the capital–output ratio is fixed in the short run, the rate of economic growth depends solely on the rate of savings (augmented by foreign savings); the rate of investment, in this broadly Keynesian model, is assumed to be identical to the rate of savings.

The simple version of the Harrod–Domar model hardly carries much weight today, but the logic expressed in it is still widely adhered to and mimicked. Most developing countries are committed to the goal of building up a local manufacturing base and governments continue to direct scarce savings/investment to this end. We know now that the record of developing countries in this regard is mixed. Pessimists like Gunder Frank maintain that very few countries have industrialized with any degree of success, and that even those countries which do have a broad industrial basis have made a Faustian bargain with Western transnational corporations and bankers. Thus is dependency reinforced and reinvented. But most students of development find fault with this view. The available evidence might not support Bill Warren's claims for a generalized industrialization of the periphery after decolonization (Warren, 1980), but neither does it suggest that the only success stories are to be found in East Asia.

One of the problems that besets the debate on the scale of Third World industrialization is a reluctance to count service industries as 'real' industries, despite the fact that many developing countries are fast becoming service economies. Another problem relates to the spatial units for which data are collected by the World Bank and other agencies. Statisticians and politicians are still besotted by the so-called nation-state, or the idea of a 'country'. But in an era of rapid globalization it is becoming ever easier for some privileged peoples and places in poor countries to access the international market-place. Parts of south New Delhi are now more prosperous than parts of the Bronx in New York or Hackney in London. There are Third Worlds in the First World and vice versa, at least in terms of local maps of poverty and plenty. At the regional scale, too, it is often noted that the economy of California would rank as one of the seven most prosperous 'national' economies in the world, with the Greater Los Angeles metropolitan region alone featuring in a list of the top twenty such spatial units. By the same logic, there are now sizeable pockets of modernity, industrialization and affluence in some low income countries. The industrialization of the periphery of the world system is not just confined to East Asia and parts of Latin America. Parts of China and India can reasonably be described as newly industrialized regions or city-regions.

Of course, questions still remain about the benefits conferred by this incipient industrialization, and about the role of state policy in promoting export-oriented or import-substitution industrialization. **Reading 19**, by **Diane Elson**, is centrally concerned with these issues, as is **Reading 20** by **Robert Wade**. Elson's paper offers a sterling review of the roles played by transnational corporations (TNCs) in binding Third World economies into the modern international economy. Her Reading sets out the basic facts about TNC growth and distribution before switching to an often critical evaluation of the role of TNCs in the Third World. Elson looks at the changing relationships between TNCs, their home governments, and the governments of the host countries in which they are located. She also looks at the mechanisms by which surplus is appropriated from developing countries and returned to the First World. Transfer pricing and overpricing feature heavily in this discussion. Elson also pays attention to the appropriateness of the technologies and products transferred by some TNCs to the Third World, and to the uneven gender divisions of labour that are exploited by Western transnationals in the developing world.

If Elson is critical of the asymmetries of power built into TNC/host country relations, she is also sensitive to the very different roles played by TNCs in developing countries. This latter point is developed at length in the full paper from which Reading 19 is drawn. Most modern accounts of TNC/Third World relations eschew simple notions of the transnational as a blessing or a curse: as a source of scarce investment, skills and foreign exchange or as the unacceptable face of neo-colonialism. It makes more sense to see this relationship as a bargain of sorts (Nixson, 1988). Where TNCs are strong and occupy quasi-monopolistic or monopsonistic positions they are able to effect deals with weak Third World states that are very much to their advantage. Some of the worst horror stories about TNCs concern mining companies in sub-Saharan Africa (Lanning and Mueller, 1979). But where TNCs are competing with one another to locate in, or sell to, well-run Third World countries, the boot can be on the other foot. This might describe the semiconductor industry in countries like Taiwan and South Korea. In some cases governments can negotiate entry deals where TNCs are required to reward labour according to agreed protocols, and where foreign firms might be asked to buy a certain percentage of component goods from domestic enterprises. The local effects of TNCs – their developmental effect – will depend critically upon the nature and range of these bargains over labour, sourcing and taxation arrangements.

A not dissimilar approach to state/capital or government/market relationships is set out by Robert Wade in what has already become a classic text on the East Asian miracle and the lessons to be learned from it (Wade, 1990). Wade accepts that Taiwan and South Korea have experienced formidable rates of economic growth, *and* that they have enjoyed a prolonged period of 'development' by any reasonable definition of that word. Some on the Left would like to discount the East Asian miracle as a case of dependent non-development, or growth without development, but it is hard to see how such an argument can be sustained without lapsing into tautology. All development must be dependent to an extent, if dependency is defined as a compulsive involvement in the global market-place and a partial reliance on foreign capital. Even the US economy is dependent to a degree, as Japanese investment houses know very well. But Wade refuses to join with those who have acclaimed the East Asian miracle as a triumph of the market, or as an advertisement for neoliberalism. Wade argues that the East Asian miracle depends crucially on the quality of state interventions in support of the market and export-oriented industrialization (EOI). He further argues that EOI in East Asia followed an earlier period of import-substitution industrialization when the state built up local industries behind tariff walls. Balanced development was also encouraged by state support for local industries that produced dominantly for local markets, before their attention was directed abroad. Many of these industries are located in rural areas or small towns, which has further discouraged permanent rural–urban migration and 'overurbanization'. Peasants in Taiwan and South Korea were also given an incentive to stay in the countryside by the radical *Land to the Tiller* land reforms of the 1950s (see Introduction to Section Two).

Wade concludes the Reading published here by setting out six lessons – or 'prescriptions' – that might reasonably be learned from East Asia, and which might judiciously be acted upon elsewhere. These prescriptions are very close to what is fast becoming the accepted wisdom on states and markets in the developing world, a point we return to in Section Six. As ever, the Guide to Further Reading that follows provides pointers to the issues raised in this section, and to related issues that have not received the attention they deserve.

Guide to further reading

Urbanization; urban design; urban problems, policies and politics

Abrams, C. 1964: *Man's struggle for shelter in an urbanizing world.* Cambridge, MA: MIT Press.

Amis, P. 1984: Squatters or tenants: the commercialization of unauthorized housing in Nairobi. *World Development* 12, 87–96.

Angel, S., Archer, R., Tanphiphat, S. and Wegelin, E. (eds) 1983: *Land for housing the poor.* Singapore: Select Books.

Armstrong, W. and McGee, T. 1985: *Theatres of accumulation: Studies in Asian and Latin American urbanization.* London: Methuen.

Bahl, R. and Linn, J. 1992: *Urban public finance in developing countries.* Oxford: OUP.

Baross, P. and van der Linden, J. (eds) 1990: *The transformation of land supply systems in Third World cities.* Aldershot: Avebury.

Berry, B. 1961: City-size distributions and economic development. *Economic Development and Cultural Change* 9, 573–87.

Bhalla, A. 1990: Urban–rural disparities in India and China. *World Development* 8, 1097–110.

Burgess, R. 1985: The limits of self-help housing programmes. *Development and Change* 16, 271–312.

Castells, M. 1972: *Imperialisme y urbanization en America Latina.* Barcelona: Gustavo Gili.

Chant, S. 1985: Single-parent families: choice or constraint? The formation of female-headed households in Mexican shanty towns. *Development and Change* 16, 635–56.

Dowall, D. and Leaf, M. 1991: The price of land for housing in Jakarta. *Urban Studies* 28, 707–22.

Drakakis-Smith, D. 1987: *The Third World city*. London: Methuen.

Eckstein, S. 1990: Poor people versus the state and capital: anatomy of a successful community mobilization for housing in Mexico City. *International Journal of Urban and Regional Research* 14, 274–96.

Fischer, C. 1982: *To dwell among friends: Personal networks in town and city*. Chicago: University of Chicago Press.

Forbes, D. and Thrift, N. (eds) 1987: *The socialist Third World: Urban development and regional planning*. Oxford: Blackwell.

Gilbert, A. 1983: The tenants of self-help housing: choice and constraint in the housing markets of less-developed countries. *Development and Change* 14, 449–77.

Gilbert, A. and Gugler, J. 1992: *Cities, poverty and development: Urbanization in the Third World* (2nd edn). Oxford: OUP.

Gilbert, A. and Varley, A. 1991: *Landlord and tenant: Housing the poor in urban Mexico*. London: Routledge.

Gilbert, A. and Ward, P. 1985: *Housing, the state and the poor: Policy and practice in three Latin American cities*. Cambridge: CUP.

Harriss, B. and Harriss, J. 1984: 'Generative' or 'parasitic' urbanism: some observations on the recent history of a South Asian market town. *Journal of Development Studies* 20, 82–101.

Holston, J. 1989: *The modernist city: An anthropological critique of Brasilia*. Chicago: University of Chicago Press.

Hoselitz, B. F. 1957: Generative and parasitic cities. *Economic Development and Cultural Change* 5, 278–94.

Irving, R. 1981: *Imperial summer: Lutyens, Baker and imperial Delhi*. New Haven: Yale UP.

Jones, G. and Ward, P. (eds) 1994: *Methodology for land market and housing analysis*. London: UCL Press.

Journal of Development Studies 1993: Beyond urban bias: special issue, Volume 29.

King, A. 1984: *The bungalow: The production of a global culture*. London: Routledge and Kegan Paul.

King, A. 1990: *Urbanism, colonialism and the world-economy*. London: Routledge.

Lewis, O. 1959: *Five families: Mexican case studies in the culture of poverty*. New York: Basic Books.

Lipton, M. 1977: *Why poor people stay poor: A study of urban bias in world development*. London: Maurice Temple Smith.

Lloyd, P. 1979: *Slums of hope: Shanty towns of the Third World*. Harmondsworth: Penguin.

Mabogunje, A. 1983: The case for big cities. *Habitat International* 7, 21–31.

Malpezzi, S. and Mayo, S. 1987: User cost and housing tenure in developing countries. *Journal of Development Economics* 25, 197–220.

Mangin, W. 1967: Latin American squatter settlements: a problem and a solution. *Latin American Research Review* 2, 65–98.

Metcalf, T. 1989: *An imperial vision: Indian architecture and Britain's* RAJ. Berkeley: University of California Press.
Meyer, D. 1986: System of city dynamics in newly industrializing nations. *Studies in Comparative International Development* 21, 3–22.
Murphey, R. 1972: City and countryside as ideological issues: India and China. *Comparative Studies in History and Society* 14, 250–67.
Onibokun, A. 1990: Poverty as a constraint on citizen participation in urban redevelopment in developing countries. *Urban Studies* 27, 371–84.
Redfield, R. and Singer, M. 1954: The cultural role of cities. *Economic Development and Cultural Change* 3, 53–73.
Richardson, H. 1989: The big, bad city: mega-city myth? *Third World Planning Review* 11, 355–72.
Roberts, B. 1978: *Cities of peasants*. London: Arnold.
Rondinelli, D. 1983: *Secondary cities in developing countries: Policies for diffusing urbanization*. London: Sage.
Rushdie, S. 1982: *Midnight's children*. London: Jonathan Cape.
Schuurman, F. and van Naerssen, P. (eds) 1989: *Urban social movements in the Third World*. London: Routledge.
Simon, D. 1989: Colonial cities, post-colonial Africa and the world economy. *International Journal of Urban and Regional Research* 13.
Sovani, N. 1964: The analysis of 'over-urbanization'. *Economic Development and Cultural Change* 12, 113–22.
Stokes, S. 1991: Politics and Latin America's urban poor: reflections from a Lima shantytown. *Latin American Research Review* 26, 75–102.
Turner, J. 1967: Barriers and channels for housing development in modernizing countries. *Journal of the American Institute of Planners* 33, 167–81.
Turner, J. 1968: Housing priorities, settlement patterns and urban development in modernizing countries. *Journal of the American Institute of Planners* 34, 354–63.
Turner, J. 1976: *Housing by people*. London: Marion Boyars.
Ward, P. (ed.) 1982: *Self-help housing: A critique*. London: Mansell.

The informal sector/economy

Bromley, R. (ed.) 1978: The urban informal sector: critical perspectives. Special Issue of *World Development* 6, 1031–198.
Bromley, R. and Birkbeck, C. 1988: Urban economy and employment. In Pacione, M. (ed.), *The geography of the Third World*. London: Routledge, 114–47.
Chant, S. 1991: *Women and survival in Mexican cities: Perspectives on gender, labour markets and low-income households*. Manchester: Manchester UP.
de Soto, H. 1989: *The other path: The invisible revolution in the Third World*. New York: Harper and Row.
Holmstroem, M. 1984: *Industry and inequality: The social anthropology of Indian labour*. Cambridge: CUP.
Lehmann, D. 1990: Modernity and loneliness: popular culture and the informal economy in Quito and Guadalajara. *European Journal of Development Research* 2, 89–107.
Moser, C. 1984: The informal sector reworked: viability and vulnerability in urban development. *Regional Development Dialogue* 5, 135–78.

Noponen, H. 1991: The dynamics of work and survival for the urban poor: a gender analysis of panel data from Madras. *Development and Change* 22, 233–60.
Portes, A. Castells, M. and Benton, L. (eds) 1989: *The informal economy: Studies in advanced and less developed countries*. Baltimore: Johns Hopkins UP.
Quijano, A. 1974: The marginal pole of the economy and the marginalised labour force. *Economy and Society* 3, 393–428.
Rodgers, G. (ed.) 1989: *Urban poverty and the labour market: Access to jobs and incomes in Asian and Latin American cities*. Geneva: ILO.
Santos, M. 1979: *The shared space: The two circuits of the urban economy in underdeveloped countries*. London: Methuen.
Sicula, D. 1991: Pockets of peasants in Indonesian cities: the case of scavengers. *World Development* 19, 137–61.
Standing, H. 1991: *Dependence and autonomy: Women's employment and the family in Calcutta*. London: Routledge.
Ward, P. (ed.) 1989: *Corruption, development and inequality*. London: Routledge.

Industrialization: TNCs, NICs and the New International Division of Labour

Amsden, A. 1989: *Asia's next giant: South Korea and late industrialization*. Oxford: OUP.
Amsden, A. 1990: Third World industrialization: 'Global Fordism' or new model? *New Left Review* 182, 3–21.
Bello, W. and Rosenfield, S. 1992: *Dragons in distress: Asia's miracle economies in crisis*. Harmondsworth: Penguin.
Chenery, H. Robinson, S. and Syrquin, M. 1986: *Industrialization with growth: A comparative study*. Oxford: OUP/World Bank.
Christiansen, S. (ed.) 1989: Special issue on 'Privatization'. *World Development* 17 (5).
Cline, W. 1982: Can the East Asian model be generalized? *World Development* 10, 81–90.
Eden, J. 1990: Race and the reproduction of factory labour in Malaysia. *Society and Space* 8, 175–90.
Elson, D. and Pearson, R. 1981: Nimble fingers make cheap workers: an analysis of women's employment in Third World export manufacturing. *Feminist Review* 7, 87–107.
Emmott, B. 1992: *Japan's global reach*. London: Century.
Felix, D. 1989: Import-substitution and late industrialization. *World Development* 17, 1455–69.
Frobel, F. Heinrichs, J. and Kreye, O. 1980: *The new international division of labour*. Cambridge: CUP.
Gwynne, R. 1990: *New horizons? Third World industrialization in an international framework*. Harlow: Longman.
Haggard, S. 1990: *Pathways from the periphery: The politics of growth in the newly industrializing countries*. Ithaca: Cornell UP.
Harrison, D. 1994: Tourism, capitalism and development in less developed countries. In Sklair, L. (ed.), *Capitalism and development*. London: Routledge, 232–57.

Henderson, J. 1989: *The globalization of high technology production: Society, space and semiconductors in the restructuring of the modern world*. London: Routledge.

Jenkins, R. 1987: *Transnational corporations and the Latin American automobile industry*. Pittsburgh: University of Pittsburgh Press.

Jenkins, R. 1991: The political economy of industrialization: a comparison of Latin American and East Asian newly industrializing countries. *Development and Change* 22, 197–231.

Kaplinsky, A. 1984: The international context for industrialization in the coming decade. *Journal of Development Studies* 21, 75–96.

Krueger, A. and Tuncer, B. 1982: Empirical test of the infant industry argument. *American Economic Review* 72, 142–52.

Lall, S. 1984: Transnationals and the Third World: changing perceptions. *National Westminster Bank Quarterly Review May*, 2–16.

Lall, S. 1990: *Building industrial competitiveness in developing countries*. Paris: OECD.

Lanning, G. with Mueller, M. 1979: *Africa undermined*. Harmondsworth: Penguin.

Lawson, V. 1992: Industrial subcontracting and employment forms in Latin America. *Progress in Human Geography* 16, 1–23.

Little, I. 1987: Small manufacturing enterprises in developing countries. *World Bank Economic Review* 1, 203–35.

Nixson, F. 1988: The political economy of bargaining with transnational corporations: some preliminary observations. *Manchester Papers in Development* 4, 377–90.

Pearson, R. 1994: Gender relations, capitalism and Third World industrialization. In Sklair, L. (ed.), *Capitalism and development*. London: Routledge, 339–58.

Porter, M. 1990: *The competitive advantage of nations*. London: Macmillan.

Riddell, R. *et al.* 1990: *Manufacturing Africa: Performance and prospects of seven countries in sub-Saharan Africa*. London: James Currey.

Riddell, R. 1993: The future of the manufacturing sector in sub-Saharan Africa. In Callaghy, T. and Ravenhill, J. (eds), *Hemmed in: Responses to Africa's economic decline*. New York: Columbia UP, 215–47.

Syrquin, M. and Chenery, H. 1989: Three decades of industrialization. *World Bank Economic Review* 3, 145–81.

UNCTC 1988: *Transnational Corporations in world development: Trends and prospects*. New York: United Nations.

Vernon, R. 1979: The product life cycle hypothesis in a new international environment. *Oxford Bulletin of Economics and Statistics* 41, 255–68.

Wade, R. 1990: *Governing the market: Economic theory and the role of government in East Asian industrialization*. Princeton: Princeton UP.

Warren, B. 1980: *Imperialism: Pioneer of capitalism*. London: Verso.

Wolf, D. 1992: *Factory Daughters: Gender, household dynamics and rural industrialization in Java*. Berkeley: University of California Press.

World Bank 1987: *World development report, 1987*. Oxford: OUP/World Bank.

World Bank 1989: *India: An industrializing economy in transition*. Washington, DC: World Bank Country Study.

World Bank 1994: *World development report, 1994*. Oxford: OUP/World Bank.

16 Michael Lipton, 'Urban Bias and the Myths of Urbanisation'

Excerpt from: M. Lipton, *Why Poor People Stay Poor: A Study of Urban Bias in World Development*, pp. 216–26, 442–3. London: Maurice Temple Smith (1977)

If people in less-developed countries fare so much worse in villages than in towns, will villagers 'vote with their feet', by migrating townwards, until the gap between rural and urban expectations is removed? If such *urbanisation* happened, major rural–urban inequalities could not persist. This chapter examines why it happens so surprisingly little – and whether policy-makers should seek to increase or decrease it. Can really poor villagers react to their poverty by moving to the town? How far can such a process increase the proportion of persons in poor countries who gain, instead of losing, from urban bias?

Over-simple readings of the data have in most LDCs led to gross exaggerations of the rate of genuine urbanisation – permanent movement from village to town, net of movement the other way. Most of the rise in urban populations is due to natural increase – not only directly but also by pushing communities over the rural–urban borderline, and by making towns expand and 'eat up' nearby villages. Most migration in LDCs is intra-rural or intra-urban; and much of the residual migration from village to town is temporary, or in other respects does not represent a real urban commitment. As for true, permanent townward migration, it activates powerful processes, demographic and economic, limiting its capacity to raise the urban share of national population: processes already apparent in the recent population counts of such major cities as Calcutta and Colombo.

Thus urbanisation has been insufficient to reduce urban bias. Moreover, its structure has tended to increase urban–rural inequalities, for two reasons. First, urbanisation is 'epidemic', as people learn from other migrants about urban prospects. Thus villages near cities, and rural families with urban members, are the likeliest to seize on any potentially beneficial further townward migration. Such villages and families are often semi-urban before these further benefits accrue. The successful migrants who remain in the city, even before they moved, were richer and better-educated than their fellow-villagers; if the really poor villager moves at all, it is usually temporarily, to a job in the urban periphery, or even to no job at all – to crime, beggary, prostitution, all the currently romanticised delights of the 'informal sector'. Second, the villages, and especially their poorest members, also lose from the drain townwards of resources of skill and leadership. Nor is the picture even lightened by substantial *net* remittances from urban migrants to their rural kinsmen. While there is no case for opposing urbanisation as such – still less for preventing it by force – it cannot cure the depressed conditions of the rural poor, except perhaps in the very long run.

Voting with their feet

The idea of an equilibrium mechanism

Though primarily a sociologist, Hoselitz is in a major tradition of marginalist economics in arguing that voluntary migration from village to town must cancel out any serious inefficiencies or inequities arising from urban bias. Suppose a townsman's material advantages over the villager exceed any net 'psychic income' from rural life. Will not the villager then do better (at the same level of capacities and effort) by becoming a townsman? If he does so, he reduces the pressure on land and the competition to get jobs in the village, and thereby improves the position of those who remain there. He also increases the number of workers struggling to get urban jobs, and the competition for housing and other amenities in the city, so that the benefits to city dwellers decline. The process of migration, runs the argument, will go on until both villager and townsman know that nothing is to be gained by moving. That can happen only when the rewards for work, requiring the same effort and ability and conscientiousness, are the same in village and town – allowing for the risk of being out of work in either place, differences between urban and rural living costs, and any possible net 'psychic income' from living in a village rather than a town. On this reasoning, any substantial urban bias must cause urbanisation, which continues until any inefficiencies caused by the bias have been removed.

This argument, typical of neo-classical economics, assumes that people have the information and the resources to respond swiftly and rationally to any chance for advantage, and that they do so within a system with few obstacles to such response. If political and market power are not too unequal, and if access to information and training is widespread, then swift response within a fluid economy can plausibly be expected to steer it towards equilibrium and thereby enhance both efficiency and equity.

In developed countries, the argument is powerful. The reduction of regional and racial inequality in the USA, as Southerners have moved west and north since 1930, is a good illustration of it. Even in such favourable conditions, however, the march to efficiency need not do much for equality, or do it swiftly. The better-off seem frequently willing to bear 'costs of discrimination': to suffer absolute economic costs in order to retain *relative* status advantages over others upon whom they inflict greater costs. Among worse-off groups, moreover, even in rich countries, it is the more literate, informed and dynamic who migrate or are otherwise 'creamed off', leaving the mass of their colleagues with even less chance to advance. And it is part of the definition of an underdeveloped economy that the poor can move about to get richer (for example, by voting with their feet) only sluggishly, if at all.

Chasing the rainbow

Indeed, the cumulative forces making the townsman better off – political and economic pressures analysed in this book – are in most really poor countries increasing the disparity much faster than townward migration can reduce it. The slowness of 'equilibriating' urbanisation has two causes: the constraints preventing the very poor from leaving the village, and the obstacles presented by the socio-economic system to their effective and lasting settlement in the town.

In poor countries, the worst-off villagers can seldom move permanently to the town. First, they are often bound to the village, by law or custom, to work off old family debts they cannot repay. In 1950, some 10 per cent of Indian families, dependent mainly on agricultural labour for a livelihood, were 'attached', and frequently, in effect, bond-slaves; Latin American peons are similarly immobilised. Even if they defy law or custom and go, their relatives may be penalised by loss of land or jobs, so the poorest class of villagers as a whole gains little. Second, most poor countries, especially in rural areas, are far from enjoying universal schooling; and the poorest villagers are the most likely to be ignorant of urban chances, or unable to exploit them because they are illiterate. Third, a migrant, while looking for an urban job (or receiving urban education) instead of working in the village, needs money from someone – usually from his father or brother in the village; to support a migrant like this, a rural family must be well-off enough to have something in reserve. Fourth, as this shows, a family decision to 'urbanise' a member is usually a sacrifice of definite income now for possibly greater income later; the very poor can seldom afford either risk or the reduction of their slender current incomes; nor can they often meet such costs by major extra borrowings at reasonable interest rates.

Hence the poorest villagers are unlikely to be able to react to urban bias by massive, permanent townward migration – except in a rich country, where they are not all that poor, and enjoy widespread schooling, social security in the event of unemployment, and no bond-slavery. In a poor country such as Ghana, the villager who moves to the town is usually a man who has *already* half-succeeded in joining the urban elite: 'the propensity to migrate increases with closeness to a large town, population size of the [village of origin], economic well-being of the rural household, number of relatives in the urban area, [and] the individual's level of education.' (Caldwell, 1968)*.

To the characteristics of poor villagers impeding their urbanisation – characteristics derived partly from the socio-economic system that cuts their prospects of earning (or borrowing) money – must be added 'immobilising' features of the system itself. These may involve deliberate restrictions on townward movement, in response to pressures from big farm employers who (because they provide the city with most of its food and raw materials) must be listened to; the very different forms of 'influx control' of Russia after 1861 and South Africa today exemplify that response. The system of most LDCs, however, restricts townward movement more subtly. Transport from village to town is often poor, infrequent, and too costly for the poorer villagers. Barriers of dialect, and in many countries of language (India has thirteen major languages, Papua several hundred), impede movement, and are often reinforced by the retention of colonial or 'mandarin' languages for official and commercial matters. Information about job prospects seldom reaches rural areas; labour exchanges are usually confined to bigger towns.

Policy towards the urbanising response

All these explicit and implicit barriers suggest that the reduction of urbanisation is an aim of urban-biased policy-makers, as indeed seems natural: a big net inflow of rural migrants would compete with (and drive down the rewards of) organised urban workers, congest urban roads, and render smaller, less competitive and hence costlier the supply of products to towns from rural areas.

Yet, in a sense, urbanisation – at least of the skilled, educated, better-off and hence more mobile villager – is a *response* to urban bias. Indeed, my earlier work on urban bias has been taken, by a sympathetic reader, as suggesting that poor countries are 'over-urbanised'; an 'excessive' share of development spending in the cities must, almost by definition, pull 'excessive' numbers of people into them. However, any attempt to cure 'over-urbanisation' by locking the villager into his village, so as to impede the urbanising response to urban bias by such villagers as may be mobile, would be the ultimate urban-biased assault on rural rights. The hypothesis that LDCs are 'over-urbanised', merely because their populations contain larger urban shares than did NRCs (now-rich countries) at comparable stages of development, is curiously ethnocentric, and anyway refuted by the statistical evidence that urbanisation is closely related to industrialisation in LDCs (but not in rich countries) today.

The issues can be clarified by making three distinctions. First, what is wrong with rapid urban population growth in LDCs is not that it damages the successful migrant: at existing levels of rural and urban living, the drawbacks of urban expansion have been much exaggerated, and after all it continues to show the preference of the migrants. What is wrong is that it aggravates the bias against the villager, despite the theoretical expectation that it would correct such bias. Second, the remedy is not to confine the artificial advantages of city life to the present beneficiaries by rendering urbanisation difficult, but to remove the arbitrarily assigned advantages that render urbanisation artificially attractive: to neutralise the pricing, investment, educational, medical and other policies that are currently transferring income from villages to towns, and encouraging the ablest villagers to follow. Linking these two distinctions is the third: many poor countries are 'over-urbanised' not in the sense that cities become undynamic or outpace industrialisation, but in the sense that urban economic dynamism confers less and less welfare (partly because its cost rises and partly because it chokes off immigration later on), increasingly takes place at the expense of rural areas, and is linked with an output structure – in building as well as in industry – that employs few, benefits mainly the well-off, and rests on arbitrary price and investment advantages conferred by public policy and secured by private monopoly.

False and true cases against urbanisation

Recent surveys of slum-dwellers and squatters in South and South-east Asia are surprisingly optimistic. Laquian sums up that 'they seem to be [quite] closely integrated with the economic and social system', to have 'many opportunities for saving and capital formation', and to reveal 'economic mobility' and 'high social and political participation' (Laquian, 1971, p. 201)*. McGee points out that 'for the Indian rural migrant about to move to Calcutta, the city is identified as a city of "hope" for the future', and asks, 'Why else would he move?' – though the limited numbers that do move, and (as we shall see) the high proportions that return, are significant here (McGee, 1971, p. 15)*. Nor should one too readily accept that a city is too big for further expansion of industry to remain economic: 'industry continues to expand in these centres, showing that many industrialists themselves still think that the major concentration retains numerous advantages.' It is the facts that these trends *do* eventually choke off urbanisation and urban industrialisation – and that much of the latter is made privately 'economic' only by socially uneconomic, urban-biased policies – that are worrying, not the natural urbanising response of migrants and businessmen to mistaken policies.

Urbanisation increases inequalities, both intra-rural and rural–urban. As both Marshall and Kautsky realised, it selects out those who could lead the village away from poverty. Successful townward migrants stay in the urban area and reinforce its pressures for extra resources. Meanwhile, they set up economic and demographic forces rendering further migration increasingly difficult, so that the villagers left behind – especially if unrelated to the early migrants – have little chance to benefit from the urban bias that early migration accentuates.

Damming the flow

There are far too many unknowns for a policy of artificially damming the urban flow to be justifiable on efficiency grounds (quite apart from the inequity of compelling even the mobile villagers to stay at home and accept urban bias). First, though the growth of urban unemployment (not mainly among immigrants) renders doubtful the case for urbanisation as a source of labour supply, it does not render 'irrelevant' the argument that if rural labour shifts to the towns it will contribute more to national output; towns in poor countries are still dominated by underutilised capacity, by actual labour shortages in unskilled manual jobs spurned by the educated unemployed, and by a generally greater output/labour ratio. Past urban overinvestment there does justify heavy labour-inputs, now, to make more of this urban capital. If the disparities of Table 16.1 justify raising the rural share in capital, they also justify lowering the rural share in labour, especially as man/land ratios rise. It suits urban trade unionists and businessmen to go for capital-intensive techniques, confining the benefits of urban growth to existing gainers (plus a few relatives), and persuading the politicians to adjust taxes and subsidies so as to make this profitable; but it is better, for development, employment and equality, to reverse these priorities.

Second, urbanisation could be needed for both low-cost industrialisation and rising levels of administrative capability. The case for small-scale and rural industry often smacks of special pleading. The worst duplication, overcentralisation and administrative confusion in early development are frequently rural, especially where agriculture, irrigation and other ministerial fiefdoms overlap. Both the high cost of rural industry and the low capacity of rural administration spring partly from urban-biased pricing and investment decisions; but to rule out the possibility that the high costs of dispersed rural activity bear some of the blame, and to delay urbanisation accordingly, would be to push policy far ahead of what can be inferred from the available research.

A third argument against preventing urbanisation is that – to counter the threat to rural jobs, incomes and nutrition from rising man/land ratios (especially where water shortages preclude major agricultural innovation, 'green-revolutionary' or other) – one must pursue two paths. One is suggested in this book: much more investment in the intensification of farmland use, especially through irrigation. But the path of labour-intensive industrialisation must be followed too, and probably in the towns if it is to be efficient. Poverty means both 'over-urbanisation' and 'over-ruralisation' – both town and country have too few resources, given their distribution and use, to provide the residents with adequate living standards. Hence (while slowing population growth and accelerating the creation of resources of skills and capital) policy should seek to place more of each sort of resource where it yields most: capital in rural areas; labour, up to a point, in urban areas. Of course, urbanisation of labour through urban-biased incentives is not justified by this argument.

Urban responses to the costs of urbanisation

It is noteworthy that many city-dwellers have sought to artificially restrain further urbanisation. Their arguments rest heavily on the conditions of filth and disease, congestion and deprivation of privacy, transport noise and transport costs, above all of the poorest and most recent arrivals (largely rural floaters rather than true migrants). Such conditions must indeed depress humane observers. But why should they constitute a practical or a moral case for stopping urbanisation? Practically, they cannot form direct costs to the urban leadership, and do not give their victims any political power. Morally, they are

Table 16.1 Capital stock: average allocation, productivity, endowment

1 Country	Argentina		Colombia	India		Japan	Mexico	South Africa	Yugoslavia
2 Year	1950	1955	1953	1950	1960–1	1955	1950	1955	1953
Percentage in Agriculture									
3 DRAC*	10.1	9.6	25.2	23.6	—	4.2	7.8	13.8	13.0
4 DRAC + livestock	—	—	35.6	36.4	—	5.6	12.7	20.1	17.4
5 DRAC + livestock + inventories ('capital')	—	—	—	30.7	27.4	6.0	13.3	6.2	19.2
6 GDP (current factor cost)	15.6	17.1	37.9	51.2	45.7	23.0	22.4	15.2	31.5
7 Labour force	21.6	19.4	52.5	70.6	73.0	40.4	57.9	33.0	66.8
Agricultural/non-agricultural ratios: productivity									
8 DRAC	1.65	1.94	1.81	3.40	—	6.81	3.41	1.12	3.08
9 DRAC + livestock	—	—	1.10	1.83	—	5.04	1.98	0.71	2.18
10 'Capital'	—	—	—	2.37	2.23	4.68	1.88	2.71	1.94
Agricultural/non-agricultural ratios: endowment per worker									
11 DRAC	0.41	0.44	0.30	0.13	—	0.06	0.06	0.33	0.08
12 DRAC + livestock	—	—	0.50	0.24	—	0.09	0.11	0.51	0.10
13 'Capital'	—	—	—	0.18	0.14	0.10	0.11	0.13	0.12

Notes:
Row 8 is $\frac{\text{row 6}}{\text{row 3}} \cdot \frac{(100-\text{row 3})}{(100-\text{row 6})}$; row 9 is $\frac{\text{row 6}}{\text{row 4}} \cdot \frac{(100-\text{row 4})}{(100-\text{row 6})}$; row 10 is $\frac{\text{row 6}}{\text{row 5}} \cdot \frac{(100-\text{row 5})}{(100-\text{row 6})}$.
These rows show the ratio, agriculture to non-agriculture, of output per unit of DRAC, DRAC + livestock, and total 'capital' respectively.
Row 11 is $\frac{\text{row 3}}{\text{row 7}} \cdot \frac{(100-\text{row 7})}{(100-\text{row 3})}$; row 12 is $\frac{\text{row 4}}{\text{row 7}} \cdot \frac{(100-\text{row 7})}{(100-\text{row 4})}$; row 13 is $\frac{\text{row 5}}{\text{row 7}} \cdot \frac{(100-\text{row 7})}{(100-\text{row 5})}$.
These rows show the ratio, agriculture to non-agriculture, of per-worker DRAC, DRAC + livestock, and total 'capital', respectively. For earlier analogues to rows 9 and 10, see C. Clark, *The Conditions of Economic Progress*, 3rd ed. Macmillan, 1960, pp. 527–9.
Sources: GDP and labour force data are latest available from UN, *Yearbook of National Accounts Statistics* and ILO, *Yearbook of Labour Statistics*, respectively. Capital stocks from R. Goldsmith and C. Saunders, eds, *The Measurement of National Wealth*, Income and Wealth Series VIII, Bowes & Bowes, especially Th.D. van der Velde, pp. 9–11, 15–16, 18, and papers in text. (I have corrected some small errors in pp. 8–10, such as item A.I.1 (b) for S. Africa and A.I.3(a) for India; and for India I have shifted half the value of capital from agriculture to industry for tea plantations, and vice versa for other plantations.) Capital stock for India, 1960–1, from *Report of the Committee on Distribution of Income and Levels of Living*, pt. 1, Planning Commission, Delhi, 1964, p. 79. *DRAC is Directly Reproducible, Allocable Capital.

after all preferred to village life by the migrants. But they do mean that urbanisation imposes, upon the settled urban community, external costs that are high and rising.

The costs are *external* in that they arise out of the actions of *A*, but are borne by *B*; if *A* moves to an already overcrowded area, in or near which *B* already lives, much of the extra unpleasantness is transferred to *B* and his like. The costs are *high* because, if existing townspeople are to be spared damage, many facilities (health, drainage, police) have to be granted to the newcomers. The cost per migrant is *rising* because each extra person or vehicle, using a congested facility, increases congestion more than proportionately.

To protect already-resident townsmen against such costs, the city acts against the migrant in several ways. Its elite refuses to raise local taxation for spending on 'foreigners', or compels migrants to bear the costs of their own contribution to pollution and disease by limiting the zones where they may reside. Or it organises political parties against them, such as Bombay's Shiv Sena, which was directed mainly against South Indian immigrants. Or it prevents their entry as best it can – difficult even in a totalitarian state (Stalin did badly at it), but eased if migrants have skins of different colour, as with influx control and the associated Pass Laws in South Africa. Or they just stop providing the job chances that attract the migrant. The understandable eagerness of city elites to restrict urbanisation should caution us against seeing such restriction as a cure for urban bias.

The facts of pseudo-urbanisation

This section shows how little urbanisation has taken place in the really poor countries of the world – those of Africa and South and East Asia. Much of what has happened (after demographic 'optical illusions' are allowed for) is temporary, or in other ways fails to change the migrant's life-style in ways that prepare him for modern industrial development. Furthermore, several mechanisms limit the impact, on the shared population in urban areas, of such townward movement as has happened.

Before demonstrating these facts, we must recall that historically the vast mass of townward migration has not been caused by rural 'push' or urban 'pull', but has been involuntary, a response to physical threats against life, limb or land. These are the great disaster treks, such as that following the partition of British India in 1947. As rural man/land ratios rise with population growth, refugees will find it increasingly difficult to find jobs, land or welcome in rural areas. The townsman is in practice unable to put up the shutters against such disaster movements, and they outweigh in importance any *voluntary* urbanising migrant response to urban bias.

In any event, the latter is not substantial. A cross-section comparison of agriculture's share in the labour force, in LDCs at different income-levels, led Kuznets to expect an observed decline in that share accompanying the income increases of the 1950s; the actual decline was a good deal less. If we take all the pairs of years between 1950 and 1967 for which estimates of that share are available in the same country, we find in Africa six cases in which it has risen, six in which it has fallen and three in which it has been static; in East and South Asia, there are four cases each of rising and falling agricultural population shares, and one static share. Only in the richer parts of the Third World, where immobilising constraints on the rural poor matter less – West Asia and Latin America – are falling shares of farm population clearly predominant. For India, 'it appears fairly certain that the final [1971] Census estimate will show a rise in the [proportion of workforce in] agriculture . . . between 1961 and 1971'. The close link in LDCs between urbanisation and industrialisation suggests that if the latter has been slow the former has also; and so it turns out.

Urbanisation: a demographic illusion?

From 1950 to 1960, the share of rural persons leaving for urban areas was 0.3 per cent yearly in South Asia, 0.6 per cent in East Asia, and 0.7 per cent in Africa – far less than rural population growth. In the poorest regions of the world, South Asia and Africa, the urban share in population rose far more slowly than is generally believed (from 16 to 18 per cent and from 14 to 18 per cent of total population respectively). In China, it proved impossible to sustain the rapid rise in the urban population share during the 1950s with surpluses of food – perhaps the last frontier that constrains urbanisation; the urban share of population actually fell from 19 per cent in 1960 to 12–13 per cent in 1971. The proportion of people in urban agglomerations (over twenty thousand persons) rose much more slowly in poor countries than in rich ones, both in 1920–50 and during the 1950s. Urban population growth has slowed down in the very places that, not long ago, were thought to pose the most uncontrollable threats of 'megalopolitan' congestion, explosion or decay: throughout 1951–71 the rate of growth of population in the Calcutta urban agglomeration was considerably lower than in rural West Bengal. A marked decline in urbanisation rates in poor countries began in the 1960s and is expected to sharpen in the next two decades.

The number of true, permanent urban immigrants is itself habitually and grossly exaggerated. Ashish Bose has provided a valuable picture of this for India. In 1961 one Indian in three was born outside his or her place of residence. But most of this migration comprises village brides, moving to their husbands' villages; only 4.2 per cent of Indians were rural-born townsmen. If we deduct the 1.1 per cent who were urban-born but lived in villages, we find that *net* townward migration covered only 3.1 per cent. Half of these at very least – more probably two-thirds – were 'turnover migrants', staying in urban areas for less than ten years before returning to rural life (Bose, 1967)*.

Several demographic 'optical illusions' help to foster the myth of mass urbanisation. The commonest, that rising proportions of urban residents must mean *permanent* streams of net townward migrants, hardly needs refuting, but there are others. The first takes place when an expanding city comes to abut on a village. Usually the village's population (and its natural increase) thereafter is counted as 'urban', even if its pattern of life has scarcely changed: '[In] Kuala Lumpur, where such boundary extension incorporated genuine suburban development, [this means true] urban growth; [not so] in other cases such as the Chartered Cities of the Philippines (which often include large rural populations).' (McGee, 1971, op.cit.)*. Bombay and Delhi, as I have seen, also contain long-submerged, but still largely rural, villages. The phenomenon is little researched, but must account for significant parts of urban growth in LDCs, except perhaps in South America.

Another 'optical illusion' of urban townward migration is created by the expansion of areas across some nominal borderline into a technically urban status that again need involve no change in economic behaviour. Their population and its natural increase then overnight become urban by classification. This, combined with migrants' own natural increase, makes the increase in the 'urban' population share a very misleading indicator of urbanisation, especially over long periods. Thus in Peru from 1940 to 1961, while national population grew by only 61 per cent, the population in agglomerations of over twenty thousand rose by 220 per cent, suggesting massive urbanisation. Yet in reality the population in cities of over twenty thousand *in 1940* grew by 175 per cent only, and the population in places of over twenty thousand *in 1961* by 173 per cent. Hence about a quarter of urban growth was due to reclassification of places as cities because they crossed the urban borderline. A third of the rest (about 61 per cent out of about 173 per cent) was due to natural increase of the *1940* population, and at least another tenth to natural increase of 1940–61 migrants. In Iran from 1951 to 1961, at least one-fifth of the apparent 'growth' in the urban share of the population was due to the reclassification of thirty-nine places that crossed the five-thousand border line. Much of Ghana's urban 'growth' has the same source. The effect also created a substantial illusion of urbanisation elsewhere.

These two effects – 'eating' of one community by another, and 'borderline-crossing' of a community owing to natural increase – separately increase the *classified* urban population. The effects can even combine; between censuses, two villages of three thousand, each swollen to four thousand by natural increase, can expand their built-up areas until they meet to form a single village of eight thousand, which is classified across an urban borderline (typically five thousand) – and, for some, this creates a statistical illusion of eight thousand townward migrants!

A third, subtler, illusion is created by the changing age-composition of populations; here the figures do not mislead us about the facts of urbanisation, but about the permanence of its upward tendency. Many LDCs have long featured a life-cycle pattern of *temporary* urbanisation of 10 to 30 per cent of rural males, for a few years between the ages of fifteen and thirty. Between 1945 and 1955, malaria control, in one country after another, slashed mortality rates, above all up to the fifth year of life. Hence rural children, who in earlier generations would have died, are surviving to enter the age-group fifteen to thirty. This swells the proportion of rural people in the age-groups of the traditional 'life-cycle' migrants – and this process will last from about 1960 (or 1945 + 15) to 1985 (or 1955 + 30). Even with no increase in either the proportion of the fifteen to thirty age-group who migrate or their average stay in the towns, the fact that this 'migration-intensive' age-group is a growing share of the population will raise the urban proportion of population. The rise in the 'migration-intensive' proportion of rural people, however, will be reversed as more of those saved from early malaria death pass into the post-migrant (and return-migrant) age-groups; it indicates neither permanent urbanisation nor a shift in preferences.

---

17 Alan Gilbert,

'Latin America's Urban Poor: Shanty Dwellers or Renters of Rooms?'

Reprinted in full (without photographs) from: *Cities* 4, 43–51 (1987)

Two images dominate writing about the housing of the poor in Latin American cities. The first is of flimsy, unserviced shanties covering the mountainsides or riverbanks on the edges of the urban areas. The second is of centrally located slums full of rooms rented by recently arrived migrants or poor natives. Both images are accurate in so far as many among Latin America's poor occupy these kinds of accommodation. Many families certainly build shanty accommodation in the environmentally least desirable areas of Latin America's cities. While many of these dwellings are slowly improved and serviced and should perhaps be removed from the category of 'shanty', there is no doubt that self-help accommodation is a long, hard struggle for the families involved. Despite the problems facing the self-help family, however, the poor almost universally prefer owner occupation to renting. Nevertheless, many families lack the resources to engage in self help and must stay with kin or rent. Many tenants live in rooms in inner cities in crowded and insanitary conditions. Large, purpose-built tenements sometimes house dozens of families, each family occupying a room and sharing the services available in a central courtyard or patio.

Neither of these images, of the poor owner occupier living on the outskirts of the city and of the poor renter occupying an inner city tenement, are inaccurate. Indeed, our understanding of the self-help process, in particular, has become much more refined in recent years as a result of numerous in depth studies. Work started by Abrams, Mangin and Turner has examined how effective self help can be in allowing the poor to mobilize resources and improve their housing conditions. If it is equally clear that self help is a hard and difficult process, few writers would argue that under current political and economic circumstances there is any real alternative. It has also become clear that the people who occupy such housing are in no sense alienated from or marginal to the urban community. They are not enmeshed in some kind of culture of poverty, nor are they in any sense excluded from urban life. Rather, numerous studies have shown that the poor participate fully as consumers and producers in the urban economy, that they participate in politics and local decision making and that their problem is less a lack of integration than the form that integration takes. As a result of this re-evaluation of self-help housing and the position of the poor, most governments have begun to legalize and service most self-help settlements. Settlement upgrading and site-and-service schemes are the recommended order of the day.

If we now know a great deal about self help, however, we still seem to know rather little about tenants. For if the world of the shanty dweller has been extensively studied, that of the tenant has been largely neglected. Perhaps this is because the central areas of most Latin American cities still contain forms of housing that were described and roundly criticized years ago. Many Mexican families, for example, still live in conditions that remind one of Oscar Lewis's descriptions of Tepito in the 1940s.[1] Recent visits to Guadalajara and Puebla[2] have shown me that the *vecindades* are alive, if anything but well. In these cities, the *vecindad* continues to house large numbers of families both in the central areas and in the older parts of the cities. . . . The narrow gateway to the street [often] leads on to a central patio which in [one] case contains 22 separate rooms. Washing and toilet facilities are available in the patio and electricity is also provided. The caretaker lives in another room on the street. . . . the problems involved in terms of noise and overcrowding can be imagined.

Perhaps the world of the tenant has been neglected because most writers assume that rented accommodation is in decline. This is certainly true in the central cities where the returns on investment are no longer competitive and major investors seem to have directed their funds towards better sources of profit. Government intervention and urban renewal have been partly responsible for this decline in rental accommodation; but so too have been the growth of new opportunities for investment such as the development of suburban housing areas. But, rental accommodation, while housing a smaller proportion of urban households than in the past, has still been expanding rapidly. This is not something to which most of the literature draws attention. Indeed, most accounts are busy demonstrating how the proportions of households living in irregular, self-help housing have increased dramatically in most cities over recent years. While this expansion cannot be denied, it errs in one important respect. For if the proportion of households in self-help settlements is increasing

[1] O. Lewis, *The Children of Sánchez*, Random House, New York, 1957.

[2] These were linked to a project financed by the Overseas Development Administration on rental housing in those two cities. This project began in 1985 and is codirected by Dr Ann Varley.

fast, the expansion in owner occupation is far more restricted. In fact, what appears to be happening is that new rental accommodation is being created in the consolidating shanty towns. As owners manage to improve their homes and extend the accommodation, they let rooms to other families. This provides an additional source of income for the owner and helps to supplement an inadequate basic income. It also provides the funds for further investment in the accommodation. Indeed, it could well be that the payments of tenants are one of the main sources of funds through which self-help accommodation is extended.

In certain cities, such as Bogotá in Colombia, rental accommodation develops rapidly as a self-help settlement becomes established. One settlement in which I carried out a survey in 1978 was only four years old; and yet 25% of the families surveyed were tenants.[3] Even though the settlement lacked adequate services, particularly water and drainage, the shortage of housing was such that tenants were willing to take rooms in the settlement. In older parts of the city, the proportion of tenant households rose rapidly. In most settlements over ten years old over half of the families are tenants.

Who are the landlords?

The housing stock in Latin American cities is expanding through self-help construction. At the same time it is creating a new source of rental accommodation and a new kind of landlord. At present little research has been conducted into the nature of landlords in the consolidated low income settlements. But current work seems to suggest that landlordism is very much a small scale business, few landlords owning more than a handful of rooms. Often, indeed, the owner lives on the same plot as the tenants. The precise situation clearly varies both between settlements and between cities. In Guadalajara, older self-help settlements seem to contain some sizable *vecindades* with absentee landlords, whereas in Bogotá it seems as if the *inquilinatos* are typically rather smaller.[4] The majority in either city, therefore, do not fit the category of exploitative, Rachmann-like landlords. Indeed, the typical landlord in Guadalajara seems to be either elderly or a widow. In Mexico, the combination of rent controls and the difficulty of evicting tenants has made life very difficult for most landlords. Although it is clearly possible to make profits from renting, the tactics that need to be employed range from the dubious to the downright nasty. To judge from the number of complaints we have heard, few landlords in Guadalajara seem prepared or able to use such tactics. As a result, most

[3] The settlement of Britalia was one of five settlements in Bogotá studied in a comparative project on land, housing and servicing in Bogotá, Mexico City and Valencia conducted in 1978–81. For a more detailed description of this settlement and the project in general see A. G. Gilbert and P. M. Ward, *Housing, the State and the Poor: Policy and Practice in Latin American Cities*, Cambridge University Press, Cambridge, 1985.

[4] Multioccupancy rental accommodation has a different local name in most countries: *conventillos, cortijos, inquilinatos, vecindades* all describe similar kinds of accommodation.

landlords seem to be regarded relatively sympathetically by the tenants. As in Bogotá, the landlords seem to be drawn from a similar social stratum to the tenants. While large scale landlordism seems more common in other cities – Puebla, for example, still has extensive areas of central *vecindades* – the nature of landlordism seems to have changed dramatically in recent years. The new landlord has few properties and is not so very different in terms of class or indeed income from the tenants.

Who are the tenants?

My work in low income settlements in Bogotá and that of Peter Ward in Mexico City suggests that the incomes of owners living in self-help settlements are fairly similar to those of the tenants.[5] Certainly, the incomes earned by heads of households are very similar. What does differ is the number of wage earners in tenant and owner households and the sources of income. In general, Table 17.1 seems to suggest that the main difference between the two groups is that the home owners are older than the tenants. As a result, they have older families

Table 17.1 Economic and social variables according to tenure in Bogotá and Mexico City

	Bogotá			*Mexico City*			
	Total	*Owners*	*Renters*	*Total*	*Owners*	*Renters*	*Sharers*
Head's monthly income[a]	45.3	46.4	42.9	49.2	49.4	46.8	50.5
Other sources of income[a]	13.7	13.8	11.8	17.9	19.6	13.5	11.9
Income of other earners[a]	44.7	47.7	35.3	55.3	57.7	45.5	45.6
Families with no other income (%)	83.3	76.3	95.1	86.8	85.5	87.1	93.8
One active worker in household (%)	58.0	54.4	64.8	65.8	63.3	78.2	64.2
Two active workers in household (%)	23.8	23.2	25.4	21.4	24.1	13.9	16.0
Age of household head	38.6	42.0	32.7	39.2	41.9	34.4	30.5
Persons in household	5.5	5.8	5.0	5.8	6.2	5.2	4.5
Head's employment – independent (%)	22.0	25.6	15.8	19.4	22.1	9.0	17.9
Head's employment – manual (%)	54.5	52.1	59.2	64.5	64.2	66.0	64.1

[a] Local pesos (hundreds).
Source: A. G. Gilbert, 'The tenants of self-help housing: choice and constraint in the housing markets of less developed countries, *Development and Change*, Vol. 14, 1983, pp. 449–77.

[5] *Op. cit.* (footnote 3).

which means that the wife is more likely to contribute financially, as well as one or more grown up children.

Different conclusions may be drawn from this description of tenancy. One interpretation is that the youth of tenants supports the idea put forward by John Turner[6] that tenancy is a temporary step on the path to ownership. In his early work he suggested that newly arrived migrants rented homes until they obtained jobs and became firmly established in the city. A similar pattern would also be typical of new households established by natives of the city. Certainly, this process is supported by my research in Bogotá where most owner occupiers began their residential career in rented accommodation. Nevertheless, although this is the typical kind of residential move, it is difficult to accomplish in many cities. Even though most families are establishing a home through self-help, limited incomes still make this a difficult step. Indeed, calculating the age at which families in five settlements in Bogotá first obtained a plot of land suggests that the transition to ownership, even in a self-help settlement, is very difficult indeed. People born in Bogotá first obtained a plot of land when they were, on average, 31 years old. This is a much more advanced age than is typical of first home owners in, say, the UK.

The mature age of first owners has little to do with aspirations because most tenants in Latin American cities readily declare their wish to become home owners. This desire for ownership is partly a recognition of economic realities, partly a consequence of the wish to be independent; it is partly an expectation created by rural life where legal and *de facto* ownership is much more common than in the cities. But it is also an attitude stimulated by the whole ideology of capitalism, reinforced continuously by television advertising and soap operas. Certainly, recent interviews in Guadalajara revealed a widespread desire for ownership among the tenant population. Indeed, it was only the very poor who did not declare their wish for home ownership: this group, it seems, realize that such an aspiration is way beyond their means.

The determinants of owner occupation

If most families wish to become home owners, even if it involves them in self-help construction, why is it that more than half of the households in certain Latin American cities are tenants (Table 17.2)? Low incomes are obviously a major problem for the majority, but it is quite clear from the tenure pattern in developed countries that tenure is not determined by income levels alone. In western Europe, for example, more families own their own homes in the UK than they do in more affluent France or FR Germany. Clearly, residential tenure depends on a range of factors linked to the economic and social structure

[6] John Turner, 'Barriers and channels for housing development in modernizing countries', *Journal of the American Institute of Planners*, Vol. 33, 1967, pp 167–81.

of the country and to the policy of the state.[7] Similarly in Latin American cities, the proportion of owner occupiers varies considerably from country to country and is by no means in direct relationship to income per capita. In addition, there are major differences between the proportions of families who rent accommodation in different cities within the same country.

Table 17.2 Proportions of homes rented in selected Latin American cities[a]

City	*Country*	*Year*	*% dwellings rented*
Barranquilla	Colombia	1977	29
Bogotá	Colombia	1977	43
Bucaramanga	Colombia	1977	44
Cali	Colombia	1977	40
Manizales	Colombia	1977	52
Medellin	Colombia	1977	38
Pasto	Colombia	1977	35
Guadalajara	Mexico	1980	48
Mexico City	Mexico	1980	46
Monterrey	Mexico	1980	32
Tijuana	Mexico	1980	48
Veracruz	Mexico	1980	48
Barquisimeto	Venezuela	1972	24
Caracas	Venezuela	1972	39
Maracaibo	Venezuela	1972	18
Maracay	Venezuela	1972	23
San Cristóbal	Venezuela	1972	32

[a] Data for Mexico record numbers of non-owning households and therefore slightly exaggerate the proportions of renters. Data for Colombia and Venezuela record proportions of rented dwellings. Data in each case are for the metropolitan area.
Sources: Colombia – *Alquileres de vivienda en Cinco Cuidades 1977: Encuesta Nacional de Hogares Etapa 16*, DANE, Bogotá; Mexico – *X Censo General de Población y Vivienda, 1980*, SPP, Mexico DF; Venezuela – *X Censo de Población y Vivienda, 1972*, Ministerio de Formento, Caracas.

The most plausible explanation for these differences, in an environment where most families are prepared to engage in some kind of self-help construction, is that the availability of land for poor people varies greatly from city to city. For while it is relatively easy to obtain cheap or even free land in certain cities, in others even the poor must pay the full market price for a plot. Cheap or free land is usually associated with the process of land invasion and there are numerous cities in Latin America where this has been the normal process of land alienation. In Venezuela, for example, most self-help settlements have been formed through large-scale occupations of land because most of the democratic governments in power since 1958, and even certain military governments before that date, have effectively turned a blind eye to such

[7] Michael Harloe, *Private Rental Housing in the United States and Europe*, Croom Helm, Beckenham, Kent, 1984.

invasions. Similarly, in many Peruvian cities, the authorities have tolerated or even encouraged the process of land invasion for most of the period since the late 1940s. At times, too, the situation in Brazilian cities such as Brasília, Rio de Janeiro and Salvador has been favourable to this form of land occupation; in Colombian cities such as Barranquilla and Cúcuta it has long been common practice and in Santiago and other Chilean cities land invasions were frequent in the late 1960s and early 1970s. In such invasion-tolerant environments, the cost of land has obviously been very cheap for those participating in the invasion. The process has certainly not been without risk or even danger; but in most of these cities the invasion was eventually authorized or tolerated. Even here, however, many families were required to purchase land. Once an invasion has been established market values are soon placed on plots and new families buy land from some of the early occupiers. Nevertheless, even for these later settlers the cost of land for self-help housing is generally much cheaper than it is for settlers in cities where land invasion is discouraged.

In cities where the authorities prevent land invasions, all land is allocated by the market. The prices charged for unserviced plots on land without planning permission are cheaper than for land used for higher income residential developments. Nevertheless, prices per square metre on the urban fringe are rarely low and it is a major struggle for low-income families to purchase a plot. The precise mechanisms through which land reaches the market vary considerably between countries and are important in so far as they seem to affect the price of land. In some Mexican cities, for example, *ejiditarios* on the urban fringe sell land as a more profitable business than farming. Since Mexican legislation does not permit the alienation of *ejidal* land, there is some risk linked to the purchase; this is reflected in the relatively low prices. By contrast, in many Brazilian, Colombian and Mexican cities, land subdivision rests in the hands of illegal urbanizers. These developers subdivide peripheral land and sell it, despite the lack of the requisite services and planning permission. These illegal subdividers are generally tolerated by the authorities who perceive the alternatives, such as land invasion, to be still more problematic. The consequence for the purchaser, however, is one of higher costs. Even if prices are much lower than in high income residential areas, costs per hectare are still very high. As a result, plot sizes in cities with illegal subdividers tend to be quite small relative to cities where invasions are common. In addition, there is a clear limit on the ability of many poor people to buy a plot of land.

While no comprehensive account of housing and urban development throughout Latin America is available to serve as the basis for a comparison of land costs, forms of land acquisition and levels of home ownership in different cities, there seems to be an approximate match between the costs of land access and the proportion of households living in rented accommodation. In Bogotá and Mexico City, for example, where land access is difficult, a low proportion of households own their homes. By contrast, in several cities where land invasions have been common, levels of ownership are much higher. A more detailed discussion of this argument is presented in Gilbert and Ward.[8]

It is interesting to speculate whether the recent liberalization of government attitudes towards self-help housing is likely to improve the chances of poor people obtaining their own home. There can be little doubt that governments are much more willing to regularize and service low income settlements than they were even a decade ago. This change in attitude reflects pressure from organizations such as the World Bank, as well as changes in thinking among architects and planners about the wisdom of condemning large parts of the urban population to illegal living. Whether, however, such a change in attitude has improved the prospects for home ownership depends on other developments in the city. Most significantly, it depends on what has happened to land values. For the cost of land continues to be the major barrier to self-help housing in every city where invasions are prohibited. Indeed, the barrier seems to be getting higher each year as land prices on the urban fringe rise faster than inflation. I know of no account which has found that land prices are rising slowly. It is indeed possible that the increasing willingness to legalize low-income areas is linked to these increasing land values. For as prices for new building land rise throughout the city, construction companies are facing increasing problems in marketing their houses. They are therefore eyeing the cheaper, illegal parts of the city. Should new areas of land be legalized then middle-income residential housing developers have a new area to develop. It is also clear that legalization brings benefits to the local authorities in so far as self-help settlers are thenceforth incorporated into the tax base of the city. Legalization may be desirable for the majority of low income settlers; but it may also increase their housing costs. In fact, the distributive consequences of liberalization are anything but clear. Besides raising taxes, more rational land development may well lead to cheaper and more effective infrastructure provision and thereby improve the quality of servicing for the majority of the population. Local variations will obviously determine the effects on the poor.

Government policy towards renting

There can be little doubt that rented accommodation is a necessary element in the housing stock of any city, whether it be provided by the private or the public sector. Short-tenure accommodation is always required by certain groups in the population and there is always likely to be some part of the population which does not wish for the responsibilities associated with ownership or which wishes to use its limited capital for some other purpose, such as establishing a business. In Latin America, it appears that most governments have played a rather negative role in encouraging and improving rented accommodation. In so far as there is a common policy, it has been to introduce rent controls. These

8 *Op. cit.* (footnote 3).

have often been ineffective; but where they have worked, they have had the long-term effect of discouraging landlords from investing in such property. As a result, one group of tenants have benefited from low rents but have seen the physical fabric of their homes gradually deteriorate. A new, and ever larger, group have been penalized by the lack of investment in new accommodation for renting. Certainly, the rent controls introduced in Mexico City in the late 1940s discouraged large investors from continuing to put money into rented housing. In turn, this has led to worsening housing conditions as landlords have refused even to repair accommodation that yields them so little return. Rightly or wrongly, the effect of most government legislation has been to discourage large-scale investment in rental housing.

Rather few Latin American governments have sought to increase the supply of rented accommodation. It has only been in the last couple of years, for example, that the Mexican state has begun to consider ways in which it might help tenants and landlords to renovate the decaying tenement blocks. Indeed, the absence of a coherent government policy towards rented housing in most Latin American cities is rather surprising, given the millions of people who occupy this form of tenure. It seems almost as if governments have washed their hands of this area of activity; it is too politically explosive.

As a consequence, the housing stock for rent has expanded through innumerable investments by self-help owner occupiers. By improving their accommodation and adding more rooms, settlers have become landlords. It is conceivable that this is the most appropriate method of increasing the housing stock for, arguably, landlords living on the premises are more likely to maintain the accommodation. It is also arguable that this rental housing market has thrived and developed because it has been largely untouched by government regulations! But the truth is, we still know so little about renting, the attitudes and characteristics of tenants and landlords, and the whole structure of the rental housing market, that it is difficult to know precisely what policies are most appropriate. Studying landlords and tenants has not been fashionable in recent years. But, given the danger that large numbers of aspirant self-help settlers will be unable to obtain cheap land in the future, this would seem a critical area for more effort. Certainly, it is to be hoped that during this International Year of Shelter for the Homeless, the tenant will not continue to be forgotten.

18 Mario Vargas Llosa,

Foreword to 'The Other Path: The Invisible Revolution in the Third World' by Hernando de Soto

Reprinted in full from: H. de Soto, *The Other Path: The Invisible Revolution in the Third World*, pp. xiii–xxii. New York: Harper and Row (1989)

Economists occasionally tell better stories than novelists. *The Other Path* by Hernando de Soto is a perfect case in point. His story, based entirely on Peruvian reality, reveals an aspect of life in the Third World that is traditionally obscured by ideological prejudice.

Unlike good literature, which teaches us indirectly, *The Other Path* preaches an explicit lesson about contemporary and future Third World reality. And unlike run-of-the-mill economic and sociological essays on Latin America, which seek to be abstract and end up distanced from any specific society, *The Other Path* never strays from the real world. It focuses on a hitherto little-studied and even less-understood phenomenon – the informal economy – and then offers a solution to the economic plight of underdeveloped countries. This solution is completely different from the economic projects the majority of Third World governments, progressive or conservative, have devised, but it is – and this is the book's main thesis – the very solution the poorest sectors of Third World societies have already put into practice.

The Other Path is an exhaustive study of the informal economy, or black market, in Peru and reveals startling information about its magnitude and complexity. But the book is much more than an exposé: After describing the magnitude and complexity of economic activities carried out outside the law in Peru, Hernando de Soto – with the assistance of his researchers at the Institute for Liberty and Democracy – examines the origins of social injustice and economic failure in Latin America. As he delineates the problems of underdevelopment, he explodes many myths about the Third World that pass for scientific truths.

The informal economy

The 'informal economy' is usually thought of as a problem: clandestine, unregistered, illegal companies and industries that pay no taxes, that compete unfairly with companies and industries that obey the law and pay their taxes promptly. Black-marketeers are brigands who deprive the state of funds it might use to remedy social problems and strengthen the very structure of society.

That kind of thinking, as Hernando de Soto proves, is totally erroneous. In countries like Peru, the problem is not the black market but the state itself. The informal economy is the people's spontaneous and creative response to the state's incapacity to satisfy the basic needs of the impoverished masses. It is, of course, paradoxical that this study, carried out by an institute that defends economic freedom, constitutes an indictment of the Third World state unrivaled in its severity and force, while at the same time it reduces most radical or Marxist critiques of underdevelopment to mere rhetorical posturing.

When legality is a privilege available only to those with political and economic power, those excluded – the poor – have no alternative but illegality. This is why the informal economy comes into being, as Hernando de Soto demonstrates with incontrovertible proof. To find out just what the 'cost of legality' is in Peru, the Institute for Liberty and Democracy set up a fictitious clothing factory and went through the procedure – the bureaucratic maze – of establishing it legally. The institute decided to pay no bribes, except when not paying them would bring the entire process to a standstill.

On ten occasions bribes were solicited, but the institute was obliged to pay only twice. To register the imaginary factory took 289 days and required the full-time labor of the group assigned to the task, as well as $1,231 (including expenses and lost wages). At that time – summer 1983 – the amount was the equivalent of 32 minimum monthly wages. This means that the process of legally registering a small industry is much too expensive for any person of modest means. It is certainly no coincidence that this is the kind of person who founds 'informal' industries in Peru.

If setting up a legal shop is a costly and time-consuming task for the poor, it is even more expensive and difficult for them to obtain legal housing. The institute found out that if a group of low-income families petitioned the state to cede them a vacant lot on which they might build, they would have to work their way for six years and eleven months through ministries and municipal offices and spend approximately $2,156 (56 times the minimum monthly wage at the time) per person. Even to get a license to open a street kiosk or sell from a pushcart is a task of Kafkaesque proportions: forty-three days of commuting between bureaucrats and $590.56 (15 times the minimum monthly wage).

The statistics that accompany the study are devastating and bolster the institute's analyses with implacable logic. The kind of country we see in these numbers is tragic and absurd: tragic because the legal system seems designed to favor those already favored and to punish the rest by making them permanent outlaws; absurd because a system of this kind condemns itself to underdevelopment. It will never progress and will slowly drown in its own inefficiency and corruption.

The Other Path's description of the origins and extent of injustice in a Third World nation is pitiless, but it does not leave us demoralized and skeptical about a remedy for these problems. The informal economy – a parallel and in many ways more authentic, hardworking, and creative society than the one that hypocritically calls itself legitimate – appears in its pages as an escape hatch from underdevelopment. Many of the victims of underdevelopment have already begun to take advantage of it and are revolutionizing the nation's economy. Curiously enough, the vast majority of those who write and theorize about the backwardness and iniquity of life in the Third World do not seem to be aware of its existence.

All over Latin America, the poor have fled from the countryside to the cities. When these poor people, driven off the land by drought, flood, overpopulation, and the decline of agriculture, reached the city, they found that the system had already closed its doors to them. They had no money and no technical training. They had no hope of getting credit, no chance to obtain insurance, and could expect no protection from the police or the judicial system. They knew their businesses would always be threatened from all sides. All they had was their will, their imagination, and their desire to work.

To judge by the four areas studied by the Institute for Liberty and Democracy – business, manufacture, housing, and transportation – these entrepreneurs have not done badly. To begin with, they are infinitely more productive than the state. The statistics in *The Other Path* are shocking. In Lima alone, the black market (excluding manufacture) employs 439,000 people. Of the 331 markets in the city, 274 have been built by the black-marketeers (83 percent). It is no exaggeration to say that it is thanks to them the citizens of Lima are able to get around, because 95 percent of public transportation belongs to them. The black-marketeers have invested more than a billion dollars in vehicles and vehicle maintenance. As to housing, the figures are equally impressive. Half the population of Lima lives in houses built by black-marketeers. Between 1960 and 1984, the state constructed low-income housing at a cost of $173.6 million. During the same period, the black-marketeers managed to construct housing valued at the incredible figure of $8,319.8 million (47 times what the state spent). Economic freedom existed only on paper before the poor of our nations began to put it into practice independently.

These numbers speak eloquently of the productive energy that restrictive legality has pushed into the black market. But that vigor also reflects the true nature of the Third World state, which is almost always grotesquely caricatured. In this respect, Hernando de Soto offers some evidence that will shatter myths.

Underdevelopment and mercantilism

One of the most widely accepted myths about Latin America is that its backwardness results from the erroneous philosophy of economic liberalism adopted in almost all our constitutions when we achieved independence from Spain and Portugal. This opening of our economies to the forces of the market made us easy prey for voracious imperialists and brought about internal inequities between rich and poor. Our societies became economically dependent (and unjust) because we chose the economic principle of laissez-faire.

Hernando de Soto attacks that fallacy head on and rebuts it. The institute's thesis is that Peru never had a market economy, and that it is only now, because of the black market, beginning to get one – a savage market economy, but a market economy nevertheless. This concept applies to all of Latin America and probably to the majority of Third World nations. Economic freedom is a principle emblazoned in our constitutions that has no more reality than the principle of political freedom, to which our politicians, especially our dictators, traditionally render hypocritical tribute. De Soto calls our economic system, which has been masquerading as a market economy for generations, mercantilist.

The term is confusing, since it defines a historical period, an economic school, and a moral attitude. Here, 'mercantilism' means a bureaucratized and law-ridden state that regards the redistribution of national wealth as more important than the production of wealth. And 'redistribution,' as used here, means the concession of monopolies or favored status to a small elite that depends on the state and on which the state is itself dependent.

The state, in our world, has never been the expression of the people. The state is whatever government happens to be in power – liberal or conservative, democratic or tyrannical – and the government usually acts in accordance with the mercantilist model. That is, it enacts laws that favor small special-interest groups – the study calls them 'redistributive combines' – and discriminates against the interests of the majority, which has marginal power or token legality. The names of the favored individuals or consortia change with each new government, but the system is always the same: not only does it concentrate the nation's wealth in a small minority but it also concedes to that minority the *right* to that wealth.

'The system' includes not only that hybrid monster I mentioned earlier – the state government – but also the entrepreneurs who work within the law. *The Other Path* does not pull any punches in its criticism of this entrepreneurial class which, instead of favoring an egalitarian and dynamic system in which the law would guarantee free competition and reward creativity, has adapted itself to the mercantile system and dedicated its best efforts to obtaining monopolies. Even today, when the comfortable house that class has been inhabiting for generations is falling down around it, it continues to view industrial activity as a sinecure instead of a way to create wealth.

This system is not only immoral but inefficient. Within it, success does not depend on inventiveness and hard work but on the entrepreneur's ability to gain the sympathy of presidents, ministers, and other public functionaries (which usually means his ability to corrupt them). In chapter 5, on the cost of legality and informality, Hernando de Soto reveals that, for the majority of formal or legal businesses, the single greatest expense, in both money and time, is bureaucratic maneuvers. This blights our economic life at its very roots.

Instead of favoring the production of new wealth, the system, owned, in effect, by the closed circle of those who benefit from it, discourages any such effort and prefers merely to recirculate an ever-diminishing amount of capital.

In that context, the only kinds of activity that proliferate are nonproductive, parasitic activities – our elephantine bureaucracies. To justify their own existence, these monsters decree, for example, that in order to register a small-scale factory, a citizen has to fight for ten months through eleven different ministerial and municipal departments and, just to keep things moving, bribe at least two people. Is it any wonder that most Third World businesses are technologically backward and have tremendous difficulty competing in international markets?

At the same time that the mercantilist system condemns a society to economic impotence and stagnation, it imposes relations between citizens and between citizens and the state that reduce or eradicate the possibility of democratic politics. Mercantilism, as described by Hernando de Soto, is based on laws that mock the most elementary democratic practices.

The legal tangle

It is said that the number of laws and executive orders – decrees, ministerial resolutions, procedures, and so forth – in Peru exceeds half a million. This is only an approximate figure because, in point of fact, there is no way to determine the exact number. We live in a legal labyrinth in which even a Daedalus would get lost. This cancerous proliferation of laws reflects the bizarre ethical conditions that prevail in our legislative process. Laws work for special interests, not for the general interest. A logical consequence of this uncontrolled growth is that for every law there is another law that emends, attenuates, or negates it. In other words, anyone involved in this morass of legal contradictions, like it or not, is at some time or another breaking the law. By the same token, anyone deliberately breaking the law can find a law that will render his or her actions lawful.

Who legislates these laws and decrees? Hernando de Soto's study shows that only the tiniest number – 1 percent – of our laws emanate from the body created to make them: the Parliament. The other 99 percent derive from the executive. That is, they flow from government departments that conceive them, draw them up, and have them promulgated – with no interference, no debate, no criticism, and often without even the knowledge of those affected by them. Laws presented in Parliament are publicly discussed, so it is conceivable that the media may inform the public about them and that their beneficiaries or victims may possibly influence their final form.

But this never happens with the majority of our laws. They are cooked up in bureaucratic kitchens (or in the private chambers of certain lawyers) in accordance with the dictates of the redistributive combine whose interests they serve. They are promulgated at such a rate that not only the ordinary citizen but even lawyers are unable to keep abreast of them and react appropriately.

Whenever a Third World nation returns to democracy, it holds more or less honest elections and permits freedom of the press. Political life takes shape and is carried on without too many impediments. But behind that façade, and particularly with regard to legal and economic life, democratic practices are conspicuously absent. The reality behind the façade is a discriminating, elitist system run by the smallest of minorities.

Black-marketeering is the masses' response to the system, which has traditionally made them victims of a kind of legal and economic apartheid. The system invents laws to frustrate the legitimate desires of the people to hold jobs and have a roof over their heads. What should the masses do? Stop living, in the name of a legality which in many ways is unreal and unjust? No. They have simply renounced legality. They go out on the streets to sell whatever they can, they set up their shops, and they build their houses on the hillsides or in vacant lots. Where there are no jobs, they invent jobs, learning in the process all they were never taught. They turn their disadvantages into advantages, their ignorance into wisdom. In politics, they act in a purely pragmatic way: they turn their backs on fallen idols and hitch their wagon to any rising star. In Peru, they were behind General Odría when he was in power, behind Prado when he governed, with Belaunde when he ran things, and firm supporters of Velasco when that general was their leader. Now they are – simultaneously – Marxists with Mayor Barrantes and followers of the Popular Revolutionary American Alliance (APRA) with President Alan García.

Hernando de Soto's book shows clearly what they really are, despite these transitory, tactical alliances: men and women who through almost superhuman hard work and without the slightest help from the legal country (in fact, in the face of its declared hostility) have learned how to create more jobs and more wealth in the zones in which they have been able to function than the all-powerful state. They have often shown more daring, effort, imagination, and dedication to the country than their legal competitors. Thanks to them, our throngs of thieves and unemployed are not larger than they are. Thanks to them, there are not more hungry people wandering our streets. Our social problems are enormous, but without the black-marketeers our situation would be infinitely worse.

But what we should thank them for most is that they have shown us a practical and effective way of fighting against misfortune which belies the preachings of Third World ideologues who cling to their worthless doctrines with baffling tenacity. The path taken by the black-marketeers – the poor – is not the reinforcement and magnification of the state but a radical pruning and reduction of it. They do not want planned, regimented collectivization by monolithic governments; rather, they want the individual, private initiative and enterprise to be responsible for leading the battle against underdevelopment and poverty.

Who would have said it? If we listen to what these poor slum dwellers are telling us with their deeds, we hear nothing about what so many Third World revolutionaries are advocating in their name – violent revolution, state control of the economy. All we hear is a desire for genuine democracy and authentic liberty.

These are the ideas Hernando de Soto convincingly defends in *The Other*

Path. The concept of liberty, in all its senses, has never been seriously applied in our countries. Only now, in the most unexpected way, through the spontaneous action of the poor, is it beginning to gain ground, showing itself to be a more sensible and effective solution than any undertaken by our conservatives and progressives as ways of overcoming underdevelopment. Extremists of both persuasions, despite their ideological differences, agree to the strengthening of the state and its interventionist practices, which does nothing but perpetuate the system of corruption, incompetence, and nepotism that is the recurring nightmare of the entire Third World.

Freedom as an alternative

That freedom should be the alternative the poor choose in their struggle with the elite will probably surprise many. One of the most commonly accepted truisms of recent Latin American history is that liberal economic ideas are characteristic of military dictatorships. Didn't the 'Chicago Boys' put them into practice with Pinochet in Chile and Martinez de Hoz in Argentina, with the catastrophic results we know so well? Didn't those politicos make the rich richer and the poor poorer in both countries? And didn't they precipitate both countries into unprecedented disasters from which they still have not recovered?

There is only one kind of liberty and it is obviously incompatible with authoritarian or totalitarian regimes. The economic liberalism they can bring about – or rather *impose* from above – will always be relative and will always be weighed down, as in Chile and Argentina, by a complementary lack of political freedom. But it is precisely political freedom that permits the evaluation, perfection, or rectification of any measure which does not work in practice. Economic freedom is the counterpart of political freedom, and only when the two are united – two sides of a single coin – can they really function.

No dictatorship can be really liberal in economic matters, because the basic principle of economic liberalism is that it is not the politically powerful but the independent and sovereign citizens who have the right to take action – to work and sacrifice – to decide in which kind of society they are going to live. The function of political power is to guarantee that all obey the rules of the game, so that action will be fairly and freely chosen. That requires a consensus, the support of the people who desire those principles, and it can take place only in a democracy.

Within liberalism, there exist extremist tendencies and dogmatic attitudes. They are usually expressed by those who refuse to change their ideas when those ideas fail the acid test of all political programs: reality. It is natural that, in a Third World country with huge economic inequities, no cultural cohesion, and tremendous social problems, like so many Latin American nations, the state has a redistributive function. Only when those huge differences have been reduced to reasonable proportions is it possible to talk about truly impartial rules of the game, identical for all. With the imbalances we have today between poor and rich, urbanites and peasants, Indians and those who live in the Western tradition, the best conceived and purest measures tend in practice to favor the few and harm the majority.

It is essential that the state remember that before it can redistribute the nation's wealth, the nation must produce wealth. And that in order to produce wealth, it is necessary that the state's actions not obstruct the actions of its citizens, who, after all, know better than anyone else what they want and what they have to do. The state must restore to its citizens the right to take on productive tasks, a right it has been usurping and obstructing. The state must limit itself to functioning in those necessary areas in which private industry cannot function. This does not mean that the state will wither away and die.

By the same token, a large government is not necessarily a strong state, as the majority of Latin American nations shows. Those immense organisms that in our countries drain the productive energies of society to maintain their own sterile existence are in fact giants with feet of clay. Their gigantism makes them torpid, and their inefficiency deprives them of the respect and authority without which no institution can function well.

The Other Path does not idealize the informal market – to the contrary. After showing their accomplishments, de Soto describes the limitations that living outside the law imposes on informal businesses: they cannot grow, they cannot plan for the future, they are vulnerable to theft, extortion, and any crisis. The report also shows the desire for legality that many of the actions of the black-marketeers betray: the street vendor's desire to get a stall in a market, the neighborhood group that improves sanitary and aesthetic conditions as soon as it gains legal title to its homes. But the study, even though it does not embellish or overvalue the informal economy, does allow us a glimpse of the black-marketeers' spirit and imagination. It shows us what might be hoped for if all that productive energy could be put into practice legally in an authentic market economy in a government which, instead of harassing the black-marketeers, would protect and stimulate them.

By calling this book *The Other Path*, Hernando de Soto challenges the movement that sprang up in the Andes in 1980 and proclaimed a Maoist utopia. It counters that program with a social project which, although totally contrary to Marxist–Leninist fundamentalism, requires a transformation of society no less radical than the one demanded by the Shining Path. It means uprooting an ancient tradition which, through the inertia, greed, and blindness of a political elite, has blended with the institutions, customs, and traditions of the official nation. But the revolution this study analyzes and defends is in no way utopian. It is already under way, made a reality by an army of the current system's victims, who have revolted out of a desire to work and have a place to live and who, in doing so, have discovered the benefits of freedom.

19 Diane Elson, 'Transnational Corporations: Dominance and Dependency in the World Economy'

Excerpts from: D. Elson, 'Dominance and dependency in the world economy', pp. 264–87 of B. Crow, M. Thorpe *et al.* (eds), *Survival and Change in the Third World*. Cambridge: Polity/Open University (1988)

Third World economies are part of an international economy. Each Third World country is integrated into the international economy through trade, investment, financial flows and migration. The most important international economic links of the majority of Third World countries are with First World countries. In general, such links are quantitatively and qualitatively more important than links with other Third World countries, and more important than links with Second World countries.

Such links are not new, but their significance has increased with commercialization and industrialization in the Third World. Today, there are very few parts of the Third World where people make a living untouched by international economic links.

According to orthodox economic theory, these are mutually beneficial links that promote development and are evidence of harmonious interdependence between First and Third Worlds. Many people in the Third World reject that view, however. They see the links as means by which the First World asserts its dominance, and confines the Third World to a subordinate position in the world economy. One school of thought, often called the dependency school, sees international links as presently constituted as a barrier to development. Far from promoting development, international trade and investment are seen by many dependency theorists as means for perpetuating underdevelopment.

In this chapter, we analyse trade and investment links between First and Third World countries, to see to what extent they are equally beneficial to both parties or whether Third World countries would be better off without them, as some dependency theorists have suggested. We need to bear in mind that there is a third possibility: the links may be extremely unequal, and vehicles for domination of the international economy by the First World, and yet they may be necessary to Third World countries and, at least in some cases, permit development. We do not have to agree with either the interpretation given by orthodox neo-Classical economic theory or the interpretation given by dependency theory.

Transnational Corporations (TNCs)

The integration of Third World countries into the international economy takes place to a very large extent under the auspices of First World firms. The channels of international trade and finance are organized by First World banks, trading houses and shipping lines. The activities of transnational corporations have been seen as especially significant by many Third World writers. Radical economists have described the TNCs as a modern form of imperialism.

Many of today's TNCs grew out of the small family businesses of nineteenth-century Europe and the USA. (Japanese transnationals have a different history and structure, but limitations of space will prevent us from considering their differences in detail here.) The first stage was the development at a national level from a family business, typically tightly controlled by a single entrepreneur or small family group who possessed all the information and made all the decisions about the firm, to a national business corporation with a formal administrative structure, with division of function (finance, personnel, purchasing, production, sales, etc.) and a hierarchical division of responsibility and power (field offices to manage local operations, head office to supervise field offices). The second stage was movement abroad. As large national firms, they had developed the capacity to go abroad, in the form of a suitable administrative structure, and financial strength. What pushed them into using this capacity was the competitive struggle with other large national corporations. Competition became a much more complex affair once a firm was competing against a limited range of known alternative suppliers, rather than against a large number of unknown alternatives. It was much more feasible to collaborate with competitors, for instance to organize cartels to keep up prices. On the other hand, there were very rich pickings for the corporation that could outwit its competitors by controlling sources of supply of inputs, or sales outlets, or by developing new products that would make the old obsolescent. Some firms invested abroad to get the security of control over their raw material requirements; some to control marketing outlets; some invested abroad as a pre-emptive measure to forestall other corporations gaining control of raw materials or markets. US corporations led the way.

The growth in TNCs

The first wave of direct US foreign investment occurred around the beginning of the twentieth century, followed by a second wave in the 1920s. This movement slackened during the depression of the 1930s, and resumed again after the Second World War, accelerating rapidly.

Between 1950 and 1969, direct foreign investment by US firms expanded at a rate of about 10 per cent per annum. It is in this twenty-year period that direct foreign investment by transnational corporations became a world economic phenomenon of great importance. Before 1914, between two-thirds and three-quarters of the value of all private foreign investment took the form of portfolio investment, that is, it took the form of purchases of financial securities, such as bonds issued by foreign institutions, governments or business firms. The City in London played a key role in these transactions. Although a few large American and European companies, such as Lever, Singer, General Electric, Courtaulds and Nestlé already owned sizeable foreign manufacturing plants in 1914, these were the exceptions rather than the rule. However, by the mid-1960s, the greater part of private foreign investment took the form of direct investment by transnational corporations.

US TNCs remain the most important, though their share in the total stock of direct foreign investment was estimated to have falled from about 54 per cent in 1967 to about 48 per cent in 1976 (United Nations Centre for Transnational Corporations, 1978, Table III-32, p. 236). A recently released report by the US Department of Commerce estimates that, in 1977, 3450 US transnationals (in all industries except banking) owned or were connected with 24,666 foreign affiliates, who controlled US$490.02 billion in assets and employed 7.2 million people.

An important feature of foreign investment by TNCs is the disproportionate role of a quite small number of very large firms. For instance, United Nations estimates published in 1973 showed over 70 per cent of direct US foreign investment to come from only 250–300 firms; in the case of Britain, over 80 per cent was controlled by 165 firms. Eighty-two firms controlled over 70 per cent of direct foreign investment by the Federal Republic of Germany. One way of grasping the sheer economic size of the largest transnational corporations is to compare their 'value added' with the gross national product of national economies ('value added' means the value of sales minus the value of purchases of material inputs). A United Nations report estimated that, in 1971, the value added by each of the top ten TNCs was greater than US$3 billion, which was greater than the Gross National Product of over eighty Third World countries. The value added of all TNCs was estimated to be US$500 billion, which was approximately 20 per cent of the combined Gross National Product of the First and Third World (United Nations Centre for Transnational Corporations, 1973, ch. 1).

Perhaps the single most important characteristic of TNCs, and the key to their power and influence, is their ownership and control of knowledge, including the technology of production, as well as organizational systems, marketing systems, and financial systems. Though *basic* knowledge tends to be produced by government-financed research and training centres, the applied development of technology is generally undertaken by business firms, and a large part of commercialized technology is in the hands of the TNCs. For instance, nearly all of the world's patents are held by the TNCs.

The international distribution of investment by TNCs

Table 19.1 shows the distribution of flows of direct foreign investment by TNCs. This makes clear that, since 1965, about three-quarters of foreign direct investment has, on average, gone to First World countries. Typically, the pattern within the First World is one of cross-investment, with US firms

Table 19.1 Direct foreign investment in selected country groups, 1965–83

Country group	*Average annual value of flows (US$ billions)*[a]				*Share of flows (%)*			
	1965–9	*1970–4*	*1975–9*	*1980–3*	*1965–9*	*1970–4*	*1975–9*	*1980–3*
Industrial countries	5.2	11.0	18.4	31.3	79	86	72	63
Developing countries	1.2	2.8	6.6	13.4	18	22	26	27
Latin America and the Caribbean	0.8	1.4	3.4	6.7	12	11	13	14
Africa	0.2	0.6	1.0	1.4	3	5	4	3
Asia, including Middle East	0.2	0.8	2.2	5.2	3	6	9	11
Other countries and estimated unreported flows	0.2	−1.0	0.6	4.8	3	−8	2	10
Total[b]	6.6	12.8	25.6	49.4	100	100	100	100

Notes: [a] Figures converted from billions of SDR to billions of US dollars based on average IMF exchange rates.
[b] Total includes IMF estimates for unreported flows.
Source: World Bank, 1985, p. 126.

investing in Canada and Western Europe, and Western European firms investing in each other's economies, and in North America. Japan does not follow exactly the same pattern. There is very little direct foreign investment in Japan itself, and Japanese firms have, until recently, concentrated the bulk of their overseas investment in the Third World, especially in Asia. But Japanese direct investment in North American and European countries rose rapidly in the 1970s, so that at the end of the decade the largest stock of Japanese direct investment was held in the USA. Indonesia and Brazil were the next largest recipients, but in fourth place was Britain, followed by Australia in fifth place. An important factor leading to Japanese investment in the USA and Britain has been the desire to jump over the trade barriers (such as voluntary restrictions on car exports) which have been increasingly erected against Japanese exports to other developed countries.

TNCs and the Third World

Direct foreign investment within the Third World has been highly concentrated in a limited number of countries. As the Brandt Report (1980) points out, 70 per cent of the direct foreign investment in the Third World has been in only fifteen countries. Over 20 per cent is in Brazil and Mexico alone, and much of the rest is in other middle-income countries of Latin America (such as Argentina, Peru and Venezuela), and in South East Asia (such as Malaysia, Singapore, Hong Kong). About one-quarter is in oil-exporting developing countries. In the poorer countries, direct foreign investment is mainly in minerals and plantations, or in countries with large internal markets, like India. But this concentration of direct foreign investments in the richer developing countries does *not* mean that the activities of TNCs are only of interest to these few. There is *some* investment by TNCs in virtually *all* Third World countries, and though in the case of many countries the amount of such investment is a drop in the ocean by world standards, it is highly significant in the context of their small domestic economies.

Until recently, investment in the Third World by transnational corporations has been mainly in oil, mining and plantations. But manufacturing now accounts for about half the current flow of foreign direct investment in the Third World. There is a strong tendency for transnationals to invest in the areas where the cultural and political influence of their 'home' countries has been greatest. The exception to this is recent large-scale Japanese investment in Latin America, but most Japanese investment in the Third World is still in Asia.

Nationalization: ownership and control?

In any serious conflicts with Third World governments, TNCs tend to rely on their home governments' ability to exert leverage to change the policy of Third

World governments. Where TNCs have been in dispute with independent governments of Third World countries, they have often been able to persuade aid agencies to cut back on aid flows. For instance, the attempts by Peru and Chile in the later 1960s to exert greater control over American transnationals were met by very large cuts in the aid channeled to them by the Inter-American Development Bank.

In the last decade, outright conflicts have in any case become far less common. The hallmark of the 1970s has been increased cooperation between Third World governments and TNCs. Joint ventures, in which an enterprise is owned jointly by a Third World state agency and a transnational corporation, are one important expression of this cooperation. Some joint ventures in the mining and oil sector were created through a process of partial nationalization. Typically, Third World governments bought 51 per cent of the shares of companies owned and operated by TNCs. For instance, the Zambian government bought 51 per cent of the shares in the companies operating the copper mines, previously wholly owned by two giant multinational mining corporations, Anglo-American and Amax.

A recent report on TNCs and world development argues that Third World governments increasingly possess the capacity to ensure a more equitable relation with the TNCs, and have achieved some success in persuading the corporations to accept more participation in decision making, and in minimizing the influence of the corporations on their domestic political process. But other researchers are less optimistic and claim that many developing countries have learned by painful experience that nationalization does not put an end to foreign control.

When the Zambian government nationalized the copper mines in 1969 it did not possess the knowledge necessary to run the mines. It, therefore, negotiated a management contract with the TNCs that had owned the mines – Anglo-American and Amax – whereby they continued to supply all managerial, financial, commercial, technical and other services needed to run the mines. In return, they got 0.75 per cent of the sales proceeds and 2 per cent of the consolidated profits (before income tax, but after mineral tax). They also continued to market the copper. The contract contained no legal obligations to train and employ Zambian personnel nor to use the profits to diversify into other industries.

This kind of arrangement, of 51 per cent ownership of shares and a management contract with the existing operators, is often unsatisfactory since it is expensive and does not give control over the most vital decisions. In the Zambian case, the government agreed to pay a total of US$296.5 million in foreign currency for the shares it nationalized. Payment was to be out of future profits on the government's 51 per cent share. This meant that the government's ability to pay depended on the mines being run at a profit, which in turn depended to a considerable extent upon management, which it did not control. After five years, the Zambian government concluded that it did not have sufficient say in the development of copper production and the management contract was terminated on 15 November 1974, at enormous cost, at a time when the industry was facing difficulties.

. . . .

State support for TNCs

Increasingly, independent Third World governments have been drawn into deploying their state power to aid, as well as to control, the business operations of TNCs. This is particularly the case in the manufacturing sector. A study of the role of TNCs in production of manufactures for export from the Third World suggested that there are two basic factors that influence the choice of which Third World country to invest in: political stability and labour docility (Nayyar, 1978). Many governments in Third World countries have shown themselves quite ready to oblige by restricting or banning the organization of trade unions and outlawing strikes, if necessary implementing these measures through the use of force. There is a good deal of evidence to support the view that many independent Third World governments have, to a great extent, taken over the role colonial governments used to play with respect to the labour force. The TNCs have no need to directly mobilize the power of their 'home' governments when they can rely on Third World government to smooth the path of their operations.

Mechanisms of surplus appropriation

The core of the dependency school's economic case against the operations of TNCs in the Third World is that they extract a surplus from the Third World and transfer it to the First World. The most obvious way in which TNCs withdraw surplus from the Third World is by remitting profits back to their head offices in the First World. For instance, it has been estimated that, between 1960 and 1968, US TNCs took, on average, 79 per cent of their declared net profits out of Latin America (Barnet and Muller, 1974, pp. 153–4).

Third World governments can retain some of the surplus by requiring payments of rent, royalties and taxes (as we saw in the case of Angola and Gulf Oil). Rent is mainly important for mineral and agricultural products where national resources are a key factor. So are royalty products by TNCs to Third World governments. Whereas rent is a payment for access to the natural resource, royalties are payments related to *utilization* of the resources. For instance, royalties paid by oil companies are typically a specified percentage of the price per barrel of oil produced, whereas rent has to be paid irrespective of the level of production. The term is also used for payments for the utilization of other resources. TNCs also get paid royalties *by* Third World governments for the use of their technology and trade marks. Taxes can be levied on company profits, or on the value of company sales or the value of company exports. The nominal tax rate of a certain percentage of profits, sales, or export value, is often not indicative of the amount of tax actually paid because there may also be all kinds of tax concessions, such as allowances for new investment, allowances for depletion of mining assets, accelerated depreciation and tax holidays.

In many cases, Third World governments have successfully raised the rate of royalty and tax payments though not always without a struggle, as is shown by the so-called 'Banana War' of 1974 between the Central American countries of Panama, Costa Rica, and Honduras and the TNCs engaged in banana production for export, including Del Monte, Castle and Cook, and United Brands. The three countries attempted to introduce a tax of US$1 per box on all banana exports (a box containing 40 pounds net weight of bananas is the physical unit in which most international trade in bananas is carried out). The TNCs resorted to a number of different tactics to try to get the tax removed. Initially, they argued on two basic grounds: firstly that the taxes were illegal because the companies were operating under long-term contracts with the governments concerned, in which there were specific clauses stating no further fiscal measures could be introduced within the time limit of these contracts; and, secondly, that the taxes were counter-productive because they would price bananas from Panama, Costa Rica and Honduras out of the market.

When these arguments failed to get the taxes removed, they resorted to other measures, such as cutting back on exports, or even stopping them altogether, and bribing high government officials. General Oswaldo Lopez Arellano, head of the Honduran military government, was forced to resign in 1975, when it became known that he had accepted a US$1.25 million bribe from United Brands in return for lowering the export tax from US$1 a box to 25 cents a box, with provision for a rise to 50 cents by 1979. The governments of Costa Rica and Panama also bowed before the pressure and reduced their taxes; in the case of Costa Rica to 25 cents a box in 1974, raised to 45 cents in 1975; and in the case of Panama to 35 cents a box in 1974, raised to 40 cents per box in 1976. Besides the immediate pressure from the companies, there was also the fear that perhaps there was something in their arguments that the tax would make banana production in their countries uncompetitive. To judge this, the governments needed information on the costs of production which only the companies had.

Lack of alternative sources of information on costs of production is often a critical factor in negotiations about the level of payments a TNC should make to a Third World government. Typically, what a government wants to do is get more of the golden eggs without killing off the goose. That is, it does not want to price the country's exports out of the market; nor render production so unprofitable that no company will undertake it; nor remove all incentive for new investment. But the government rarely possess the knowledge required to judge the rate of payment which will secure these goals. Generally, it has neither sufficient *technical* knowledge to judge accurately the physical potential of a mine or a plantation; nor sufficient *economic* knowledge to judge the assets' profitability.

Transfer pricing

The difficulty of making an assessment of the economic value of the operations of a TNC in a particular country is, to a large extent, the result of the fact that there is often no 'open market' price for the inputs used or the outputs produced. Frequently, the inputs are supplied by different branches of the same firm; and/or the outputs are bought by different branches of the same firm. The inputs and outputs are not bought and sold between independent sellers and buyers who transact at 'arms length'. Instead they are intra-firm transfers, and the prices at which they are transacted are known as 'transfer prices'. A large TNC has considerable discretion in deciding the level at which it sets its internal transfer prices, and can manipulate such prices in order to reduce the payments it makes to governments. For instance, it can transfer funds from a particular country by raising the prices it charges for inputs to its subsidiary in that country; or by lowering the price it pays for outputs from that subsidiary.

This ability to decide, within rather wide limits, the pattern of internal transfer prices, means that a company can afford to concede higher rates of royalty payment or profit tax. It can then minimize the impact of the higher rates by setting internally transferred output prices low and input prices high.

It is much easier to estimate the *potential* for manipulation of transfer pricing than it is to estimate the actual extent of such manipulation. This is because it is often difficult to establish a market or arms-length price of a particular product with which to compare the transfer price: the product in question may be specific to one particular corporation, with nothing else exactly like it produced by other sources.

To gain some idea of the potential for transfer pricing, we need to look at the proportion of international trade that is intra-firm trade: that is, the proportion of imports or exports that are transferred internationally *between* countries but *within* one TNC. Unfortunately, in most countries, data on intra-firm trade is only sporadically collected. Only the USA regularly collects and reports two different sets of data indicating the extent and composition of US intra-firm imports. These are data on imports from majority-owned affiliates of US companies, and data on imports from 'related parties', i.e. from firms in which 5 per cent or more of the shares are owned by the importing firm. In 1975, nearly one-third of all US imports originated from majority-owned affiliates. The proportion was somewhat higher for US imports from developing countries (35 per cent from majority-owned affiliates) than from developed countries (28 per cent from majority-owned affiliates). However, these figures do include oil imports. If oil is excluded, then the figure for imports from developing countries drops to 11 per cent.

The data for related-party imports shows that, in 1977, nearly 54 per cent of US imports from developed countries came from related parties, and about 43 per cent of US imports from developing countries. If oil is excluded, the figure for imports from related parties in developing countries falls to 28 per cent. If we consider only international transactions with majority-owned affiliates, then we leave out all those joint ventures where TNCs have a minority of the shares. On the other hand, some 'related-party' transactions are not between a US transnational and its overseas affiliates, so the proportion of US non-oil imports from developing countries which are 'intra-firm' is somewhere between 11 per cent and 28 per cent (Helleiner, 1981).

To what extent is the potential for manipulation of transfer prices actually used to the detriment of Third World countries? It *is* difficult to get evidence on transfer prices and their appropriateness, but the verdict of Indian economist Sanjaya Lall is that the few investigations that have taken place have *all* shown that transfer prices are used against Third World countries (Lall, 1981, p. 62). One of the most thorough studies is that carried out by Constantine Vaitsos on the impact of transfer pricing in Colombia, Peru, Chile, Bolivia and Ecuador. He found that manipulation of transfer pricing, especially the over-pricing of imported inputs, is used extensively by TNCs as a way of repatriating profits from those countries. Here is an example: in the pharmaceutical industry in Colombia, Vaitsos found that reported profits accounted for 3.4 per cent of effective returns; royalties for 14.0 per cent; and over-pricing for 82.6 per cent (Vaitsos, 1973, p. 319). He estimates that the loss of surplus to Colombia in 1968 could have been as much as US$20 million (Vaitsos, 1974).

. . . .

Payments to TNCs

There are some other, less obvious, ways in which a transfer of surplus can take place. Payments to TNCs for patents, product and technology licences, brand names, trade marks, and management, marketing and technical services, have all been growing rapidly. A United Nations estimate puts the growth rate at 20 per cent a year (Colman and Nixson, 1978, p. 229). One of the reasons for this rapid growth is the trend to nationalization in the mining and agricultural sectors. The other is industrialization in the Third World orientated towards producing the kind of products, with the kind of technology, that TNCs have developed. But why should payments for patents, licences, trade marks, brand names and management marketing and technical services, be regarded as transfers of surplus rather than as payments for a necessary input? This depends on the indeterminacy of the price of such items; the fact that such prices are not determined on an open market with independent and competing buyers and sellers. Particularly important is the fact that, in many cases, the buyer is not in a position to know enough about what is being purchased, nor how necessary it really is for the production and marketing process. As Vaitsos has put it:

> a prospective buyer needs information about the properties of the item he intends to purchase so as to be able to make appropriate decisions. Yet in the case of technology, what is needed is information about information which could effectively be one and the same thing. Thus, the prospective buyer is confronted with a structural weakness intrinsic to his position as a purchaser, with resulting imperfections in the corresponding market operations. (Vaitsos, 1975, p. 190)

The same argument applies to management and marketing contracts. From the point of view of the TNC, an important aspect of income from these new sources is the way they diminish risk. There are many circumstances that can have an adverse effect on profits – bad weather, low demand, geological difficulties, labour unrest – so income in the form of profit is subject to considerable risk. But payments for licences, management contracts, etc. have to be made whatever the level of profits; and know-how is not subject to nationalization in the same way as physical assets. So there is considerable evidence that TNCs withdraw a surplus from the Third World, and are capable of finding ways of circumventing some of the efforts of Third World governments to retain more surplus within their countries. But does that mean that the activities of TNCs in Third World countries lead to a polarization with underdevelopment in the Third World (or periphery) and development in the First World (or centre)?

TNCs and development

Many economists who accept that there are costs to the operations of TNCs argue that there are also benefits; they claim that TNCs make contributions to the creation of a surplus, as well as withdrawing a good part of it, once it has been created; and that in the process of surplus creation there are 'spin-offs' that aid development, such as wages and salaries paid to Third World workers, dividends paid to Third World shareholders, foreign exchange from exports, physical assets – mines, factories, plantations – and some transfer of technology. They make the same kind of point that Marx made about British investment in colonial India that foreign investment does contribute to the development of the productive forces of Third World countries.

How valid is this argument? What kind of contributions might TNCs make? One obvious suggestion is that they are a source of finance, that they bring finance into the Third World when they set up their operations there, helping to close the gap between finance required for investment and the savings available within the Third World. However, TNCs do not necessarily *transfer* large quantities of capital *to* the Third World, even though they may *own* large quantities of capital *in* the Third World. The activities of TNCs within the Third

World may, to a considerable extent, be financed from local sources within the Third World. TNCs borrow from local banks, sell shares to local shareholders and take over already established local firms. They do reinvest some of their profit within the Third World, but these reinvested earnings have often been derived from investment financed by local sources. One United Nations study estimates that, in Latin America, during the period 1957–65, US TNCs financed 83 per cent of their total investment from local sources, either from reinvested earnings (59 per cent) or local borrowing (24 per cent). In the case of the manufacturing sector, a larger proportion of finance came from local borrowing (44 per cent) than from reinvested earnings (38 per cent) (Colman and Nixson, 1978, p. 224). In 1973, the US Senate Committee on Finance reported that less than 15 per cent of the total financial needs of US-based manufacturing subsidiaries abroad originated from US sources (United States Senate, Committee on Finance, 1973, p. 38).

The most important contribution that TNCs make is the provision of know-how, whether in the form of 'hardware' such as machines and components, or 'software', such as managerial and marketing skills. Not all this know-how is necessary and appropriate for the people of the Third World. We may doubt whether the knowledge of transnational food corporations about how to produce and market dried milk for bottle feeding babies helps mothers and babies in the Third World. Lack of sanitation and clean water, lack of refrigeration, illiteracy, and low incomes, result in over-diluted and contaminated bottle feeds, with terrible effects on the health and survival of babies receiving them.

However, we must not forget that elite groups within the Third World play an important role in selecting what kind of know-how TNCs contribute to their countries, particularly in their choice of the type of goods they wish to consume. It has been argued that TNCs cannot be directly blamed for the lack of development (or the direction that development is taking) within the Third World. Their prime objective is global profit maximization and their actions are aimed at achieving this objective, not developing the host Third World country. If the technology and the products that they introduce are 'inappropriate', if their actions exacerbate regional and social inequalities, if they weaken the balance of payments position, in the last resort, it is suggested, it is up to Third World governments to pursue policies that will eliminate the sources of these problems (Colman and Nixson, 1978).

Some Third World governments have done just that, carefully regulating the role that TNCs are allowed to play in their development. . . . So, there seem to be some reasons to doubt a simple argument that the activities of TNCs in the Third World are a barrier to development. Frequently, the relationship between the corporation and the country is very unequal; frequently the TNCs play a dominant role. But the quality of the relationship depends a great deal on the policies of Third World governments and the skills at their disposal. In the case of many of the poorer countries, it has been suggested that their problem is not too much, but too little, investment from TNCs. For, in allocating their resources, TNCs certainly obey the biblical maxim: to them that hath shall be given. If we want to understand the barriers to development, we should look not just at transfers of surplus out of Third World countries, but also at the factors that prevent the investment in the first place.

There is certainly a correlation between high rates of growth and a high share of TNC investment in the Third World. Most of the investment in recent years has gone to the newly industrializing countries (or NICs) of Latin America and Asia. The NICs include large countries – Brazil, Mexico and India – city states like Hong Kong and Singapore, and medium-size countries like South Korea and Taiwan. With the exception of India, their industrialization over the last fifteen years has been export-oriented. They have rapidly expanded their exports of manufactures to the First World, leading to talk of a new international division of labour replacing the old international division of labour, in which Third World countries exported mining and agricultural products.

Most of the manufactures exported to the First World are labour intensive products: textiles and garments, footwear, toys, televisions, radios and watches. The labour intensive stages of production of high technology goods like semi-conductors (more familiarly known as micro-chips) are also carried out in many NICs. Some of the NICs are now competitive in the export of cars, and even of heavy industrial goods such as iron and steel and ships. South Korea's shipbuilding industry ranks in the world second only to that of Japan.

The role of the TNCs in the newly industrializing countries has been variable, both across industries and across countries. The TNCs have been very important in high technology industries like electronics. But in more traditional industries they have been far less important. Locally owned firms, especially in Taiwan, South Korea and Hong Kong, have produced most of the exports of clothing and textiles, though often operating under sub-contract to First World firms. In general, the transnationals have played a more dominant role in Latin America than in East Asia (Lall, 1981, p. 218).

Selected references

Barnet, R. and Muller, R. 1974: *Global reach: The power of the multinational corporation*. New York: Simon and Schuster.

Brandt, W. 1980: *North–South: A programme for survival*. London: Pan.

Colman, D. and Nixson, F. 1978: *Economics of change in less developed countries*. Oxford: Phillip Allan.

Helleiner, G. K. 1981: *Intra-firm trade and the developing countries*. London: Macmillan.

Lall, S. 1981: *Developing countries in the international economy*. London: Macmillan.

Nayyar, D. 1978: Transnational corporations and manufactured exports from poor countries. *Economic Journal* 88, 349.

United Nations Centre for Transnational Corporations 1973: *Multinational corporations in world development*. New York: UN.

United Nations Centre for Transnational Corporations 1978: *Transnational corporations in world development: A re-examination*. New York: UN.

United States Senate, Committee on Finance 1973: *Implications of multinational firms for world trade and investment and for US trade and labor*. Washington, DC.
Vaitsos, C. 1973: Bargaining and the distribution of returns in the purchase of technology by developing countries. In Bernstein, H. (ed.), *Underdevelopment and development*. Harmondsworth: Penguin, 315–22.
Vaitsos, C. 1974: *Inter-country income distribution and transnational enterprises*. Oxford: OUP.
Vaitsos, C. 1975: The process of commercialization of technology in the Andean Pact. In Radice, H. (ed.), *International firms and modern imperialism*. Harmondsworth: Penguin.
World Bank 1985: *World development report, 1985*. Oxford: OUP.

20 Robert Wade, 'Lessons from East Asia'

Excerpts from: R. Wade, *Governing the Market: Economic Theory and the Role of Government in East Asian Industrialization*, chapter 11. Princeton: Princeton UP (1990)

The debate about the role of the state in economic development demonstrates the power of infinite repetition as a weapon of modern scholarship. The issue is normally posed in terms of the 'amount' of state intervention or the 'size' of government. The neoclassical side says that more successful cases show relatively little intervention in the market, while less successful cases show a lot (Brazil and Mexico compared to East Asia; or sub-Saharan Africa at the bottom). It uses this evidence to urge governments to shrink the size of the state and remove many of its interventions in the market. The political economy side says that the neoclassicals have their facts wrong: the most successful cases show 'heavy' or 'active' intervention. It concludes from this evidence that governments *can*, in some circumstances, guide the market to produce better industrial performance than a free market, even in the absence of neoclassical-type market failure. But neither side has been noticeably enthusiastic to specify just what evidence would be consistent with its position and what would not. Both have exercised a selective inattention to data that would upset their way of looking at things. So the debate about the role of the state is less a debate than a case of paradigms ('parrot-times') talking past each other.

I have shown for Taiwan – and suggested for Korea and Japan – that ample evidence is available in support of both the free market/simulated free market (FM/SM) and the governed market (GM) theories. This poses an identification problem. How important are those facts which are consistent with the FM/SM interpretation, and how important are those which are consistent with the GM approach? My argument is simply that the GM facts are too important to ignore in an explanation of Taiwan's (and Korea's and Japan's) superior performance. This challenges economics to deploy – or invent – theories which will make the non-neoclassical facts of East Asia analytically tractable. But does it also support the prescription that other middle-income countries should try to govern the market in a broadly similar way (with appropriate adjustment for national circumstances)?

That depends on the answers to three questions. First, are the conditions of the international economy as favorable to a rapid, forced, and export-dependent industrialization today as they were for Taiwan and Korea? Second, is there a general economic rationale for GM policies? Third, can governments significantly improve their administrative and political capacity to govern the market?

Conditions of the world economy

Both the FM/SM and the GM interpretations of East Asian success emphasize the importance of domestic factors, in particular 'right' policies – though they differ on what constitutes 'right.' Implicitly, they assume that the trajectories of states are parallel and theoretically independent, each separately subject to the same economic tendencies. Development is a kind of marathon race, in which each runner's position is a function of his internal resources and in which all runners could in principle cross the finish line at the same time.

Yet it is clear from what has been said that a good part of the reason for East Asian success has to do with international factors. These created opportunities for relatively low-cost industrial production sites to be integrated into the world economy. In the 1960s several conditions came together to produce at one and the same time relatively favorable access to industrial country markets, dramatically increased access to international finance, and increasing relocation of production by multinational corporations to low-wage sites. These conditions created opportunities, but did not determine which countries would seize them. Which countries seized them can be explained partly in terms of domestic factors – including the existence of an industrial base resulting from prior import-substitution and the existence of a hard state pursuing GM policies. Location and geopolitical importance are also relevant. The United States 'invited' Taiwan, Korea, and Japan to become economically strong because of their location on the West's defense perimeter (which made them more strategic than, say, the Philippines, Indonesia, or Brazil). Japan, the most dynamic economy of the postwar era, had special ties with Taiwan and Korea derived from proximity and colonial history. Hence, part of the success of GM policies in East Asia is due to the favorable historical and international conditions in which they were implemented. To the extent that these factors are different at other times and places, this throws doubt on the possibilities for other countries at other times to emulate East Asian success.

A central difference between the world economy of today and that of the

1960s, when Taiwan, Korea, and Japan made big inroads into Western markets, is that it is no longer in an expansionary phase. There has also been a dramatic fall in the demand for unskilled labor and raw materials per unit of industrial production. Consequently, developing countries in the 1980s face an external environment more hostile than in any previous decade since the Second World War. They are doubly squeezed on trade and on capital. Growth in world output slowed from 4.1 percent in 1970–79 to 2.6 percent in 1980–87. Terms of trade for nonfuel exports from developing countries deteriorated from a 1.1 percent per year decline in 1970–79 to a 1.7 percent per year decline in 1980–87. Protection in developed country markets has increased since the early 1970s, accelerating in the early 1980s to the point where by 1986 21 percent of manufactured goods imported into the United States and Europe were restricted by quantitative barriers. This protection is being applied with special discrimination against developing countries. Eighteen percent of manufactured imports from developed countries were covered by quantitative restrictions, and 31 percent of manufactured imports from developing countries. Yet manufactured imports from developing countries account for a mere 1.5 percent of manufactured consumption in developed countries. Meanwhile the microelectronics revolution has reduced the advantage of cheap labor sites, slowing the inflow of foreign direct investment to developing countries at large. The debt service burden for indebted developing countries increased sharply in the 1980s, while voluntary private lending to developing countries almost stopped (US$3.5 billion in 1987 compared to $73.4 billion in 1980).

On top of these trends has come a sharp increase in the volatility of the international economy, and therefore much more uncertainty facing developing country governments and producers. With the internationalization and deregulation of financial markets, financial capital is ricocheting around the world in amounts thirty to forty times bigger than trade flows. The relationships between exchange rates and trade, interest rates and investment, and fiscal and monetary policies have become unhinged (Drucker, 1986). Governments' ability to control as fundamental a parameter of economic activity as money supply is diminished, and long-term investment is depressed. Dealing with currency fluctuations 'is like changing the handicap in golf on every hole,' protested the president of Sony recently. 'Wouldn't you lose interest in playing golf eventually? If money scale expands or shrinks every day in different currencies, how can we make up our minds to invest?' (*Toronto Globe and Mail*, 1 June 1987). If Sony finds long-range investment difficult in current conditions, think of the predicament of would-be exporters and investors in developing countries with free trade and capital movements. They are forced to adjust and readjust to signals from the international economy which are essentially short-term. These adjustments to price signals that turn out to be misleading guides to economic fundamentals may cause high costs. They are 'distortionary' in a sense different from but as important as the conventional meaning in economics, of price misalignments which arise when an economy has not adjusted sufficiently to international price signals (Bienefeld, 1988).

The implications are ominous for those developing countries that would seek to 'follow the NICS.' If many countries are to succeed in increasing their exports of light manufactures, world trade would have to expand fast; but all the signs are it will not. The new protectionism is directed especially at the light manufactured goods which the next tier countries are urged to make their leading export sectors. Moreover, the East Asian four are 'stretching' their industrial structures as they expand into more advanced sectors, using technology to remain competitive in light manufactures and thereby only slowly vacating these sectors for others to enter. It will therefore be more difficult for the others (such as Mexico, Brazil, Thailand, Poland, and Hungary) to use industrial and trade policies 'successfully.'

Underlying these ominous trends are shifts in technology which imply potentially far-reaching effects on the competitive position of nations. It is difficult to forecast these effects, for the implications appear to be mixed. The increase in the capabilities of machines to perform the tasks of unskilled labor may facilitate a shift in the location of production back to the developed countries; but it may also enable some developing countries to become more competitive through better quality control and cheaper engineers. The reduced importance of unprocessed inputs worsens the export prospects of raw material producers; but the new technologies enable them to process the raw materials in-country, and free some regions from some of nature's constraints (biotechnologies can make deserts bloom).

Faced with these new dangers and opportunities, what broad lines of economic policy should developing countries follow? We can be fairly sure that policies to impart an East Asian kind of directional thrust will have a smaller effect than they did in East Asia, if for no other reason than the less favorable conditions of the international economy. On the other hand, this does not mean that FM/SM policies are the better alternative.

We have seen that the confidence with which the neoclassical school prescribes liberalization and privatization cannot be grounded in theory, for the theory which shows how liberalization and privatization generate growth is scarcely developed. We have also seen that it cannot be grounded in the experience of the East Asian NICS, for with the partial exception of Hong Kong they have pursued a mix of policies – many of which are inconsistent with neoclassical prescriptions. Hong Kong, though historically much wealthier than Taiwan and Korea, has not been doing as well in terms of income growth and industrial transformation. But what about the other 98 percent of countries? Does the experience of a broad cross-section of countries provide solid grounds for the neoclassical confidence? My review of the cross-sectional evidence suggests not.

Given this, we might usefully deploy a more inductive approach to policy and policy-making. We can ask what policies the most successful countries actually adopted, and then construct a rationale for why those policies may have helped their growth. Due recognition has to be given to the Darwinian or Malinowskian fallacy in this exercise – the assumption that because something exists it must be vital to the survival of the organism or society in which it exists. Translated into East Asian terms, this leads one to argue that because these most successful countries used high protection, tightly controlled financial systems, and the like, such measures must have been vital to their success. But what could be called the Ptolemaic fallacy is more prevalent and more inhibiting of learning: the assumption that only those features of economic policy consistent with neoclassical prescriptions could have contributed to superior economic performance, so that everything else can be safely ignored. It seems useful to err for a change on the side of the former.

A distinction must be made between what is consistent with neoclassical theory and what is consistent with neoclassical prescription. There is room within the confines of neoclassical theory for practically any mix of markets and intervention. Most neoclassical economists argue the costs of government regulation and industrial targeting and the virtues of very wide choice by individual market agents on (often implicit) empirical grounds. The empirical significance of market failure has been exaggerated, they say, and government efforts to repair such failure are likely to make matters worse because government failure is empirically more acute than market failure. In what follows I use ideas that are familiar in neoclassical analysis, as well as some that are unfamiliar, to reach different conclusions about the possible economic benefits of governed market policies. These conclusions have the merit of being consistent with what the governments of very successful economies – the East Asian ones – actually did. I state the argument in the form of six prescriptions for micro- and mesoeconomic interventions.

The argument is most relevant to the circumstances of newly industrializing or newly industrialized countries with per capita incomes in the middle-income range. For the most part I duck the question of its relevance to the industrialized market economies. I am unclear about how a world economy would work in which the leading economies, especially the United States, adopt the kinds of actions endorsed here. Past experience suggests it could be benign: the laissez-faire world of the 1920s had disastrous economic consequences, while the postwar era of stable but negotiable exchange rates and national controls over capital movements generated steady expansion. Keynes used his understanding of the prewar boom and bust to argue passionately in favor of import controls and central bank control of international capital movements as instruments of postwar economic management. But in any case, the late comer industrializers, which constitute only a tiny part of world income and world markets, can use different principles from those appropriate to the older industrialized market economies. They can, as have many latecomers before them, free ride on the (more or less) liberal norms embraced by the latter out of self-interest. After all, the per capita income of the fifty-two middle-income countries is still, after four decades of self-conscious development, only 10 percent of the per capita income of the 'industrialized market economies' (2 percent for the low-income countries; World Bank, 1988, table 1).

My argument is also relevant mainly to noncrisis conditions, when a longer-term view can be taken. Sadly, that currently excludes many countries of sub-Saharan Africa, the Caribbean, and Latin America. Fully two-thirds of middle-income countries had negative growth of gross domestic investment in 1980–86, compared with only 2 percent in 1965–80; one-third had negative growth of GDP (World Bank, 1988, table 4). Many have not been investing enough to maintain essential infrastructure. Many are unable to obtain even basic economic statistics, because so much activity has moved into grey and black markets.

I also assume benign political leaders, whose concerns go beyond using state power to support the affluence of a small group. Some rulers, it is true, are predatory, in the sense that their efforts to maximize the resource flow under their control erode the ability of the resource base to deliver future flows. In these cases enhancing the power and autonomy of the state could be disastrous. But states vary in terms of the benignness or maliciousness of their leaders, and the more any particular case is toward the benign end of the spectrum the better the argument applies.

Prescription 1: Use national policies to promote industrial investment within the national boundaries, and to channel more of this investment into industries whose growth is important for the economy's future growth. We must note at the outset that the objective of such policies is not efficient allocation of resources in a Pareto-optimum sense, but growth and innovation. This means that the theorems of welfare economics about the conditions of market failure are largely irrelevant for development purposes. They judge failure in terms of the allocation of resources to the most efficient uses, rather than in terms of the generation of new resources. Theories about market failure in a growth context are not well developed. Here I shall do no more than sketch out a plausible rationale for national industrial policies.

The first step is to consider why the government should take steps to intensify the level of investment and reinvestment within the national boundaries. The point of interest is not the causal effect of investment on growth, for that is well established theoretically and empirically (Romer, 1987). So also is the connection between high investment and growth, on the one hand, and high labor demand; and between high labor demand and high wages. The point of interest, rather, is the need for political power to focus the investment process on the national territory, which means channeling the options of both domestic capitalists and foreign capitalists by means of import restrictions, domestic content requirements, foreign exchange controls, conditions on the admission of foreign investment, export incentives, technology incentives, and the like. The reason is that as capital becomes internationally mobile, its owners and managers have less interest in making long-term investments in any specific national economy, and hence less interest in the overall development of any specific economy – including their home base. As wages rise, they may be inclined to relocate their assets abroad, or divert them into short-term speculative uses at home, or use their influence over state power to keep labor costs lower than otherwise. From the perspective of a national interest, however, they should be encouraged or cajoled to reinvest at home, and specifically to invest in technological improvements as a way of remaining internationally competitive despite higher wages. For the domestic workforce is not internationally mobile, and its rising real wages are a primary indicator of developmental success. This argument has to be qualified in several ways – by the desirability of some outward foreign investment in terms of a national interest test, and by the problems of overcoming purchasing power constraints on the investment cycle if large export markets are not available. But the qualifications do not change the basic point. Empirically, the work of Alexander Gerschenkron (1962) and Dieter Senghaas (1985), among others, supports the proposition that the more successful European latecomers used a political mechanism to channel the competitive process in the direction of higher-wage and higher-technology activities. (The United States is an exception, where the same result occurred because of the scarcity of labor in relation to land and capital. Hong Kong is also a partial exception. Wages have risen and the benefits of growth have been widely diffused even with capital operating wholly in terms of an international perspective partly because of its very small size and partly because of its role as a regional service center. And to repeat, in the past decade or so it has been less successful in transforming its industrial base than Taiwan and Korea.)

The next step is to consider why government efforts to concentrate investment in selected industries may help overall growth and productivity. One reason involves economies of scale and learning. Whereas neoclassical analysis normally assumes rising cost curves, in many manufacturing processes a doubling of production volume per unit of time gives rise to a substantial fall in unit costs, commonly on the order of 20 percent. But the size of plant or firm required to achieve these economies of scale is typically large in relation to the existing assets of firms in developing countries. The risks confronting potential investors are therefore high, and the investment process will be slowed if the risks are not partially lifted. If domestic producers are given assistance to enable them to compete against foreign suppliers in the domestic market despite higher costs, they may be able to expand their production volume to the point where, thanks to economies of scale and the transactions cost of imports, they can compete without further assistance.

. . . .

A third justification, in addition to scale and learning economies and capital market imperfections, comes from that most elastic of concepts, 'externalities' or 'spillovers.' In the general sense, external costs or benefits are those which are created by a firm or other economic agent but which do not bear on or accrue to it. Simultaneous externalities occur where a firm's potential gains from an investment are contingent upon complementary decisions by other firms. Even if all the parties know they would gain by coordinating their investments to capture the externalities, they may face inherent contradictions of interest, as in a Prisoner's Dilemma game. Hence market prices may not adequately signal the interdependence that exists among these investment decisions, and uncoordinated firms may invest at suboptimal levels from a national perspective. A big push, involving simultaneous expansion of several industries, can insure the profitability of each investment, even though each on its own would be unprofitable. One important reason is that such simultaneous expansion helps to overcome the constraint of a small domestic market, when entry and participation in world trade entails significant costs.

There is also a second kind of externality, sequential rather than simultaneous. Sequential externalities occur where a large upstream plant would, if built, induce the entry of downstream firms to make use of new profit opportunities created by the upstream firm but not appropriable by it. The upstream plant brings greater social benefit, in the form of induced downstream growth, than is reflected in its private profit.

. . . .

Another justification for governing the market has to do with the adverse effects of market instabilities on long-term investment. Any moderately complex economic system encounters a source of instability arising from the uncertainty inherent in the attempt to match present supply decisions with future demand decisions. One would expect that if prices and quantities are left wholly to the instabilities of the market, investment in industries or technologies which require a large commitment of time or capital may not be made, and a higher than desirable proportion of the economy's investment will go into quick return projects. Individual firms on their own may be more inclined to stick within a narrow range of familiar product lines than branch out into new industries and products. It may well be that, within limits, price *instability* has a more adverse effect on growth than price *distortions* as conventionally defined. A context of deliberately created stability, achieved by risk-spreading mechanisms such as protection or subsidies, can facilitate industrial deepening, export expansion, and political compromises to share adjustment costs.

Again, a role for industrial targeting can be warranted by the fact of differences between industries in prospects for long-term growth in output, profits, and wages. Unassisted entrepreneurs may not have either the foresight or the access to capital to follow long-term potential. Their decisions may lock the country into specialization in industries with inferior prospects (an issue beyond the scope of comparative advantage theory). Given a world of technical change, falling cost curves, and differential rates of growth across industries, it can be rational for a government to select from within the plausible industries those which have high growth potential and to use the powers of government to supplement those of the market in marshalling resources for entry and successful participation. This means diverting resources from currently profitable activities into ones that might be fast-growing and/or profitable in the future – which is risky. But any successful large company follows a strategy of diversifying from currently profitable activities into new ones, on the assumption that the future will probably be different from the present. Governments at the national level can aim to carry through a parallel strategy of diversification. The scarcer the supply of capital and the higher the entry barriers of the new industries, the stronger the case for selective assistance.

But can comparative advantage really be modified, made, or achieved in this way? Traditional theory takes comparative advantage as exogenous, largely determined by 'factor endowments.' In a gross way these considerations are still relevant: Burundi should not go in for computer production just yet. But as Bela Gold says,

> Virtually all empirical findings of comparative advantage represent no more than ex post facto rationalizations of past trade patterns, often reflecting *market interventions* rather than substantial differentials in efficiency and costs. Moreover, even the demonstrable comparative advantages prevailing in a given period have frequently been undermined and even reversed thereafter through determined efforts to advance technologies, shift input requirements, alter transport costs, and develop new markets. . . . The very identification of current comparative disadvantages often represents the first step in developing means of overcoming them.
>
> (1979, pp. 311–12, emphasis added)

William Cline reaches a similar conclusion: 'Increasingly, trade in manufactures appears to reflect an exchange of goods in which one nation could be just as likely as another . . . to develop comparative advantage, and the actual outcome is in a meaningful sense arbitrary' (1982, pp. 39–40). In place of 'arbitrary' I would say 'subject to strategies of firms and governments.' Talk of 'revealed comparative advantage' (measured by the relative preponderance in a country's exports of product × compared to its preponderance in the trade of the world as a whole) is hence misleading, for the export pattern may reveal government assistance as much as factor endowments. And factor endowments, it should be remembered, can themselves be arranged on a spectrum from unalterable to alterable, with sunshine at one end and knowledgeable brains at the other. The classic case of Portuguese wine and British sheep reflects unalterable natural endowments; the modern case of British whiskey and Japanese electronics reflects human capital build-up, long-term horizons, and other acquired advantages. Government assistance can create new advantages of the acquired kind, some of which are industry-specific.

The popular belief that governments cannot 'make winners' rests on remarkably little empirical research into the record of different governments in selective industrial promotion. Many governments, especially in small countries, routinely target industrial assistance at specific industries and even at specific firms, particularly where economies of scale call for a minimum level of subsidy per firm. Yet we do not have systematic data on the performance record of different governments which would allow us to distinguish those with one failure out of four from those with seven failures out of eight. (No failure is itself failure, because it means that the targeters are not taking enough risk.) Research on this question has to balance the record of government failure against the record of failure by private business; and examine, too, what happens to economies where few transformation projects are attempted because the government declines to take an initiative and private business declines to take the risk.

In short, several considerations – economies of scale and learning, capital market imperfections, externalities, market instabilities, and differential growth potential – give grounds for state assistance to industry and to some industries more than others.

. . . .

Prescription 2: Use protection to help create an internationally competitive set of industries. Two of the things which economists disagree least about are that protection, whether for restraining the demand for imports or for promoting domestic industries, is always second-best, and that quantitative restrictions (QRs) are always inferior to tariffs. When unrestrained demand for imports leads to balance-of-payment difficulties, the solution is devaluation plus restrictive expenditure (fiscal and monetary) policy. If for some reason it is deemed necessary to promote specific industries, credit subsidies should be used.

These prescriptions are backed by an impressive body of theoretical reasoning. But once one moves beyond a concern for what is logically consistent with the theoretical system of neoclassical economics, they are not compelling guides to decision-making in the real world. As regards devaluation, the first problem is that experts often disagree by large margins as to what the 'desirable' exchange rate should be, not only in developing countries but in industrialized countries as well (notably the United States). Second, even where experts agree that the exchange rate is substantially overvalued, 'markets' often seem to be poor at correcting the imbalances. Third, the policy instrument is the nominal exchange rate, but there may be no close connection between changes in the nominal rate and changes in the real rate except in the very short run; and it is the real exchange rate which counts for resource allocation. Fourth, the neoclassical argument recognizes no limits on how far the exchange rate can be made to fall. But a fall in the real exchange rate means a fall in the price of noninternationally tradable goods and services relative to the price of tradables. The most important nontradable is labor, so a fall in the real exchange rate tends in practice to cause a fall in the real wage. Workers may revolt. More generally, inflexibilities of import-dependent production processes and consumption patterns may mean that the needed fall in the exchange rate is not possible without disruption of production, inflation, social unrest, and political conflict, fear of which may induce a well-meaning government to find other methods of maintaining external balance.

The argument to replace protection with credit subsidies as a means of assisting particular industries is also not compelling. First, there can be no presumption that the subsidies needed for infant industries to compete equally against foreign suppliers would match the finance available. Unless a close connection is assumed between the revenue-raising capacity of government and the amount of subsidies needed, the subsidies may exceed the capacity. Second, the advantages of subsidies cannot be presumed to outweigh the 'distortionary' effects of raising revenue through the existing tax system. Third, protection through tariffs raises revenue in an administratively simple way, compared to the difficulties of raising revenue through direct taxes; and is probably no more difficult to administer effectively than a subsidy program. Fourth, subsidies are generally a more visible means of transferring resources and may therefore generate more political conflict than protection, which transfers resources more invisibly. (Whether this is desirable depends on whether the pattern of protection makes national sense.) Finally, insofar as changes in subsidies are more contested politically than changes in protection, subsidies are unlikely to be changed enough to buffer short-term external fluctations.

There are indeed many cases where protection has not had any noticeable innovation- or investment-enhancing effect (e.g., India). This reflects the failure to integrate protection with a wider industrial policy, or link it to export performance, or make the quid pro quo conditions credible, or to maintain macroeconomic stability. If protected producers know that in the foreseeable future protection will be much reduced or that government will pressure them to enter export markets, then protection may give them breathing space in which to undertake the necessary investment and innovation. They can use higher than normal profits in the domestic market to subsidize their entry into export markets, practicing discriminatory pricing. The same effect may be induced by awarding import licenses for targeted products only on evidence that the product could not be obtained from domestic producers within some reasonable margin of the import price. Such an 'approval' mechanism or 'law of similars' at least forces would-be importers to obtain full information about domestic supply capability. It also helps to stabilize demand for domestic producers of import substitutes, thereby lowering their risk and encouraging them to invest enough for economies of scale. But at the same time, the price criterion means that international competitive pressures are brought to bear on domestic producers, though in a modulated way.

There is, of course, a tension between stimulating demand for nationally made products by protection (or domestic content requirements, or government procurement) and stimulating the international competitiveness of users of those nationally made products. Supply-side measures of assistance to the domestic producers can help to reduce the conflict. But in any case, it is important that exporters be exempt from most import restrictions, the exemption being greater the bigger the price and quality differential between imports and domestic substitutes. The government can, however, use its import-restricting ability to encourage users of imported inputs to negotiate with local suppliers for upgraded production or lower prices in return for guaranteed sales. Repeated across many products, this mechanism can nudge the production structure of the country upwards.

Notice that the mechanism uses QRs rather than tariffs. QRs (and domestic content requirements) have merit when the acquisition of technological capacity and subsequent adaptive innovation depend on extensive interaction between users and suppliers. However, QRs have the costly consequence of amplifying the volatility of price signals, because with changes in domestic prices the tariff-equivalent of any given QR also changes. But where macroeconomic stability prevails, as in East Asia, this familiar cost of QRs is much less significant. The East Asian experience supports the argument that QRs have lower costs in stable than in unstable macroeconomic conditions.

The desirable degree of import liberalization is much affected by country size. For most small countries – most of the time and in most industries – a relatively liberal trade regime is a necessity because of the lack of domestic economies of scale. Bigger countries have a wider latitude of choice. In general, the wider the latitude of choice, the more the overall degree of trade freedom should emerge as the result of calculations of the appropriateness of lowering protection to particular industries, bearing in mind that domestic competition can substitute for foreign competition, as in Japan, and that domestic competition, even between oligopolists, can be stimulated by government policies. Taiwan and Korea show how liberal trade policies in some industries can be combined with import substitution policies for other industries, resulting in different incentives to different industries. They also show how a rapidly industrializing country can soften pressures from its trading partners to open its markets or face retaliation, by a judicious combination of camouflage, statements of intent, and real liberalization.

Some developing countries, particularly in sub-Saharan Africa, are unable to earn enough foreign exchange to cover import demands at any politically viable exchange rate, because of the limited supply of internationally saleable products. Here it makes no sense to talk of protection only as a temporary measure to assist the emergence of infants able within five to ten years to compete against international competition with no protection. Protection has to be seen as a part of longer-term measures to gain experience of industry and large-scale organization. In its absence resources may remain largely unemployed or confined to very small-scale production. The trick is to use such longer-term protection in a way which does not eliminate all competitive pressures.

. . . .

Prescription 3: If the wider strategy calls for heavy reliance on trade, give high priority to export promotion policies. East Asia's fast growth and equitable distribution was undoubtedly helped by the rapid growth of exports. Exports faced less of a demand constraint than output in general; they provided a channel for technical assistance from buyers; and they gave more scope for labor utilization than the manufacturing sector as a whole or the existing import-substituting industries in particular. However, export growth is not the only important reason for fast and equitable growth, and other countries with different natural resource endowments and larger economic size may be able to achieve 'good' growth and distribution with less reliance on exports. Indeed, they may not have much choice in the matter, because Western countries will probably intensify protection to avoid big (especially China-scale) increases in competing imports from developing countries.

Where, nevertheless, heavy reliance is to be placed on trade, the government must recognize that successful exporting of manufactured goods to richer countries is not just a matter of getting the exchange rate right and keeping labor cheap, even in the absence of protection. This is because many kinds of manufactured exports to richer countries are only saleable as complete packages meeting all buyer specifications, including packaging, labeling, colors, raw materials, finishes, and technical specifications. Costs rule out the option of importing an incomplete or defective package and correcting the defects in a subsequent stage of manufacturing. Thus, marketing, transmission of information, and quality control turn out to be key activities for export success. Buyers can supply some of these services; but especially because of the externalities the government also has an important role. The government can arrange for information about foreign markets and about domestic suppliers to be easily and freely available; it can directly help the promotion of some products (e.g., through trade fairs); and it can help to curb the tendency of firms without brandnames to compete by producing shoddy goods, spoiling the country's reputation for other producers. Very importantly, the government can also inspire producers to seek out export markets as a normal part of their operations.

All this holds even in the absence of protection. If the economy is protected, cheap labor and a proexport exchange rate are still less likely to be sufficient. Without quick and automatic access to imported inputs at world market prices, free of customs duty, quantitative restrictions, and indirect taxes, would-be exporters will be handicapped in world markets by being forced to pay more than competitors for the same inputs or by being forced to use inferior domestic substitutes. Since manufactured exports from developing countries are normally sold in intensely competitive markets, producers in a country without a scheme for duty drawback and relaxation of quantitative restrictions are unlikely to obtain big export orders. Buyers for industrialized countries will simply pass them by. However, even once export sales have near-free trade conditions producers of manufactured goods may still face net incentives to sell on the protected domestic market, and exports may still be uncompetitive because the costs of nontradable inputs (especially labor) are raised by demand for those same inputs from the protected and hence larger-than-otherwise domestically oriented industries. An export subsidy scheme may be needed to make export sales as attractive as domestic market-related sales.

Combining this discussion of export promotion with the preceding discussion of import protection, we see how misleading it is to present import substitution and export promotion as mutually exclusive strategies, as in Anne Krueger's claim that 'export promotion outperforms import substitution' (1981, p. 5). They are mutually exclusive only if defined to refer to the overall balance of incentives between domestic and foreign sale. But at the individual industry level, import-substituting incentives and export-promoting incentives can be complementary. On the one hand, development of the supply side through import substitution may be a prerequisite for the demand-side growth of exports. On the other hand, export growth can be helpful for the further development of industries that are nearing the limits of import substitution. Likewise, export promotion in one industry can complement import substitution in another by providing foreign exchange, for example. At any one time export promotion and import substitution should coexist, reflecting the different development stages of different industries.

We also see how misleading is the common assumption that policy-induced neutrality (as when export incentives 'counteract' the effects of import controls) is equivalent to free trade. It is not clear how the many kinds of incentives for export- and import-substituting industries can be commensurated (effective protection rates are hardly adequate). It is fairly clear that the structure of relative prices at the time when 'neutrality' is achieved reflects the prior rounds of intervention, and differs from that of an economy with untrammeled prices and exchanges throughout. Therefore we cannot presume that relative prices and resource allocations would be unchanged if the entire array of incentives and protection were eliminated at a stroke.

Prescription 4: Welcome multinational companies, but direct them toward exports. Multinationals are the primary source of knowledge about technology and production and an important source of knowledge about marketing. No country is going to get far in knowledge-intensive manufacturing and services without their help. Hence the government of a newly industrializing country should establish attractive policies for foreign capital, whether as subsidiaries, joint ventures, or licensors. However, foreign firms should be under pressure to direct their sales toward exports and their input purchases toward local suppliers. For if their products dominate the domestic market the developmental consequences of the protection system may well be worse than if domestically owned firms dominate the domestic market.

. . . .

Prescription 5: Promote a bank-based financial system under close government control. A closely regulated bank-based financial system has several advantages in industrializing country conditions. First, it permits higher investment than would be possible if investment depended on the growth of firms' own profits or on the inevitably slow development of securities markets. In a capital market-based system, the decentralized preferences of the public largely determine the allocation of potential savings into productive investment, financial speculation, or consumption. In a bank-based system, in which enterprises depend heavily on banks for finance and less on a broad public of shareholders, the long-term growth preferences of government officials and/or bank executives have more weight. Investment decisions are hence more insulated from the preferences of the public. Credit can be more cheaply provided for productive investment, in the context of a long-term approach to the economy's investment activity. In a capital market-based system, on the other hand, government attempts to stimulate investment by tax cuts and deregulation may have only a modest effect on investment, as in the economic reforms of President Reagan and Prime Minister Thatcher.

Second, a bank-based system encourages more rapid sectoral mobility and permits the government to guide that mobility insofar as it can influence the banks. Even small changes in the discount rate or in concessional credit supply between sectors can have a significant effect on resource allocation (provided the use of credit is controlled enough to prevent unlimited fungibility), because the effect of such changes on firms' cash flow position is greater than where firms have smaller debt/equity ratios. Where the government is trying to foster key sectors, a bank-based financial system gives it a powerful mechanism for inducing firms to enter sectors they otherwise would not. Where, on the other hand, capital is allocated mainly in decentralized markets, the government's ability to extend a visible and vigorous hand in the functioning of the industrial economy is limited, because firms are less susceptible to state influence.

Third, a bank-based system can help to avoid the bias toward short-term company decision-making inherent in a stock market system. The creditor needs the borrowing company to do well: it is concerned about the company's market share and ability to repay loans over the long term, which depend on how well the company is developing new products, controlling costs and quality, and so on. So these become the criteria which managers are concerned with, rather than stock market quotations.

The fourth advantage is more directly political. Industrial strategy requires a political base. Control over the financial system, and hence over highly leveraged firms, can be used to build up the coalitions needed to support the government's objectives – thus helping to implement the industrial strategy. Firms are dissuaded from opposing the government by the knowledge that opponents may find credit difficult to obtain. Of course, such a practice is easily abused. If it becomes common to allocate credit for 'loyalty' rather than for economic performance or potential the legitimacy of the administrative discretion will be impugned. Sparing but well-publicized use may reap the political gains without the legitimacy costs.

These are four potential advantages of a bank-based, administered-price financial system. However, such a system contains certain imperatives for government action which have far-reaching implications for the government's role in the economy.

The first is that the government must help to ease the downside risk of debt-financing. Higher deposit interest rates can increase the flow of financial savings; but at the new rates the private sector may not be prepared to borrow the savings unless the government intervenes to socialize some of the prospective private losses. Even if in the short run the savings are translated into loans, the higher savings and investment made possible by the higher rates may not be sustainable in the longer run without measures to spread risk. This is because highly indebted (or leveraged) firms are vulnerable to decline in current earnings to below the levels required by debt repayment, repayments on debt being fixed (whereas payments on equity are a share of profits). With firms vulnerable in this way, so are the banks which carry the 'nonperforming' loans. So where debt/equity ratios are high, there is an ever-present danger of financial instability in the economy: bankruptcies, withdrawal of savings, a fall in real investment, and slower growth. To ease such dangers, firms are likely to borrow less and banks to lend less than if the government were to underwrite some of the risks to which lenders and high debt/equity producers are exposed. If the government does bear some of the risk of private losses, the supply and demand for loanable funds will be greater, so investment, technical change, and hence growth can be higher.

The need to socialize risk applies especially in the case of highly correlated risks, to which most firms in major sectors are exposed. So it applies especially to interest rate changes, or major recession, or changes in major export markets, or political risks. Therefore the impetus for government to shoulder some of the risks of investment and saving in an economy with high debt/equity ratios is especially strong in countries which are trade-dependent and/or under external threat (like Taiwan, Korea, and Japan). The impetus is reinforced in industries where both entry and exit take a long time.

This impetus then leads the government to provide a battery of ways to reduce the risks of financial instability – not only in the form of deposit insurance and lender-of-last-resort facilities, but also in the form of subsidies to banks imperiled by loan losses, product and credit subsidies to firms in financial difficulties, banks' share-holding in companies, government share-holding in banks and in lumpy projects, and even government ownership of banks. Government can also, of course, control interest rates and exchange rates to dampen firms' exposure to market fluctuations in these two important sources of correlated risk.

The second imperative is for the supplier of credit to become involved with company management. The supplier of credit may for this purpose be the government (Korea), or the banks (Germany), or some of both (Japan). In any case, the reason for involvement with management is that the creditor cannot simply withdraw when a company runs into difficulties by selling the securities in the secondary capital market; the secondary capital markets are too thin.

Given that the 'exit' option of the capital market is not available, the alternative is the 'voice' option, to try to restructure company management so as to make it more competitive and to take the long-term view.

Nevertheless the government and/or the banks must, third, develop an institutional capacity to discriminate between responsible and irresponsible borrowing, and to penalize the latter. Firms which borrow without due commercial caution and run into trouble must not expect the government or the banks to continue to bail them out (the so-called moral hazard problem). The government must also develop mechanisms of bank supervision to curb the tendency for banks faced with big loan losses to conceal them in the 'performing' part of the balance sheet while making even riskier loans in the hope of getting back enough to offset the losses. This is the path that turns good bankers into bad ones, solvent banks into insolvent ones.

Once market signals are blunted by administered pricing and socialized risk, the government must, fourth, create a central guidance agency capable of supplementing market signals by its own signals as to which sectors will be most profitable – but in a way which allows plenty of scope for private pursuit of opportunities not seen by the guidance agency.

Finally, the government must maintain a cleavage between the domestic economy and the international economy with respect to financial flows. Without control of these flows, with firms free to borrow as they wish on international markets and with foreign banks free to make domestic loans according to their own criteria, the government's own control over the money supply and cost of capital to domestic borrowers is weakened, as is its ability to guide sectoral allocation. Speculative inflows seeking exchange rate gains can precipitate accelerating movements in exchange rates, with damaging consequences for the real economy. Uncontrolled outflows can leave the economy vulnerable to an investment collapse and make it difficult for government to arrange a sharing of the burden of adjustment to external shocks between the owners of capital and others; 'the others' are likely to be made to take the burden, with political unrest, repression, and interrupted growth as the likely result. More generally, foreign exchange controls are needed to intensify the cycle of investment and reinvestment within the national territory, with outflows only where they can be shown to meet national economic priorities. Otherwise domestic interest rates come to be determined in large part by US interest rates, and therefore make the economy subject to the kind of macro-economic mismanagement of the US economy seen during the 1980s. Although presented here as just one in a list of several requirements, this cleavage between the domestic financial system and the international financial system is a prior condition for all the others.

. . . .

Prescription 6: Carry out trade and financial liberalization gradually, in line with a certain sequence of steps. Many neoclassical analysts urge large-scale and quick liberalization, to get a whole package of reforms in place before opposition builds up. And many urge that comprehensive import liberalization should be carried out before export earnings increase, so as to flush away the inefficiencies generated by protective barriers and enable a subsequently better response to export demand. By contrast, the East Asian experience is consistent with a prescription for more gradual change and a different sequence. It suggests the following: (1) macroeconomic stabilization should come before trade liberalization; (2) substantial external financial assistance greatly eases the transition from stabilization to liberalization; (3) liberalization of imports of export inputs should come before deprotective competition-providing import liberalization; (4) import liberalization of the latter type is not a prior condition for successful exporting; it should follow the growth of exports; (5) successful exporting requires a large promotional role for public agencies; (6) gradual trade liberalizations can be sustained; and (7) financial liberalization should come late in the queue, after a substantial measure of import liberalization.

. . . .

The argument for economic liberalization – whether in trade, finance, or other spheres – also needs to address the question of what kinds of private sector groups will gain from the change. It cannot be assumed that they will wish to be entrepreneurial investors rather than luxury consumers. Nor can it be assumed that they will wish to place limits on the arbitrary actions of the state and discipline the state to provide effective services. Liberalization may lead to the capture of economic power by less accountable cliques around the power-holders, Marcos-style. The analytical dichotomy between 'state' and 'economy' can lead us to overlook the point that the same people or groups may have feet planted firmly on both sides of the divide, in which case a shrinkage of the state and expansion of the private sector may further remove economic power formerly in the hands of the state from some degree of accountability. It may further erode a 'center' – a cohesive organizational structure – where collective interests can be articulated and followed.

These are six broad economic prescriptions supported by the experience of Taiwan, Korea, and Japan. But we must note another lesson to do with differences rather than similarities. While the three East Asian states all governed the market, they used somewhat different methods for doing so. Taiwan used large upstream public enterprises and selected foreign firms to provide 'unbalanced' pushes in certain sectors, arms-length incentives to steer the response of myriad small downstream firms, and stable prices and real effective exchange rates. Korea used huge private business groups as the spearheads, steering them with massive credit subsidies and more direct cajoling (recently switching to more of a negotiation mode). It obtained more of its technology under license than through direct foreign investment, and sacrificed some macroeconomic stability for faster industrial transformation. Japan, which already had huge private business groups in place in the 1930s, pioneered the route that Korea was later to follow, except that consultative decision-making procedures linking government and business were in place from much earlier on. So there is more than one way to govern the market effectively.

Selected references

Bienefeld, M. 1988: The significance of the newly industrializing countries for the development debate. *Studies in Political Economy* 25, 7–39.

Cline, W. 1982: *Reciprocity – A new approach to world trade policy?* Washington, DC: Institute for International Economics.

Drucker, P. 1986: The changed world economy. *Foreign Affairs* 64, 768–91.

Gerschenkron, A. 1962: *Economic backwardness in historical perspective.* Cambridge, MA: Harvard UP.

Gold, B. 1979: *Productivity, technology and capital.* Lexington: Lexington Books.

Krueger, A. 1981: Export-led industrial growth reconsidered. In Hong, W. and Krause, L. (eds), *Trade and growth of the advanced developing countries in the Pacific basin.* Seoul: Korea Development Institute.

Romer, P. 1987: Crazy explanations for the productivity slowdown. *Macroeconomics Annual* 2, 163–202.

Senghaas, D. 1985: *The European experience.* Dover, NH: Berg.

World Bank 1988: *World development report, 1988.* Oxford: OUP/World Bank.

SECTION FIVE

THE GLOBAL POLITICAL ECONOMY

Editor's Introduction

Developing countries have long been integrated into the world economy. Some early modernizers liked to treat the Third World as a place filled by 'people without history', but the relative absence of modern icons in the Third World was a product of the West's imperialism as much as anything else. Gunder Frank was close to the mark when he declared that all countries had once been undeveloped, but only Third World countries had been underdeveloped by others.

Fifty years later the scars of colonialism still exist in many places, in terms of meaningless 'national' boundaries, an excessive reliance on primary commodity exports, and a continuing imbalance in local space-economies. Some countries have also discovered that their sovereignty has to be shared with institutions like the World Bank and the IMF if they are to gain funds for development, or even repay past debts. Robert Jackson has written of a world of quasi-states in Africa (Jackson, 1990). And yet the nature of the global political economy is fast changing, and with it the different trajectories of low- and middle- income economies within it. A world system drawn together by international trade in the 1950s and 1960s is now dominated by private capital flows. Distances, too, are shrinking, at least when expressed in travel times. As late as 1930 the best average speed of steam locomotives was 65 miles per hour, while steam ships managed just 36 miles per hour (Dicken, 1992). In the 1990s most developing countries have their own airlines, and the world is increasingly wired together by telephone and fax machines and by satellite TV and electronic mail networks. In one sense, perhaps, we have no Others.

But the 'end of geography' thesis can be overdone. People in the developing world are less able to command space than their counterparts in the developed world, and face-to-face contacts still matter even in the cyberspace worlds of foreign exchange dealers and futures traders (Agnew and Corbridge, 1995). On a day-to-day basis what matters most to trading agents in developing countries are the rules and institutions that govern international flows of labour, commodities and money. The regulations that prevent a free and fair trade in people are not dealt with in this Reader. It must suffice to say that the problems of refugees and stateless people are getting worse, and that often racist barriers to international migration sit uncomfortably with the free trade rhetoric trumpeted by Western countries and the Bretton Woods institutions. **Readings 21–23** are concerned instead with three of the staple topics of international political economy: trade, aid and structural adjustment. But even here there is a subversive edge. **Readings 21 and 22** have been chosen in part because they cut against the grain of orthodox views on trade and aid respectively.

The orthodox view of trade as an engine of growth takes its cue from certain versions of comparative advantage theory. Briefly stated, these hold that countries should specialize in the production of those goods and services that make the best use of local skills and resources, and then trade the surpluses which result with other countries. Total world production will increase as a consequence and with it global welfare. In a world of perfect competition and frictionless free trade it matters not at all if Japan ends up producing cars and computers, while Bangladesh produces rice and jute. So long as each country is playing to its strengths, Japan and Bangladesh can trade their cars and computers for rice and jute respectively.

This view of international trade has been growing in popularity since the 1970s and underpins recent endorsements of trade liberalization and outwardly-oriented development strategies. Its proponents often shout hard because they feel their vision of free trade is undermined by development economists who have lent their support to import-substitution industrialization and protectionism (Balass *et al.* 1982; Haberler, 1961). In fact, there is more common ground between these two camps than is commonly recognized. The 'orthodoxy' of the 1950s and 1960s did suggest that developing countries might need to build up their manufacturing base behind tariff walls *in the short run*. The premiss was that countries specializing in the export of primary commodities would suffer a secular decline in their terms of trade (Prebisch, 1950; Singer, 1950, 1989). But even in the 1950s and 1960s it was generally supposed that protectionism and import-substitution industrialization would be temporary affairs. Given time, developing countries were expected to build up their manufacturing base sufficiently to compete freely and fairly in the international market-place. Tariff barriers would then come down and trade would be installed as an engine of growth worldwide.

The jury is still out on these 'two' orthodoxies. World Bank data confirm that outwardly-oriented developing economies have out-performed inwardly-oriented developing economies since about 1970 (World Bank, 1987). But the question of causality is not so easily settled. It might still be the case that successful outwardly-oriented countries first need to build up a manufacturing or services base, and that this is best effected behind tariff walls and with public subsidies. A country like Bangladesh might also want to think twice about taking the logic of some comparative advantage models too far, and placing all its eggs in the basket of primary commodity production. Economic theory might find it difficult to explain why the terms of trade should move against non-oil primary commodities, but economic history suggests this often happens. In any case, countries might have a legitimate interest in wanting to produce a wide range of goods and services, including some goods and services in which they appear to lack a comparative advantage. The historical geography of international trade is more complicated than some economic theories suggest, and for good reasons.

Just such a thought is echoed by **James K. Boyce** in **Reading 21**. Boyce is less concerned with formal theories of international trade than with the specifics of the world coconut market, where 'the Philippines is king'. He notes that while the Philippines accounts for more than half of world coconut exports, its 'ability . . . to act as a "price-maker" in the world coconut market is severely constrained by the existence of natural and synthetic substitutes'. Despite rhetoric to the contrary, the Philippines is not the 'Saudi Arabia of coconut oil'. The prices of many coconut oil substitutes are regulated by governments in the developed world, and Filipino producers face declining terms of trade for their product.

Boyce makes two further points. In the first part of the Reading he provides an account of the development of coconut production for export in the Philippines under Spanish and American rule. Boyce notes that: 'Preferential trading arrangements with the United States played a key historical role in the growth of Philippine coconut exports in the face of the competitive and politicized world market for edible oils'. This system of quasi-imperial preference continued until the early 1970s. The larger part of Boyce's Reading is then given over to an account of the local relations of production and exchange under which coconut oil is produced for the world market, a topic largely excluded from standard presentations of world trade issues. The Filipino 'coconut farms' that Boyce describes are trapped in a series of *local* trading relationships that are far from free and fair, and in production systems that were and are dominated by cronies of the ruling political elite. The surplus income generated by small farmers in the 1970s and early 1980s was effectively redistributed to 'a handful of politically powerful individuals'. But this elite was ineffectual when it came to pressing demands for higher coconut oil prices on foreign buyers. Some associates of Marcos did try to bid up prices by means of UNICOM, a coconut oil milling cartel, but it seems that the IMF asked for UNICOM to be dismantled as a condition of a standby credit in the mid-1980s. UNICOM was abolished in January 1985.

Boyce concludes his paper with a brief review of 'two limitations of comparative advantage as a guide to trade strategy'. A first limitation relates to declining terms of trade and price instability. These problems could be relieved by attacks upon the subsidies offered to competing oilseed producers in the developed world and by a stabilization fund. A second limitation relates to the fact that, 'within a country, one person's advantage can be another's disadvantage'. Programmes of land reform and market reform in the Philippines could ensure a much fairer distributional outcome from the coconut production that currently obtains (see also **Reading 7**); an unequal 'distributional outcome . . . is not inherent in export agriculture'. The wider import of Boyce's argument is that trade alone cannot always serve as an engine of growth, nor is it the unitary phenomenon that it sometimes appears to be in economics textbooks. Goods and services for export have to be produced with regard to particular and imperfect relations of production and exchange. Injunctions to free up international trading relations are not without merit, but they are surely strengthened by corresponding injunctions to attend to local production arrangements and trading systems. Reform is often needed on a broad front, and with close attention to the sequencing of reforms: a lesson being learned painfully in the 1990s by the transitional economies of the ex-Soviet bloc.

Boyce's account of agricultural markets in the Philippines is at odds with some other studies which have praised Third World trading networks and the entrepreneurs that make them efficient. **Peter Bauer** was one of the first development economists to argue in favour of private marketing arrangements in the rural Third World, and against the anti-producer bias of state marketing boards. His account of *West African Trade* is a classic of its type (Bauer, 1954). More recently, Bauer has been a leading light in the counter-revolution in development studies, although his 'dissent on development' (Bauer, 1972) owes more to his mentor Hayek than is common amongst neoliberals. In the United Kingdom, Bauer was made a Baron in 1983 by Mrs Thatcher and he was promptly dubbed Lord Anti-Aid by the *Observer* newspaper. Such attention reflects not only the importance of Bauer's often singular views, but also the fact that his strongly worded attacks on foreign aid have appeared in the popular press as well as in academic publications.

Reading 22 presents a pithy statement of Bauer's views on foreign aid. Bauer rejects the usual arguments for giving official development assistance. Aid given for humanitarian reasons is dismissed for pandering to 'a widely articulated feeling of guilt in the West'. Aid given for economic reasons is likewise given short shrift. The main economic argument for giving aid is that it will help to fill a gap which is presumed to exist in the developing world: usually a savings gap (as per the Harrod–Domar model), but sometimes a foreign exchange gap or a technology gap. Aid can be given either as a gift, or more usually as a loan with a strong element of concessionality attached to it. Aid can also be tied to purchases from the donor country, and can be linked to specific development projects or given as general programme assistance. Whatever its form, Bauer finds little to commend in the practice of aid giving. Like many on the Left (ironically), he believes that aid traps people in a culture of dependency. He also claims that the aid industry has a logic of its own, serving the interests of some privileged groups in the North and South at the expense of Northern taxpayers and the poor in the South. Bauer further points out that the developed world managed to develop without foreign aid; it follows, he suggests, that aid cannot be a precondition for development, whatever its proponents maintain. Finally, Bauer develops a startling political argument about aid. Instead of seeing aid as a response to certain problems in the Third World, Bauer maintains that the Third World is an invention of the aid industry itself. 'The concept of the Third World and the policy of official aid are inseparable. The Third World is merely a name for the collection of countries whose governments, with occasional and odd exceptions, demand and receive official aid from the West'.

Bauer's views on foreign aid are nothing if not controversial, and it needs to be clearly stated that his views are not widely shared by students of development (a point that does not worry Bauer). Readers of this book are strongly encouraged to read Riddell's intelligent and balanced assessment of changing aid practices for a contrary point of view, and for a reasoned defence of foreign aid (Riddell, 1987). Riddell provides good empirical evidence for suggesting that the effectiveness of Western aid has slowly been improving, even if total disbursements remain low. He also makes recourse to a modified Rawlsian theory of justice – there but for the grace of God go I – to challenge the philosophy of just deserts that underpins Bauer's attack on foreign aid. Bauer maintains that individual countries are responsible for their own fortunes, and that the West has no duty to 'aid' the developing world or to make amends for an earlier history of colonialism. (Bauer also denies that aid can serve such a purpose.) The Rawlsian challenge suggests that we are not the sole authors of our fortunes or misfortunes (Rawls, 1972). A person born into poverty in Bangladesh does not have the same life-chances as a person born into plenty in the United States, *through no fault of his or her own*. It is unreasonable to oppose a redistribution of assets or incomes on the grounds that it is a coercive attack on the fairly won rewards of sovereign individuals (though see Nozick, 1974).

Some proponents of 'development ethics' have suggested that a Rawlsian theory of justice can be extended to the international stage, and be made to underwrite policies that facilitate a transfer of resources from rich countries to poor (Gaspar, 1992). Quite how this transfer might be effected is a matter for policy that will not be discussed here. The broader claim is that people in rich countries would not accept the conditions of absolute poverty that exist today in many parts of the developing world. Since the same people could easily have been born into absolute poverty themselves (the accident of birth . . .), they should support policies designed to provide for the basic human needs of people everywhere. If foreign aid can help meet this end it should be supported.

In practice things are not this simple, and the new sub-discipline of development ethics is still in its infancy (Crocker, 1991). Some will argue that the best way to help poor people is to set them free, and to increase the general size of the cake before worrying about its distribution (Kirzner, 1989). This is the position of most neoliberals, many of whom argue that developing countries must continue to adjust the structures of their economies to a changing global economy if they are to maximize their potential for development.

Paul Streeten is not opposed to all aspects of this argument. In **Reading 23** he points out that structural adjustment is synonymous with development in a broad sense, and that to oppose it *a priori* is to oppose change itself. Streeten also notes that structural adjustment in the narrow sense of an adjustment to major disruptions is bound to hurt some poor people in the short run, 'if only because the poor are so numerous and average income is so low in low-income countries'. Here is an implicit rebuke to romantics on the Left. Development is fraught with dilemmas, and it is a delusion to believe otherwise. (A tendency to self-delusion is equally to be found among 'pricist fanatics who say that the correct prices will by themselves induce the right innovation and even the correct action by the government in the public sector.') Nevertheless, the question of who benefits and who loses from structural adjustment is a serious one, and Streeten is careful not to duck it. His paper is a model of careful analysis. Arguments that too often are run together are properly picked apart and exposed to close scrutiny.

Streeten is anxious to insist on the need for a global response to problems of structural adjustment that some others see as the singular responsibility of a country in difficulty. He is critical of a fallacy of composition which suggests that all countries can grow out of debt or trading deficits by running export surpluses, a position that the IMF and the World Bank seem to maintain. The fallacy of composition is sometimes a ploy by which responsibility for a 'common crisis' (Brandt, 1983) is dumped on those countries forced to undertake programmes of structural adjustment according to certain conditions laid down by the Bretton Woods institutions. The pre-history of the debt crisis illustrates how unjust this approach is, and how ostensibly 'economic' approaches to structural adjustment are guided by geopolitics. The debt crisis emerged partly as a result of poor policies in some developing countries, and partly as a result of a global recession in the early 1980s that was induced by quasi-monetarist policies in the developed world. The crisis was also encouraged by the aggressive actions of some private banks in the 1970s and 1980s, and by new opportunities for capital recycling opened up by the Euromarkets. Yet the banking-cum-debt crisis that broke in Mexico in

August 1992 was coded as a debt crisis pure and simple, and was policed throughout the 1980s in such a way that the costs of adjustment were borne almost exclusively by the indebted developing world (and poorer households in particular). America's large money-center banks were protected at the cost of a lost decade of development in Latin America.[1] Had international justice prevailed over geopolitical strength, it is doubtful that a similar distribution of pain and pleasure would have ensued. The Age of Empire might be over, but our interdependent world economy continues to be an asymmetrical world economy. Real power remains in the hands of a decentred network of hegemonic actors in the OECD world economy; with the United States, the European Union and Japan, and with the IMF, the World Bank and the private capital markets. The Third World, so-called, lacks a presence and lacks a voice.

Guide to further reading

General: interdependence and globalization

Aglietta, M. 1982: World capitalism in the eighties. *New Left Review* 136, 25–36.
Agnew, J. and Corbridge, S. 1995: *Mastering space: Hegemony, territory and international political economy*. London: Routledge.
Beenstock, M. 1984: *The world economy in transition* (2nd edn). London: George Allen and Unwin.
Brett, E. 1985: *The world economy since the war: The politics of uneven development*. London: Macmillan.
Brookfield, H. 1975: *Interdependent development*. London: Methuen.
Calleo, D. 1987: *Beyond American hegemony*. New York: Basic Books.
DerDerian, J. and Shapiro, M. (eds) 1989: *International/intertextual relations*. New York: Lexington.
Dicken, P. 1992: *Global shift: The internationalization of economic activity*. London: Paul Chapman.
Gilpin, R. 1987: *The political economy of international relations*. Princeton: Princeton UP.
Goldgeier, J. and McPaul, M. 1992: A tale of two worlds: core and periphery in the post-Cold War era. *International Organization* 46, 467–91.

[1] The reference here to the Latin American banking-cum-debt crisis is not meant to occlude an equally serious debt crisis in sub-Saharan Africa and other regions. Ratios of debt outstanding to exports or GNP are often higher in Africa than in Latin America. The main difference between the 'two' debt crises is that African debt is largely owed to official lenders and not private banks. The debt crisis in Africa also threatens to last longer than the crisis in Latin America, where positive net private capital flows have been restored since 1992. In the context of international trade, too, the position of large parts of Africa is not a happy one. In 1965 Africa accounted for about 5 percent of world trade; by 1990 the corresponding figure was under 2 percent. There are justifiable fears that large parts of Africa no longer 'matter' to the OECD world economy, except as a source of raw materials, occasional debt repayments, and as a dumping ground for toxic waste. On this 'hemmed in' Africa, and possible responses to Africa's economic decline, see Callaghy and Ravenhill (1993).

Harris, N. 1986: *The end of the Third World*. London: Penguin.
Hoogvelt, A. 1982: *The Third World in global development*. London: Macmillan.
Jackson, R. 1990: *Quasi-states: Sovereignty, international relations and the Third World*. Cambridge: CUP.
Kennedy, P. 1993: *Preparing for the twenty-first century*. New York: Random House.
Knox, P. and Agnew, J. 1994: *The geography of the world economy*. London: Edward Arnold.
Krasner, S. 1985: *Structural conflict: The Third World against global liberalism*. Berkeley: University of California Press.
Krueger, A. 1993: *Economic policies at cross-purposes: The United States and developing countries*. Washington, DC: Brookings Institution.
Lash, S. and Urry, J. 1994: *Economies of signs and space*. London: Sage.
Lipietz, A. 1987: *Mirages and miracles: The crises of global Fordism*. London: Verso.
Ruggie, J. (ed.) 1993: *Multilateralism matters*. New York: Columbia UP.
Sklair, L. 1991: *Sociology of the global system*. Hemel Hempstead: Harvester Wheatsheaf.
Sklair, L. (ed.) 1994: *Capitalism and development*. London: Routledge.
Spybey, T. 1992: *Social change, development and dependency*. Cambridge: Polity.
Strange, S. 1988: *States and markets: An introduction to international political economy*. London: Pinter.
Thrift, N. 1989: The geography of international economic disorder. In Johnston, R. and Taylor, P. (eds), *A world in crisis?* (2nd edn). Oxford: Blackwell.
Wallace, I. 1990: *The global economic system*. London: Routledge.
Walter, A. 1993: *World power and world money*. Hemel Hempstead: Harvester Wheatsheaf.
World Bank 1985: *World development report, 1985*. Oxford: OUP/World Bank.

Trade and exchange

Balassa, B. and associates 1982: *Development strategies in semi-industrial economies*. Baltimore: Johns Hopkins UP.
Bauer, P. 1954: *West African trade*. Cambridge: CUP.
Bauer, P. 1972: *Dissent on development*. Cambridge, MA: Harvard UP.
Bhagwati, J. 1978: *Foreign trade regimes and economic development*. Cambridge, MA: Billinger.
Bhagwati, J. 1989: Is free trade passé after all? *Weltwirtschaftliches Archiv* 125, 17–44.
Bliss, C. 1989: Trade and development: theoretical issues and policy implications. In Chenery, H. and Srinivasan, T. (eds), *Handbook of development economics, Volume II*. Amsterdam: North Holland, Chapter 23.
Dornbusch, R. 1992: The case for trade liberalization in developing countries. *Journal of Economic Perspectives* 6, 69–85.
Emmanuel, A. 1972: *Unequal exchange*. New York: Monthly Review Press.
Evans, D. 1989: *Comparative advantage and growth: Trade and development in theory and practice*. Hemel Hempstead: Harvester Wheatsheaf.

Folke, S. Fold, N. and Enevoldsen, T. 1993: *South-south trade and development*. Basingstoke: Macmillan.
Furtado, C. 1964: *Development and underdevelopment*. Berkeley: University of California Press.
Grimwade, N. 1989: *International trade: New patterns of trade, production and investment*. Basingstoke: Macmillan.
Haberler, A. 1961: Terms of trade and economic development. In Ellis, H. (ed.), *Economic development for Latin America*. London: Macmillan.
Heckscher, E. 1935: *Mercantilism*. London: George Allen and Unwin.
Krueger, A. 1980: Trade policy as an input to development. *American Economic Review* 70.
Krugman, P. (ed.) 1986: *Strategic trade policy and the new international economics*. Cambridge, MA: MIT Press.
Lewis, M. 1989: Commercialization and community life: the geography of market exchange in a small-scale Philippine society. *Annals of the Association of American Geographers* 79, 390–410.
Morgan, D. 1979: *Merchants of grain*. New York: Norton.
Prebisch, R. 1950: *The economic development of Latin America and its principal problems*. New York: ECLA.
Rodrick, D. 1992: The limits of trade policy reform in developing countries. *Journal of Economic Perspectives* 6, 87–105.
Salvatore, D. and Hatcher, T. 1991: Inward-oriented and outward-oriented trade strategies. *Journal of Development Studies* 23, 7–25.
Singer, H. 1950: The distribution of gains between investing and borrowing countries. *American Economic Review* 40, 473–85.
Singer, H. 1989: Terms of trade and economic development. In Eatwell, J. Milgate, M. and Newman, P. (eds), *The new Palgrave: Economic development*. London: Macmillan, 323–8.
Singer, H. and Gray, P. 1988: Trade policy and growth of developing countries: some new data. *World Development* 16, 395–403.
Tarrant, J. 1985: A review of international food trade. *Progress in Human Geography* 9, 235–54.
Wood, A. 1991: How much does trade in the South affect workers in the North? *World Bank Research Observer* 6, 17–31.
World Bank 1987: *World development report, 1987*. Oxford: OUP/World Bank.

Aid

Ayres, R. 1983: *Banking on the poor: The World Bank and world poverty*. Cambridge, MA: MIT Press.
Bauer, P. 1984: *Reality and rhetoric: Studies in the economics of development*. Cambridge, MA: Harvard UP.
Cassen, R. *et al*. 1986: *Does aid work? Report to an intergovernmental task force*. Oxford: OUP.
Chenery, H. 1989: Foreign aid. In Eatwell, J. Milgate, M. and Newman, P. (eds), *The new Palgrave: Economic development*. London: Macmillan, 137–44.
Griffin, K. 1991: Foreign aid after the Cold War. *Development and Change* 22, 645–85.

Hayter, T, 1971: *Aid as imperialism*. Harmondsworth: Penguin.
Krueger, A. Michalopoulos, C. and Ruttan, V. (eds) 1989: *Aid and development*. Baltimore: Johns Hopkins UP.
Mosley, P. 1981: Aid for the poorest? *Journal of Development Studies* 17, 214–25.
Mosley, P. Harrigan, J. and Toye, J. 1991: *Aid and world power* (2 volumes). London: Routledge.
Riddell, R. 1986: The ethics of foreign aid. *Development Policy Review* 1, 35–46.
Riddell, R. 1987: *Foreign aid reconsidered*. London: James Currey/ODI.
Rudner, M. 1989: Japanese aid to Southeast Asia. *Modern Asian Studies* 23, 73–116.
Shaw, J. and Clay, E. (eds) 1993: *World food aid*. London: James Currey.
Streeten, P. 1983: Why development aid? *Banco Nazionale del Lavoro Quarterly* 147, 379–85.
White, J. 1974: *The politics of foreign aid*. London: Bodley Head.

Debt, structural adjustment and development ethics/justice

Bierstaker, T. (ed.) 1993: *Dealing with debt: International financial negotations and adjustment bargaining*. Boulder: Westview.
Brandt, W. (Chair) 1983: *Common crisis north–south: Cooperation for world recovery*. London: Pan.
Branford, S. and Kucinski, B. 1988: *The debt squads: The US, the banks and Latin America*. London: Zed.
Buiter, W. and Srinivasan, T. 1987: Rewarding the profligate and punishing the prudent and poor: some recent proposals for debt relief. *World Development* 15, 411–17.
Callaghy, T. and Ravenhill, J. (eds) 1993: *Hemmed in: Responses to Africa's economic decline*. New York: Columbia UP.
Congdon, T. 1988: *The debt threat*. Oxford: Blackwell.
Corbridge, S. 1993: *Debt and development*. Oxford: Blackwell.
Corden, M. 1991: The theory of debt relief: sorting out some issues. *Journal of Development Studies* 27, 133–45.
Cornia, A. Jolly, R. and Stewart, F. 1987: *Adjustment with a human face*. Oxford: OUP.
Crocker, D. 1991: Towards development ethics. *World Development* 19, 457–83.
Diaz-Alejandro, C. 1984: Latin American debt: I don't think we are in Kansas any more. *Brookings Papers on Economic Activity* 2, 335–89.
Dissent 1992: Africa today: crisis and change (special issue). Summer, 293–416.
Dornbusch, R. and Edwards, S. (eds) 1991: *The macroeconomics of populism in Latin America*. Chicago: University of Chicago Press.
Eichengreen, B. and Lindert, P. (eds) 1989: *The international debt crisis in historical perspective*. Cambridge, MA: MIT Press.
Gaspar, D. 1992: Development ethics – an emergent field? A look at scope and structure with special reference to the ethics of aid. *Working Paper 134*. The Hague: Institute of Social Studies.
George, S. 1989. *A fate worse than debt*. Harmondsworth: Penguin.
Griffith-Jones, S. 1986: *Managing world debt*. Hemel Hempstead: Harvester Wheatsheaf.

Killick, T. (ed.) 1984: *The quest for economic stabilisation: The IMF and the Third World*. London: Heinemann.
Kirzner, I. 1989: *Discovery, capitalism, and distributive justice*. Oxford: Blackwell.
Kuczynski, P.-P. 1988: *Latin American debt*. Baltimore: Johns Hopkins UP.
Lustig, N. 1990: Economic crisis, adjustment and living standards in Mexico, 1982–1985. *World Development* 18, 1325–42.
McKinnon, R. 1991: *The order of economic liberalization: Financial control in the transition to a market economy*. Baltimore: Johns Hopkins UP.
Mosley, P. and Smith, L. 1989: Structural adjustment and agricultural performance in sub-Saharan Africa 1980–1987. *Journal of International Development* 1, 321–55.
Nelson, J. (ed.) 1990: *Economic crisis and policy choice: The politics of adjustment in the Third World*. Princeton: Princeton UP.
Nozick, R. 1974: *Anarchy, state and Utopia*. Oxford: Blackwell.
O'Neill, O. 1991: Transnational justice. In Held, D. (ed.), *Political theory today*. Cambridge: Polity, 276–304.
Rawls, J. 1972: *A theory of justice*. Oxford: OUP.
Sachs, J. (ed.) 1989: *Developing country debt and the world economy*. Chicago: University of Chicago Press/NBER.
Taylor, L. 1988: *Varieties of stabilisation experience: Towards sensible macroeconomics in the Third World*. Oxford: OUP.
Woodward, D. 1992: *Debt, adjustment and poverty in developing countries* (2 volumes). London: Pinter/Save the Children.

21 James K. Boyce, 'Of Coconuts and Kings: The Political Economy of an Export Crop'

Reprinted in full from: *Development and Change* 23 (4): 1–25 (1992)

'The time has come,' the Walrus said,
'To talk of many things:
of shoes – and ships – and sealing wax –
Of cabbages – and – kings –
And why the sea is boiling hot –
And whether pigs have wings.'

Carroll (1922, p. 186)

Introduction

In the world coconut market, the Philippines is king. The country accounts for more than half of world exports, and hence is sometimes termed the 'Saudi Arabia of coconut oil', a label which understates its market share but overstates its market power. The ability of the Philippines to act as a 'price-maker' in the world coconut market is severely constrained by the existence of natural and synthetic substitutes. The prices of these substitutes – and therefore the price of coconut oil – are strongly influenced by policies of other governments, including protection for domestic oilseed producers and environmental regulation of the production and consumption of synthetics. In the world of international commerce, the coconut king wields little power. Moreover, to say that the Philippines is the coconut king is not to say that all Filipinos in the coconut industry dine at the royal table. On the contrary, the relations of production and exchange in the Philippine coconut industry have brought poverty to many and fortunes to a few.

This article investigates the political economy of coconuts in the Philippines. Section I provides a brief historical overview, followed by an examination of the terms of trade on the world market and the competition from natural and synthetic substitutes in Section II. Section III looks at the relations of production in coconut agriculture. Section IV describes the relations of exchange, manipulation of which became the primary vehicle for concentration of the income generated in the coconut sector during the Marcos era. Section V offers some concluding remarks.

A central theme of the paper is the importance of linkages between wealth and power in determining the size and distribution of income in export agriculture. In the simplified world of economics textbooks – where endowments and technology are exogenous, markets are perfect and externalities do not exist – interactions between wealth and power are of no importance. In the real world of coconuts, however, they matter a great deal.

I. The coconut sector: an overview

In the Philippines, as in many Asian, African and Latin American countries, export agriculture has historically been the single most important locus of interaction with the world economy. In the mid-1960s, sugar, coconuts and forestry accounted for 80 per cent of Philippine export earnings; coconuts alone accounted for 33 per cent (see Table 21.1). Their share declined over time owing to adverse price trends and export diversification, but in the mid-1980s coconut products remained the Philippines' single most important export, accounting for 12 per cent of total earnings.

Table 21.1 Major Agricultural and Forestry Exports from the Philippines, 1962–85 (US$ Million FOB Value)[a]

Product	*1962–6*	*1967–71*	*1972–6*	*1977–81*	*1982–5*
Coconut products:	239	217	444	850	614
Copra	153	107	148	101	13
Coconut oil	56	78	230	575	461
Other[b]	30	32	66	174	140
Sugar and sugar products	150	175	470	451	320
Forestry products	184	267	317	426	316
Total exports	724	973	2120	4537	5012
Coconut share in total (%)	33.0	22.2	20.8	18.7	12.2

[a] Quinquennial averages; [b] dessicated coconut and copra meal or cake.
Sources: Calculated from data in National Economic and Development Authority (1976; p. 423; 1986, pp. 362–3).

Growth in agro-forestry exports has long been a central element of the country's economic development strategy. The World Bank (1973; p. 19) placed 'expansion of agricultural exports' alongside foodgrain self-sufficiency as the major goals for Philippine agriculture. Juan Ponce Enrile, who served as

Defence Minister and as a senior coconut industry official under former President Ferdinand Marcos, predicted in 1980 that '25 per cent of Philippine growth in the next twenty to thirty years will come from coconuts'.[1]

Coconut production in the Philippines beyond domestic needs dates from 1642, when a Spanish edict required each 'indio' to plant 100–200 coconut trees to provide caulk and rigging for the colonizers' galleons. Large-scale exports of copra (dried coconut meat from which oil is extracted) began in the late nineteenth century, in response to demand from European and North American manufacturers of margarine and soap. The first mills for extraction of coconut oil in the Philippines were established early in the twentieth century. The United States became the largest market for Philippine coconut oil, and gave it preferential tariff treatment until 1974. Copra continued to predominate in trade with European countries, owing to tariffs on oil which were imposed to protect European millers.[2] After the Second World War, the primary uses of coconut oil in the world economy shifted from edible to non-edible industrial products, such as soap, detergents, cosmetics, explosives and pharmaceuticals.[3]

In the early 1960s, two factors boosted the Philippine coconut industry. The first was the devaluation of the peso in 1962 by almost 100 per cent, which brought windfall profits to agro-exporters (Legarda, 1962; Treadgold and Hooley, 1967). The second was the introduction of large ocean tankers for the transport of coconut oil, a technological breakthrough which cut shipping costs and set the stage for the Philippines to move up the processing ladder from the export of copra to the export of oil.[4] At the same time, the 'green revolution' in rice agriculture saved the country's land frontier for non-rice uses.[5]

The increased profitability of coconut exports stimulated rapid acreage growth. From one million ha in the 1950s, the area planted to coconut rose to 2 million ha by 1971, and to 3 million ha by 1979 (see Table 21.2). Much of this growth occurred on the land frontier, notably on logged-over virgin lands in Mindanao (Tiglao, 1981, p. 58). Coconut yields stagnated, however, at about one metric tonne per ha (see Table 21.2), reflecting low input use and a lack of investment in the replanting of ageing trees.[6]

[1] 'Enrile Cites Scheme for Coconut Industry', *Business Day* (Manila), 14 April 1980, p. 3, cited by Tadem (1980: 43).

[2] Copra is allowed to enter Europe duty free, while coconut oil is subject to tariffs ranging from 5 to 15 per cent (Canlas and Alburo, 1989: 84). Hicks (1967: 89) reported similar rates in the 1960s.

[3] For details, see Tiglao (1981) and Hawes (1987: 59–68). On uses of coconut oil, see also Woodruff (1979: 112–23).

[4] The freight rate for Philippine coconut oil to the US Pacific coast, for example, dropped from US $26 per ton to US $9 (Hicks, 1967: 160–1; see also Tiglao, 1981: 24).

[5] Some observers like Hooley and Ruttan (1969), Barker (1978) and Kikuchi and Hayami (1978) saw the 'closing of the land frontier' in the Philippines in the 1950s and 1960s. But while rice acreage levelled off after 1960, acreage under export crops grew rapidly. For discussion, see Boyce (forthcoming: Chapter 3).

[6] The reliability of the official data on coconut yields is open to question. Alternative estimates by the Philippine Coconut Authority indicate that average yields fell from 1.2 mt/ha in the 1960s to 0.7 mt/ha in the 1980s (Galang, 1988: 72).

Table 21.2 Area, Yield and Output of Coconut, 1962–85 (Output in Raw Nuts)[a]

	1962–6	*1967–71*	*1972–6*	*1977–81*	*1982–5*
Area (000 ha)	1475	1880	2253	2996	3210
Yield (mt/ha)	1.0	0.9	1.1	1.4	1.0
Output (000 mt)	1523	1704	2461	4244	3264
Value (m pesos)	626	1053	2367	6513	8512
Coconut share in total crop output value (%)	20.4	16.8	14.9	19.0	14.4

[a] Quinquennial averages.
Sources: Calculated from data in National Economic and Development Authority (1976, pp. 134–53; 1986, 266–75).

Coconut yield stagnation is sometimes attributed to the low prices received by growers, but the fact that farmgate prices were attractive enough to stimulate acreage expansion suggests that the explanation lies elsewhere. One possible factor is inadequate investment in coconut research and extension. Hicks and McNicoll (1971; pp. 205–6) stated that 'practically no basic research has been directed toward developing a higher yielding coconut palm'. They attributed this to the fact that, unlike other tree crops in tropical agriculture, coconuts 'have always been a smallholders' crop, and few growing interests are large enough to justify expensive research'. Precisely for this reason, however, most crop improvement research throughout the world is conducted by public sector institutions rather than by private growers. Relatively low investment in coconut research and extension in the Philippines therefore may reflect the political weakness of small growers and the existence of other political priorities among the larger growers.

In a review of Philippine agricultural research, Evenson *et al.* (1980, p. 26) noted that in the case of coconuts, 'Few varietal advances appear to have been made over the past 50 years or so.' In its 'Green Paper' for the new Aquino government, the Agricultural Policy and Strategy Team of the University of the Philippines (1986, p. 249) reported that 'no more than two agronomists with a doctorate degree are working on coconut plant breeding in the country'. The Team also reported (p. 499) that research expenditures on coconut from 1974 to 1984 were, on average, only 28 per cent of those on sugar-cane.[7] The primary coconut crop improvement initiative under the Marcos government was a programme to replace older trees with the new, higher-yielding, 'Mawa' (Malayan × West African) hybrid, touted as the coconut equivalent of IRRI rice. It was not a great success. The Mawa hybrid, although shorter than traditional varieties, also has a shorter root system which makes it unable to withstand typhoons. The traditional tall varieties bend with the high winds, but

[7] This may overstate the effective investment in coconut research, since much of the expenditure was channelled through the Philippine Coconut Authority, which tended to devote its research to 'buildings, public relations and funds' (Evenson *et al.*, 1980: 27).

Mawa topples and dies, a shortcoming which greatly limits its potential geographic range in the Philippines.

Critics perceived the replanting programme as a vehicle for private gain rather than for the public good. A special tax levied on coconut growers (described in more detail in Section IV) was used to finance the purchase of Mawa seed from a farm owned by 'coconut king' Eduardo Cojuangco, a close associate of President Marcos.[8] The seedlings were then distributed to the growers, who received subsidies (financed by the levy) for replanting costs. According to Jose V. Romero Jr, chairman of the Philippine Coconut Authority under the Aquino government, 'In many cases the growers just banked the money and threw away the seedlings. Cojuangco's aim was just to get the government to buy his production, and then let the government dump it' (pers. comm., J. V. Romero, Manila, 25 January 1989). In marked contrast to the spread of IRRI rice, the Mawa variety remains a rare sight in the Philippine countryside.[9] The Agricultural Policy and Strategy Team of the University of the Philippines (1986, p. 249) claims that government funding for coconut breeding research was 'abruptly discontinued by the Philippine Coconut Authority when the MAWA variety was first earmarked for exclusive use in the replanting program, for now obvious reasons'.

The 1970s saw a boom in coconut oil milling in the Philippines, and a shift from copra to oil in the composition of the country's coconut exports (see Table 21.1). The milling boom was encouraged not only by reduced shipping costs, but also by government policies. These included higher export tariffs on copra (Tiglao, 1981, p. 30), and investment incentives which in the end led to the creation of substantial excess capacity in the milling industry.[10]

II. Terms of trade

Movement in the external terms of trade – the prices of exports relative to the prices of imports – is a crucial aspect of any nation's interactions with the world economy. For the Philippines, this trend has been quite adverse. Power (1983, p. 9) remarks, 'Few countries in the world have suffered as much from the

[8] The farm was on Bugsuk island, off the southern tip of Palawan. Cojuangco acquired most of the island in a trade with the Philippine government, giving up one ha in Tarlac province for every three in Bugsuk. 'In retrospect', a US Embassy (1980: 5–6) cable remarked, 'the trade appears particularly attractive, since most of the rice land in Tarlac subsequently came under the President's land reform program.'

[9] This may be fortunate, as the prospect of widespread adoption raised serious concerns about genetic vulnerability to crop disease epidemics (Banzon and Velasco, 1982: 43–4; Sangalang, 1987: 226–7). For further discussion of the hybrid replanting programme, see Tiglao (1981: 61, 85, 95–6), Sangalang (1987) and Habito (1987a: 195–202).

[10] Hawes (1987: 63–8) attributes this overcapacity to 'faulty planning by technocrats in government', and notes that a number of the new mills ultimately 'reneged on their loan payments and closed down'. An additional reason may have been the opportunities for profit and capital flight afforded by the procurement of milling equipment; see Boyce (1990: 40–1, 71–2).

movements of international prices.'[11] Terms of trade for coconut oil are reported in Table 21.3. In 1985 each barrel of coconut oil exported by the Philippines bought less than half as much in imports as it had in 1962. The decline in the terms of trade was by no means smooth, however, as prices fluctuated greatly, transmitting instability from the world economy to the Philippine economy.

Table 21.3 Terms of Trade for Philippine Coconut Oil, 1962–85

Year	*(1972 = 100)* Coconut oil export price index	Import price index	*Terms of trade*
1962	119.9	71.4	167.9
1963	133.7	76.2	175.5
1964	145.4	76.8	189.3
1965	159.8	78.1	204.6
1966	134.2	79.4	169.0
1967	142.3	81.2	175.2
1968	166.9	80.7	206.8
1969	140.1	82.7	169.4
1970	160.4	93.5	171.6
1971	143.9	95.5	150.7
1972	100.0	100.0	100.0
1973	198.8	128.8	154.3
1974	508.1	211.6	240.1
1975	208.7	219.6	95.0
1976	192.4	217.2	88.6
1977	296.8	241.2	123.1
1978	338.7	245.8	137.8
1979	512.6	270.1	189.8
1980	342.6	358.6	95.5
1981	284.3	398.6	71.3
1982	241.5	340.5	70.9
1983	286.8	342.4	83.8
1984	547.2	386.7	141.5
1985	295.7	363.8	81.3

Sources: Calculated from data in National Economic and Development Authority (1976, pp. 426–8, 434; 1986, pp. 364–5, 377).

To a certain extent, the decline and instability of coconut and other agricultural export prices can be understood as an outcome of market forces. An overabundance of agricultural commodities on the world market and intense competition among producing countries have been general features of the

[11] Measures of changes in terms of trade are notoriously sensitive to choice of the time period. Josling (1984: 10) reports that in the period 1970–9, the decline in the purchasing power of Philippine agricultural exports was about average for 79 developing countries.

post-war era. Moreover, agricultural commodities have long served as the textbook examples of the boom-and-bust price cycles which competitive markets can engender. The markets for Philippine agricultural exports are also characterized, however, by pervasive 'imperfections', and the unfavourable price movements reflect the interplay of economics and politics on a world scale.

The terms of trade for coconut were relatively stable in the mid-1960s, declined from 1968 to 1972, and then rose sharply in 1973 and 1974 – notwithstanding the petroleum import price increase of those years – thanks to a boom in the coconut market. This initiated several years of extraordinary price instability. Nominal coconut prices dropped to less than half the 1974 level in 1976, then doubled by 1979, and then collapsed again.[12] Meanwhile the Philippines faced an inexorable rise in nominal import prices. Hence, in spite of several ups and downs, the overall trend in the terms of trade was negative.[13]

This deterioration continued an earlier trend. Coconut prices moved generally downward after the Second World War: the world price of coconut oil had fallen from 26 US cents per pound in 1948 to 11 cents by 1962 (UNECAFE, 1969; p. 86). Between 1950 and 1962, the ratio of Philippine export prices to Philippine import prices had declined by 24 per cent.[14] Price instability is also a longstanding feature of the Philippine engagement with world markets. The price of coconut oil fell an astonishing 90 per cent in the post-war depression of 1921–2, and again by 70 per cent between 1928 and 1934 with the onset of the Great Depression (Tiglao, 1981, pp. 3, 6).

Although it is by far the world's largest coconut exporter, the Philippines exercises little market power owing to the availability of close substitutes. 'Coconut oil is a minor oil locked in a competitive battle', Hicks (1967, p. 212) observed, 'not so much with other producers of coconut oil but with other sources of oil.' Soyabean oil, groundnut oil, cottonseed oil, lard and tallow are among the major competitors in edible uses such as margarine and cooking oil. Petroleum-based synthetics are the major competitors in non-edible uses such as the manufacture of detergents, cosmetics, explosives and pharmaceuticals. Demand for coconut oil, and for the copra from which it is extracted, is therefore highly price-elastic.

The downward trend and instability of coconut prices can be explained primarily in terms of the world supply of competing substitutes. Producers of edible oils in the industrialized countries receive subsidies and other protection

[12] These fluctuations are even more dramatic in the monthly price data. Between January 1973 and December 1976, for example, coconut oil soared from US $168 per metric tonne to a peak of US $1,138 (in April 1974), and then dropped to US $305 (Hawes, 1987: 70).

[13] Between 1962 and 1985, the exponential trend for coconut oil prices (deflated by the import price index) was *minus* 3.6 per cent per year; for copra it was *minus* 4.8 per cent (growth rates estimated by ordinary least squares).

[14] Calculated from data in NEDA (1976: 434); see also Tryon (1967). This figure refers to all Philippine exports, but as noted above these were predominantly agricultural and forestry products.

from their respective governments. The effect of these policies is to increase the world supply of edible fats and oils, eroding the competitive position of Philippine coconut oil, and depressing world prices for all freely-traded fats and oils. The introduction of petroleum derivatives as substitutes for coconut oil in industrial uses began on a large scale in the late 1960s. In this case, an important constraint on the competitiveness of coconut oil has been the failure of market prices to capture the negative externalities in the production and use of the synthetics. Even within the Philippines, there are complaints that coconut oil-based soap 'is being replaced by petro-based detergent which is polluting our waterways' (Abadilla, 1987, p. 4). Here, as in a number of other sectors of the world economy, neglect of the social costs of pollution confers an illusion of efficiency upon the products of the petrochemical industry.[15] In addition, Hicks (1967, p. 201) noted 'an alarming contrast between the resources, research, and investment applied to the problem of reducing the cost of synthetic oils and the general neglect of the really substantial cost-reducing potential of natural oil'. This contrast reflects profound differences in the size and power of their respective producers. Large firms in the petroleum sector have the capacity not only to finance research and development, but also to internalize much of the resulting benefit streams. Natural oil producers must rely on public sector research, which has often been minimal.

Preferential trading arrangements with the United States played a key historical role in the growth of Philippine coconut exports in the face of the competitive and politicized world market for edible oils. The terms of Philippine access to the US market have reflected the balance among the competing interests of US edible oil producers, US firms engaged in the milling and export of coconut oil in the Philippines and US industrial consumers of coconut oil, all of which generated conflicting demands (Hicks, 1967, p. 135). The resulting tariff policies protected US edible oil producers from foreign competition, but applied substantially lower duties to Philippine oils than to those of other countries (Hicks, 1967, pp. 52–80; UNECAFE, 1969; pp. 56–7).

US tariff and commercial policy has been described as 'the most important factor in stimulating the expansion of the Philippine coconut industry' in the colonial period (Hawes, 1987, p. 61). The Philippines continued to receive preferential access to the US market after independence under the United States–Philippine Trade Agreement of 1946 and the Laurel–Langley Agreement of 1955. This special treatment was gradually reduced, however, and finally terminated in 1974 with the expiration of the Laurel–Langley Agreement. Stripped of this protection, the Philippine coconut industry became fully exposed to competition from producers of substitutes in the world market.

[15] Commoner (1990: 53) remarks, 'Nearly all the products of the petrochemical industry are substitutes for perfectly serviceable preexisting ones.'

III. Relations of production

'Coconut farms' are a distinct entity in Philippine agriculture, since coconut is often a farm's primary or sole crop, rather than one component of a diversified crop mix. A 1978 survey of coconut farms by the Ministry of Agriculture found an average total cropped area of 4.9 ha, of which 4 ha were under coconuts.[16] In recent years the intercropping of coconut with other crops, such as coffee, abaca and lanzones, has been increasing, but monoculture remains the rule.

The 4 ha average conceals significant farm-size disparities among coconut farms. The 1971 agricultural census found that the largest 10 per cent of coconut farms, with 10 ha or more, accounted for 42 per cent of total coconut area (Cornista and Pahm, 1987, p. 23). The expansion of coconut acreage sparked by the 1962 decontrol of foreign exchange occurred primarily in the larger farm-size classes. A survey conducted by Nyberg (1968, pp. 51–2) found that 84 per cent of new plantings were on farms of 10 ha or more, and that 'essentially no new plantings were being made on farms smaller than two hectares'.

The census data on farm-size distribution obscure the degree of concentration in coconut lands, since they refer to operational holdings rather than to ownership. For example, Hayami *et al.* (1990, p. 118) report that in an upland coconut village in Quezon a single landlord owned all the land and cultivated it by means of share tenants; the agricultural census would record each tenant's holding as a separate farm. Estimates of the extent of tenancy vary widely, from 22 to 68 per cent of all coconut farms (Cornista and Escueta, 1983, pp. 4–5). Tenancy is reportedly more widespread in the coconut regions of southern Luzon than in the newer, and typically larger, coconut farms of Mindanao. Coconut landlords include middle-class professionals, teachers, managers and military officers, many of them absentees (Putzel and Cunnington, 1989, pp. 13–15). Share tenancy is most common, with considerable variation in both the share and the responsibilities of the tenant. Variants include a 50–50 division of the copra with the tenant bearing all expenses in Quezon, or a 60–40 split in favour of the landowner if the latter bears certain expenses; a 2/3–1/3 division in favour of the landowner in Albay, again with the tenant bearing all expenses; and, in Laguna, a one-seventh share of the green nuts to the tenant whose job is simply to protect the trees and to clear the land between them (Cornista and Escueta, 1983, pp. 5–6; pers. comm., L. B. Cornista, Manila, 24 January 1989).

Although family and exchange labour remain important on the smaller coconut farms, Cornista and Escueta (1983, p. 2) report an increasing use of hired labour in production activities. Government surveys conducted in 1974–8 indicate that hired labour accounted for 61 per cent of total labour inputs on

[16] Cited by Hawes (1987: 57). In sandy, coastal zones this tendency towards monoculture reflects the special characteristics of soils, but in other places food crops such as rain-fed rice or white corn would be among the feasible alternatives.

coconut farms (Tiglao, 1981, pp. 38–40). David (1977) similarly estimates that in 1976 hired labourers accounted for 65 per cent of the 1,550,000 persons 'engaged in the actual act of farming on a coconut farm', and that of the remainder, 32 per cent were tenants and 3 per cent were owner-operators.[17]

Hired labourers are paid mainly on a piece-rate basis, which reduces the need for supervision and facilitates the use of family helpers by the hired worker (Tiglao, 1981, p. 44). Official data indicate that their *daily* wages tend to be slightly higher than those for agricultural labourers in rice, corn and sugar-cane cultivation (World Bank, 1985, p. 29). This may reflect longer working days, the use of family helpers by coconut workers and perhaps higher daily earnings to compensate for fewer days of employment, since the average *annual* earnings of coconut labourers are among the lowest of any occupational group in the country (Cornista and Escueta, 1983, pp. 10–11).

Unlike hired labourers in the other major export crops, coconut workers are generally unorganized, with 'no channel for the articulation of their interests and needs'. Nevertheless some coconut labourers have developed 'means of sharing poverty' at the local level, including the rotation of limited employment, the formation of work groups in which remuneration is shared equally among members and the inclusion in such groups of non-members in dire need of employment (Cornista and Escueta, 1983, pp. 12–13). In addition, Hawes (1990) reports that in some areas the coconut workers have been organized by the National Democratic Front.

Tiglao (1981, p. 51) estimates that gross farm income from copra production is, on average, partitioned as follows: 50 per cent to the owner of the land as ground rent; 11 per cent to the farm operator as profit; 37 per cent to wage and family labour; and 2 per cent to cover non-labour production costs. These must be regarded as rough approximations, which would vary from farm to farm and from year to year depending, for example, on the farmgate price of copra.[18] There is little doubt, however, that returns to land typically exceed returns to labour in coconut production. The pattern of ownership of coconut land is therefore a critical determinant of income distribution in the sector.

There is no evidence of economies of scale in coconut cultivation. Agricultural census data reveal no correlation between farm size and the number of nuts per tree.[19] Data published by the Philippine Coconut Authority (1988, p. 102) indicate that yields in Mindanao, where average holdings are larger, tend to be

[17] Cited by Guerrero (1985: 23–5). Mangahas (1985: 212–13) presents unpublished national sample survey data indicating that hired labourers accounted for a somewhat lower proportion, 48 per cent of the labour force in coconut farming in 1975.

[18] Tiglao's estimates are derived from data for the mid-1970s. Annual data on real wages of coconut labourers presented by the World Bank (1985: 29) for the years 1970–82 show a near-zero growth trend and much less instability than copra prices. The elasticity of the real wage with respect to the real price of copra for these years is −0.01 (estimated using the ratio of the nominal copra price index to import price index as reported in Table 21.3). There is thus little evidence that copra price movements 'trickle down' to coconut labourers.

[19] For 1980 data, see Hayami *et al.* (1990: 117).

higher than in Luzon, where average holdings are smaller, but Habito (1987b, p. 2) points out that environmental conditions are more favourable in Mindanao, and that the trees there are generally younger and hence more productive. On the basis of interviews in coconut-growing areas of Luzon, Hayami et al. (1990, p. 117) report that intercropping is more common and more intensive on small family farms than on larger wage-labour farms. This may reflect lower costs of labour and supervision on smaller farms.[20] One constraint on the spread of inter-cropping, however, has been the fear of landlords 'that if they allow farmworkers to inter-crop these areas with vegetables and other food crops, they may lose control of the land' (Putzel and Cunnington, 1989, p. 35). In some cases, according to Habito (1987a, p. 206), 'landlords actually prohibit their tenants from inter-cropping'. Because of these tensions over property rights, as well as the more widely recognized incentive problems, share tenancy poses 'a barrier to increased productivity on coconut farms' (Habito, 1987a, p. 220).[21]

In spite of its image as a smallholder's crop, then, coconut production in the Philippines is typically characterized by a dichotomy between ownership of land and labour on it. Land ownership in the coconut sector is more widely dispersed than in other major export crops in the Philippines, but most coconut labourers do not own the land on which they work. The result is not only inequitable, but also possibly inefficient.

IV. Relations of exchange

The overt aim of the Marcos regime's development strategy for export agri-culture was growth in output and export earnings. Behind the scenes, however, the regime aggressively pursued another agenda: the redistribution of income to favoured individuals. Most notably, Marcos deployed state power to put control of the country's top agricultural exports, coconut and sugar, securely into the hands of presidential 'cronies'. The result was a dramatic redivision of the agro-export income pie, with bigger slices for the privileged few and smaller slices for the rest.

Most copra in the Philippines is purchased from the growers by barrio (village) traders, who then sell it to town-based buyers, who in turn sell to the coconut mills and copra exporters. In the 1960s, according to Cernohous, there were more than 250,000 producers, roughly 10,000 barrio buyers, 3,400 town

[20] Ofreneo (1980: 104) remarks that many small owner-operators in the Philippines continue to grow coconuts even though they 'would be registering net losses if their unpaid labour and that of their families were included in the computation' of farm income when valued at the market wage. The same finding in Indian farm management studies in the 1950s sparked the development of the theoretical literature on the inverse relation between farm size and land productivity (Sen, 1962).

[21] However, in Laguna and Cavite, where proximity to the Manila market opens profitable opportunities for intercropping with fruit trees, landowners have planted lanzones using hired labour, often with no involvement of the coconut tenant (pers. comm., L. B. Cornista, Manila, 24 January 1989).

buyers, and fewer than 50 exporters and oil crushers. Superimposed on this pyramid was a network of credit ties, binding individuals at each layer to those above them. 'The trade is financed from the top of the channel down, with many a town buyer actually being a mere agent of either an exporter or crusher, and most barrio buyers being sub-agents of town buyers.' The result was a pattern of market power in which 'middlemen are essentially price takers *vis-à-vis* the group immediately above them, while at the same time being price makers *vis-à-vis* the group immediately below them' (Cernohous, 1966, p. 74).

The mass of producers, of course, are simply price takers. Nyberg (1968, p. 52) reported that large coconut plantations often received prices 10–20 per cent higher than those received by small growers, perhaps by virtue of scale economies in marketing and their ability to bypass the lowest link in the marketing chain. In addition, small coconut farmers often have relatively little bargaining power owing to their lack of storage facilities, their need for cash, and their indebtedness to the buyer. The growers' lack of timely price information also enhances the monopsony power of the buyer, though Cernohous (1966, p. 76) noted, 'even where the information reaches the farmer, in the absence of an actual competitive bid, it probably does not significantly alter his relative bargaining position.'

In some barrios there is only one buyer. Where there are several, the ability of growers to choose freely among them is often constrained by *suki* relationships, personalized ties between the grower and the barrio trader-creditor, which are social as well as economic. These ties do not completely subordinate the grower to the buyer, but neither is the grower completely free. Cornista (1981, p. 350) reports, on the basis of fieldwork in two Laguna villages, that a grower is obliged to sell to the *suki* buyer once the latter has extended credit.[22] If the buyer consistently offers an unfavourable price, however, the grower 'would seek out new buyers after paying his loan'. The extent of market power hence would hinge on the grower's ability to repay any accumulated debts to the copra buyer.

The sharp increase in coconut acreage in the 1960s and 1970s brought new buyers into the market, perhaps resulting in a diminution of local-level market power. Improved price information reduced the buyer's advantage in bargaining with individual growers. In some cases, a new practice has emerged in which several potential buyers submit sealed bids to the grower, who then can select the highest offer. Cornista reports that this practice, known in Laguna as *subasta*, is confined to big coconut growers and remains less prevalent than individual bargaining. Moreover:

> To counteract the effects of *subasta* the buyers agreed to allow one of them (usually the biggest *kapitalista*) to outbid the rest. The profits which accrued from the

[22] Similarly, in a 1977–8 survey of coconut marketing in southern Mindanao, the Ministry of Agriculture found that 50 per cent of producers obtained cash advances from copra buyers, 'thereby depriving them of the opportunity to sell at higher prices offered by other buyers' (Valiente *et al.*, 1979: 2). Also see Tiglao (1981: 67).

transaction were divided equally among them. In a way, a monopsonistic situation resulted. (Cornista, 1981: 340)

In the two villages studied by Cornista, this collusion eventually broke down as 'a number of coconut buyers started to act independently'.

The major change in coconut marketing during the Marcos era, however, was a dramatic concentration of market power at the top. This was achieved by the open exercise of political muscle. In the name of 'vertical integration' and 'rationalization', the Philippine coconut industry was consolidated under a single entity with effective control over virtually all copra purchases and over the production and sale of coconut oil on the domestic and export markets. The takeover was engineered by a series of Presidential Decrees.[23] The three key steps were the imposition of levies on all coconut production, the creation of a bank in which these monies were deposited interest-free and the purchase by that bank of the bulk of the country's oil-milling capacity. In theory, all of this was done for the benefit of the coconut growers, the nominal 'owners' of the assets purchased with the levies. In practice, the main beneficiaries were a few close political associates of President Marcos, notably 'coconut king' Eduardo Cojuangco and Defence Minister Juan Ponce Enrile.

The first levy, imposed by legislative action in 1971, was to be used to provide credit to growers, to invest in the industry, and to finance the Philippine Coconut Producers' Federation, known as COCOFED, an association of large coconut landowners which had lobbied for passage of the law. A second, much larger levy was imposed by Presidential Decree in 1973. Known as the Coconut Consumers Stabilization Fund (CCSF) levy, its initial rationale was to subsidize domestic consumption of coconut products at a time of unprecedented high prices. In the next two years further decrees expanded the uses of the CCSF levy to include the establishment of Cojuangco's hybrid coconut seed farm and the acquisition of a commercial bank 'for the benefit of the coconut farmers'.

The United Coconut Planters' Bank (UCPB), with Cojuangco as president and Enrile as chairman of the board, soon became one of the largest commercial banks in the Philippines.[24] The interest-free deposit base provided by levy funds gave the bank a unique advantage. In January 1979, UCPB acquired Legaspi Oil Company, which milled a quarter of all Philippine coconut oil exports, in 'one of the largest corporate takeovers in Philippine history' (Tiglao, 1981, p. 88). Later in the same year UCPB bought Granexport Corporation from the US firm Cargill. These and other mills were placed under the control of the newly formed United Coconut Oil Mills (UNICOM), which by

[23] For more detailed accounts, see Tiglao (1981: 80–92), Sacerdoti (1982), Clarete and Roumasset (1983: 14-21) and Hawes (1987: 68–80).

[24] A US Embassy (1980: 7) cable ranked UCPB fourth among domestic private banks in liquid assets, third in deposits, fourth in net worth, and first in total investment.

1980 had cornered more than 80 per cent of the country's entire oil-milling capacity.[25]

Overcapacity in coconut milling, which had been stimulated by government Board of Investment incentives in the 1970s, contributed to the willingness of firms to sell out to UNICOM. An added push came from a 1978 Presidential Decree, which provided that subsidies funded by the CCSF levy would be provided only to mills owned and controlled by 'the coconut farmers', that is, by the UCPB and UNICOM. According to a US Embassy cable, for the Filipino owners of relatively small mills, 'It was sell or else!' Large foreign mills were not subjected to 'direct pressure', but 'were simply put in a position where the owners believed that it was in their best interest to sell' (US Embassy, 1980, p. 11). President Marcos characterized these developments as a historic triumph for the coconut growers: 'For half a century, the coconut farmers were the forgotten men of the country. Now you are no longer just coconut planters, you are bankers, owners of a coco mill complex' (quoted by Tiglao, 1981, p. 92).

A rather different picture emerges from other sources, among them US government cables obtained under the Freedom of Information Act. A May 1980 cable, marked 'confidential', offered the following candid assessment:

> Since martial law was declared in September 1972, the coconut industry has been steadily brought under the influence of a small group of people, chief among whom are [name deleted] and Eduardo 'Danding' Cojuangco, both men long and close political associates of President Marcos. The prime motivation appears to be near total control of the coconut industry. There are four reasons which might explain why President Marcos would implicitly support, and even aid and encourage [names deleted], men whom he trusts in this effort.
> – First, control of the coconut industry provides President Marcos [deleted] with additional political and financial leverage to remain in power.
> – Second, control of the industry by men close to the President denies that control to anyone else.
> – Third, Marcos appears to use this method to reward his associates in the business community, the military and the bureaucracy.
> – Finally, control of the industry has allowed the Philippines to attempt to better its terms of trade for coconut oil sold on the world market. (US Embassy, 1980, pp. 1–2)[26]

The total 'surplus' extracted from coconut producers by the Marcos–Cojuangco–Enrile combine was substantial indeed. A 1984 US Embassy cable reported that total collections levied on coconut producers since 1973 amounted

[25] This control was subsequently consolidated further. A 1984 US Embassy cable reported: 'UNICOM mills, those which have toll [oil?] crushing contracts with UNICOM, and those which "cooperate" with UNICOM, account for more than 85 per cent of the country's crushing capacity. Even the "independent mills" are constrained by PCA [Philippine Coconut Authority] regulations to course their coconut oil exports through UNICOM. Copra exports are prohibited completely' (US Embassy, 1984: 4). Canlas and Alburo (1989: 93) report that by 1983 UNICOM directly controlled 93 per cent of milling capacity.

[26] The deleted name is evidently former Defence Minister Juan Ponce Enrile, whose name inadvertently was not blacked out on page 6 of the same document.

to 9.26 billion pesos, equivalent to more than US $1.1 billion at prevailing rates of exchange. In addition, UNICOM used its control over prices to establish profit margins of two or three pesos per kilo of copra, compared to a 'good' normal margin of fifty centavos; the Embassy calculated that in 1983 alone, each peso of margin netted UNICOM an extra US $214 million. At the same time, the Embassy reported that UNICOM officially undervalued its coconut oil exports, the difference between the actual and stated value being 'deposited in dollar accounts abroad or used to fund various Cojuangco projects'. The cable reported that, on top of these exactions, 'Cojuangco has found many indirect methods of profiting from the monopoly'. For example, equipment and materials purchases by UNICOM were routed through a company operated by Cojuangco's son, who 'takes a ten per cent commission on all purchases' (US Embassy, 1984, pp. 11, 14, 15).

The Embassy estimated that the income personally accumulated by Cojuangco through the coconut industry ranged 'from several hundred million dollars to over a billion' (US Embassy, 1984, p. 16). It also reported that:

> Cojuangco supports the President in many different ways. It is generally believed that Cojuangco shares the spoils of the coconut monopoly with the President, although details of amounts and methods of payment are lacking. It is rumored that he handles some of the President's own investments. As a regional KBL chairman [Marcos's political party] in Central Luzon, Cojuangco is responsible for keeping the party faithful in line and for promoting KBL victories at the polls . . . On a more personal level, Cojuangco offered Leer [sic] jets as wedding presents for Irene Marcos [the President's daughter] and her spouse when they were married last year. President and Mrs Marcos eventually received the gifts when the newlyweds declined them. (US Embassy, 1984, p. 20)

Some of the wealth extracted from the coconut sector was used to enter other industries via takeovers of existing firms. For example, Cojuangco purchased large blocks of stock in the San Miguel Corporation, the largest private corporation in the Philippines and the largest food and beverage firm in Asia, and became its vice-chairman (US Embassy, 1984, p. 17).[27] A substantial fraction was undoubtedly transferred abroad, a component of the estimated US $20 billion in capital flight from the Philippines from 1962 to 1986 (Boyce, 1990, p. 51).

The coconut cabal was much less successful in wielding market power to secure higher prices for coconut oil abroad, owing to the ease of substitution among competing oils. An attempt to establish a 'COCOPEC' cartel failed completely; in the words of the US Embassy (1984, pp. 13–14) cable, 'All UNICOM got for its efforts was anti-trust suits in the US.'[28] The takeover of the coconut industry thus did nothing to enlarge the Philippine economic pie, but only changed the way it was sliced.

Within the Philippines, there was little organized resistance to the takeover.

[27] With annual sales of US $653 million, San Miguel ranked 125th among Third World public and private corporations in 1984 (*South*, 1984).

[28] For details, see Bonner (1987: 326–30).

This was due in part to the poverty and geographic dispersion of the coconut growers, but outright intimidation also played a role. In July 1982, former Vice-President Emmanuel Pelaez, the leading critic of the coconut monopoly in the Philippine legislature, was seriously wounded and his driver killed in an ambush.[29]

The most influential opposition came instead from the World Bank, which pressed for a 'return of market forces' to the coconut industry (US Embassy, 1984, p. 25). The International Monetary Fund likewise is reported to have demanded the dismantling of UNICOM as a condition for a standby credit. In response, a January 1985 Presidential Decree abolished UNICOM and replaced it by a 'cooperative endeavor' with sole legal authority to export coconut oil (Espiritu, 1987). The very limited nature of this reform is evident from an August 1985 US Embassy memorandum which reported that 'current members of the coconut exporters cooperative have agreed to give Danding Cojuangco 32 centavos per kilogram of coconut oil exported and 20 centavos per kilogram of other coconut product exports', and that the money 'will be used for the coming elections'. The memorandum added, 'Cojuangco has sole control of the money and no audit is made' (US Embassy, 1985, p. 2).

In sum, the market structure of the Philippine coconut industry underwent a major concentration at the top during the 1970s and early 1980s. The primary motive for this transformation, and its major effect, was redistribution of the income generated by the country's leading agricultural export. The beneficiaries were a handful of politically powerful individuals. The losers included millers and traders driven out of the industry, and above all the coconut growers, who experienced an intensification of the monopsonistic environment in which they sold their product.[30]

V. Concluding remarks

The story of the Philippine coconut industry reveals two limitations of comparative advantage as a guide to trade strategy. The first was discussed in Section II: the dynamic phenomena of declining terms of trade and price instability. The second was discussed in Sections III and IV: within a country, one person's advantage can be another's disadvantage.

On the world stage, the coconut king wields little power, owing to competition from natural and synthetic substitutes. In recent decades, the size of

[29] The US Embassy (1984: 23) reported to Washington, 'It was universally assumed that Cojuangco was behind the attempt on Pelaez's life.'

[30] After the collapse of the Marcos regime, the spread between farmgate and mill-gate copra prices dropped, suggesting that by the mid-1980s monopsony had intensified at the local level, too. Philippine Coconut Authority (1988: 138) data indicate that the average spread declined from 1.70 pesos per kilogram in 1983–5 to 58 centavos in 1987. With total annual production of 2 million metric tonnes, this difference would amount to 2.25 billion pesos, or more than US $100 million, per year. See also Lopez (1987).

the Philippine coconut-earnings pie has dwindled. Governmental support to competing oilseed producers in other countries, failure to internalize the external costs of synthetics, and relative neglect of research and development in coconuts and coconut products have contributed to this result.

With little leverage in the international arena, the coconut king in the Marcos era turned his entrepreneurial talents to extracting more tribute from his subjects at home. 'It was our best and brightest, Harvard-trained lawyers, against the no-read-no-writes, the fourth-grade dropouts', explains Philippine Coconut Authority Chairman Jose V. Romero Jr (pers. comm., Manila, 25 January 1989). Like the walrus in Lewis Carroll's *Through the Looking Glass*, the architects of the coconut monopsony cloaked their predations with sanctimonious expressions of concern for the welfare of their victims. But here, as elsewhere, the Marcos regime demonstrated that the pursuit of self-interest does not necessarily advance the public interest.

Of course, the distributional outcome recounted above is not inherent in export agriculture. Land reform and market reform could redirect income to those whose labour produces the crops. However, the external constraints imposed by the world market would remain. Hence a case can be made not only for redistribution, but also for diversification via allocation of land and labour to other uses.[31]

Coconut 'rent seeking' in the Marcos era not only exacerbated economic and political inequalities, but also can be understood, at least in part, as a *consequence* of these inequalities. Unchecked by a more equitable distribution of wealth and power, the ruling élite was free to pursue its self-interest even at the expense of the public interest. A comparison between the Philippines and its East Asian neighbours may be instructive in this regard. China, Japan, Korea and Taiwan are by no means models of pure egalitarianism, but after the Second World War each had a redistributive land reform which broke the power of the landed oligarchy. In the Philippines this historic task has yet to be accomplished.

The Marcos regime has passed into history, but the entrenched interests in the Philippine coconut economy have proven quite resilient. Six years after Marcos's downfall, export agriculture in the Philippines remains virtually untouched by land reform. Although the Aquino government quickly moved to dismantle the coconut monopsony, the fate of the coconut levy funds, now valued at 30 billion pesos (more than US $1 billion), remains undecided (Malabed, 1991, p. 6). Juan Ponce Enrile and Eduardo Cojuangco Jr are reported to be 'back into the coconut business' in the provinces of Batangas and Quezon, respectively (Cloa, 1991, p. 5). In the looking-glass world of the political economy of coconuts, Humpty Dumpty may indeed be put together again.

[31] In its 'Green Paper' for the Aquino government, the Agricultural Policy and Strategy Team (1986, p. 263) concluded: 'Now is the time to introduce diversification in the Philippine agricultural sector to correct the overdependence on a few export crops and to avoid a position of vulnerability.'

Notes

This article draws on my book, *The Philippines: The Political Economy of Growth and Impoverishment in the Marcos Era* (forthcoming, Macmillan). I am grateful to the Development Centre of the Organisation for Economic Co-operation and Development which commissioned that book; to the Joint Committee on Southeast Asia of the Social Science Research Council and American Council of Learned Societies for additional support; and to Craig Nelson of the National Security Archive in Washington, DC, for assistance in obtaining documents released under the US Freedom of Information Act. For thoughtful comments on earlier drafts, I am indebted to Howarth Bouis, Gary Hawes, José E. R. Ledesma, James Putzel and three anonymous referees of this journal. Responsibility for the views expressed here and for any errors is mine alone.

References

Abadilla, D. C. 1987: The Coconut Jigsaw Puzzle. *Malaya*, 6 July: 4–5.

Agricultural Policy and Strategy Team 1986: *Agenda for Action for the Philippine Rural Sector*. Manila: University of the Philippines at Los Banos (UPLB) Agricultural Policy Research Program and Philippine Institute for Development Studies. (The 'Green Paper'.)

Banzon, J. A. and Velasco, J. R. 1982: *Coconut Production and Utilization*. Manila: Philippine Coconut Research and Development Foundation.

Barker, R. 1978: Barriers to Efficient Capital Investment in Agriculture, in T. W. Schultz (ed.) *Distortions in Agricultural Incentives* 140–60. Bloomington: Indiana UP.

Bonner, R. 1987: *Waltzing with a Dictator: The Marcoses and the Making of American Policy*. New York: Times Books.

Boyce, J. K. 1990: *The Political Economy of External Indebtedness: A Case Study of the Philippines*. Makati: Philippine Institute for Development Studies, Monograph Series No. 12.

—— (forthcoming) *The Philippines: The Political Economy of Growth and Impoverishment in the Marcos Era*. London: Macmillan.

Canlas, D. B. and Alburo, F. A. 1989: Trade, Protectionism, and Vertical Integration in Agro-Industrial Processing: The Philippine Coconut Oil Industry, in J. T. L. Hoe and S. Sharma (eds) *Trade, Protectionism, and Industrial Adjustment in Vegetable Oils: Asian Responses to North America*, 75–94. Singapore: Institute of Southeast Asian Studies.

Carroll, L. 1922: *The Complete Works of Lewis Carroll*. New York: Random House.

Cernohous, Z. 1966: The Marketing of Agricultural Products in the Philippines. *Philippine Economic Journal* 5(1): 64–94.

Clarete, R. L. and Roumasset, J. A. 1983: An Analysis of the Economic Policies Affecting the Philippine Coconut Industry, PIDS Working Paper 83–08. Makati: Philippine Institute for Development Studies.

Cloa, J. 1991: Policies in the Coconut Industry: From Marcos to Aquino. *Farm News and Views*. Quezon City: Philippine Peasant Institute 3 (6), 4–5.

Commoner, B. 1990: *Making Peace with the Planet*. New York: Pantheon.

Cornista, L. B. 1981: Social Dynamics of Coconut Farming: The Case of Two Philippine Villages, unpublished PhD dissertation, University of Wisconsin at Madison.

—— and Escueta, E. F. 1983: The Structure of the Coconut Farming Industry. Los Banos: UPLB, Agrarian Reform Institute, Occasional Paper No. 10.

—— and Pahm, E. A. 1987: Directions of Agrarian Reform in the Coconut Sector. Los Banos: UPLB, Institute of Agrarian Studies, Occasional Paper No. 17.

David, V. 1977: The Barriers in the Development of the Philippine Coconut Industry, unpublished MBA dissertation. Ateneo de Manila University.

Espiritu, C. E. 1987: Monopoly Still Controls Local Coconut Industry. *Business Star* (Manila), 4 August: 1, 2.

Evenson, R. E., Waggoner, P. and Bloom, P. 1980: Investing in Technology for Philippine Agriculture, report for the USAID Asia Bureau/University of Minnesota Asian Agricultural Research Review (mimeo).

Galang, J. 1988: The Coconut Crisis. *Far Eastern Economic Review*, 24 November: 72–3.

Guerrero, S. H. 1985: A Review of Welfare Issues in the Coconut Industry. PIDS Working Paper 85–01. Makati: Philippine Institute for Development Studies.

Habito, C. 1987a: Policy Issues in the Coconut Industry: A Survey, in UPLB Agricultural Policy Working Group. *Policy Issues on the Philippine Rice Economy and Agricultural Trade* 193–224. College, Laguna: UPLB, Center for Policy and Development Studies.

—— (1987b): Should Scale Economies Stand in the Way of Philippine Agrarian Reform? paper presented to the Philippine Economic Society, Manila (29 May).

Hawes, G. 1987: *The Philippine State and the Marcos Regime: The Politics of Export*. Ithaca, NY: Cornell UP.

—— 1990: Theories of Peasant Revolution: A Critique and Contribution from the Philippines. *World Politics* 42 (2), 261–98.

Hayami, Y., Quisumbing, M. A. and Adriano, L. 1990: *Toward an Alternative Land Reform Paradigm: A Philippine Perspective*. Quezon City: Ateneo de Manila UP.

Hicks, G. L. 1967: The Philippine Coconut Industry: Growth and Change, 1900–1965. Washington, DC: National Planning Association, Center for Development Planning, Field Work Report No. 17.

—— and McNicoll, G. 1971: *Trade and Growth in the Philippines: An Open Dual Economy*. Ithaca, NY: Cornell UP.

Hooley, R. and Ruttan, V. W. 1969: The Philippines, in R. T. Shand (ed.) *Agricultural Development in Asia*, 215–50. Canberra: Australian National UP.

Josling, T. 1984: Agricultural Prices and Export Earnings: The Experience of Developing Countries in the 1970s. Rome: FAO Economic and Social Development Paper No. 3.

Kikuchi, M. and Hayami, Y. 1978: Agricultural Growth against a Land Resource Constraint: A Comparative History of Japan, Taiwan, Korea, and the Philippines. *Journal of Economic History* 38 (4), 839–64.

Legarda, B. 1962: Foreign Exchange Decontrol and the Redirection of Income Flows. *Philippine Economic Journal* 1 (1), 18–27.

Lopez, M. E. 1987: Has the 'Coconut Monopoly' Really Been Cracked? Manila: Center for Research and Communication Staff Memo No. 2: 2–5.

Malabed, J. 1991: Coco Levy: One Big Rip-off. *Farm News and Views* (Quezon City: Philippine Peasant Institute) 3 (6), 6–9.

Mangahas, M. 1985: Growth, Equity and Public Policy in the Rural Sector: A Philippine Case Study with Subnational Data, draft report submitted to the Food and Agriculture Organization of the United Nations, November.

National Economic and Development Authority (NEDA) 1976: *Statistical Yearbook of the Philippines 1976*. Manila: NEDA.

——, 1986: *1986 Philippine Statistical Yearbook*. Manila: NEDA.

Nyberg, A. J. 1968: Growth in the Philippine Coconut Industry, 1901–1966. *Philippine Economic Journal* 7 (1), 42–52.

Ofreneo, R. E. 1980: *Capitalism in Philippine Agriculture*. Quezon City: Foundation for Nationalist Studies.

Philippine Coconut Authority 1988: *Coconut Industry Yearbook*. Manila: PCA.

Power, J. H. 1983: Response to Balance of Payments Crises in the 1970's: Korea and the Philippines. PIDS Staff Paper Series No. 83–05. Makati: Philippine Institute for Development Studies.

Putzel, J. and Cunnington, J. 1989: *Gaining Ground: Agrarian Reform in the Philippines*. London: War on Want.

Sacerdoti, G. 1982: Out of a Cooking-oil Crisis – A Multipurpose Giant. *Far Eastern Economic Review*, 8 January: 44–8.

Sangalang, J. B. 1987: The Coconut Replanting Program, in UPLB Agricultural Policy Working Group, *Policy Issues on the Philippine Rice Economy and Agricultural Trade*, 225–45. Los Banos: UPLB, Center for Policy and Development Studies.

Sen, A. K. 1962: An Aspect of Indian Agriculture. *Economic Weekly* 14 (4–6), 243–6.

South 1984: South 500, July: 50–64 (London).

Tadem, E. C. 1980: *Mindanao Report: A Preliminary Study on the Economic Origins of Social Unrest*. Davao City: Afrim Resource Center.

Tiglao, R. 1981: *Looking into Coconuts: The Philippine Coconut Industry, Export-Oriented Agricultural Growth*. Davao City: Alternate Resource Center.

Treadgold, M. and Hooley, R. W. 1967: Decontrol and the Redirection of Income Flows: A Second Look. *Philippine Economic Journal* 6 (2), 109–28.

Tryon, J. L. 1967: Internal and External Terms of Trade in Post-War Philippines. *Philippine Economic Journal* 6 (2), 189–209.

United Nations Economic Commission for Asia and the Far East (UNECAFE) 1969: *The Coconut Industry of Asia*. Regional Plan Harmonization and Integration Studies Series No. 1. Bangkok and New York: UNECAFE.

US Embassy 1980: Concentration of Power in the Philippine Coconut Industry: Implications for Investment and Trade. Cable No. A-47 from US Embassy, Manila, to Department of State, 9 May.

——, 1984: The Philippine Coconut Monopoly. Cable No. 05607 from US Embassy, Manila, to Secretary of State, Washington, DC, March.

——, 1985: Confidential Memorandum of Conversation, Subject: Meeting with [deleted], Participants: [deleted], John H. Penfold, Economic Counselor, AmEmbassy, Manila, John H. Andre II, Economic Officer, AmEmbassy, Manila, 8 August.

Valiente, A. M., Jr *et al.* 1979: Coconut Socio-Economic and Marketing Study: Southern Mindanao. Manila: Ministry of Agriculture (mimeo).

Woodruff, J. G. 1979: *Coconuts: Production, Processing, Products* (2nd edn). Westport, CT: AVI Publishing.

World Bank 1973: Current Economic Position and Prospects of the Philippines. Vol. I. Report No. 78-PH, 20 April. Washington, DC: World Bank.
——, 1985: The Philippines: Recent Trends in Poverty, Employment and Wages. Report No. 5456-PH, 20 June. Washington, DC: World Bank.

22 Peter Bauer, 'Foreign Aid: Central Component of World Development?'

Excerpt from: P. Bauer, *The Development Frontier: Essays in Applied Economics*, pp. 38–49. Hemel Hempstead: Harvester Wheatsheaf (1991)

In the United States, *foreign aid* has become the accepted term for the policy of official subsidies in the form of grants and soft loans from governments of relatively prosperous countries to those of less prosperous ones. In Europe, *development aid* is the term most often used.

Foreign aid and *development aid* are misleading expressions. To call official wealth transfers 'aid' promotes an unquestioning attitude. It disarms criticism, obscures realities, and prejudges results. Who can be against aid to the less fortunate? The term has enabled aid supporters to claim a monopoly of compassion and to dismiss critics as lacking in understanding and sympathy. To paraphrase Thomas Sowell, aid is a major example of a policy which allows intellectuals and politicians to be on the side of the angels at a low apparent cost. The term also clearly implies that the policy must benefit the population of the recipient countries, which is not the case. If these transfers were generally known as government-to-government subsidies or subventions, terminology would encourage more systematic assessment.[1]

Unfortunately, the term *foreign aid* is now so widely used that it is not possible to avoid it in a book intended for both academic and wider readership. I shall use interchangeably the terms *transfer, subsidies*, and *aid*, and occasionally refer to *aid-recipient countries*. But it should be remembered

[1] The familiar terms *bilateral* and *multilateral aid* are also misleading in implying reciprocity between donors and recipients. There is no such reciprocity since donors give and recipients receive. Under bilateral programs the resources go direct from donor to recipient governments, and under multilateral programs they go through international organizations for subsequent reallocation.

that the recipients of official aid are always governments. They are not the poor, destitute, or starving people shown in aid propaganda.[2]

There are presumably also other reasons for the prevalent uncritical attitude toward the policy of official subsidies to governments of poorer countries. One is the familiar but facile identification of governments with the population at large. Perhaps more important is the belief that rich Western countries can readily afford to give away a small fraction of their income, which may do some good to the peoples of the recipients and cannot possibly harm their prospects. This latter belief ignores the political, social, and cultural repercussions of these subsidies.[3]

There are other forces favoring an uncritical attitude, such as a widely articulated feeling of guilt in the West, and the operation of political, administrative, and commercial interest groups. Altogether, a climate of opinion has emerged in which only supporters of these subsidies are regarded as expert, and dissenters can be readily dismissed as lacking in both understanding and compassion. The extension of these subsidies to Eastern European governments underlines the importance of a more thoughtful attitude to this policy.

Unquestioning support for aid helps to explain many anomalies. Western aid has gone to governments explicitly hostile to the West, such as those of Ethiopia, Vietnam, and Cuba; to governments at war with each other (which has enabled recipients to claim that the donors support their enemy); to governments whose policies have created refugees; and to governments of other countries who have taken them in, at high cost and with resultant tensions. Examples include the flow of refugees from Vietnam to other Southeast Asian countries, such as the Philippines, Malaysia, and Indonesia.[4] The governments of all these countries have received Western aid simultaneously throughout the relevant periods. A notable anomaly occurred at the height of the Falklands War in 1982, when Britain was supplying aid to the Argentine government under the United Nations Development Program at a time when that government was deploying expensive, sophisticated weaponry against British forces. This episode evoked no protest in Britain.

[2] In some instances, subsidies go through governments rather than to them. In our context the distinction is immaterial because the direction and use of the funds require government approval in the recipient countries.

[3] This disregard accords with much modern macroeconomic theory, including unrefined Keynesian methodology. Here is a key passage from *The General Theory of Employment, Interest and Money* (London, 1936), p. 245: 'We take as given the existing skill and quantity of available labor, the existing quality and quantity of available equipment, the existing technique, the degree of competition, the tastes and habits of the consumer, the disutility of different intensities of labor and of the activities of supervision and organization, as well as the social structure.' It is debatable whether such assumptions are helpful for the analysis of fluctuations in output and employment in advanced industrial economies. They are plainly inappropriate for the examination of the factors behind the economic performance and advance of entire societies. For this purpose, acceptance of these assumptions deprives the analysis of all predictive power.'

[4] In the 1980s, multi-million-dollar official relief funds were set up to help refugees who had fled as a result of persecution by aid-recipient governments.

Aid has gone and still goes to governments that severely restrict the inflow of capital, the shortage of which is said to be the reason for aid. More generally still, these subsidies go to governments pursuing policies that plainly retard economic advance and damage the interests of their poorest subjects.[5]

'Foreign aid is the central component of world development.' This was said by Professor Hollis B. Chenery in 1981 when he was vice-president of the World Bank in charge of economic research.[6] He could not have been right. Large-scale development takes place in many parts of the world without foreign aid, and did so long before this policy was invented some forty years ago.[7]

Though evidently not a central component of development, since the Second World War foreign aid has been the centerpiece of academic and public discussion on the economic prospects of Asia, Africa, and Latin America and on the relation between the West and these regions. Whatever its impact on development, foreign aid has had far-reaching results. For instance, it has brought into existence as concept and collectivity the Third World, also called the South. Foreign aid is the source of the North-South conflict, not its solution. The primary significance of aid lies in this important political result. It has also promoted the politicization of life in the recipient countries. These results have been damaging both to the West and to the peoples of the less developed world. The amount of money spent by the West in no way measures these effects.

Discussion on foreign aid envisages the world as being one-third rich (the West) and two-thirds poor (the Third World or South). In this picture, extreme poverty is the common and distinguishing characteristic of the Third World. But there is a continuous range in the per capita incomes of countries. The absence of a distinct break in the series undermines the concept of a Third World demarcated from the West on the basis of per capita incomes. The line of division between rich and poor countries is quite arbitrary. One could equally well say that the world is two-thirds rich and one-third poor.

The picture is misleading also in that many groups or societies in Third World countries, especially in the Far East, the Middle East, Southeast Asia, and Latin America, are richer than large groups in Western countries. Nor is the Third World stagnant. Both before and after the Second World War, many Third World countries grew rapidly, including South Korea, Taiwan, Thailand, Malaysia, Singapore, the Ivory Coast, Mexico, Venezuela, Colombia, and Brazil.

[5] Further and detailed examples will be found in P. T. Bauer, *Reality and Rhetoric: Studies in the Economics of Development* (London and Cambridge, Mass., 1984), chap. 3.

[6] *New York Times*, 1 March 1981.

[7] In accordance with standard practice, foreign aid in this essay refers to official economic aid, that is, gifts from donor governments to recipient governments, both directly and also indirectly through international organizations. It includes the grant element in subsidized loans. It excludes military aid, private investment, and the activities of charities; the West includes Japan, Australia, and New Zealand – that is, it refers to the OECD countries.

It is not sensible to lump together and average the incomes of the very different societies of the Third World or South, which comprise at least two-thirds of mankind. What is there in common between, say, Thailand and Mozambique, Nepal and Argentina, India and Chad? Their societies live in widely differing physical and social environments and display radically different attitudes and modes of conduct. The Third World includes millions of aborigines and pygmies, and also peoples with ancient and sophisticated cultures, and others employing highly advanced methods of business and technology. It is both misleading and condescending to treat the richly varied humanity of the majority of mankind as if it were much of a muchness, an undifferentiated, uniform, stagnant mass, a mass, furthermore, that could not emerge from this state without external subsidies.

Nor is brotherhood a common characteristic of the Third World, as is evident from the persistent hostility and even armed conflict between many Third World countries, including India and Pakistan, Iraq and Iran, and many others, not to mention the numerous civil wars. It is not surprising that attempts at organizing economic cooperation within the Third World have failed, except for collective bargaining with the West over aid and related matters. Moreover, this collective bargaining is often organized and financed by the West.

The common characteristic of the Third World is the receipt of foreign aid and not poverty, stagnation, exploitation, brotherhood, or skin color. The concept of the Third World and the policy of official aid are inseparable. The Third World is merely a name for the collection of countries whose governments, with occasional and odd exceptions, demand and receive official aid from the West. This is the only bond joining its diverse and often antagonistic and warring constituents, which have come to be lumped together since the late 1940s as the underdeveloped world, the less developed world, the nonaligned world, the developing world, the Third World, or the South.

The Third World is, moreover, a progeny of the West. Aid was introduced and has always been organized by the West. It began with President Truman's Point Four Program of 1949. He urged bold measures to help the less developed countries, where, he said, over half of mankind was living in sickness and wretchedness.

In creating the Third World, the West has created an entity hostile to itself. Some individual Third World countries have been neutral or even friendly to the West, but the organized and articulate Third World is critical or hostile.

Foreign aid also encouraged the notion of the West (or North) as a single economic decision-making entity, a homogeneous aggregate with identical interests capable of imposing its will on the Third World. In fact, Western governments do not cooperate in setting market prices, and foreign suppliers compete for business in the Third World markets. Obvious examples include suppliers of manufactured products such as cars, trucks, and chemical products, and construction companies. Manufacturers and traders compete vigorously in the purchase of exports from the Third World.

Although the case for aid goes largely unquestioned, various arguments are often heard, ranging from restitution of alleged wrongs to preservation of the African elephant. The three most persistently influential are the promotion of development, the relief of poverty, and the political and economic interests of the donors. Official Western aid is therefore envisaged as being simultaneously a moral, political, and economic imperative.

Since its inception in the early postwar years, the central argument for foreign aid has been that without it Third World countries could not progress at a tolerable rate, if at all. In fact, external donations have never been necessary for the development of any society anywhere. (As shown later in this chapter Marshall aid to postwar Europe provides no exception to this statement.) Economic achievement depends on personal, cultural, social, and political factors, that is, people's own faculties, motivations, and mores, their institutions, and the policies of their rulers. In short, economic achievement depends on the conduct of people and that of their governments.

It diminishes the people of the Third World to suggest that, although they crave material progress, unlike the people of the West they cannot achieve it without subsidies. Much of the Third World progressed rapidly long before foreign aid – witness Southeast Asia, West Africa, and Latin America. Large parts of these regions were transformed in the hundred years or so before aid. There are, of course, Third World societies that have not progressed much over the last hundred years. This lack of progress reflects factors that cannot be overcome by aid, and are indeed likely to be reinforced by it. The notion that poor countries cannot progress without subsidies derives from the hypothesis of the vicious circle of poverty and stagnation, a major theme of development economics since the Second World War. It was concisely formulated by the Nobel laureate Paul Samuelson: 'They [the backward nations] cannot get their heads above water because their production is so low that they can spare nothing for capital formation by which the standard of living could be raised.'[8] This hypothesis is refuted by every individual family, group, community, or country that has emerged from poverty without subsidies, and that has often done so within a short space of time. Indeed, if the hypothesis were valid, the world would still be in the Old Stone Age.[9]

If a hypothesis conflicts with empirical evidence, this means that the model behind it is defective in that the variables specified or implied are either relatively unimportant as determinants or do not interact in the fashion

[8] Paul A. Samuelson, *Economics: An Introductory Analysis*, 2nd edn (New York, 1951), p. 49.

[9] The world is a closed system which has not received resources from outside. All developed countries began as underdeveloped. Academic readers will recognize the model underlying the vicious circle of poverty and stagnation. The crucial variables and relationships of this model are the following: the growth of income is a function of the rate of capital accumulation, that is, of investment; investment depends on saving; and saving is a function of income. Hence the growth of income depends on the growth of capital, and the growth of capital depends on the growth of income. The model behind the thesis of the vicious circle of poverty pivots on the notion that the low level of income itself prevents the capital formation required to raise income.

implied. Both defects mar the hypothesis of the vicious circle. The growth of income does not depend on the volume of saving and investment, and poor people can save and invest sufficient amounts to emerge from poverty.

The argument for aid as necessary for development rests on the belief that possession of capital is critical for economic advance. If this were so, how is it that large numbers of very poor people could have become prosperous within a few years without donations, as they have done the world over? Evident examples include very poor immigrant communities in North America and Southeast Asia. To have capital is the result of economic achievement, not its precondition.

Poor people can and often do save and invest sufficient amounts to emerge from poverty. They can save enough from small incomes for direct investment in trading, agriculture, and other purposes. They can also work harder or longer or redeploy their resources more productively to improve their lot.

And much recent research by leading scholars, including Simon Kuznets, has confirmed that increase in capital played a minor role in the economic advance of the West in recent centuries.[10] Moreover, these findings refer to capital formation. They apply even more to the volume of investible funds because much spending, conventionally termed *investment*, does not result in productive capital formation. Furthermore, the findings refer to capital formation in societies where social and political conditions were helpful to economic achievement.

The contribution of capital formation to development can be secured without aid. To begin with, direct investment from abroad is likely to flow into areas in which capital can be employed productively. Externally financed plantation, mining, trading, and other commercial enterprises have been established in many parts of the less developed world in apparently unpropitious conditions in the absence of a hostile social and political climate. The inflow of this type of capital has been accompanied by an inflow of technical and administrative skills, and it encouraged new ideas and methods of production. As part of their operations these enterprises, notably banking and trading companies, have also financed small-scale local farmers and traders. Foreign direct investment played a large part in the economic progress and transformation of much of the less developed world in the nineteenth and twentieth centuries.

Moreover, enterprises and governments in the Third World capable of using capital productively can readily borrow commercially abroad as well as at home.

Both before the Second World War and since, the thousands of enterprises in the less developed world that advanced from very modest beginnings to considerable size and prosperity readily found the necessary capital. Even in very poor Third World countries, small-scale producers have had access both to domestic and to external funds. Western and Levantine trading firms routinely lend substantial amounts of money to trustworthy African

[10] References to the work by Kuznets and others are in P. T. Bauer. *Equality, The Third World and Economic Delusion* (London and Cambridge, Mass., 1981), pp. 242 and 280.

borrowers, primarily traders, who in turn lend to farmers. Indeed, lending by trading firms to local people is generally a prerequisite for doing business. Ability to borrow does not depend on the level of income, but on responsible conduct and the ability to use funds productively.

Third World governments have also been able to borrow readily, even too readily, from international banks. This also applies to borrowing for facilities that do not yield a directly appropriable return. If such spending, often known as infrastructure spending, is productive, it increases national income (and thus taxable capacity) so that the loans can be readily serviced.

Since Third World governments and enterprises that can use funds productively and conduct their finances responsibly can secure external funds, it follows that the maximum contribution of aid to development cannot exceed the avoided cost of borrowing, that is, interest and amortization charges payable to creditors as a percentage of GNP. The most that aid can do for development is to reduce the cost of a resource that is not a major independent factor in the development process. For large Third World countries these benefits must be modest, even minimal, far too small to affect any macroeconomic aggregate. For instance, for India in the early 1980s this benefit would have been of the order of 0.25 to 0.5 percent of recorded GNP.

It may be thought that even such a modest contribution to development is worthwhile: the recipient countries must derive some benefit from the inflow of resources, while the donors can afford to give away a small proportion of their incomes. This prima facie plausible reasoning overlooks the adverse repercussions of official wealth transfers. Most of these repercussions arise because the subsidies go to the governments; some others would arise even if they went to the private sector.

Unlike manna from heaven, aid does not descend indiscriminately on the population at large, but goes directly to the government. Because aid accrues to the government it increases its resources, patronage, and power in relation to the rest of society. The resulting politicization of life enhances the hold of governments over their subjects and increases the stakes in the struggle for power. This result in turn encourages or even forces people to divert attention, energy, and resources from productive economic activities to concern with the outcome of political and administrative processes and decisions. This sequence provokes tension that often erupts in armed conflict, especially in countries comprising different ethnic, tribal, and cultural groups. Such sequences must inhibit economic advance because the deployment of people's energies and resources necessarily affects the economic performance of a society. Foreign aid has not been the sole cause of the politicization of life in the Third World, but it has contributed significantly to the process.

Foreign aid has also enabled many recipient governments to pursue policies that plainly retard growth and exacerbate poverty. The long list of such policies includes persecution and sometimes the expulsion of the most productive groups, especially ethnic minorities; suppression of private trade, and at times the destruction of the trading system; restriction on the inflow of foreign capital and businesses, depriving the country not only of capital but also of enterprise and valuable skills; extensive confiscation of property, including forced collectivization; takeover of foreign enterprises, which uses up scarce capital and deprives the country of valuable skills; price policies that discourage agricultural production; expensive forms of state support of unviable activities and projects, including uneconomic import substitution; and the imposition of economic controls, which, among other adverse effects, restricts external contacts and domestic mobility and so retards the spread of new ideas and methods often crucial for economic advance. Many aid recipients regularly pursue several of these policies simultaneously; the Ethiopian government has pursued all of them.

Such policies, when pursued singly, and much more so when pursued together, impoverish people and even cripple the economy. Vietnam, Burma, Ethiopia, Tanzania, and Uganda are conspicuous examples.

These policies often directly provoke conflict, even armed conflict, the effects of which are exacerbated by the neglect of public security and the protection of life and property, which in the Third World often goes hand in hand with ambitious, far-reaching policies.

Most aid recipients, possibly all, severely restrict private inward investment. This policy is both anomalous and damaging. It is evidently anomalous because shortage of capital is the basis, the rationale of official aid. The inflow of official aid makes it easier for governments to restrict inward private investment. These restrictions serve the political purposes of the government. They reinforce the hold of the rulers over their subjects, whose economic opportunities are extended by the inflow of private capital. The restrictions often also benefit special interest groups by shielding them from competition, and the beneficiaries are usually politically effective or useful supporters of the government. The restrictions on the inflow of private capital are very damaging because, as already noted, such investment from abroad is often accompanied by the inflow of administrative, technical, and commercial skills, the inflow of which in turn brings with it new ideas, crops, and methods of production.

Such policies are pursued because they accord with the purposes and interests of the ruling group, including promotion of governmental power, satisfaction of politically effective pressure groups, and provision of financial rewards to politicians, civil servants, and their allies. The pursuit of many of these policies is therefore not irrational according to the accepted meaning of the term, although if carried too far they may prove counterproductive and undermine the position of those in power. To describe them as irrational or misguided is to imply that the policymakers are engaged in the single-minded pursuit of increasing general economic and social welfare, so that policies which plainly go counter to this objective are only honest, well-intentioned errors. It is evident that the damaging, destructive policies pursued for years, even decades, cannot be attributed to well-intentioned mistakes. A number of governments pursuing such policies, most obviously in Africa and Asia, could not have survived without Western aid, which in some cases has even been increased substantially in the face of such destructive policies, notably in Ethiopia, Sudan, and Tanzania in the early 1980s.

This situation throws into relief a conspicuous and persistent anomaly. Recorded per capita income is a major factor in the allocation of much Western aid.[11] On this criterion, governments pursuing destructive policies can qualify for more aid. Paradoxically, aid allocated on the basis of per capita incomes rewards policies of impoverishment.

This anomaly applies evidently to program aid, that is, subsidies for the general purposes of government. But it applies also to project aid, that is, official subsidies for particular projects whether in the public or private sector. Such subsidies will be sanctioned by the recipient government only if the project fits with general government objectives. And generally it enables the recipient government to spend more in pursuit of its policies.

Actual or anticipated balance-of-payments difficulties also affect the allocation of aid. This criterion has also encouraged policies adverse to development. Confidence that balance-of-payments difficulties will attract further aid encourages financial profligacy, notably inflationary policies. These policies are often accompanied by exchange and other controls. Inflation, payments difficulties, and controls engender tension and insecurity, even a crisis atmosphere, thereby inhibiting domestic saving and productive investment and encouraging a flight of capital.

There are other untoward effects of aid. Official transfers have often biased development policy toward unsuitable external models. Subsidized import substitution, construction of petrochemical complexes, and state ownership of airlines are familiar examples. Adoption of external prototypes in development policy has often gone hand in hand with attempts at more comprehensive modernization. Such efforts have included attempts to transform people's mores, values, and institutions and have in turn invited backlash and conflict. Governments that engage in ambitious programs and projects of these kinds have simultaneously neglected the basic functions of government, including the protection of life and property.

The advocacy of official subsidies as necessary for economic advance also encourages or reinforces widely prevalent attitudes and inclinations in poor countries, especially in Asia and Africa, that are uncongenial to economic advance, such as belief that economic improvement depends on circumstances and influences outside one's control. These subsidies go to governments and encourage them to engage in beggary and blackmail rather than in exploring the potentialities of economic improvement at home. And such inclinations are apt to spread outwards from the rulers to the population.

The inflow of aid funds drives up the real rate of exchange and adversely

[11] Examples include British aid and the large amounts of aid distributed by the International Development Association, an affiliate of the World Bank. These are allocated explicity on the basis of recorded per capita income.

affects foreign trade competitiveness. This effect can be offset to the extent that the subsidized transfers enhance the overall productivity of resources. For various reasons, in the case of official aid such enhancement is unlikely in practice. In any case, it can occur only after a time lag of years. Meanwhile the higher real exchange rate makes for continued dependence on external assistance. This effect of external subsidies applies whether they go to the government or to the private sector.

An indirect effect of these subsidies also damages the economic prospects of recipients. It is a familiar paradox that the donors erect severe barriers against imports from the recipients of their largesse. The granting of subsidies diminishes effective resistance to these barriers. The recipient governments are less inclined to protest for fear that the complaints may endanger the inflow of subsidies. And within donor countries, criticism of import barriers is muted because many people feel that they are doing enough for the recipients and do not wish to face the dislocation and opposition that accompany liberalization of imports. Since external trade and the contacts it promotes are important instruments in the economic advance of LDCs, the restriction of imports from these countries is correspondingly damaging. The subsidies serve to some extent as conscience money for the damage inflicted by the trade restrictions.

Trade restrictions are imposed because they benefit special interest groups in the donor countries. The subsidies also benefit important commercial, political, and administrative interests there. The presence of influential lobbies behind both the trade restrictions and the subsidies resolves the paradox of providing subsidies to LDC governments and simultaneously restricting the trading opportunities of the recipient countries.

Finally, the adverse effect of one type of aid needs to be noted. Official aid to Asian and African governments often takes the form of free or heavily subsidized food. Even in the absence of underpayment of producers by special taxation and the operation of government buying monopolies, this policy depresses farm incomes in the recipient countries and thereby promotes otherwise uneconomic migration from rural areas to the cities with attendant social, political and economic costs.

23 Paul Streeten, 'Structural Adjustment: A Survey of the Issues and Options'

Excerpts from: *World Development* 15 (12): 1469–82 (1987)

1. Introduction

The essence of development is structural adjustment: from country to town, from agriculture to industry, from production for household consumption to production for markets, from largely domestic trade to a higher ratio of foreign trade. While the advanced industrial countries also have to adjust to a changing world and to new technologies, their structures are more stable and less subject to change. In this very general sense, development is synonymous with structural adjustment and a paper on structural adjustment would be a paper on development.

For the purpose of this paper, however, a somewhat narrower definition is chosen. In the pursuit of their development objectives – growth, equity, poverty eradication, self-reliance, environmental protection, cultural values – countries are at times faced with major disruptions to which they have to adjust. Adjustment is then adaptation to sudden or large, often unexpected changes. These changes may be favorable or unfavorable to the set of objectives pursued by a government. In the case of a favorable change (e.g., unexpected improvement in the terms of trade, additional foreign capital available for investment or greater benefits from the international division of labor), the challenge is to derive the maximum benefits from it; in the case of an unfavorable change, to adapt with the minimum social costs, that is to say, the lowest sacrifice of the objectives and the minimum undesirable side-effects.

The adaptation to unfavorable change may be forced upon a country, 'too little and too late,' and therefore its social costs may be very high; or it may be anticipated and prepared for, and therefore its costs minimized. Even when the change is not anticipated, there are methods that provide for adaptability and flexibility, thereby reducing, usually at a cost, the cost of adjustment.

Change may be small and slow or large and sudden. For slow and gradual change, the price mechanism is one of the best instruments of adaptation. It combines a decentralized system of signals and incentives, to buyers and sellers, for the allocation and redeployment of resources in response to changes in demand and supply, and it avoids some of the drawbacks of bureaucratic controls, such as ignorance, inefficiency and corruption. If with Alfred Marshall, we think that nature does not make jumps, we shall rely heavily on the price mechanism and the market.

For a large and sudden change, the price mechanism is less suited, at least by itself, and has often to be supplemented or replaced by other non-price measures.

The main issues discussed in this paper are the implications for structural adjustment in an economy that is engaged in world trade, world capital flows and other relations with the outside world. The structural changes to which such an economy has to adjust may come from outside. Structural changes or shocks originating within the country also have repercussions on the balance of payments and require correcting actions and adjustments.

A special case is adjustment from a set of wrong policies to better policies. This comprises adjustments both for a given set of objectives that had been pursued in a misguided manner and for a set of different objectives, more in line with a reformed social welfare function. We shall call these the problems of transition.

The balance of payments of a developing country both reflects attempts of domestic adjustments and imposes the need for domestic adjustments to changes in the rest of the world. Thus, a changeover to an export-oriented strategy, an attempt to grow more food at home, a land reform, a tax reform or a redistribution of income, may lead to a temporary balance of payments deficit. On the other hand, the rise in the price of an important import, global inflation, a fall in demand for a country's exports, or policy changes in foreign countries may also cause a deficit, to which the developing country must adjust. It is now generally agreed that the use of exchange rates to equilibrate payments at each moment ('clean floating') is not acceptable and would inflict unecessary damage, even if effective. Large speculative capital movements can lead to 'overshooting' and be counterproductive to the adjustment process. But the exchange rate, often in conjunction with other measures, more fully discussed later, is a powerful instrument to bring about more long-term changes.

2. Adjustment for what?

The first and most fundamental question to ask is: adjustment for what purpose? Any adjustment must have some end in view. Sometimes constraints are considered as if they were objectives of policy, and, of course, they can take the form of intermediate objectives. Thus the elimination of a deficit in the balance of payments or in the budget or in public enterprises, though a constraint, can become an overriding short-term objective. Or we may wish to adjust from a strategy of import substitution to one of export orientation, or to growing debt service, or to more food production. Or we may wish to correct other distortions in order to improve the allocation of resources. Or we may wish to reduce high rates of inflation. These are, at best, intermediate objectives.

Among the objectives of structural adjustment are usually cited (1) the reduction or elimination of a balance of payments deficit, (2) the resumption of higher rates of economic growth, and (3) the achievement of structural changes that would prevent future payments and stabilization problems. One of the most important purposes of structural adjustment is to make the economy less vulnerable to future shocks. This can be done by increasing flexibility and adaptability. The success of a structural adjustment program depends largely on the absence of rigidities. But it may also be its aim to reduce such rigidities. Unless they can be removed, structural adjustment can be very costly, or altogether out of reach. Growing flexibility is therefore both a condition for and an objective of adjustment policy.

Flexibility can be applied to the market for products or for factors of production. If it is confined to products, but factors remain inflexible, large rents will arise which have no economic function.

There are more fundamental objectives, such as the elimination of hunger and malnutrition, or the alleviation of poverty, or cultural autonomy or self-reliance or greater national strength and military power. Some would say that accelerating economic growth is also such a fundamental objective, although growth is simply the time dimension of other goals of policy such as consumption, or poverty alleviation, or reduced inequality, or more employment. Whatever the technical intricacies of the adjustment process, it is useful to bear its purposes in mind, if only because some of these objectives may conflict with one another.

3. Adjustment to what?

The next question to be asked is adjustment *to what?* Adjustments may be to an unexpectedly favorable turn of events, or, more normally, to an unfavourable turn of events. Adjustments to a favorable turn may be to an exceptionally good harvest, to an improvement in the terms of trade, to a rise in export prices, such as that experienced by oil-exporting countries in the 1970s or in response to a rise in the price of coffee of the Ivory Coast or Colombia, to unexpected inflows of financial resources, or to a sharp drop in the price of oil for oil-importing countries, such as that in 1986. Inability to adjust to such favourable turns can lead to opportunity losses and can be almost as important as inability to adjust to unfavorable turns of events.

The large literature on the 'Dutch disease' that arose from the bonanza of natural gas discoveries in Holland testifies to the fact that a large rise in the supply of foreign exchange can be, at best, a mixed blessing, and, at worst, a curse. The exchange rate appreciates, exports of other goods and services than the one the price of which has risen, decline, competitive imports flood into the country, domestic employment declines while inflation rises as the demand for non-tradables increases. Income distribution may become more unequal and the poor may be particularly hard hit. The removal of the foreign exchange and savings constraints can bring to the fore other obstacles. Adjustment policies should then be devoted to their identification and removal. Among these will be promotion of exports other than the commodity whose price has risen, some control of imports, both possibly through a dual exchange rate, sterilization of some of the inflowing foreign exchange, control of domestic inflation, and creation of alternative productive assets. Even a favorable turn of events can cause serious adjustment problems. One is reminded of Bernard Shaw: 'There are two tragedies in life; one is not to get your heart's desire; the other is to get it.'

Much more common, unfortunately, is the need to adjust to an unfavorable turn of events. Adjustments may be in response to shocks that are expected to last a long time, or a short time, or they may have their origins in macroeconomic or in microeconomic factors.

A widely accepted distinction is that between the need to adjust to shocks caused by external factors, or exogenous shocks, such as a drastic deterioration of the international terms of trade, a reduction in the demand for export volumes, resulting, for example, from a world recession, or a rise in interest rates for countries with large debts, and those caused by domestic events or policies, or endogenous shocks, such as an excessively lax fiscal and monetary policy, price distortions, losses on the part of public enterprises, excessive protection, excess foreign borrowing or domestic up-heavals, including revolutions. On the face of it, the distinction is clear enough. Foreign reductions in demand, due to a new technology, to a change in tastes, to rising competitors, or to a recession abroad, or a sudden and large rise in the prices of essential imports, are very different phenomena from domestic harvest failures, strikes, corrupt policies, mis-management or political change. Yet, on closer inspection, the distinction becomes blurred. If prices of a country's exports drop and its terms of trade deteriorate, it is a matter of good policy to have foreseen this event or at least its possibility, and to be ready to move out of the declining export trade into more profitable lines. Thailand, Malaysia and the Philippines diversified their crops and raised productivity in existing crops in response to declining price prospects (for example rubber in Malaysia), while other countries failed to do this (for example Tanzania for sisal).

. . . .

Some of the major changes arising mainly from the world economy to which adjustments by the developing countries are called for are these:

(1) growing debt service, combined with fewer loans and higher interest rates;
(2) deteriorating terms of trade, whether resulting from rises in import prices, such as that of oil, or drops in export prices, such as that of major export crops;
(3) high levels of inflation in the world;
(4) slower growth in the OECD countries;
(5) technical innovations such as those in electronics that change the location of industries.

In addition, there are

(6) continuing high rates of population growth;
(7) urbanization;
(8) scarcities of land and certain raw materials;
(9) scarcities of foodgrains;
(10) policies adopted by the developed countries: protection of industry, agriculture and services;
(11) environmental pollution;
(12) international migration;
(13) natural disasters, such as prolonged droughts;
(14) man-made disasters, such as the arms race or wars.

. . . .

So far we have discussed adjustments as mainly responses to shocks, whether external or internal. But adjustments may be required as a result of more active initiatives for a change in strategy. A government may wish to change from import-substituting industrialization to export-orientation, or from excessive urban bias to more agricultural production both for home consumption and export, or from a conventional concentrated growth strategy to a more egalitarian one, or to institute a land reform, or a tax reform, or to make poverty eradication one of its principal targets. The adjustment problems that such transitions create are therefore often inadequately treated and sometimes misunderstood. These will include sectoral imbalances in supply and demand, manifesting themselves in unemployment combined with inflation, and disturbances in the balance of payments, and will raise the time period for financing adjustments. As income is redistributed to the poor, the supply of food will be inelastic in the short run and prices will rise or more imports will burden the balance of payments. As expenditure on luxury goods is reduced, unemployment in these trades may rise. Owners of capital will try to move their capital abroad. If the reformist government replaces a repressive dictatorship, previously oppressed groups will assert their claims to higher incomes with inflationary results. There may be strikes and even coups d'état. Some of these adjustment problems are mistaken for manifestations of mis-management, which, of course, especially for inexperienced reformist governments, may independently add to their difficulties.

Economists have been better in analyzing comparative statics and comparative dynamics, than the transition path from one type of strategy to another. We completely lack a handbook for reform-minded Prime Ministers and Presidents, who would like to know how to manage the transition to a better society. The international community, in turn, should be ready to assist such reform-minded leaders with Radical (or Reformist) Adjustment Loans (RALs) to ease the transition, to help overcome the dislocations, to provide more flexibility and elbow room. When President Duarte in El Salvador undertook a land reform in 1980–82, he was promised US aid to compensate landowners. But Senator Jesse Helms introduced legislation preventing this. In addition, the prices of the country's commodity exports fell, and the reformer found himself in disgrace.

4. Adjustment of what?

Next there is the question: adjustment of *what*? In developed countries adjustment sometimes means the revival, with new technology, of old industries, like phoenix rising from the ashes, and at other times it means the creation of new industries, at the frontiers of technological knowledge. The textile industry is an example of either. The question then is often posed whether labor should move to the industries or capital and entrepreneurs to the labor. Developing countries have a much smaller industrial base and the question is normally a different one. Should there be adjustment of *policies*, particularly pricing policies, say from protection to freer trade or from keeping food prices down to raising them to the level of world prices, or should there be adjustment of *institutions*? These may refer to a land reform or to population control or to the administrative system or to education and training.

A vast amount of writing and exhortation has recently been devoted to advocating higher agricultural prices where they are below border prices. This, it is said, would stimulate agricultural production – both export crops and food for domestic consumption – and help poor rural people at the expense of the urban middle class.

When it is pointed out that, quite apart from the fact that there are many poor buyers of food, both in towns and in the country, who should be protected from hunger and starvation caused by higher food prices, other measures besides higher prices are needed only to get a substantial supply response, the advocates of higher prices would agree. Yes, they would say, we need roads to get the produce to the market, technology to apply to agricultural production, irrigation, credit, efficient marketing, etc. But the proposition 'other things are also necessary' is open to two diametrically opposed interpretations. It may mean that other things help, but to 'get prices right' by itself is better than nothing: it is a step in the right direction. Or, alternatively, it may mean that only if price policies are combined with non-price measures do we get a supply response. Price measures by themselves can be either quite ineffective, or actually counterproductive, even if we have only total production in mind, and pay no attention to poverty or income distribution.

Let us first briefly look at the 'other things' and then illustrate the futility, in some situations, of applying only price measures. It so happens that the six prongs which are necessary to achieve an agricultural supply response each begin with 'In,' so that we can call them the six 'Ins.' Prices are the first In, because they serve as Incentives. Second, we need Inputs: water, fertilizer, equipment. Third, we need Innovation: technology, such as high-yielding varieties, irrigation, etc. Fourth, we need Information: the technological knowledge must be diffused through extension services. Fifth, we need Infrastructure. Unless there are roads and harbors to get the crops to the market, even the best incentives will not get production and sales up. And sixth, we need Institutions: credit institutions, efficient marketing institutions that do not cream off too high a margin from the final price, and in some cases, land reform. Also adequate health and education services for farmers. These are the six Instruments that provide a mnemotechnic framework for agricultural policy. Many of these, particularly infrastructure and innovation, are normally carried out in the public sector.

There are some pricist fanatics who say that the correct prices will by themselves induce the right innovation and even the correct action by the government in the public sector. There are others, who say that the Ins other than Incentives are already in place in many developing countries. If these two groups of people were right, no independent public sector action would need to be taken. Unfortunately, it is obvious that this is often not the case. In Tanzania, roads are so inadequate that even if farmers produced more in response to higher prices, the crops could not be transported: a failure of Infrastructure. Or the marketing boards would appropriate the gains, leaving nothing to the farmers, a failure of Institutions.

In addition, complex and difficult choices arise within each of the six Ins. Consider infrastructure, normally provided at least partly at public expense. There are choices between physical, legal, human, social and producer-specific types of infrastructure to be made; choices between centralized and decentralized types of infrastructure; choices between infrastructure for producers and for consumers, choices between maintenance of existing projects and new projects, and choices between different methods of financing infrastructure. The same is true for institutions and for information. (For example, should extension workers concentrate on a single line of conveying information or should they combine several?) All these compete for scarce resources with directly productive investment in agriculture.

Now let us illustrate how higher agricultural prices without action on the other five fronts can be ineffective or counterproductive. The illusion that higher prices by themselves will lead to a large response in supply derives from an illegitimate application of what happens if the price of one crop rises relatively to others, to the case where the agricultural terms of trade as a whole improve. All the evidence shows that supply response is much lower for total output.

Moreover, raising agricultural prices may not be capable of improving the terms of trade. In the effort to raise producer prices as an incentive to agricultural production, the benefits can be undone if the higher prices are communicated to the rest of the economy through proportionately higher money wages and higher markups on industrial goods. If this were to happen, the attempt to raise agricultural prices would only set off an inflationary movement, would not have improved agricultural-industrial terms of trade, would not have achieved its objective of raising agricultural production, and would have undesirable side effects.

The second example is from Bangladesh. There, a large proportion of the rural population are tiny farmers who produce some food for themselves on their little plots of land, but not enough, so that they have to work for bigger landlords and buy food with their wages. When the price of food rises, these deficit farmers pledge their little plots against consumption loans. When they cannot repay the loans, they forfeit their land. We are not here concerned with the fact that these people may be driven to starvation, but only with the effect on output. Since the output per acre on large farms tends to be lower than on small farms, the transfer, caused by the initial rise in food prices, will lower total agricultural production.

A third example is inspired by Tanzania. Assume there is a scarcity of consumer goods in the villages, their prices are controlled and there is no black market. As agricultural prices are raised, the farmers can buy the limited amount of consumer goods at controlled prices for a smaller supply. If we also assume that they are not interested in storing money for the future, they will produce and sell less. Their elasticity of supply becomes minus one. As supply prices rise, they reduce supply proportionately, because they need only a given total of receipts to buy the rationed and price-controlled goods. Let us further assume that foreign exchange earnings are derived from agricultural crops and are spent on imports of consumer goods. And let us assume, not unrealistically, that a certain absolute amount of these imported consumer goods are reserved for the urban population. The reduced agricultural supply will mean reduced imports. Since a larger proportion of these are preempted for the towns, the farmers will get even less. Their supply will further shrink and we shall witness a contractionist spiral, fewer consumer goods leading to less agricultural supply, leading to fewer consumer goods for farmers, etc. (Bevan *et al.*, 1987).

The conclusion is that correct pricing policies often (clearly not always) work best in conjunction with action in the public sector.

. . . .

5. Adjustment by whom? Who benefits and who loses?

When we consider the question of adjustment *by whom*, the candidates are labor or capitalists, the rich or the poor, men or women and children. UNICEF has recently started to think about, and to urge upon the IMF, what they call 'adjustment with a human face.' How can we bring about the required adjustment with the minimum harm to the most vulnerable groups? Some cuts in expenditure are required. Much investment is productive investment and we do not want to cut this, for it would undermine long-term growth. In low-income countries the rich are few and some consumption cuts are bound to fall on the poor. But it will be possible to cushion the poorest and most vulnerable groups, such as children and women, against having to suffer cuts in their consumption. In wartime Britain, adjustment to the war economy has led to improved standards for children and mothers, by programs of rationing, free issues of certain products such as orange juice, and redistribution of real income.

In considering the impact of adjustment policies on the poor, three things should be clear to start with. First, any form of macroeconomic adjustment is bound to affect the distribution of incomes, if only because its purpose is to raise incomes in some sectors and reduce them in others. It can therefore not be excluded from any full analysis of the effects of adjustment policies. Second, what should be of main concern is the impact of adjustment on the poor, not on relative income distribution. If some groups benefit without anybody being hurt, this should be no cause for complaint, although in fact conflict over the division of benefits can be as serious as conflicts over gains and losses. Third, any cut in living standards, imposed by the need to correct a balance of payments deficit, is bound to hit some of the poorer sections (though not necessarily the poorest of those groups, the meeting of whose needs contributes to human capital formation), if only because the poor are so numerous and average income is so low in low-income countries. This is particularly so if productive investment that contributes to economic growth and to safeguarding living standards in the future is to be protected. These reductions in living standards of the poor are likely to be more serious in countries with low average incomes per head, partly because incomes are already low, and even across the board cuts will drive many below the poverty line, and partly because these countries are liable to be less flexible and responsive to shocks and therefore have to deflate by more for any given correction in the balance of payments.

Some of the austerity in the much bewailed austerity programs is therefore inevitable. In particular, employed wage earners in the organized sector, including those in public enterprises, not normally among the poorest, may have to suffer. The question is whether stabilization policies inflict an undue share of the burden upon the poor, for instance by a sudden elimination of food subsidies, or by cuts in social services, or by price increases of essentials.

. . . .

It is not inevitable that stabilization policies should hit the poor exclusively. Adjustment normally consists of two parts: expenditure-reduction and expenditure-switching. Expenditure-reduction, in a low-income country, will tend to hit the poor sections, unless they are specifically protected or compensated, but expenditure-switching need not harm them. The reallocation from domestic sectors to tradables will tend to raise incomes in these sectors. If traded goods are mainly agricultural goods and if many poor are engaged in agriculture, they may benefit from the reallocation. (On the other hand, some non-tradables, such as construction and government services, are labor-intensive, and a switch to the production of tradables may reduce employment and income of the poor.) Analysis of the impact of balance of payments adjustments on income distribution and poverty is sometimes conducted in terms of a two-sector neoclassical model, in which tradables have different factor proportions from non-tradables. A switch to tradables then benefits the owners of the factor of production that is used more intensively in the production of tradables, not only in the tradables sector but in both sectors. It also benefits the consumers of non-tradables because their relative prices fall. But the difficulty with this type of analysis is that incomes of people do not always follow the functional lines of capital and labor, and the two-sector model is of limited use. Within each sector there may be, for example, an informal, labor-intensive sub-sector, in which producers own a bit of capital but operate on a small scale, and a formal, capital-intensive large-scale sub-sector, with wage employment. As expenditure is switched to the tradables sector, workers in the formal sector may benefit but the informal sub-sector poor may be harmed. If there is substantial mobility between the sub-sectors, this would not matter, for people would move from the informal into the formal sector. But if mobility is limited, the distinction is crucial. It has already been argued that more important than the impact of stabilization policies on income distribution is the impact on absolute poverty.

The higher prices of imports of necessities caused by the devaluation could be offset by subsidies, targeted towards the poor. The restrictions on investment could be on luxury housing and durable consumer goods for the upper class. In fact, however, many stabilization policies have hit the poor hard. The price rises that follow devaluation and the removal of controls frequently affect particularly the necessities consumed by the poor. Reductions in government expenditures are often on labor-intensive public works, food subsidies and social services. High interest rates tend to encourage concentration of wealth if lenders are richer than borrowers (but this is not always so), though lower rates combined with credit rationing favoring the rich often do the same. Monetary and fiscal restrictions raise unemployment and reduce the bargaining power of unskilled labor. In all cases it is the poor who suffer more than they would have suffered in a situation of repressed inflation, where demand for labor is high, the prices of necessities are controlled, and social services are more generous.

Although we do not have a rigorous model by which to assess the impact of adjustment policies on vulnerable groups, including children and women, it is possible to lay down certain guidelines. *First*, maintain a floor for minimum nutrition, health and education levels, and do not permit cuts in expenditure to affect these. *Second*, restructure production within the productive sectors – industry, agriculture, services and foreign trade – so that the producers in the small-scale informal sectors are not discriminated against but are made to complement the organized sector (manufacturing components or spare parts, conducting repair, subcontracting, providing ancillary services, etc.); and give them access to credit, domestic markets and sources of supply. *Third*, restructure the social services – education, health, water and sanitation – so that the vulnerable groups are not deprived of minimum services. And *fourth*, provide international support for these types of adjustment, both financial and technical assistance, and particularly ease debt service conditions, so as to permit the flexibility and adaptability such programs call for.

. . . .

6. Adjustment – how?

What means are to be employed to bring about the adjustment? It is in the nature of adjustment policies that they comprise measures additional to stabilization, such as the reduction of tariffs, trade liberalization, the elimination of controls on wages and prices, the creation of institutions to facilitate export credits, and the improvement in infrastructure, all of which aim at increasing the capacity to export.

The analysis can be conducted in three stages.

1. How severe will be the adjustment problem – the disease – as registered in the balance of payments deficit? (A secondary, more important but more difficult calculation would be the costs imposed by alternative corrective measures, such as deflation, devaluation, import restrictions, tariffs, etc.)
2. What range of medicines is available? E.g., exchange rate flexibility increases the number of medicines, while pegging exchange rates reduces the cupboard by one medicine; forswearing increases in tariffs by another. The possibility of retaliation complicates matters.
3. How effective is any given medicine? E.g., with a more slowly growing volume of trade, demand elasticities will be lower and exchange rate adjustments less effective. On the other hand, within a free trade area or common market, elasticities may be expected to be higher.

The task for policy is to combine financing deficits and correcting them in such a way as to minimize reductions in employment, output and growth, and the standard of living of the poor (and any other objectives such as income distribution). Appropriate methods of financing deficits, combined with the right type of conditionality, do not frustrate the process of adjustment, but facilitate it, and can reduce its costs.

7. Adjustment through the market or government intervention?

The virtue of allowing the market to bring about adjustments is that it combines signals and incentives in response to change. Nobody maintains that markets work perfectly, but then nobody maintains that government intervention can always produce the best results. We live in an imperfect world, and it is a matter for skillful judgment to decide which imperfections – market failures or government failures – are least bad. Governments can, of course, intervene in the free play of market forces by using prices as instruments of policy. Indirect taxes and subsidies can attempt to use prices and market forces for the purposes of government policy, where the free play of market forces otherwise would conflict with social objectives.

Fortunately, there are not only the much discussed market failures and government or bureaucratic failures, but also market successes and government successes. The success stories in structural adjustment show neither the success of laissez-faire, nor of centralized bureaucratic control, but show governments (a) that knew in which areas to intervene and which to leave alone, and (b) how to conduct interventions efficiently. The failures illustrate not only excessive government intervention, but also unwise and inefficient intervention in some areas, and inadequate intervention in others.

When adjustment is needed to a change, such as a change in supply or demand for a commodity which has been produced and sold in the past, consequential price changes normally signal the right responses and produce the right incentives. In response to a fall in demand, prices, profits and investment will tend to fall, and labor will leave the activity, or new workers will be discouraged from entering. But some exceptions should be noted.

(1) When workers have to be retrained for new jobs, or have to move to new locations, capital is needed to finance these adjustments. If capital markets were perfect, workers could borrow and repay loans out of their future higher earnings. But capital markets are not perfect, and there is a role for governments to finance adjustment assistance.
(2) Adjustments may be impeded by lack of knowledge, or by inertia. At a minimum, government may, in such situations, have to provide information, and perhaps again some subsidies to the moves. On the other hand, the information available to the government may not be much better than that available to individual workers, or their reluctance to move may be justified because their subjective costs exceed probable benefits.
(3) If real wages are sticky and do not fall in response to a decline in demand, unemployment will rise. Sometimes such unemployment may bring about the adjustment more effectively than reductions in real wages, particularly if there is excess demand for labor in other sectors, and this is known. But if this does not happen there may again be a case for government intervention.
(4) One of the most common justifications for government intervention is the need to correct an undesirable impact on income distribution. As we have seen, in any adjustment process, some people are likely to gain, others to lose. If the losses of the losers are undesirable, the government will try to step in.

Price policies are not the same as free market policies. Prices can be used as an instrument of government intervention, in order to achieve public objectives. Tariffs, indirect taxes and subsidies permit prices to affect the allocation of resources by cutting the links between buyers and sellers that would prevail in their absence. If, in response to an increase in demand, an indirect tax is imposed on a product, this will reduce the demand for the product, and therefore not permit an increase in the supply, unless the revenue is spent on additional supplies. A subsidy can be used to lower the price to consumers, who otherwise could not buy the product. It raises supply, without raising prices to buyers. The limits to such policies are set by fiscal constraints. Direct controls do not call for taxes or subsidies, but call on scarce administrative capacity. Both fiscal and administrative constraints are severe limitations in most developing countries, and choices between the two are difficult.

It is sometimes said that if we always pursue efficient policies, such as those indicated by market forces, while there will be occasionally losers, in the long run everyone will be better off. There is therefore no case for government intervention to correct for income distribution. Such intervention is only likely to reduce efficiency. But this defense is untenable. First, there is no reason to believe that the gains would be spread randomly in the long run. Many forces such as increasing returns, higher savings out of higher incomes, etc., work for cumulative gains – 'unto those who have shall be given' – and the same group of people can be consistent losers. Second, in the long run the identity of individuals changes, and it is illegitimate to compare the gains of one generation with the losses of another. Even if the compensatory gains were spread over the same generation, the individuals' tastes, wants and needs will change over time and it is then not clear that a future gain is always an adequate compensation for a current loss.

8. Fallacies of aggregation: the need for a global view

One of the criticisms of the Fund and the World Bank has been the absence of a global view on adjustment policies. Not all countries can have export surpluses. New gold apart, the surpluses of some must be matched by the deficits of others. Similar problems could arise in structural adjustment. If several countries were to reduce their dependence on a single export crop, the price of which is likely to fall in world markets, by diversifying into crops that are supplied by other countries, but are also in surplus, they would aggravate the deterioration in their terms of trade. Similar problems can arise for manufactured exports, against which protectionist barriers are likely to be raised if the total exceeds what other countries think they can absorb. It has often been said, for example, that the experience of export-led growth of the East Asian countries cannot be repeated by all developing countries because the additional exports would be so large that protectionist barriers would go up or the terms of trade would deteriorate.

In its simplest form there are several things wrong with this argument. First, in principle, the foreign exchange earned by the exports of the developing countries will be spent on extra imports and, problems of composition (calling for adjustment) aside, need not give rise to protectionism. Second, the phasing of trade liberalization and export-orientation will be different for different countries, and not all exports will be dumped simultaneously. Third, as a result of a better international division of labor, incomes will rise in the developing countries and out of these higher incomes more will be spent on the exports from the developed countries, as well as from other developing countries. In the context of accelerated growth, the adjustment problem of switching resources from import-competing to new export industries should be eased. Fourth, the commodity composition and the export/GDP ratios will be different for different liberalizing countries. Many developing countries will continue to export many primary products. The experience of the Gang of Four (Taiwan, South Korea, Singapore, Hong Kong) should not be held up as a model. These four countries are labor-rich, natural resource-poor, while other developing countries, such as Brazil and Argentina, are likely to be more resource-rich and labor-poor, and will therefore export only a smaller proportion of their GDP in the form of labor-intensive manufactured products, thus reducing the risk of swamping the market. According to these and other characteristics, different developing countries will export different goods (as well as at different times). Fifth, some exports would be directed to other Third World countries, in which vested interests clamoring for protection are less than in the developed countries. Sixth, as a result of trade liberalization some counter-protectionist pressure groups in the developed countries are likely to arise or be strengthened, for example agriculture in the United States.

In spite of these arguments there are bound to be adjustment problems in the industrialized countries if all developing countries were to liberalize their trade and adopt more outward-looking policies. These will apply with particular severity to some product groups and some countries. A lot depends on what is assumed about growth rates in the late 1980s and 1990s, and how successful industrialized countries' adjustment policies will be. The conclusion of this discussion is that, although adjustment problems in the developed countries in response to an export drive on the part of all developing countries would be serious, and recommendations along these lines suffer from some degree of illegitimate aggregation, they need not be as serious as some observers have anticipated.

Reference

Bevan, D. L., Bigsten, A., Collier, P. and Gunning, J. W. 1987: Peasant supply response in rationed economies. *World Development* 15 (4) (April).

SECTION SIX

NEW DIRECTIONS IN DEVELOPMENT STUDIES

Editor's Introduction

The study of development is bound to be in a constant state of flux, if only because the worlds that it seeks to describe are fast-changing themselves. Development studies also takes a cue from the more general intellectual climate that surrounds it, as witness the recent movement away from metatheories, and the current interest in empowerment, institutions, democracy, feminism, postmodernism and privatization. But development studies also builds on past achievements. The empirical studies that have accumulated over the past fifty years have done a lot to provide us with viable working hypotheses about the relationships between fertility and the changing position of women, for example, or about the price responsiveness of farmers, the skills of informal sector workers, the merits of self-help housing, the effectiveness of official development assistance, and the benefits of outwardly-oriented development strategies.

Some of the most important large-scale research studies have been conducted by the World Bank. The Bank has managed long-standing research programmes on issues like population, the environment, poverty, industrialization, trade and capital flows, as well as supervising a growing number of country studies. It is fashionable in some quarters to write off the World Bank as a purveyor of neoliberal economic orthodoxies, and the IMF for 'Imposing Misery and Famine'. But this is too easy: the Bretton Woods institutions are not unitary bodies, and what goes on at a Board of Directors level is not always reflected in the academic studies commissioned by the Bank and the IMF. The annual *World Development Reports* of the World Bank did suffer from an excess of neoliberal zeal in the mid-1980s, in my view, but these Reports are an indispensable guide not only to the 'facts of development' (the 'world development indicators' at the end of each Report), but also to the main topic chosen for discussion in any year.

Some of these Reports are simply outstanding: for example, the 1984 Report, which deals with population and development, and the 1990 and 1992 Reports, which are focused on poverty and the environment respectively. The chosen theme of a Report also gives a clue to certain shifts in development thought and policy. In this respect, it is significant that the text of the 1994 *World Development Report* is devoted to the topic of *Infrastructure for Development*. The 'new' development studies which is emerging in the 1990s is greatly concerned with infrastructure for development, and the *institutions* through which this infrastructure is best provided. Institutional economics is also in favour in the 1990s, as witness the recognition of the work of Douglass North by the award of a Nobel Prize in 1993 (see North, 1991). Some other issues that are commanding attention include: the greening of development; gender hierarchies in development thought and practice; privatization and deregulation; democratization and development; and citizenship, empowerment and development. These issues by no means exhaust the list of topics now in fashion. Nor are they being researched to the exclusion of some of the staple topics of development studies which have been discussed elsewhere in this Reader. Development studies is, or should be, a cumulative enterprise.

The final section of this Reader takes up several of these 'new' directions in development studies, where 'new' refers to their likely incorporation into the mainstream of development studies and policy. **Reading 24** is taken from the **World Bank**'s *World Development Report* for 1994. It is the most optimistic of the four Readings which close this book, and the most indebted to development economics. The 1994 *World Development Report* begins by noting that: 'Developing countries invest $200 billion a year in new infrastructure – 4 percent of their national output and a fifth of their total investment'. It further notes that monies spent on transport, power, water, sanitation, irrigation and telecommunications have markedly improved the lives of many poor households in the developing world. In stark contrast to the claims of some anti-developmentists, the World Bank points out that: 'During the past fifteen years [from the late 1970s], the share of households with access to clean water has increased by half, and power production and telephone lines per capita have doubled. Such increases do much to raise productivity and improve living standards'.

These are important points and they deserve to be put at the top of the World Bank's annual Report in this way. People *are* generally empowered by development, providing this development addresses the basic needs and employment prospects of ordinary men and women. But the World Bank goes on to warn against complacency: 'One billion people in the developing world still lack access to clean water – and nearly 2 billion lack adequate sanitation'. So what is to be done, and what accounts for this lamentable state of affairs fifty years into the Age of Development? The gist of the World Bank's answers to these questions is as follows. The Bank first argues that too much attention has been paid to the quantity of infrastructure stocks, and too little to the macro-economic efficiency of a given developing country and the quality of infrastructure services delivered by different agencies. To this extent, the Bank is adapting the 'orthodox' views with which it was associated in the 1980s. The Bank accepts that infrastructure can deliver major benefits in economic growth, poverty alleviation, and environmental sustainability, but it is critical of delivery systems that have been inefficient and failed to reward the providers of infrastructure. The key lies in providing incentives to 'ensure efficient, responsive delivery of infrastructure'. Such incentives will come from new forms of commercial management, increased competition, and stakeholder involvement.

The question of stakeholder involvement in turn is linked to the second main theme that emerges from the World Bank's Report. In the final part of the Reading published here, we find the Bank commending new forms of public–private partnerships to provide infrastructure and to manage common property and common pool resources. This links through to a new emphasis on the role of institutions and institutional incentives in promoting sustainable development (Ostrom, Schroeder and Wynne, 1993). There is a growing awareness that new technologies – gas turbine generators, long-distance cable networks, etc. – are opening up new opportunities for the competitive provision of infrastructures even in rural areas of the developing world. This is allied to a growing willingness in some developing countries to grant a greater role to the private sector in infrastructure provision. The speed with which telephones are connected now in India compared to just a few years ago is seen by some as evidence of the greater efficiency of a public–private partnership, as compared to the bureaucratic torpor of many public-sector enterprises.

The World Bank is also signing up for the view that common pool resources are best managed by means of cooperative institutional arrangements involving state actors and local resource users (Sengupta, 1991). A further example from India illustrates this point. Until the mid-1980s India's Reserved and Protected Forests were managed by the state in such a way that local forest-dwellers were turned into criminals if they grazed their animals on government-owned land, or collected more than a specified amount of fuelwood. Timber contractors, meanwhile, were often allowed to mine the forests in an unsustainable way. Lacking a stake in 'their' forests, and yet needing access to the forests on a daily basis, many villagers were driven to exploit India's Protected Forests in a covert, hurried and damaging way. Trees were hacked down instead of being properly coppiced, and India's forest cover dropped alarmingly towards 15 percent of the country's land area (with much of this being badly degraded). But now things promise to be different. Since 1989 the Government of India has encouraged all the States in the Republic to enter into a system of Joint Forest Management with local forest users. In the State of Bihar a radical plan has been approved whereby Protected Forests will be run by Village Forest Protection and Management Committees (VFPMC) in conjuction with the Forest Department. The Forest Department will provide advice and saplings, but the new trees planted will be chosen largely by villagers, possibly ending a recent bias to eucalyptus. When trees are suitable for coppicing (on a ten-year rotation of coppice-with-standards, say), the timber will be harvested and marketed by a local cooperative organized by the Forest Department. One third of the proceeds will be returned to the VFPMC for forest development schemes, a further third will be ploughed into village improvement schemes like road-building, while the last third will be 'profit' for participating members of the VFPMC. In return, villagers are expected to join with the local Forester in protecting 'their' forest against intruders, and guarding against the removal of timbers not previously agreed to by the committee of management of the VFPMC (Corbridge, 1994)

Quite how well India's experiment with Joint Forest Management will work remains to be seen. On the plus side, institutions are being set up which call for the participation of local people (including women and representatives of the Scheduled Castes and Scheduled Tribes), and which provide them with incentives and rewards on a household and village basis. Set against that, there is the perennial fear that such institutions will be captured by members of the local elite. Some approaches to institutional incentives and sustainable development have rightly been criticized for being too formal, and for being insensitive to the local contexts within which real institutions have to work (Khan, 1994). The proper design of institutions is not just a technical matter or a topic only for development economists. It is also about communications, language and power.

Empirical research will illuminate these and other concerns. Such research will possibly also open our eyes to a darker, and less obviously 'economic', side to development than the World Bank is sometimes prepared to acknowledge. (It is significant that World Bank forecasts of growth in the world economy have consistently been too optimistic, a point admitted by the World Bank in its *World Development Report for 1991*, p. 28; see also Cole, 1989.) **Readings 25–27** challenge some aspects of the liberal orthodoxy that is emerging in development studies, while focusing directly on three issues that are now much in vogue (and rightly so): the environment; democracy, and the citizenship–justice–gender constellation.

Reading 25 by **Susanna B. Hecht** exemplifies the virtues of an interdisciplinary approach to development studies. At one level her paper can be read for its detailed account of the conversion of tropical forests into agricultural land in the Eastern Amazon Basin. Hecht takes her readers through the history of land conversions in the region, before considering the available scientific evidence on the ecological effects of land clearances and cattle ranching. Particular attention is paid to the changing properties of local soils, in terms of calcium, magnesium and potassium concentrations and soil pH. Hecht concludes that soil physical changes *are* adversely affecting the productivity of agricultural land, and that this in turn affects the profitability of ranches *as ranches* (given increased costs of weed control, etc.).

But this is only one part of the story that Hecht tells, and by no means the most disturbing. The wider point that Hecht makes is that land conversions are continuing in spite of the ecological damage that is already apparent to some ranchers. Although land is still used for cattle, the entrepreneurs/speculators who started to buy up land in the Eastern Amazon Basin in the 1960s depend for their profits '*not on the annual productivity of the land, but on the rate of return to investment*' (emphasis in the original). This finding (which is well documented by Hecht) in turn calls into question the main models that are commonly used to explain land degradation in the humid tropics. Hecht quickly dismisses the Malthusian model for its 'ludicrous [description] . . . of environmental degradation in this situation as only a function of demographics'. She also discounts 'commons models and externalities frameworks . . . because, in fact, there has been a rapid shift to private property [in the Eastern Amazon Basin] which has been associated with the increase, rather than the decline of environmental degradation'. In place of these and other models Hecht commends an approach to environmental degradation that begins with 'the role of large government subsidies in the creation of land markets and speculation in the Brazilian Amazon'. The governments of Brazil created a situation where land in the Amazon was turned into a commodity, the value of which bears little relation to its value for production. In such a situation, cautious land management becomes irrelevant. A possible solution to the environmental crisis hitting this part of the Amazon basin has to begin with controls over the private land market. Whether the government of Brazil will take such action is a moot point, and a matter of politics and political will. The title of Hecht's paper is instructive here, and offers an interesting contrast to books and papers that purport to deal with 'the environment' without also dealing with local patterns of capital accumulation and the dynamics of political systems. A development economics – even an environmental development economics – that is inattentive to power relations is unlikely to be very helpful. The environment might well need to be priced – that is, not treated as a free good – but the manner in which it is priced matters greatly.

Power and governance are certainly at the heart of **Reading 26** by **Adrian Leftwich**. The Reading published here is only part of a much longer article, but some important points still emerge. Leftwich is concerned to examine a 'new orthodoxy [that] dominates official Western aid policy and development thinking' (Leftwich, 1993, p. 605). This new orthodoxy asserts not simply that 'good governance' and democracy are desirable in themselves, and the likely outcomes of development (roughly the view of modernization theory in the 1960s), but that 'democracy is a necessary *prior or parallel* condition of development, not an outcome of it' (ibid., emphasis in the original). Leftwich rejects this assertion. In the Reading published here he begins by tracing the recent history of the claim that a lack of good governance underlies the development problems of a number of developing countries, and most obviously countries in Africa. Leftwich proceeds to argue that the model of good governance proposed by the World Bank is unobjectionable to the point of being vacuous; few people are going to speak against open and honest government or independent and fair legal systems. He further suggests that 'the Bank's analysis is naive . . . because it entirely ignores that good governance is not simply available on order, but requires a particular kind of *politics* both to institute and sustain it' (emphasis added). The real question, Leftwich contends, is 'whether democratic politics . . . can promote development in the Third World and elsewhere'. In the final part of his paper he suggests that the most likely answer to this question is 'no'. What is required, Leftwich concludes, is not necessarily a democratic state (though this might be desirable), but a strong developmental state that can pursue 'market-friendly' policies in such a way that even poor households are empowered to participate in development.

Leftwich's conclusion will be deemed 'unfashionable and uncomfortable' in some quarters. Some will agree with his charge that 'unlimited liberal democracy and unrestrained economic liberty may be the last thing the developing world needs' if the elimination 'of poverty and misery is the *real* target' (emphasis in the original), but many others will recoil from the illiberalism that is latent in this statement. Leftwich is raising an important issue for analysis, but it is possible that his conception of poverty and misery is too narrowly economic. Certainly, the rather aggregated view of poverty and misery that Leftwich is discussing is far removed from the very bodily conceptions of poverty and misery that are discussed by **Nancy Scheper-Hughes** in her disturbing account of violence, citizenship and justice in contemporary Brazil (**Reading 27**).

There are several reasons for closing this volume with the Scheper-Hughes Reading, but not the least of them is the breathtaking scope (and alarming subject matter) of the book from which this short excerpt is taken: her account of *Death Without Weeping: The Violence of Everyday Life in Brazil* (Scheper-Hughes, 1992). Development studies has thrown up many classic monographs since the 1940s, but this book surely ranks with the best of them. In some respects it is a successor to James Scott's account of the *Weapons of the Weak* (Scott, 1985), except that Scheper-Hughes is rightly critical of Scott for running the risk of 'romanticizing human suffering or trivializing its effects on the human spirit, consciousness, and will' (Scheper-Hughes, 1992, p. 533). Scheper-Hughes is also present in her text in a way that Scott is not present in his, although her accounts of her own place and work in the town of Bom Jesus de Mata, in the Brazilian north-east, are disarmingly honest and never lapse into narcissism or solipsism. The worst excesses of the postmodern turn are neatly sidestepped in favour of a committed ethnography. In still other respects, Scheper-Hughes's book is a successor to Josué de Castro's classic, but neglected, accounts of *The Geography of Hunger* (de Castro, 1952) and *Death in the Northeast* (de Castro, 1969).

Notwithstanding these antecedents, *Death Without Weeping* is a work without obvious parallel which deserves to be read in its entirety, even though it features here only in the form of a telling excerpt. Scheper-Hughes draws on her vast experience as a paramedic and anthropologist in Bom Jesus to alert us to the constructions of life and death, and mothering, that are held to by many poor women in a Brazilian town where violence and grinding poverty are the stuff of everyday life. **Reading 27** also confronts us with the obscene bodily violence that too often must be 'taken for granted' by the abused residents of the Alto do Cruzeiro, the hillside shanty town that forms part of Bom Jesus. Her accounts of the abduction, murder and mutilation of the *moradores* [residents] of the Alto, and what she calls the 'political tactic of disappearance', hold up a disturbing mirror to the languages of citizenship and justice that are now confidently paraded by a bourgeois elite in 'democratic' Brazil. The *moradores* of the Alto 'speak of bodies that are routinely violated and abused, mutilated and lost, disappeared into anonymous public spaces – hospitals and prisons but also morgues and the public cemetery'. They also refer to their 'invisibility' and 'anonymity'; to the fact that 'they are lost to the public census and to other state and municipal statistics'.

These are not people without history, but they are a collection of men, women and children whose homes and streets are often missing from maps, and whose presence is a matter of little significance to the powers that be (whether military or 'democratic'). Democracy may or may not matter to development at a macro-scale (**Reading 26**), but the absence of 'Western' conceptions of democracy, citizenship and the modern state from the Alto is of direct concern to the human *bodies* that are at the centre of Scheper-Hughes's narrative. Development means little if it has no respect for the integrity of the human body; likewise, development studies is the poorer when it fails to marry the 'big picture' of economics and macrosociology to the life stories of men, women and children bound up in wider processes of development and change. *Death Without Weeping* is a classic precisely because, in the words of its dust jacket, 'It spirals outward, taking the reader from the wretched huts of the shantytown into the cane fields and the sugar refinery, the mayor's office and the legal chambers, the clinics and the hospitals, the police headquarters and the public morgue, and finally, the municipal graveyard of Bom Jesus'. Scheper-Hughes forces us to attend to the different worlds of practice and meaning that join together, and separate, the worlds of the elite and state functionaries in Brazil and the worlds of the *moradores* of the Alto. Her sensitivity to sameness *and* difference is a model for a 'new' development studies.

Guide to further reading

Pointers to further reading have already been given in the Guides to Further Reading at the end of Sections One to Five. This Guide lists material referred to in the editorial matter; it also provides some further references on institutions, environment and development, democratization, and citizenship/gender/international justice.

Consolidated guide

Adelman, I. and Thorbecke, E. 1989: Special issue on the role of institutions in economic development. *World Development* 17 (9).

Apthorpe, R. (ed.) 1987: Institutions and policies: special issue. *Public Administration and Development* 6 (4).

Baddeley, O. and Fraser, V. 1989: *Drawing the line: Art and cultural identity in contemporary Latin America*. London: Verso.

Bardhan, P. 1989: The new institutional economics and development theory. *World Development* 17, 1389–95.

Booth, D. 1994: Rethinking social development: an overview. In Booth, D. (ed.), *Rethinking social development*. Harlow: Longman, 3–34.

Callaghy, T. and Ravenhill, J. (eds) 1993: *Hemmed in: Responses to Africa's economic decline*. New York: Columbia UP.

Clapham, C. 1985: *An introduction to Third World politics*. Beckenham: Croom Helm.

Clapham, C. 1992: The collapse of socialist development in the Third World. *Third World Quarterly* 13, 13–26.

Cole, S. 1989: World Bank planning forecasts and planning in the Third World. *Environment and Planning A* 21, 175–96.

Comaroff, J. and Comaroff, J. (eds) 1993: *Modernity and its malcontents*. Chicago: University of Chicago Press.

Corbridge, S. 1994: Forest struggles and forest policies in the Jharkhand, India, 1980–1993. *Mimeo*: University Lecture – University of Wisconsin, Madison, 18 March 1994.

de Castro, J. 1952: *The geography of hunger*. Boston: Little, Brown.

de Castro, J. 1969: *Death in the Northeast*. New York: Random House.

Diamond, L. 1992: *The democratic revolution*. New York: Freedom House.

Dunkerley, J. 1988: *Power in the Isthmus*. London: Verso.

Elson, D. (ed.) 1991: *Male bias in the development process*. Manchester: Manchester UP.

Fox, J. 1994: The difficult transition from clientalism to citizenship. *World Politics* 46, 151–84.

Friedmann, J. 1992: *Empowerment: The politics of alternative development*. Oxford: Blackwell.

Granovetter, M. 1992: Economic institutions as social constructions. *Acta Sociologia* 35, 3–11.

Grant, J. 1993: *The state of the world's children*. Oxford: OUP/UNICEF.

Hawthorn, G. 1993: Sub-Saharan Africa. In Held, D. (ed.), *Prospects for democracy*. Cambridge: Polity, 330–54.

Huntington, S. 1993: The clash of civilizations. *Foreign Affairs* 72, 22–49.

Karl, T. 1990: Dilemmas of democratization in Latin America. *Comparative Politics* 23, 234–67.

Khan, M. 1994: State failure in weak states: a critique of new institutionalist explanations. *Mimeo*: Department of Land Economy, University of Cambridge.

Leftwich, A. 1993: Governance, development and democracy in the Third World. *Third World Quarterly* 14, 605–24.

Lehmann, D. 1990: *Democracy and development in Latin America: Economics, politics and religion in the postwar period*. Cambridge: Polity.

Lele, S. 1991: Sustainable development: a critical review. *World Development* 19, 607–21.

Leonard, J. 1989: *Environment and the poor: Development studies for a common agenda*. New Brunswick: Transaction Books.

Manor, J. (ed.) 1991: *Rethinking Third World politics*. Harlow: Longman.

Marglin, F. and Marglin, S. (eds) 1990: *Dominating knowledge: Development, culture and resistance*. Oxford: Clarendon.

Momsen, J. and Kinnaird, V. (eds) 1993: *Different places, different voices: Gender and development in Africa, Asia and Latin America*. London: Routledge.

Nelson, J. 1992: *Encouraging democracy*. Washington, DC: Overseas Development Council.

North, D. 1990: *Institutions, institutional change and economic performance*. Cambridge: CUP.

North, D. 1991: Institutions. *Journal of Economic Perspectives* 5, 97–112.

O'Donnell, G. 1994: Delegative democracy. *Journal of Democracy* 5, 55–69.

Ostrom, E. 1990: *Governing the commons: The evolution of institutions for collective action*. Cambridge: CUP.

Ostrom, E. Schroeder, L. and Wynne, S. 1993: *Institutional incentives and sustainable development: Infrastructure policies in perspective*. Boulder: Westview.

Pearce, D. and Warford, J. 1993: *World without end: Economics, environment and sustainable development*. Oxford: OUP/World Bank.

Phillips, A. 1993: *Democracy and difference*. Cambridge: Polity.

Potter, D. 1993: Democratization in Asia. In Held, D. (ed.), *Prospects for democracy*. Cambridge: Polity, 355–79.

Radcliffe, S. and Westwood, S. (eds) 1993: *Viva! Women and popular protest in Latin America*. London: Routledge.

Redclift, M. 1987: *Sustainable development: Exploring the contradictions*. London: Methuen.

Rowe, W. and Schelling, V. 1991: *Memory and modernity: Popular culture in Latin America*. London: Verso.

Rushdie, S. 1991: *Imaginary homelands*. London: Granta.

Said, E. 1993: *Culture and imperialism*. New York: Knopf.

Scheper-Hughes, N. 1992: *Death without weeping: The violence of everyday life in Brazil*. Berkeley: University of California Press.

Scott, J. 1985: *Weapons of the weak: Everyday forms of peasant resistance*. New Haven: Yale UP.
Sengupta, N. 1991: *Managing common property: Irrigation in India and the Philippines*. New Delhi: Sage.
Swanson, T. 1994: *The international regulation of extinction*. Basingstoke: Macmillan.
Taussig, M. 1993: *Mimesis and alterity*. London: Routledge.
Thompson, J. 1993: *Justice and world order: A philosophical inquiry*. London: Routledge.
Turner, R. (ed.) 1993: *Sustainable environmental economics and management*. London: Belhaven.
Whitehead, L. 1992: The alternatives to 'liberal democracy': a Latin American perspective. *Political Studies* XL, 146–59.
World Bank 1991: *World development report, 1991*. Oxford: OUP/World Bank.
World Bank 1992: *Governance and development*. Washington, DC: World Bank.
World Bank 1994: *World development report, 1994*. Oxford: OUP/World Bank.

24 World Bank, 'Infrastructure for Development'

Excerpts from: World Bank, *World Development Report*, 1994, pp. 1–8. Oxford: OUP/World Bank (1994)

Developing countries invest $200 billion a year in new infrastructure – 4 percent of their national output and a fifth of their total investment. The result has been a dramatic increase in infrastructure services – for transport, power, water, sanitation, telecommunications, and irrigation. During the past fifteen years, the share of households with access to clean water has increased by half, and power production and telephone lines per capita have doubled. Such increases do much to raise productivity and improve living standards.

These accomplishments are no reason for complacency, however. One billion people in the developing world still lack access to clean water – and nearly 2 billion lack adequate sanitation. In rural areas especially, women and children often spend long hours fetching water. Already-inadequate transport networks are deteriorating rapidly in many countries. Electric power has yet to reach 2 billion people, and in many countries unreliable power constrains output. The demands for telecommunications to modernize production and enhance international competitiveness far outstrip existing capacity. On top of all this, population growth and urbanization are increasing the demand for infrastructure.

Coping with infrastructure's future challenges involves much more than a simple numbers game of drawing up inventories of infrastructure stocks and plotting needed investments on the basis of past patterns. It involves tackling inefficiency and waste – both in investment and in delivering services – and responding more effectively to user demand. On average, 40 percent of the power-generating capacity in developing countries is unavailable for production, twice the rate in the best-performing power sectors in low-, middle-, and high-income countries. Half the labor in African and Latin American railways is estimated to be redundant. And in Africa and elsewhere, costly investments in road construction have been wasted for lack of maintenance.

This poor performance provides strong reasons for doing things differently – in more effective, less wasteful ways. In short, the concern needs to broaden from increasing the *quantity* of infrastructure stocks to improving the *quality* of infrastructure services. Fortunately, the time is ripe for change. In recent years, there has been a revolution in thinking about who should be responsible for providing infrastructure stocks and services, and how these services should be delivered to the user.

Against this background, *World Development Report 1994* considers new ways of meeting public needs for services from infrastructure (as defined in

Box 1) – ways that are more efficient, more user-responsive, more environment-friendly, and more resourceful in using both the public and private sectors. The report reaches two broad conclusions:

- Because past investments in infrastructure have not had the development impact expected, it is essential to improve the effectiveness of investments and the efficiency of service provision.
- Innovations in the means of delivering infrastructure services – along with new technologies – point to solutions that can improve performance.

Box 1 What is infrastructure?

This Report focuses on *economic infrastructure* and includes services from:

- Public utilities – power, telecommunications, piped water supply, sanitation and sewerage, solid waste collection and disposal, and piped gas.
- Public works – roads and major dam and canal works for irrigation and drainage.
- Other transport sectors – urban and interurban railways, urban transport, ports and waterways, and airports.

Infrastructure is an umbrella term for many activities referred to as 'social overhead capital' by such development economists as Paul Rosenstein-Rodan, Ragnar Nurkse, and Albert Hirschman. Neither term is precisely defined, but both encompass activities that share technical features (such as economies of scale) and economic features (such as spillovers from users to nonusers).

This Report marshals evidence in support of these conclusions – identifying causes of failure and examining alternative approaches. The main messages and policy options are summarized in Box 2.

Infrastructure's role and record

The adequacy of infrastructure helps determine one country's success and another's failure – in diversifying production, expanding trade, coping with population growth, reducing poverty, or improving environmental conditions. Good infrastructure raises productivity and lowers production costs, but it has to expand fast enough to accommodate growth. The precise linkages between infrastructure and development are still open to debate. However, infrastructure capacity grows step for step with economic output – a 1 percent increase in the stock of infrastructure is associated with a 1 percent increase in gross domestic product (GDP) across all countries. And as countries develop, infrastructure must adapt to support changing patterns of demand, as the shares of power, roads, and telecommunications in the total stock of infrastructure increase relative to those of such basic services as water and irrigation.

The kind of infrastructure put in place also determines whether growth does all that it can to reduce poverty. Most of the poor are in rural areas, and the

Box 2 Main messages of *World Development Report 1994*

Infrastructure can deliver major benefits in economic growth, poverty alleviation, and environmental sustainability – but only when it provides services that respond to effective demand and does so efficiently. Service is the goal and the measure of development in infrastructure. Major investments have been made in infrastructure stocks, but in too many developing countries these assets are not generating the quantity or the quality of services demanded. The costs of this waste – in forgone economic growth and lost opportunities for poverty reduction and environmental improvement – are high and unacceptable.

The causes of past poor performance, and the source of improved performance, lie in the incentives facing providers. To ensure efficient, responsive delivery of infrastructure services, incentives need to be changed through the application of three instruments – commercial management, competition, and stakeholder involvement. The roles of government and the private sector must be transformed as well. Technological innovation and experiments with alternative ways of providing infrastructure indicate the following principles for reform:

- *Manage infrastructure like a business, not a bureaucracy.* The provision of infrastructure needs to be conceived and run as a service industry that responds to customer demand. Poor performers typically have a confusion of objectives, little financial autonomy or financial discipline, and no 'bottom line' measured by customer satisfaction. The high willingness to pay for most infrastructure services, even by the poor, provides greater opportunity for user charges. Private sector involvement in management, financing, or ownership will in most cases be needed to ensure a commercial orientation in infrastructure.
- *Introduce competition – directly if feasible, indirectly if not.* Competition gives consumers choices for better meeting their demands and puts pressure on suppliers to be efficient and accountable to users. Competition can be introduced directly, by liberalizing entry into activities that have no technological barriers, and indirectly, through competitive bidding for the right to provide exclusive service where natural monopoly conditions exist and by liberalizing the supply of service substitutes.
- *Give users and other stakeholders a strong voice and real responsibility.* Where infrastructure activities involve important external effects, for good or bad, or where market discipline is insufficient to ensure accountability to users and other affected groups, governments need to address their concerns through other means. Users and other stakeholders should be represented in the planning and regulation of infrastructure services, and in some cases they should take major initiatives in design, operation, and financing.

Public-private partnerships in financing have promise. Private sector involvement in the financing of new capacity is growing. The lessons of this experience are that governments should start with simpler projects and gain experience, investors' returns should be linked to project performance, and any government guarantees needed should be carefully scrutinized.

Governments will have a continuing, if changed, role in infrastructure. In addition to taking steps to improve the performance of infrastructure provision under their direct control, governments are responsible for creating policy and regulatory frameworks that safeguard the interests of the poor, improve environmental conditions, and coordinate cross-sectoral interactions – whether services are produced by public or private providers. Governments also are responsible for developing legal and regulatory frameworks to support private involvement in the provision of infrastructure services.

growth of farm productivity and nonfarm rural employment is linked closely to infrastructure provision. An important ingredient in China's success with rural enterprise has been a minimum package of transport, telecommunications, and power at the village level. Rural enterprises in China now employ more than 100 million people (18 percent of the labor force) and produce more than a third of national output.

Infrastructure services that help the poor also contribute to environmental sustainability. Clean water and sanitation, nonpolluting sources of power, safe disposal of solid waste, and better management of traffic in urban areas provide environmental benefits for all income groups. The urban poor often benefit most directly from good infrastructure services because the poor are concentrated in settlements subject to unsanitary conditions, hazardous emissions, and accident risks. And in many rapidly growing cities, infrastructure expansion is lagging behind population growth, causing local environments to deteriorate.

In developing countries, governments own, operate, and finance nearly all infrastructure, primarily because its production characteristics and the public interest involved were thought to require monopoly – and hence government – provision. The record of success and failure in infrastructure is largely a story of government's performance.

Infrastructure's past growth has in some respects been spectacular. The percentage of households and businesses served has increased dramatically, especially in telephones and power. The per capita provision of infrastructure services has increased in all regions; the greatest improvements have been in East Asia and the smallest in Sub-Saharan Africa, reflecting the strong association between economic growth and infrastructure.

In other important respects, however, the performance has been disappointing. Infrastructure investments have often been misallocated – too much to new investment, not enough to maintenance; too much to low-priority projects, not enough to essential services. The delivery of services has been hampered by technical inefficiency and outright waste. And too few investment and delivery decisions have been attentive to meeting the varied demands of different user groups, or to the consequences for the environment.

. . . .

Diagnosing the causes of poor performance

The problems of insufficient maintenance, misallocated investment, unresponsiveness to users, and technical inefficiencies present daunting challenges for future reforms – challenges compounded by new demands and constrained resources. The solutions lie in the successes and failures of policy and in the lessons from recent policy experiments.

There is great variation both within and across countries in the efficiency of providing infrastructure services. Moreover, good performance by a country in one infrastructure sector is not necessarily associated with good performance in other sectors. Some developing countries – and not always the richer ones – perform at high levels. Côte d'Ivoire meets the 85 percent best-practice standard in water supply, while in Manila only about 50 percent of treated water is delivered to customers. In railways, the availability of locomotives is high where maintenance is good: at any given time, India has 90 percent of its locomotives available. Availability is low where maintenance is neglected: 50 percent in Romania and 35 percent in Colombia, compared with a developing country average of about 70 percent. For telephones, call completion rates are 99 percent in the best-performing countries, 70 percent in the average developing country, and far lower in some. These findings indicate that the performance of infrastructure derives not from general conditions of economic growth and development but from the institutional environment, which often varies across sectors within individual countries.

Therefore, to understand what accounts for good performance – and bad – requires understanding the institutional arrangements for providing infrastructure services and the incentives governing their delivery. This Report identifies three reasons for poor performance.

First, the delivery of infrastructure services usually takes place in a market structure with one dominating characteristic: the absence of competition. Most infrastructure services in the developing world are provided by centrally managed monopolistic public enterprises or government departments. Almost all irrigation, water supply, sanitation, and transport infrastructure is provided in this manner. Until a few years ago, telephone services in most countries were the responsibility of a state-owned post, telephone, and telegraph enterprise. The bulk of power has also been provided by a public monopoly. As a result, the pressure that competition can exert on all parties to perform at maximum efficiency has been lacking.

Second, those charged with responsibility for delivering infrastructure services are rarely given the managerial and financial autonomy they need to do their work properly. Managers are often expected to meet objectives at variance with what should be their primary function – the efficient delivery of high-quality services. Public entities are required to serve as employer of last resort or to provide patronage. They are compelled to deliver services below cost – often by not being allowed to adjust prices for inflation. The other side of the coin is that public providers are rarely held accountable for their actions. Few countries set well-specified performance measures for public providers of infrastructure services, and inefficiency is all too often compensated for by budgetary transfers rather than met with disapproval.

Third, the users of infrastructure – both actual and potential – are not well positioned to make their demands felt. When prices reflect costs, the strength of consumer demand is a clear signal of what should be supplied. Through the price mechanism, consumers can influence investment and production decisions in line with their preferences. But prices of infrastructure services typically do not reflect costs, and this valuable source of information about consumer needs is lost. For example, power prices in developing countries have generally fallen, while costs have not. As a result, prices now cover only half the supply costs, on average. Water charges and rail passenger fares typically cover only a third of costs. Excess consumer demand based on below-cost prices is not a reliable indicator that services should be expanded, although often it is taken as such.

Users can express preferences in other ways, such as local participation in planning and implementing new infrastructure investments. But they seldom are asked, and investment decisions are all too often based on extrapolations of past consumption rather than on true assessments of effective demand and affordability.

Individually, each of these three points is important. Together, they go a long way toward explaining the disappointing past performance of much infrastructure. Rival suppliers and infrastructure users might have exerted pressure for better services, but they were prevented from doing so. Governments – by confusing their roles as owner, regulator, and operator – have failed to improve service delivery.

New opportunities and initiatives

Creating the institutional and organizational conditions that oblige suppliers of infrastructure services to be more efficient and more responsive to the needs of users is clearly the challenge. But is it possible? Three converging forces are opening a window of opportunity for fundamental changes in the way business is done. First, important innovations have occurred in technology and in the regulatory management of markets. Second, a consensus is emerging on a larger role for the private sector in infrastructure provision, based in part on recent experience with new initiatives. Third, greater concern now exists for environmental sustainability and for poverty reduction.

New technology and changes in the regulatory management of markets create new scope for introducing competition into many infrastructure sectors. In telecommunications, satellite and microwave systems are replacing long distance cable networks, and cellular systems are an emerging alternative to local distribution networks. These changes erode the network-based monopoly in telecommunications and make competition possible. In power generation, too, combined-cycle gas turbine generators operate efficiently at lower output levels, while other innovations are reducing costs. New technology makes competition among suppliers technically feasible, and changes in regulations are making competition a reality by allowing competitive entry in activities such as cellular phone service or power generation. Technical and regulatory change in other infrastructure sectors – ranging from transportation to water supply and drainage and irrigation – also make them more open to new forms of ownership and provision.

Alongside such changes are new perceptions of the role of government in infrastructure. An awareness is growing in many countries that government provision has been inadequate. Brownouts and blackouts in power systems, intermittent water supplies from municipal systems, long waiting periods for telephone service connection, and increasing traffic congestion provoke strong reactions. Reforms in some industrial countries have increased the competition in telecommunications, in road freight and airline transport, and in power generation – proving that alternative approaches are possible. The poor performance of planned economies has also provoked a reassessment of the state's role in economic activity.

These developments have led governments to search for new ways to act in partnership with the private sector in providing infrastructure services. Most dramatic have been the privatizations of such enterprises as the telephone system in Mexico and the power system in Chile. Elsewhere, various forms of partnership between government and the private sector have evolved. Port facilities have been leased to private operators – the Kelang container facility in Malaysia being among the first. Concessions have been granted to private firms, particularly in water supply; Côte d'Ivoire is one of the earliest examples. Contracting out services, as Kenya has done with road maintenance, is well under way in many countries. Private financing of new investment has grown rapidly through build-operate-transfer (BOT) arrangements under which private firms construct an infrastructure facility and then operate it under franchise for a period of years on behalf of a public sector client. This approach has been used to finance the construction of toll roads in Mexico and power-generating plants in China and the Philippines.

An increasing regard for the environmental sustainability of development strategies and a deepening concern for poverty reduction after a decade of stagnation in many regions of the world also give impetus to infrastructure reform. Creating pressures for change, environmental issues are coming to the fore in transport (traffic congestion and pollution), irrigation (increased waterlogging and salinity of agricultural land), water supply (depleted resources), sanitation (insufficient treatment), and power (growing emissions). At the same time, a decade of reduced economic growth – especially in Latin America and Sub-Saharan Africa – shows that poverty reduction is not automatic and that care must be taken to ensure that infrastructure both accommodates growth and protects the interests of the poor.

Options for the future

To reform the provision of infrastructure services, this Report advocates three measures: the wider application of commercial principles to service providers, the broader use of competition, and the increased involvement of users where commercial and competitive behavior is constrained.

Applying commercial principles of operation involves giving service providers focused and explicit performance objectives, well-defined budgets based on revenues from users, and managerial and financial autonomy – while also holding them accountable for their performance. It implies that governments should refrain from ad hoc interventions in management but should provide explicit transfers, where needed, to meet social objectives such as public service obligations.

Broadening competition means arranging for suppliers to compete for an entire market (e.g., firms bidding for the exclusive right to operate a port for ten years), for customers within a market (telephone companies competing to serve users), and for contracts to provide inputs to a service provider (firms bidding to provide power to an electric utility).

Involving users more in project design and operation of infrastructure activities where commercial and competitive behavior is constrained provides the information needed to make suppliers more accountable to their customers. Users and other stakeholders can be involved in consultation during project planning, direct participation in operation or maintenance, and monitoring. Development programs are more successful when service users or the affected community has been involved in project formulation. User participation creates the appropriate incentives to ensure that maintenance is carried out in community-based projects.

These elements apply whether infrastructure services are provided by the public sector, the private sector, or a public–private partnership. To this extent, they are indifferent to ownership. However, numerous examples of past failures in public provision, combined with growing evidence of more efficient and user-responsive private provision, argue for a significant increase in private involvement in financing, operation, and – in many cases – ownership.

25 Susanna B. Hecht, 'Environment, Development and Politics: Capital Accumulation and the Livestock Sector in Eastern Amazonia'

Excerpts from: *World Development* 13 (6): 663–84 (1985)

Summary

Deforestation and environmental degradation are increasingly common themes in the literature on humid tropical rural development. This paper explores the frameworks

Funds for field research and laboratory analysis were provided by the Ford Foundation, the National Science Foundation and AID. Earlier drafts of this paper benefited from the valuable comments of Michael Storper, Dennis Mahar, Richard Norgaard, John Nicolaides. Jack Ewel and Anthony Anderson.

used to analyze environmental questions in developing economies and how well these function in the particular case of livestock development in the Eastern Amazon Basin. The paper argues that, due to the peculiarities of the state subsidies available for ranching activities that spurred a frenzy of land speculation, the exchange rather than productive value of land became paramount. In such a context, cautious land management was irrelevant and serious environmental degradation was the result. The paper suggests that models of environmental degradation that focus only on the question of production cannot capture the environmental dynamics of speculative economies.

1. Introduction

The Amazon Basin has increasingly become the focus of international attention. As the largest area of remaining tropical rainforest biome, it has a relatively unexplored resource potential, and is regarded as one of the last agricultural frontiers. Recent decades have witnessed the rapid conversion of forest to agricultural landscapes (primarily pasture) in Amazonia, stimulated by a combination of infrastructure development, fiscal incentives and colonization programs. Estimates indicate that about one million hectares are deforested each year (Tardin *et al.*, 1979; OTA, 1983). Unfortunately, much of the area converted to pasture is only ephemerally productive; and a few years after conversion, the productivity of these lands shows a pronounced decline. Estimates of the area of degraded pasture vary from about 20% (Serrao *et al.*, 1979; Toledo and Serrao, 1982), to those, based on LANDSAT data indicating deteriorated pasture levels at closer to 50%, of lands cleared (Tardin *et al.*, 1977; Hecht, 1982; Santos, de Novo and Duarte, 1979).

The conversion of substantial areas of tropical rainforest in Amazonia has been a source of considerable controversy. Biologists voice concern about species extinction (Myers, 1980; Gomez-Pompa, Vasques-Yanes and Guevara, 1972; Pires and Prance, 1977); changes in hydrological regimes (Gentry and Lopez Parodi, 1980), local and global climate modification (Salati *et al.*, 1979; Molion, 1975; Salati and Schubart, 1983) and soil resources degradation (Goodland and Irwin, 1975; Goodland, 1980). Social scientists have pointed to the intense land conflicts (Schmink, 1982; Souza-Martins, 1983), increasing peasant marginalization (Wood and Schmink, 1979; Sawyer, 1979; Santos, 1979), extinction of indigenous groups (Davis, 1977) and increased rural-to-urban migration (Martine, 1982; Aragon, 1978; Wood and Wilson, 1983) that have accompanied the process of land development through livestock expansion into the Amazon region.

In contrast to the concerns of biologists and social scientists noted above, Amazonian integration has been described as an essential national development project both in terms of its potential economic and social effects (see, for example, National Development Program – PND II, 1975; and Alvim, 1978) and the only means of assuring that millions of Brazilians are not condemned to lives of abject poverty. While ecological impacts are acknowledged in this literature, the benefits of development are thought to outweigh the irreversible costs of species lost and Indian extinction. The less catastrophical environmental issues are generally seen as temporary and resolvable. Ecological problems in this view follow a sort of Kuznets curve: as development begins, things deteriorate, but as growth accelerates, technical solutions eventually diminish the deleterious environmental effects.

This paper has several purposes. First, it will briefly evaluate the frameworks used to analyze environmental questions in developing countries, particularly those pertaining to agricultural resources in the humid tropics areas like Amazonia. This will require some discussions of the models themselves. Next, the paper will discuss how environmental questions can be inserted into an analysis of the political economy of Amazonian development, by using an historical and political economy approach to integrate development and environmental issues into one analysis. Through this approach, the dynamics of Amazonian occupation in the 1970s, its associated pattern of capital accumulation and the serious problems of environmental degradation are closely linked. Finally, this paper is a 'middle-level' analysis that focuses on the concrete expression of larger national and international processes as they unfolded and affected the eastern Amazon Basin.

2. Models of environmental degradation

> A model is a metaphor of historical process. It indicates not only significant parts of this process, but the way in which they change. In one sense, history remains irreducible; it remains all that happened. In another sense, history does not become history until there is a model: at the moment at which the most elementary notion of causation, process, or cultural patterning intrudes, then some model is assumed. (Thompson, 1965, p. 288)

The paradigms used for understanding Third World environmental issues are certainly not unique to this particular context; they pervade the environmental literature generally. How the questions are framed, the kinds of data used to support various positions, the level of abstraction and the level of generalization are major problems in environmental analysis. The way ecological problems are theorized in developing countries, however, has simply not received much scrutiny. Further, the conceptions that are used often ignore specific regional histories, their biologies, economies and social dynamics that strongly influence the form and impact of environmental issues.

This section will outline the major environmental frameworks used to understand how Third World environmental questions are posed and developed. These frameworks can be roughly separated into: (1) general biological metaphors; (2) general economic metaphors; (3) at the other end of the spectrum, technology assessment.

(a) *Biological metaphors: Malthus's inexorable population increase*

By far the most prevalent thesis about land degradation in the Third World is the Malthusian view that increasing demographic pressure results in overuse of land and use of marginal land. This kind of cultivation generates soil nutrient decline or erosion with consequent deterioration of the agricultural resource base. In reference to humid tropical areas, the analysis specifically emphasized the problems of slash-and-burn agriculture with population increase (*cf*. Myers, 1980; Sanchez *et al*., 1982).

The Malthusian framework, currently enjoying a renewed vogue, when invoked as an explanation for Third World poverty, has come under attack from several quarters. This literature is enormous, but representative critics include Durham (1980), Mamdami (1972), Murdoch (1980), Harvey (1974) and Sen (1981). These authors of diverse disciplinary background, contending with a broad array of questions, argue that increasing population numbers must be analyzed in a historical and more socially complex manner than simply invoking aggregate population parameters. Family size, they suggest, reflects rational economic decisions. The cost-benefit ratio of extra children is high for poor families in poor societies or societies where resources are extremely unequally distributed, because children contribute economically in agricultural labor or the informal economy to the household at an early age and continue to do so throughout their lives. Their opportunity cost, however, is relatively low.

The application of Malthusian perspectives in the Third World has remained central to the analysis of environmental degradation (Brown, 1982; Myers, 1980; Nicolaides *et al*., 1982). In by far the most studied area of Third World environmental issues, the Sahel, a simple Malthusian view has been sharply criticized (*cf*., Watts, 1983; Franke and Chasen, 1980; and Hoskins, 1982) because of its lack of attention to the social and historical factors underlying the shift to larger animal and human populations. These authors argue that environmental problems have their origins in the structural poverty of rural populations. Rather than seeing population increase as an expression of natural law 'beyond the reach of human ingenuity' (Malthus, 1970), these authors argue that the decision to have children is a rational response to quite concrete economic factors. The point of analysis then should not focus at the correlation between population and poverty, but rather at historical and political economic issues that underlie reproductive choice.

The problem in much of the application of a Malthusian framework is due to the strong correlation of population numbers with poverty and environmental degradation but their weak powers of explanation. Thus, as a 'theory,' Malthusian perspectives merely seek out generalized relations among various empirical objects and events themselves, and not abstractions about what produced them. As Ellen (1982) suggests, 'Crucial correlations are rarely those gross observable relationships between totalities, but the subtle, hidden connections between peculiarities.'

(b) *Economic perspectives on environmental degradation*

The economic analyses of Third World environmental issues usually fall into one of the three general categories. There is considerable conceptual relation between the three, and the ideas developed in one may be a sub-theme in another. The first can roughly be described as the 'Tragedy of the Commons'; the second, 'The Issue of Externalities' (often a function of the first); and finally, 'Dependency Perspectives'. The unifying idea of all of these centers on the concept that irrational land use leads to environmental degradation, and this irrationality can be understood through an analysis of issues associated with the economics of production.

Environmental degradation in developing countries is often perceived as a function of faulty property relations centered around the questions of common resources. In the original formulation by Hardin (1968), destruction of commons was triggered by population growth, and carried a strong Malthusian current within its analysis. The position can be roughly summarized by the idea that in the use of common resources, each economic actor seeks to maximize his individual utility. In the case of the commons, each individual garners the benefits of his additional exploitation, while the costs (in terms of environmental degradation) of each incremental increase in resource use are shared over all users, or society as a whole. Thus, there is a minor cost to the individual, but maximum potential gains. Each actor then will attempt to expand his exploitation competitively, with the result that the resource itself eventually collapses. The individual, personal rationality generates a collective tragedy. The two solutions proposed by Hardin to this situation are increased population control and the privatization of public goods to assure their proper management.

Clarke (1974) has refined and expanded the ideas of the commons, pointing out that not only common property competitive exploitation, but also private maximization of net present value of profits, are central to the destruction of 'commons resources.' By focusing on discount rates, Clarke emphasizes that 'high discount rates have the effects of causing biological over-exploitation wherever it is commercially feasible.' It is thus often rational for entrepreneurs to accelerate their level of exploitation to the point of resource destruction and to reinvest their profits elsewhere. In Hardin's view, common resource externalities are not serious until population pressures make them so. Clarke, on the other hand, specifically points out that either competitive exploitation or certain mechanisms of capital accumulation produce the destruction of the 'commons.' Further, he emphasizes the mobility of capital, which makes it possible to avoid the longer-term economic consequences of resource degradation.

The 'commons' framework is most often applied in discussions of fisheries, grazing lands and, increasingly, forest resources. Counterarguments to Hardin focused on his Malthusian and 'Lifeboat Ethics' stance, but several critiques have also suggested that resources perceived as 'commons' reflect a variety of relations that were overlooked in Hardin's original formulation. The model rests on the neoclassical ideas that with private property, individual maximizers will rationally manage the resource to use the land or resource at its best and highest use and to remain competitive within the market. This assumes that markets are always the best means of allocating natural resources and that competition necessarily leads to appropriate management – propositions that are open to debate.

. . . .

Implicit in the commons models is the idea that all the costs of production are not borne by the person who receives the benefits, but accrue to nature or society as a whole. This is the starting point of the analysis of externalities. Unlike the 'Tragedy of the Commons' approach, which essentially argues that the penetration of capitalist relations is necessary for proper natural resource management, the analysis of externalities reduces the scope of the commons issue from a historical-economic argument to a technical question, or one of mere market imperfection. The challenge is either technological or one of integrating the process into a market framework through the uses of taxes, subsidies and regulation. There is no question that some environmental questions can be resolved through only technical measures; but their implementation, almost without exception, requires the invocation of state economic manipulation through regulation. This thrusts such issues into the political realm.

Most writers on Third World environmental issues point out the high frequency of externalities (World Bank, 1975; Milton and Farver, 1968; Eckholm, 1982; Redclift, 1984; Murdoch, 1980). However, the political resolution of these questions is seen as profoundly problematic due to the structure and control of state power, the power of elites in Third World countries, the limited influence states have over both technology and investment choices made by national and foreign entrepreneurs, and finally the view that environmental deterioration is a necessary price for economic growth.

Commons/externalities frameworks focus on the *internal* organization of the resource management. By modifying internal production structure through privatization, technology and/or appropriate intervention, individuals will rationally manage their assets in a manner to produce a good for sale in competitive markets. The 'dependency' perspective, and the term is loosely used here, generally is concerned with the external factors acting on the production system that lead to environmental degradation. These factors include the orientation to and discipline by global markets; inappropriate international technologies, and the question of control over the revenues and responsibility for the consequences of the development process in intensive agriculture, or the exploitation of mineral or biotic resources (Sunkel, 1982; Mueller and Gligo, 1982).

The argument here follows roughly these lines. Historically, the major agricultural focus in international development concentrated on a few export crops; these tended to be the 'dessert crops' – cacao, coffee and sugar – or fiber crops like cotton, all of which compete for roughly the same market. Particularly in the case of short-cycle crops like sugar and cotton, the production process has become highly technical (in part because the United States is an important producer of these crops and US technology was often transferred 'wholesale'), and involves the use of a variety of fertilizer, pesticide and herbicide imports, in addition to heavy farm machinery. Environmental degradation occurs through a combination of the careless use of technologies evolved for different environments and different management regimes (Janzen, 1973; Milton and Farver, 1968) in the face of a substantially different and more complex biota. Further, the pressure for increased output often implies 'pushing' agro-ecosystems. Such an analysis is widely associated with export cropping, and can include a discussion of state promotion of particular commodities. The larger questions of control of revenues remain localized in the 'North–South' debate.

(c) *Technology assessment*

The issue of inappropriate technology, what Michael Todaro (1977) has called the 'false paradigm' model, has increasingly come to the fore in analysis of environmental impacts. In this view, planners or agencies apply land use technologies that, due to local environmental or cultural conditions, are not well adapted to the area, and thus cause undesirable ecological effects. The literature on appropriate technologies is vast; but in practice, the focus of research is heavily weighted toward technological solutions and centers around genetic improvement and husbandry techniques.

This model does not try to replace environmental issues within an economic context except as a resolvable externality. Its most salient feature is its atheoretical approach that treats ecological issues as discrete problems associated with policy mistakes and a lack of fine tuning of the physical and biological technologies. It is not usually explicitly linked with any larger theories of environmental degradation, although Malthusian assumptions are often implicit in the 'false paradigm.'

I will argue that environmental degradation in much of Amazonia is not adequately explained by any of these models. Rather than analyzing regional ecological problems as strictly endogenous, due to population increase or to the use of inappropriate technologies, one needs to examine the role that Amazonian development (especially via cattle ranching) played in Brazil after the military coup in 1964. In both substantive and ideological ways, the development of Amazonia addressed the strong national as well as international pressures that confronted the new military government. These pressures led policymakers to choose cattle ranching, the *latifundia* land use *par excellence* over all other alternative land uses, as the defining strategy of Amazonian occupation (Hecht, 1982a).

The process of forest conversion and its consequences were implicit in the choice of cattle production. The factors that triggered continued conversion, in spite of the dramatic environmental costs and productivity declines, require a new environmental analytic framework that is linked to specific patterns of capital accumulation.

3. The military coup of 1964 Amazonian policy

(a) *The question of legitimacy*

In 1964, Brazil experienced a military coup of the type Barrington Moore (1968) would call a 'revolution from above.' Diverse political factions were supporters of the coup (Stepan, 1968), but its outcome was particularly favorable to certain groups: the agro-industrial and industrial entrepreneurial elite. The coup inaugurated a variety of changes, if not in kind certainly in emphasis, in the Brazilian economic scene. These transformations were reflected in increased international investment, the strengthening of entrepreneurial capital and significant modifications in the role of the Brazilian State in national and Amazonian planning.

When the military seized power, several basic political issues had to be addressed. First, the regime had to legitimate its right to govern. Second, it had to resolve many of the pressing economic constraints that had hampered capital accumulation by national elites (such as wage demands, high inflation, import industrialization policies and lack of investment outlets). Third, it was necessary to solve, or at least foster the appearance of contending with the social and political problems of rural areas, as reflected in stagnant agricultural production, low rates of investment and rural out-migration (Knight, 1971; Wood and Wilson, 1983).

Increased economic output was seen as a solution to the questions of legitimation and economic reorganization. The economic growth policies chosen by the new regime relied on increased international borrowing, a profound wage squeeze, augmented transnational participation in the economy, repression and conventional expansionist monetary and fiscal policies (Fishlow, 1973; Taylor *et al.*, 1980; Belassa, 1979). While the role of these policies in spurring the 'Brazilian Miracle' is open to question (Taylor *et al.*, 1980; Fishlow, 1973; Belassa, 1979; Singer, 1982; Malan and Bonelli, 1977), the regime took credit for the rapid growth of the Brazilian economy during the 1968–73 period. This unusual economic performance, triggered by massive borrowing, foreign investment and an exceptionally active world market, aided in the institutionalization of the various military regimes that followed the 1964 coup.

Agricultural and agrarian questions required a profound shift in policy as well. Except for a few exports, the Brazilian agricultural sector in the late 1950s and early 1960s suffered from a lack of credit and investment capital, import tariffs that made the cost of inputs such as machinery and agricultural chemical stocks very expensive, export taxes and marked regional disparities in investment. Overvalued exchange rates made Brazilian agricultural products relatively costly on the international market, while national policies emphasized exports only as a vent for surplus production (Knight, 1971). Structural change and mechanization in Brazilian agriculture began to erode access to land for tenant farmers and sharecroppers, while the closing of the southern frontiers of Paraná and Rio Grande do Sul further reduced agricultural options for the rural poor (Foweraker, 1980).

Attempts to confront the situation in the countryside took the form of increased availability of funds for agriculture. Policy mechanisms for agricultural change included subsidized interest rates for rural modernization, mechanization, export incentives, and revaluation of the *cruzeiro*. The greater availability of financing for investments in the agricultural sector was intended to modify the production processes on the farm and in specific regions. The agro-industrialist elites, as well as urban entrepreneurs, were attracted to agriculture by the subsidies, for diversification of their investment portfolio, and to take advantage of tax credits. These initiatives would ostensibly promote greater efficiency and rational economic behavior, compared with that of semi-feudal elites and through essentially a diffusion effect, would transform agricultural production in Brazil.

The agrarian question, whose outwards symptoms were accelerated rural to urban migration and increasing peasant activism, was addressed by the new regime by repression in areas of insurrection, such as the Northeast and Parana, and the opening of a new agricultural frontier, Amazonia. Amazonian development obviated the need for land reform, and implied a national will to include the rural poor in the government's development strategy.

The occupation of Amazonia was an idea that resonated closely with other ancillary themes in the government programs. Among the most important was the military ideology of National Security. The large size of Amazonia (more than 50% of the Brazilian territory), with its sparse population and unpatrolled border shared with eight other countries, and with a history of annexation and border conflict between them, fueled a certain disquiet about the area. The geopolitical importance of Amazonia, reflected in the slogan '*Integrar para nao entregar*', is present throughout the Amazon planning documents.

The military language characteristic of the development rhetoric also contributed to the sense of Amazonian occupation as the moral equivalent of war. Ideologically, such a military focus unifies national factions around a common national goal and justifies current sacrifices in welfare for a larger (future) good. Thus, General Castello-Branco stated in 1964 that 'Amazonian occupation would proceed as though it were a strategically conducted war,' and the first post-coup body of legislation concerning the Amazon had the appropriately military title of 'Operation Amazonia.' Not only would Amazonian occupation function like a war in the ideological sense, but it would also stimulate the economy in a like manner through construction and heavy industries linked to infrastructure development.

The concept of national security was closely allied to that of national integration, another common policy theme. The latter, in turn, can be seen as a version of Manifest Destiny, an idea consistent with the orthodox economic approach to the region. '*Integracao Nacional*' implied greater economic linkages of the Amazon hinterlands to urban centers, facilitated by the development of infrastructure and the creation of investment credits. The attack on regional disparities held out the image of the developed center-south – Sao Paulo and Rio – as the achievable future of Brazil's backlands. Further, through regional occupation, the riches of Amazonia, rightfully Brazilian, could be realized and contribute to the overall welfare of Brazil.

Through national security and manifest destiny, backed by infrastructure and investment, the new agricultural frontier in the Amazon was to provide a solution to vital economic and ideological questions, and thus served important political and legitimizing functions to the new regime. The role that cattle ranching was to play in this process can only be understood through an appreciation of the internal and international pressures that influenced the new government.

(b) *Internal pressures underlying policy*

Significant internal pressures came to bear on the new military regime (Stepan, 1968; 1973). Urban unrest required police coercion, and the wage squeeze on workers characteristic of the post-1964 regime (Bacha, 1977; Fishlow, 1973; Malan and Bonelli, 1977) exacerbated dissatisfaction (Stepan, 1973; O'Donnell, 1979). Since few other concessions were granted to labor by the regime, cheap food policies (especially for beef) were important priorities (Leff, 1967; Bergsman, 1970). The beef industry in the early 1960s, however, was at a cyclic production low, with price ceilings making cattle production uneconomic for producers, while urban and international demand soared.

The capacity of traditional landed elites to respond to this crisis was perceived as dubious, while the technological orientation of agribusiness and the entrepreneurial spirit of parts of the industrial sector, coupled with the new Australian pasture technologies, seemed a reasonable solution. The apparent viability of this avenue was contingent on the extent to which long-term credit could be made available and antiquated production bottlenecks could be circumvented. These credit and production constraints could be addressed without drastic structural change through increased agricultural lending and horizontal expansion of land use through the sale of state properties to agro-entrepreneurs.

The expansion of cattle production had other attractions since it was consistent with the desire to expand exports of 'non-traditional' Brazilian products, a fundamental feature of the new regime's economic policy. Little was known about Amazonian ecologies, but the 300-year history of ranching on the island of Marajo and the existence of upland natural grasslands made it appear that only technical insufficiency limited the productivity of livestock in the region. Further, compared with the other agricultural options in the region, such as pepper, cacao and rubber plantations, ranching seemed relatively easy to implant and maintain and had low labor requirements.

. . . .

The expansion of ranching into the Amazon was also conditioned by high inflation periods of the 1960s and 1970s that increased land speculation. As the latter parts of this article show, the fiscal incentives and land concessions provided by the government in Amazonia facilitated land acquisition and contributed to the extreme increase in the value of land in the region. Those who invested through SUDAM (Superintendency of Amazonian Development) and other cattle projects could make enormous capital gains simply through the increased valorization of land.

(c) *International factors affecting policy*

International factors implicitly or explicitly played an important role in the evolution of ranching as the main development strategy for the Amazon, primarily through the expanded global demand for beef during the mid-1960s. This increased demand reflected both changes in the US production system and rising European and Middle Eastern purchases. Feder (1979) has pointed out that changes in the US beef production system were important in the expanded global beef demand and in the increased lending for the livestock sector by international agencies.

In the 1950s, the United States embarked on a program to increase the production of high-quality beef using the feedlot system. Grain-fattening of cattle was also a means of disposing of surplus wheat and corn, a serious problem in the late 1950s and early 1960s. Feedlots were capable of generating large tonnages of high-quality beef, but the success of this production system was not without some difficulties. In particular, demand began to soar for lower-quality, utility beef used in fast foods and sausages. Utility, or cutter beef, is expensive to produce by the feedlot system, and suppliers turned to international sources for beef. The rise in US demand occurred at the same time that meat consumption and demand increased in Europe, the Eastern Bloc and Japan. Since South American (as opposed to Central American) beef has traditionally been oriented to European markets, the expanded purchasing power of these countries in the mid-1960s was a major stimulus to demand.

The general international perspective for the expansion of Brazilian beef is summarized by FAO/ECLA's (1964) study called *Livestock in Latin America*. The FAO document indicated that although Brazil's existing productive capacity was rather low, it had great potential for expansion through the incorporation of new land and the rationalization of production. The FAO argued that overcoming certain bottlenecks, primarily related to credit, was essential if Brazil were to capture a sizeable market share. This document concluded that global beef markets were buoyant and would continue to expand as national and international demand increased, a tendency that was particularly strong in the early 1960s.

Finally, Brazil was seen as an appropriate area for the transference of the Australian pasture technologies. If the conditions of long-term credit and better grass varieties were met, FAO pointed out that Brazil could become one of the premier beef exporters. This influential document frequently underlay the great push toward ranching throughout Latin America in the early 1960s (see, for example, Parsons, 1976), the precise period when policy for the Amazon was being developed. The various international agencies, such as the World Bank, were able to argue that with the proper technology and better credit lines, livestock represented an excellent investment for development. As a consequence, during the mid- to late-1960s, financial resources poured into livestock projects.

. . . .

The international investment picture coincided very well with Brazil's own development ambitions and dovetailed with Brazil's geopolitical and balance-of-payments concerns. It was against the backdrop of these internal and international pressures that, after several trips to the Amazon, General Castello Branco, the first military president after the coup, laid the groundwork for the far reaching legislation that was to become known as 'Operation Amazonia.'

(d) *Operation Amazonia*

In late 1965, General Castello Branco began what he described as a new era in Amazonian planning that would set the tone of the region's development. He emphasized that planning would occur in an ambience where technical considerations would take precedence over clientelistic interests that had dominated the previous planning agency, SPVEA (Superintendency for the Economic Valorization of Amazonia). He stressed greater efficiency in planning and emphasized the enhanced role of private enterprise in regional development. The government would provide infrastructure and general funding for development, while the actual task of regional occupation would be carried out by the entrepreneurs.

The fundamental legislation for Operation Amazon was Law 5.1744 (October, 1966), in which fiscal incentives were provided for the Amazon. These incentives stipulated that 50% of a corporation's tax liability could be invested in Amazonian development projects, essentially permitting taxes to become venture capital. The projects could be new ones or the expansion of existing enterprises. Since several southern Brazilian land magnates already had substantial land investments in the Amazon, this was an attractive means for valorizing existing holdings.

The government provided exemptions of 50% of the taxes owed for 12 years, to enterprises already established in 1966, and exemptions of up to 100% for projects implanted prior to 1972. Qualifying firms were permitted to import machinery and equipment duty-free, as well as being exempted from export duties for regional products (for example, timber). The various states of the region also provided their own incentives and inducements (land concessions, usually), while international credit lines such as those from the Inter-American Development Bank provided special agricultural development credit that could be mobilized for Amazonia (Pompermeyer, 1979). These incentives differed from previous development funds in the magnitude of resources, but more importantly in that land acquisition could be stipulated as part of the development costs.

Another incentive to the private sector was FIDAM (Funda para Desenvolvimento da Amazonia), which was to receive 1% of federal tax revenues, proceeds of BASA (Banco da Amazonia) securities and fiscal incentive funds not applied to specific projects (Mahar, 1979). These would be invested by BASA in research and various private firms. The result of these various incentives was that the federal government would supply 75% of the investment capital needed for the enterprises.

The new incentive laws also raised the ceiling from 50 to 75% of capital costs and had more generous grace periods. Another important aspect of the incentive legislation was the eligibility for fiscal incentives of foreign corporations, a rider that had not existed previously. Although foreign investment in Amazonia is often discussed (Kohlhepp, 1978; Davis, 1977; Ianni, 1978), the magnitude of foreign investment in the agricultural sector in Amazonia is comparatively low compared to Brazilian national investment. Evans (1979) has aptly pointed out that, in general, the foreign investor is a poor candidate for the entrepreneurial role when information is low. Further, foreign investment in Brazil was characterized during the 1964–78 period by its involvement in the industrial sector rather than agriculture, and in the post-1975 period by its emphasis on mining when engaged in rural activities.

The combination of fiscal incentives and other credit lines resulted in an explosion of ranching in Amazonia. . .. The peak investment period was 1967–72 as investors began to implant projects prior to the 1972 cutoff date for the 12-year tax holiday. During these five years, SUDAM approved some 368 new projects. By 1978, 503 cattle projects had been approved and of these, 335 were new projects, while 168 were reformulations, or expansions of existing SUDAM projects. By 1978, about $ 1 billion of SUDAM funds had been invested in these ranches, or on the order of $ 2.6 million per ranch in direct investment. The other loans applied to these ranches represented another subsidy that is almost impossible to assess.

Throughout the 1960s, livestock production was publicized as the most promising investment to be made in the region. As the president of BASA (Banco da Amazonia), Lamartine-Nogueira (1969), put it:

> Ranching . . . is an activity that has all the necessary conditions to be transformed into a dynamic sector of the northern economy . . . the fiscal incentives and road construction have generated a remarkable preference for livestock, and for this reason, a new era in the sector is opened.

Not only was there a marked preference for ranching, but also bias to certain regions: northern Mato Grosso, southern Para and northern Goias.

The extraordinary fiscal incentives and the (seemingly) relatively low risk associated with ranching created an unparalleled opportunity for gaining control of land. As Mahar (1979) has shown, investment in crop production in the Northeast (where incentives were also available) was comparatively risky, as is most cropping in the North. But, if land values increase, then the desire for investment becomes intelligible because land tends to hold value in inflationary economics. This was certainly the case in Brazil throughout the 1960s and 1970s. Infrastructural development in an area like the Amazon also increases the value of land. The influx of incentives that permitted the acquisition of land as part of the development costs created a situation where, as Mahar has noted, the value of Amazonian land was increased at 100% per year in real terms. . . . This speculative process was driven in part by the hope of future production returns, or the value of future resources; but land, and its modifications by ranching, became the primary vehicle for capturing enormous state subsidies.

As we show later in this paper, the nature of land in the Amazonian economy began to change in a fundamental way. Land itself, and not its product, became a commodity since even lands whose productivities were declining increased in value in this speculative context. What became crucial at this juncture was the emphasis on the exchange rather than the use value of lands.

. . .

4. Conversion of forest to pasture: development or destruction

Among those who have investigated the soil effects of forest conversion to grassland in the Latin American tropics, and the subsequent performance of these pastures, there have been substantial differences in opinion. Goodland (1980) and Fearnside (1983) have argued that ranching represents the worst of all conceivable land-use alternatives for Amazonian development due to the high ecosystem losses relative to the short-term profits and low employment potential. Myers (1980), Fearnside (1978), Hecht (1982a) and others have argued, for a variety of reasons, that ranching is a relatively unstable and unproductive land use for the region. Others, notably Serrao *et al.* (1979), Alvim (1980) and Falesi (1976) disagree. Central to the forest to pasture controversy is a widely cited study by EMBRAPA (Empresa Brasiliera de Pesquisa Agropecuaria) the research arm of the Ministry of Agriculture.

The EMBRAPA study (Falesi, 1976) maintained that conversion of forest to pasture improved soil properties, particularly for Ca, Mg and pH. Increases in P were transitory, although economically questionable. These results were frequently used by the Ministry of the Interior in policy conflicts to help undermine small-scale agricultural credits. Peasant agriculturalists were perceived as ecologically damaging, while large ranchers were portrayed as environmentally rational.

Based on the EMBRAPA results, it was suggested that 'the formation of pastures on low fertility soils is a rational and economic means by which to rationalize and increase the value of extensive areas' (Falesi, 1976). Serrao *et al.* (1979) argued that 'the subsequent substitution of pasture with perennial crops would require only a small amount of P fertilizer for development . . . due to the favorable conditions of the majority of the soil's components after a long period under pasture.' In the international development literature, Cochrane and Sanchez (1982) indicated that 'The data suggest a remarkable degree of nutrient cycling and maintenance of soil fertility under pasture. . . . These data are encouraging because they indicate a very high beef production potential with minimum inputs.'

In spite of such optimism, the roughly 10 million hectares of land converted from forest to pasture in Brazilian Amazonia do not appear to be particularly stable. Estimates of the area of severely degraded pasture range from 15% (Toledo and Serrao, 1982) to 50% (Tardin *et al.*, 1977; Santos *et al.*, 1979; Hecht, 1982). The major factors involved in pasture land degradation include soil nutrient changes, compaction and weed invasion.

In the next section, I summarize data on soil changes on a clay-loam oxisol in the major cattle areas in eastern Amazonia, comparing more recent research with the EMBRAPA study. Methodological issues and the differences between the studies are presented in more detail elsewhere (Hecht, 1982a). The most important difference, however, is that of sample size. The EMBRAPA study was based on five samples per age class, while this study involves eighty samples per age class of pasture.

(a) *Results of conversion studies*

(i) *Effect on soil pH* When forests are felled and burned, an increase in pH occurs as the bases held in the biomass are transferred to soils, regardless of the land use implanted (Nye and Greenland, 1960; Sanchez, 1976). As Figure 25.1 shows, there are substantial increases in the soil pH of the sites examined in the EMBRAPA study. By contrast the clay-loam oxisol showed only a moderate pH unit rise. Though the ranges of pH in the clay loam included some values as high as those of the other sites, as the number of samples increase, the pH increases are less dramatic. This result is corroborated in another data set (presented in Sanchez, 1976) that analyzed soil changes before and after deforestation on 60 sites. These samples were analyzed at the EMBRAPA laboratories. One of the interesting aspects of the pH data is that the liming effect is maintained through time. Cochrane and Sanchez (1982) and Toledo and Serrao (1982) believe that efficient nutrient cycling on the part of the grasses is responsible for the persistence of the pH improvement. While Tietzel and Bruce (1972) have shown in Australia that *Panicum maximum* (the most widely planted grass species in the Amazon) is a reasonably effective cycler of Ca, Mg and K, there is an alternative interpretation that also merits consideration. When forests are cut and burned for pasture formation, only about 20%

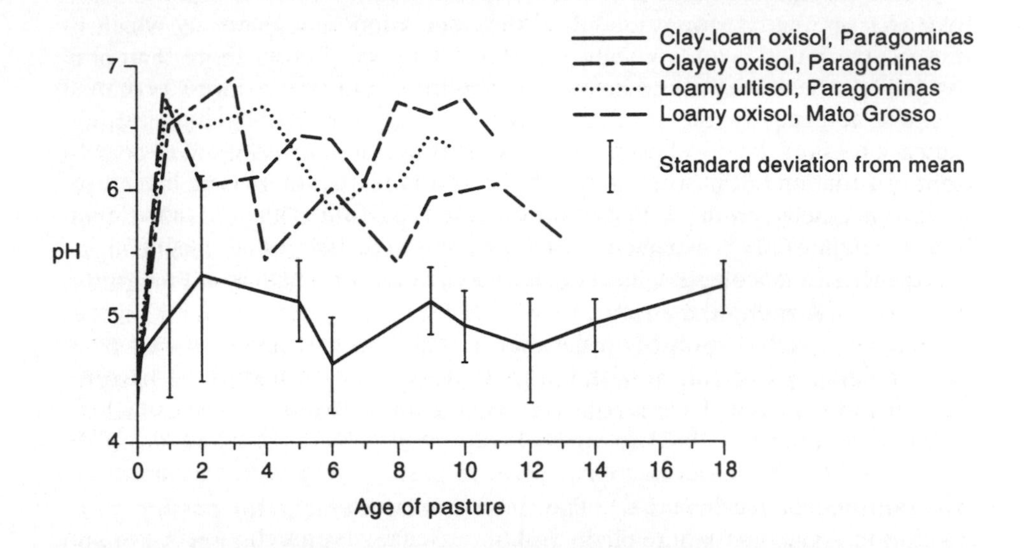

Figure 25.1 Changes in pH after conversion of forest to pasture.
Sources: Clay-loam oxisol from Hecht (1982a); clayey oxisol, loamy ultisol and loamy oxisol; from Falesi (1976).

of the slash actually combusts (Olsen, 1968). Since 80% of the total ecosystem Ca, Mg and K are stored in tree boles, their gradual decay after cutting could supply these elements at a rate that could maintain the pH. Since the maintenance of pH is recorded for other agricultural systems after forest conversion (Bandy and Sanchez, 1981), it may be that cycling is less important than slash decay. This hypothesis, of course, does not exclude the possibility of nutrient cycling by the grasses. But, in fact, it appears to. pH levels are probably maintained by inputs via slash decomposition. Once decomposition has run its course, the implication is that pH will fall.

(ii) *Soil calcium and magnesium* Closely associated with the increase in pH are additions of Ca and Mg to soil. The rise in these elements (and their variability) is most pronounced in the years immediately following clearing. Since rainforests store over one ton each of Ca and Mg per hectare (Klinge *et al.*, 1975), and the ash additions after burning supply immediately at least 100 kg of Ca (Seubert, Sanchez and Valverde, 1977), the rise in Ca and Mg after conversion is not surprising. While soils may be improved in terms of Ca and Mg contents, in fact, the overall level of these elements is below that critical for pasture production (Coordenadoria de Assistencia Tecnica Integral, 1974).

(iii) *Soil potassium* Potassium is a monovalent cation (an ion with one positive charge) that is stored mainly in the vegetation in tropical ecosystems, cycles quickly and is quite vulnerable to leaching. In general, more than one ton of K per hectare is stored in the forest biomass. Due to the mobility of this element, K values are quite erratic throughout the pasture sequence reflecting periodic burning, weed invasion and other management activities. The coefficient of variation of this element is so high, however, that there is no statistical significance between the K values of forest and pasture (Hecht, 1982b). Soil improvement of K after conversion is thus open to question as is decline in K. Deficiencies of this element have been documented for pastures in Paragominas, Brazil (Koster, Khan and Bosshart, 1977). The high value for K in the Mato Grosso oxisol probably reflects an initially high K level, as well as a greater frequency of palms, both in the native vegetation and in the pasture weed invaders. Palms have a relatively high level of K in their leaves (Silva, 1978).

(iv) *Phosphorus levels in soil* The most crucial element for pasture production in Amazonia is P (Toledo and Serrao, 1982), and 10 ppm is usually considered the minimum value for sustained production of pastures. After conversion, P values increase dramatically, . . . but after the fifth year, they tend to continuously decline to a level of about 1 ppm.

The decline of P has been identified by some researchers as the main reason for pasture instability in Amazonia (Serrao *et al.*, 1979). The high demand of Panicum for this element, coupled with losses due to erosion and animal export, as well as the competition the grass undergoes from weeds adapted to low P levels, lead to drastic drops in productivity, which often result in pasture abandonment. Although the grass responds well to fertilization (Serrao *et al.*, 1979), the high transport, application and opportunity costs, as well as erratic availability of P, make widespread fertilization uneconomic at this time.

(v) *Soil nitrogen* Soil nitrogen values reflect nitrogen-accumulating activities like N fixation, additions from the atmosphere and organic matter decay, as well as N-decreasing activities like volatization, denitrification, leaching, erosion and plant uptake. Many of these processes are mediated by the biota, and rates of loss and addition are affected by environmental factors (pH, temperature, soil moisture). Nitrogen is an element that can vary strongly from site to site. As Figure 25.2 suggests, the Paragominas ultisol shows a slight initial increase and a subsequent equilibration, indicating that differences between forest and pasture N soil storage are insignificant. In the clay-loam oxisol, soil N decreases, but when analyzed through multiple range tests, pasture soils are not significantly different from forest soils, except in the oldest pastures and during the first year after clearing (Hecht, 1982a).

The heavy clay oxisols from Paragominas and Mato Grosso both show N declines although the Paragominas samples are decidedly more erratic. The high N values in year 13 in the Paragominas heavy clay soil may reflect N fixing

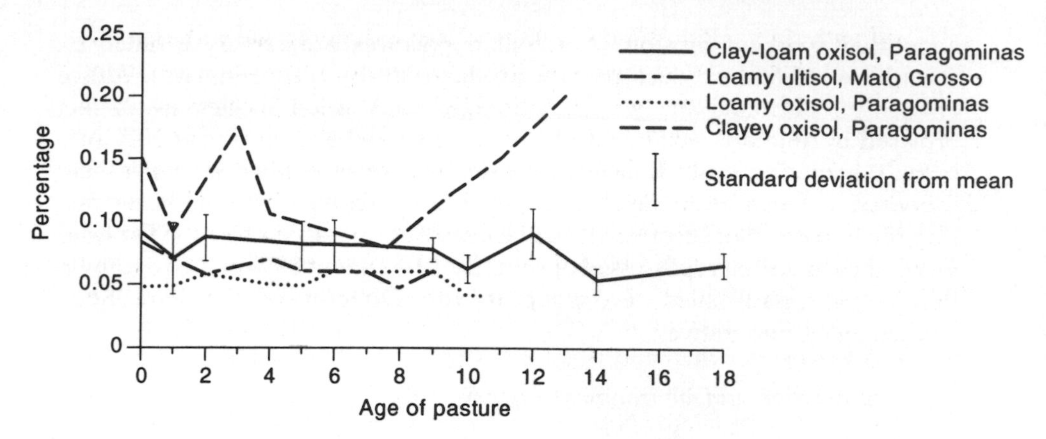

Figure 25.2 Changes in N after conversion of forest to pasture.
Source: As for Figure 25.1.

by native weedy legumes. On the other hand, the Mato Grosso site N levels decline by 50% after conversion.

(vi) *Organic carbon in soil* Soil carbon levels often drop with burning, but they can increase if there is an addition of fine charcoal, which probably occurred in clay-loam and loamy oxisols, and the loamy ultisol. Levels of C can increase due to slash decomposition, organic matter additions from the grass and heavy weed invasion, but such gains can be short-lived. In the Mato Grosso oxisol and Paragominas ultisol, C declined after clearing. In the Mato Grosso site, the value dropped to 50% of the forest level.

(vii) *Summary of soil findings* The effect of conversion of forest to pasture on soil chemical properties can be described as relatively neutral for N and K, negative for P and C, and mildly positive for Ca, Mg and pH, particularly in the first years after clearing. The widely cited 'dramatic increases' in these elements after conversion seem to be moderated when larger sample sizes are used. In any case, the absolute levels of Ca and Mg are low-to-marginal for pasture production. Soil nutrients during the first five years are adequate for animal stocking rates of one animal unit (au) per hectare, but drop to 0.25 au/hectare after five or six years. This is due not only to soil chemical changes, but also soil compaction and weed invasion.

(b) *Other factors affecting pasture productivity*

Soil physical changes also affect productivity. For example, soil bulk densities that double with increasing pasture age reduce infiltration that influences sheet and other types of erosion. Amazonian pastures are rapidly invaded by shrubby weeds that reduce pasture productivity by competing with forage grasses for nutrients and water. Although many weed species are in fact browsed by animals, weed control is expensive and absorbs about 20% of a ranch's operating cost. Ranches that did not receive fiscal incentives are squeezed between declining productivity and escalating weed and infrastructure repair costs. Not surprisingly, when livestock operations pass the five-year mark, they are often repossessed or sold. By 1978, about 85% of the ranches in Paragominas had failed, according to the Director of the Para State Cattlemen's Cooperative.

(c) *Accumulation and environmental decay*

Even as pasture productivity declined, for the reasons mentioned, the value of pasture land increased due to the infrastructure development, mineral resource potential and the generous fiscal incentives and sweetheart loans made available to investors. The creation of a land market in Amazonia where previously none had existed generated an extraordinary speculative boom that held sway throughout the 1960s and 1970s. What is essential here is that the exchange value of the land itself was far higher in this speculative context, than any commodities it could produce. Entrepreneurs depended for profits, *not on the annual productivity of the land, but on the rate of return to investment*.

Whether ranching achieved many of its hoped for results is quite dubious and requires a more extensive analysis than can be presented here (see Mahar, 1979; Pompermeyer, 1979; Hecht, 1982). However, pastureland created from forests is expensive to implant and to maintain, and the value of the animal product does not recompense its production costs even with large subsidies (Hecht, 1982a; Possio, 1984).

Declining pasture productivity was not the only consequence of converting forest to pasture. Ranching expansion played a key role in marginalizing small landholders as land increasingly came under the control of large estates. Areas where ranching dominated, such as the south of Para and the north of Mato Grosso, were characterized by land distributions, where 6.7% of the landholders controlled fully 85% of the private sector lands, while almost 70% of the remaining farms owned merely 6% of the area. Along the Belem-Brasília Highway, one of the earliest upland cattle ranching areas, Gini coefficients for land owning increased from 0.66 to 0.77 in the years between 1967 and 1975 (Santos, 1979), indicating extreme land concentration. Such marginalization produced a great deal of land conflict (Souza-Martins, 1983; Schmink, 1982).

5. Environmental models and the Amazonian case

I would like to return now to the earlier sections of this article where the main models used to explain land degradation in the humid tropics are discussed:

Malthusian, economic dependency and technology assessment approaches. I will describe why each of these models does not adequately explain the dynamics of environmental deterioration in the eastern Amazon and why a more complex framework for analysis of international environmental problems must be developed.

The Malthusian perspective, as it has been applied to Amazonia, generally asserts that while shifting cultivation may be ecologically sound at low human population densities, it becomes environmentally destructive when demographic increases reduce the land/person ratios, and fallow times are shortened so that land cannot recover its production potential after an agricultural cycle. Soil fertility declines from overuse, weeds hamper crop production, and erosion results.

There is no question that much of the shifting cultivation observable in eastern Amazonia along roads and in colonization areas follows such a pattern of land deterioration, but this phenomenon is unrelated to human population numbers or physical availability of land. Roughly 12 million people live in the Brazilian Amazon, a region that encompasses about 60% of the Brazilian national territory. Amazonia itself encompasses close to 5,000,000 km^2. It is ludicrous to describe the environmental degradation in this situation as only a function of demographics. Rather, the situation is due to the extraordinary maldistribution of land, as the figures cited in earlier paragraphs show. Further, these slash-and-burn cultivators are often harassed by large landowners or land speculators and are unable or unwilling to cultivate carefully because they have no title and even their short-term squatters rights may be violently curtailed.

Commons models and externalities frameworks do not apply well to the Amazonian situation because, in fact, there has been a rapid shift to private property which has been associated with the increase, rather than the decline of environmental degradation. The externalities: extinction, climatic and hydrologic, that occur in Amazonia remain in the realm of debate. No species has been certified as extinct, although undoubtedly many have gone out of existence. Since a large percentage of species has never been catalogued, at this time we know neither the organisms nor the value of the organisms that have passed from this earth. The existence of climatic and hydrological impacts continues to be debated.

The 'dependency' framework to explain environmental degradation in tropical agriculture centers on the role of export crop production. There are two main approaches, one based on analysis of the production process, while the other focuses on the role of the state. These two models of export stimulus and environmental decay can certainly help in developing an understanding of the dynamics of resource degradation in Amazonia, but they do not constitute a complete analysis. It is important to recognize that Amazonian beef is not an export product in the usual sense of the term. Although Amazonian development policy states that '. . . Beef (from) Amazonia should be oriented fundamentally toward export markets . . .' (MINTER, 1974), in fact, cattle were only occasionally exported to Surinam (via boats from Marajo) and walked or trucked in small numbers across the Venezuelan border. Indeed, the Federation of Exporters of the state of Para reported that in the years documented by their records, beef generated only an annual average of $ 44,000 compared to the $ 33,000,000 of export revenues from Brazil nuts (FIP, 1977). Amazonia itself remains a net meat importer. Roughly 40% of the animals slaughtered in the Para abattoirs in the late 1970s were trucked from Goias (Hecht, 1982a). Thus, strictly speaking, Amazonian beef cannot be considered an export commodity, in spite of hopes to the contrary.

The first export-oriented explanatory model argues that development occurs under the discipline of international commodity prices, which demand that exports be produced as intensively and efficiently as possible. Therefore, agronomic technologies with high productivities are employed; they are generally mechanized and imply the use of large amounts of fertilizers, pesticides and herbicides, and thus are capital-intensive. These methods can generate direct environmental effects, such as compaction, and indirect effects, like water contamination and poisoning of fauna.

If the relationship of capital investment to employment and output is analyzed for eastern Amazonia, cattle ranching is more capital intensive than is industry per job created or per unit of output (Mahar, 1979; Hecht, 1982a). Given the high levels of capital investment, the first export paradigm – a highly technical production process – might appear to apply to the eastern Amazon area, particularly since the use of mechanical clearing is fairly widespread and becoming more so. Forest conversion to agriculture and pasture with machinery in the humid tropics has been shown to undermine the productivity of soil resources and to reduce yields (Seubert, Sanchez and Valverde, 1977). Mechanical deforestation is also substantially more expensive than is manual (Toledo and Serrao, 1982). I have argued elsewhere (Hecht, 1982a) that mechanical clearing on large ranches is related to the desire to minimize the use of human labor on ranches for a variety of compelling reasons including public health, fear of squatters, violence and difficulty in controlling huge work teams. The capital intensity of eastern Amazonian cattle operations, however, is not reflected in the actual process of animal production *per se* since the technologies are quite primitive. What makes cattle ranching in Amazonia capital intensive is the very high barriers and infrastructure costs, not active investment in production technology.

The second model of the role of export production concludes that state subsidies (such as tax credits) and high international beef prices make it possible for landowners to make profits in spite of inefficient production organization and unsound land practices. This model has been applied to Central American livestock (Nations, 1980). In Central America, pasture productivity declined because of poor management and for ecological reasons, but not in a manner that completely put land out of production, as often occurs in Amazonia. Thus, animals can still be stocked in very low numbers while farmers expand grasslands into new areas to serve the international demand.

This approach contains some persuasive arguments, since the role of tax incentives and the pattern of land distribution in Amazonia are somewhat analogous to the case of Central America. But, there are several differences that make the two areas incommensurable. First, as noted, the price of beef is controlled, at least through gentlemen's agreements, in Brazil, making livestock profitability marginal even under the best of circumstances (Lattimore and Schuh, 1979). Livestock production in the Amazon is far more expensive than in other areas of the country (*O Globo*, 1980; *Conjuntura Economica*) and is usually not profitable on improved pastures even with extensive subsidies. Finally, Amazonian pastures generally go completely out of production in 10 years, a crucial difference between the Amazonian and Central American situation. Moreover, Brazilian beef has never been as closely integrated with international markets as Central American production.

In technology evaluation, the 'inappropriate technology' model argues that environmental problems result when well-intentioned planners mistakenly recommend technologies that for cultural or ecological reasons, generate resource degradation. The difficulties that occur are perceived as simple technological problems that can be resolved with a 'technological fix.' Such a point of view does have some validity in eastern Amazonia, in that the pasture species initially used were not well adapted to the low soil levels, and subsequent introductions, though more tolerant of the soil conditions, have been bothered by insect attack (i.e., *Brachiaria humidcola*). But the question really goes beyond the issue of better cultivars, since the structure of the pasture ecosystem itself very well may be inappropriate to the region (Hecht, 1982; Fearnside, 1980; Goodland, 1980). The more central issue is why certain land uses and technologies are promoted over other alternatives. I have argued in this article that ranching served important political and ideological functions, and that these, rather than varieties of *Panicum*, are what underlie the choice of pasture ecosystems for Amazonia, and their instability.

The environmental degradation associated with pasture development in eastern Amazonia is best understood as a consequence of the role of land in inflationary economies, the traditional function of livestock as a means of acquiring large areas (and the institutional rents associated with them), the stimulating effect of the physical opening of the agricultural frontiers on certain industrial sectors of the economy, and the role of large government subsidies in the creation of land markets and speculation in the Brazilian Amazon. The expansion of ranching through tax breaks, fiscal incentives and road development generated enormous untaxed capital gains for certain groups. Acquisition of land was the vehicle for capturing direct and indirect state subsidies. The productivity of the land became secondary because the land itself was the commodity. Due to the peculiarities of state subsidies, a radical disjuncture between the value of land for production and its value for exchange was created. If the productivity of the land itself has little importance, cautious land management becomes irrelevant, and environmental degradation is the inevitable result.

References

Alvim, P. T. 1978: Perspectivas de producao agricola na Regiao Amazonica. *Interciencia* 3, 243–51.

——, Agricultural production potential of the Amazon Region, in F. Barbira-Scazzochio (ed.) 1980: *Land, People and Planning in Contemporary Amazonia*. Cambridge: Center for Latin American Studies, Cambridge UP.

Aragon, L. 1978: Migration to Northern Goias, Ph.D. thesis, East Lansing: Michigan State University.

Bacha, E. L. 1977: Issues and evidence on recent Brazilian economic growth. *World Development* 5 (2), 47–67.

Bandy, D. and Sanchez, P. A. 1981: Continuous crop production on acid soils in the Peruvian Amazon, paper presented at a Workshop on Management of Low Fertility Acid Soils, University of Surinam, Paramaribo: Surinam.

Belassa, B. 1979: Incentive policies in Brazil. *World Development* 7 (2), 1023–42.

Bergsman, J. 1970: *Brazil's Industrialization and Trade Policy*. New York: OUP.

Brown, S. and Lugo, A. 1982: Storage and production of organic matter and their role in the global carbon cycle. *Biotropica* 14, 161–79.

Bunker, S. 1979: Power structures and exchange between government agencies in the expansion of the agricultural sector in Para. *Studies in Comparative and International Development* 14 (1), 56–76.

Clarke, C. 1974: The economics of overexploitation. *Science* 181, 630–4.

Clarke, W. C. 1982: *Carbon Dioxide Review*. New York: OUP.

Cochrane, T. and Sanchez, P. A. 1982: Land resources of the Amazon Basin, in S. B. Hecht (Ed.) *Land Use and Agricultural Research in the Amazon Basin*. Cali, Colombia: CIAT, 137–211.

Conjuntura Economica (Rio de Janeiro: Fundagão Getulio Vargas).

Coordenadora de Assistencia Technica Integral 1974: *Normas Para Manejo de Pastagens, Boletim Tecnico 81*. Belem, Brazil.

Davis, S. 1977: *Victims of the Miracle*. Cambridge: CUP.

Durham, W. 1980: *Scarcity and Survival in Central America*. Berkeley: University of California Press.

Eckholm, E. 1982: *Down to Earth*. London: Pluto Press.

Ellen, R. 1982: *Environment, Subsistence and System*. Cambridge: CUP.

Evans, Peter 1979: *Dependent Development: The Alliance of Multinationals, the State, and Local Capital in Brazil*. Princeton: Princeton UP.

Falesi, I. 1976: *Ecosistema de Pastagem Cultivada na Amazonia Brasileira. Boletim Tecnico* 1. Belem, Para: EMBRAPA.

FAO/ECLA 1964: (United Nations Food and Agricultural Organization/Economic Commission on Latin America), *Livestock in Latin America*. Rome: FAO.

Fearnside, P. 1978: Estimation of carrying capacity for human settlement of the TransAmazon Highway colonization area of Brazil. Ph.D. thesis, University of Michigan.

——, 1980: Effects of cattle pastures on soil fertility in the Brazilian Amazon: Consequences for beef production sustainability. *Tropical Ecology* 21 (1), 122–37.

Feder, E. 1979: Lean cows, fat ranchers: A study of the Mexican beef industry, *manuscript*. Berlin.

FIP (Federacao de Industrias do Para) 1977: *Para Exportacao*. Belem: FIP.

Fishlow, A. 1973: Brazilian size distribution of income. *American Economic Review*, 62 (2), 391–402.

Foweraker, J. 1980: *The Struggle for Land*. Cambridge: CUP.

Franke, R. and Chasen, B. 1980: *Seeds of Famine*. New Jersey: Allanheld, Osmun.

O Globo, 5 January 1980.

Gentry, A. and Lopez-Parodi, J. 1980: Deforestation and decreased flooding in the Upper Amazon. *Science*, 210, 1354–6.

Gomez-Pompa, A., Vasques-Yanes, C. and Guevara, S. 1972: The tropical rain forest: A non-renewable resource, *Science* 177, 762–5.

Goodland, R. 1980: Environmental ranking of development projects in Brazil. *Environmental Conservation*, 7 (1), 9–25.

——, and Irwin, H. 1975: *Red Hell to Green Desert?* Amsterdam: Elsevier.

Hardin, G. 1968: The tragedy of the commons. *Science*, 162, 1243–8.

Harvey, D. 1968: Population, resources and the ideology of science. *Economic Geography*, 50 (3), 256–78.

Hecht, S. B. 1982a: Cattle ranching in the Brazilian Amazon: Evaluation of a development strategy, Ph.D. thesis, Berkeley: University of Calfornia.

——, (ed.) 1982b *Land Use and Agricultural Research in the Amazon Basin*. Cali, Colombia: CIAT.

Hewitt, D. 1984: *Interpretations of Calamity*. Boston: Allen & Unwin.

Hoskins, M. 1982: Musings on technology transfer, OTA Workshop. Shepards Ford, VA.

Ianni, O. 1978: *A Luta Pela Terra*. Petropolis: Vozes.

InterAmerican Development Bank (IADB) 1971–76: *Annual Reports*. Washington, D.C.: IADB.

Janzen, D. 1973: Tropical Agro-ecosystems. *Science*, 182, 1213–18.

Jarvis, L. 1984: Livestock in Latin America. *Mimeo*. Berkeley, CA.

Klinge, H., Rodrigues Brunig, E. and Fittkau, E. J. 1975: Biomass and structure in a Central Amazonian rainforest, in F. Golley and E. Medina, (eds), *Tropical Ecological Systems*. New York: Springer-Verlag, 115–22.

Knight, P. T. 1971: *Brazilian Agriculture, Technology, and Trade*. New York: Praeger.

Kohlepp, G. 1978: Erschliessung und Wirtschaftliche Inwertsetzune Amazoniens. Entwicklungsstrategien Brasilianischer Planungs-frolitik und Privater Unfernehmen, *Geografische Rundschau*, 30 (1), 2–13.

Koster, H., Kahn, E. J. and Bossert, R. 1977: *Programa e Resultados Preliminarios dos Estudos de Pastagems na Regiao de Paragominas Para, e o Nordeste de Mato Grosso*. Belem: SUDAM/IRI.

La Marche, V., Graybill, D., Fritts, H. and Rose, M. 1984: Increasing atmospheric carbon dioxide: The evidence for growth enhancement in natural vegetation. *Science*, 225, 1019–21.

Lamartine-Noguero, F. de. 1969: *A Agropecuaria e o Processo de Desenvolvimento da Amazonia*. Belem: BASA/MINTER.

Lattimore, R. G. and Schuh, G. D. 1979: Endogenous policy determination: The case of the Brazilian beef sector, *Canadian Journal of Agricultural Economics*, 27 (2), 1–16.

Leff, N. H. 1967: Export stagnation and autarkic development in Brazil, *Quarterly Journal of Economics* (May), 286–301.

Mahar, D. J. 1979: *Frontier Development Policy in Brazil: A Study of Amazonia*. New York: Praeger.

Malan, P. S. and Bonelli, R. 1977: The Brazilian economy in the 70s: Old and new development. *World Development*, 5 (1/2), 19–37.

Malthus, T. 1970: *Essay on Population*. Harmondsworth: Penguin.

Mamdami, M. 1972: *The Myth of Population Control*. New York: Monthly Review Press.

Martine, G. 1982: Expansao e retracao de *emprego na fronteira agricola*, in N. Gligo and C. Mueller (eds), Brasilia: University of Brasilia.

Milton, M. and Farvar, M. 1968: *The Careless Technology*. St. Louis: Washington University Press.

MINTER (Brazil, Ministerio do Interior) 1975: *Programa Nacional de Desenvolvimento*. Brasilia: MINTER.

Molion, L. C. 1975: A climatic study of the energy and moisture fluxes of the Amazonas Basin with considerations of deforestation effects, Ph.D. thesis.

Moore, B. 1968: *The Social Origins of Dictatorship and Democracy*. New York: Beacon.

Moran, E. 1981: *Developing the Amazon*. Bloomington: University of Indiana Press.

Mueller, C. and Gligo, N. 1982: *Espansao da Fronteira e Meio Ambeinte*. Brasilia: University da Brasilia.

Murdoch, W. 1980: *The Poverty of Nations*. Baltimore: Johns Hopkins.

Myers, N. 1980: *Conversion of Moist Tropical Forests*. Washington, D.C.: National Academy of Sciences.

Nations, J. 1980: Report to the Tinker Foundation.

Nicolaides J., Sanchez, P., Bandy, D., Villachica, J., Coutu, A. and Valverde, C. 1983: Crop production systems in the Amazon Basin, in Moran (ed.), *The Dilemma of Amazonian Development*. Boulder: Westview Press.

Nordin, C. and Meade, R. H. 1982: *Science*, 215, 426.

Nye, P. and Greenland, D. 1960: *The Soil Under Shifting Cultivation*. Farnham Royal, Bucks, UK: Commonwealth Agricultural Bureau.

O'Donnell, G. 1979: *Modernization and Bureaucratic Authoritarianism*. Berkeley: University of California Press.

Office of Technology Assessment (OTA) 1984: *Technologies for Sustaining Tropical Forests*. Washington, D.C.: OTA.

Olsen, G. 1968: Energy storage and the balance between producers and decomposers, *Ecology*, 44, 322–31.

—— 1983: Manuscript cited in Woodell *et al.*

Parsons, J. J. 1976: Forest to pasture: Development or destruction? *Revista de Biologia Tropical*, 24 (1), 121–38.

Pires, J. M. and Prance, G. T. 1977: The Amazon forest: A natural heritage to be preserved, in G. T. Prance and T. Elias, (eds), *Extinction is Forever*. Bronz: New York Botanical Garden.

Programa Nacional de Desenvolvimento (PND) 1975: Brasilia: MINTER.

Pompermeyer, M. J. 1979: The state and frontier in Brazil, Ph. D. thesis. Stanford University.

Possio, G. 1984: An economic analysis of cattle ranching in Amazonia, M.S. thesis Berkeley: University of California.

Redclift, M. 1984: *Development and the Environmental Crisis*. New York: Methuen.

Salati, E., Dall Olio, A., Matsui, E. and Gat, S. R. 1979: Recycling of water in the Amazon Basin: an isotope study, *Water Resources Research*, 15 (5), 1250–8.

——, and Lovejoy, T. 1983: Principitating change in Amazonia, in. E. Moran (ed.), *The Dilemma of Amazonian Development*. Boulder: Westview Press.

——, and Shubart 1982: Natural resources for land use in the Amazon, in S. B. Hecht (ed.), *Land Use and Agricultural Research in the Amazon Basin*. Cali, Colombia: CIAT.

——, and Vose, P. 1984: Amazon forest, a system in equilibrium, *Science*, 225 (4658), 129–37.

Sanchez, P. A. 1976: *Properties and Management of Tropical Soils*. New York: Wiley Interscience.

—— 1979: Soil fertility and conservation considerations for agro-forestry systems in the humid tropics of Latin America, in H. Mongi and P. Hucksley (eds), *Soils Research in Agro-Forestry*. Nairobi: ICRAF.

——, Bandy, D., Villachica, J. and Nicolaides, J. 1982: Amazon soils management for continuing crop production, *Science* 211, 821–7.

Santos, A., Odeao de Novo, E. and Duarte, V. 1979: *Degradacao de Pastagens na Amazonia* Belem: Relatorio do Projeto INPE/LANDSAT.

Santos, R. 1979: Sistema de propriedade e relacoes de trabalho no meio rural Paraense, in Monteiro de Costa, (Ed.), *Amazonia: Desenvolvimento e Occupacao*. Rio de Janeiro: IPEA/INPES.

Sawyer, D. 1979: Peasants and capitalism on the Amazonian frontier, Ph.D. thesis. Harvard University.

Schmink, M. 1982: Land conflicts in Amazonia, *American Ethnologist*, 9 (2) (May), 341–57.

Sen, A. 1981: *Poverty and Famines*. Oxford: Clarendon Press.

Serrao, A., Falesi, I., Vega, J. B. and Teixeira, J. F. 1979: Productivity of cultivated pastures on low fertility soils of the Brazilian Amazon, in P. A. Sanchez and L. E. Tergas, (eds), *Pasture Production in Acid Soils of the Tropics*. Cali, Colombia: CIAT.

Seubert, C. E., Sanchez, P. A. and Valverde, C. 1977: Effects of land clearing methods on the soil properties and crop performances on an ultisol of the Amazon jungle of Peru, *Tropical Agriculture*, 54, 307–21.

Silva, L. F. 1978: *Influencia do Manejo de un Ecosistema nas Propriedades Edaficas dos Oxisols de Tabuleiro*. Itabuna Bahia: CEPLAC.

Smith, N. J. 1976: TransAmazon Highway: A cultural and ecological analysis of settlement in the humid tropics, Ph.D. thesis. Berkeley: University of California.

Souza-Martins, J. 1980: *O cativeiro da Terra*. Sao Paulo: Libraria Editora Ciencias Humanas.

Stepan, A. 1968: *The Military in Power*. Princeton: Princeton UP.

—— 1973: *Authoritarian Brazil*. Princeton: Princeton UP.

SUDAM (Superintendencia de Desenvolvimento da Amazonia) 1974: *O Problema da Occupacao Economica da Terra*. Belem: SUDAM.

Sunkel, O. 1982: Estilos de Desarrollo e el Medio-Ambiente. Brasilia: University of Brasilia.

Tardin, A. T., dos Santos, A., Moraes-Novo, E. M. and Toledo, F. L. 1977: *Relatorio Atividades do Projeto SUDAM/INPE* No. 1034, NTE. Sao Jose dos Campos, Brazil: INPE.

——, *et al.* 1979: Levantemento de Areas de Desmatamento na Amazonia Legal Atraves de Imagens do Satelite LANDSAT, INPE Report No. 411-NTE/142. Sao Jose dos Campos, Brazil: INPE.

Taylor, L., Bacha, E., Cardoso, E. and Lysy, F. 1980: *Models of Growth and Distribution for Brazil*. New York: Oxford Press.

Thompson, E. P. 1965: *Making of the English Working Class*. New York: Monthly Review.

Tietzel, J. and Bruce, R. 1972: Fertility of pasture soils in the wet tropical coast of Queensland, *Australian Journal of Experimental Agriculture and Animal Husbandry* 12, 49–54.

Todaro, M. 1977: *Economic Development in the Third World*. London: Longman.

Toledo, J. and Serrao, A. 1982: Pasture and animal production in Amazonia, in S. B. Hecht (ed.), *Land Use and Agriculture Research in the Amazon Basin*. Cali, Colombia: CIAT.

Watts, M. 1983: *Silent Violence*. Berkeley: University of California Press.

Wood, C. and Schmink, M. 1979: Blaming the victim: Small farmer production in an Amazon colonization project, *Studies in Third World Societies*, 7, 77–93.

——, and Wilson, J. 1983: The role of the Amazon frontier in the demography of rural Brazil, in C. Wood and M. Schmink, *The Frontier Development in the Amazon*. Gainsville, FL: University of Florida Press.

Woodell, G. *et al.* 1983: Global deforestation: Contribution to atmospheric carbon dioxides, *Science*, 222 (4628), 1081–6.

World Bank 1978: *Forestry Sector Paper*. Washington D.C.: World Bank.

Glossary

BASA	Banco da Amazonia
cerrado	A savanna type of vegetation, a grassland with an important woody component
EMBRAPA	Empresa Brasiliera de Agropecuaria. The Ministry of Agriculture network of research institutions
INCRA	National Institute for Colonization and Agrarian Reform
Mata fina	A lower biomass forest that often occurs near the transition to cerrado, or along the 1,200 mm rainfall isohyet
MINTER	Ministry of the Interior
PIN	National Program for Integration
PDAM	National Program for Amazonian Development
PND	National Development Program
SPVEA	Superintendency for the Economic Valorization of Amazonia
SUDAM	Superintendency for Amazonia Development

26 Adrian Leftwich, 'Governance, Democracy and Development in the Third World'

Excerpts from: *Third World Quarterly* 14 (3): 605–24 (1993)

. . . .

Good governance: emergence and meanings

The first contemporary public appearance of the notion of good governance came in a 1989 World Bank report on Africa, which argued that, 'Underlying the litany of Africa's development problems is a crisis of governance.'[1] The Bank defines governance as '. . . the exercise of political power to manage a nation's affairs'. Good governance included some or all of the following features: an efficient public service; an independent judicial system and legal framework to enforce contracts; the accountable administration of public funds; an independent public auditor, responsible to a representative legislature; respect for the law and human rights at all levels of government; a pluralistic institutional structure, and a free press.[2]

This adds up to a comprehensive statement of the minimum institutional, legal and political conditions of liberal democracy, though the Bank never stated this explicitly. Rather, these characteristics are advanced as the largely functional and institutional prerequisites of development. This reading of the Bank's initial managerial approach is confirmed by its latest definitive statement on *Governance and Development*. This publication, the product of two years of work by Bank staff,[3] treats good governance as '. . . synonymous with sound development management'.[4]

Between 1989 and 1991 there followed a flow of pronouncements on governance, democracy and the relationship of either or both to development. These pronouncements issued from all major Western governments and were especially forceful from the British, French, German, US and Nordic governments. They were supported by the main international development institutions, and a variety of cooperative, intergovernmental and regional organisations, such as the Organisation for Economic Cooperation and Development (OECD), the Organisation of African Unity (OAU), the European Communities and the Commonwealth.[5]

[1] World Bank, *Sub-Saharan Africa: From Crisis to Sustainable Growth* Washington DC: The World Bank, 1989.

[2] *Ibid*, pp. 6, 15, 60–61 and 192.

[3] M. Moore, 'Declining to learn from the East? The World Bank on "Governance and Development"', *IDS Bulletin*, 24(1), 1993, pp. 39–50.

[4] World Bank, *Governance and Development*, Washington DC: World Bank, 1992, p. 1.

[5] *IDS Bulletin*, 'Good government', 24(1), 1993, p. 7.

Their views on the relationship of governance, democracy and development were not all identical. Some stressed the democratic theme, others focused on administration and yet others emphasised human rights as either necessary or desirable components of development. But despite these differences in emphasis, a common underlying shape of the concept of democratic good governance became clear. It may be thought of as having three main components, or levels, ranging from the most to the least inclusive: systemic, political and administrative.

First, from a broad systemic point of view, the concept of governance is of course wider than that of government which conventionally refers to the formal institutional structure and location of authoritative decision making in the modern state. Governance, on the other hand, refers to a looser and wider distribution of both internal and external political *and* economic power.[6]  Governance thus denotes the structures of political and, crucially, *economic* relationships and rules by which the productive and distributive life of a society is governed. In short, it refers to a *system* of political and socioeconomic relations or, more loosely, a regime. In current usage there is no doubt that good governance means a democratic capitalist regime, presided over by a minimal state which is also part of the wider governance of the New World Order.

Second, in its more limited but explicitly political sense, good governance implies a state enjoying both legitimacy and authority, derived from a democratic mandate and built on the traditional liberal notion of a clear separation of legislative, executive and judicial powers. And, whether presidential or parliamentary, federal or unitary, it would normally involve a pluralist polity with some kind of freely elected representative legislature, subject to regular elections, with the capacity at the very least to influence and check executive power and protect human rights. This is the position which most Western governments adopt.

Finally, from a narrow administrative point of view, good governance means an efficient, open, accountable and audited public service which has the bureaucratic competence to help design and implement appropriate policies and manage whatever public sector there is. It also entails an independent judicial system to uphold the law and resolve disputes arising in a largely free market economy. This is the position of the World Bank.

In summary, in its most extensive sense, the idea of democratic good governance is not simply the new *technical* answer to the difficult problems of development, although some of its proponents sometimes like to present it in that light. Democratic good governance is better understood as an intimate part of the emerging politics of the New World Order. And clearly, the barely submerged structural model and ideal of politics, economics and society on which the contemporary notion of good governance rests is nothing less than

[6] M. J. Lofchie, 'Reflections on structural adjustment' in *Beyond Autocracy in Africa*, Inaugural Seminar of the African Governance Programme, The Carter Center, Atlanta GA: Emory University, 1989, pp. 121–125.

that of Western liberal (or social) democracy – the focal concern and teleological terminus of much modernisation theory.

Who could possibly be against good governance in its limited administrative sense? Is it not the case that any society – whether liberal or socialist – must be better off with a public administrative service that is both efficient and honest, open and accountable, and with a judicial system that is independent and fair? In this sense, at least, the World Bank's conception of good governance re-identifies precisely the principles of administration that have long been argued as being of benefit to developing countries. They are impeccably Weberian in spirit, if not letter.[7] And even under the most unpromising Third World circumstances, good governance in this sense must be better for development than its opposite, bad governance.

However, the Bank's analysis is naive (whereas Weber's was not) because it entirely ignores that good governance is not simply available on order, but requires a particular kind of politics both to institute and sustain it, as I have argued elsewhere.[8] The far more urgent question, therefore, is whether democratic politics – as part of the wider notion of *democratic* good governance discussed above – can promote development in the Third World and elsewhere. Re-stated, this boils down to the essential question: will the current wave of democratisation last, and will it hinder or enhance development? It is to this crucial set of issues that I now turn.

Democracy and development

In the 1960s it was widely assumed in comparative politics that democracy was a concomitant of 'modernity' and hence an *outcome* of socioeconomic development, not a condition of it.[9] Democracy, it was argued, required a high level of literacy, communication and education; an established and secure middle class; a vibrant civil society; relatively limited forms of material and social inequality,[10] and a broadly secular public ideology. All this, it was held, was a function of prior economic development which would yield these necessary, though not sufficient, conditions for sustainable democracy. 'The more well-to-do a nation, the greater the chances that it will sustain democracy'.[11] Recently it has been argued that a well organised working class (itself a product of

[7] M. Weber, *The Theory of Social and Economic Organization*, edited by T. Parsons, New York: The Free Press, 1964.

[8] A. Leftwich, *States of Development*, Cambridge, Polity, forthcoming. This book will develop a wider argument about the primacy of politics in development than can be discussed here.

[9] S. M. Lipset, *Political Man*, London, Heinemann, 1960; P. Cutright, 'National political development: measurement and analysis', *American Sociological Review*, 28, 1963 pp. 253–264; and M. Needler, 'Political development and socio-economic development: the case of Latin America', *American Political Science Review*, 62, 1968 pp. 889–897.

[10] R. A. Dahl, *Polyarchy*, New Haven CT: Yale University Press, 1971, p. 103.

[11] Lipset, *op cit*, note 9, pp. 49–50.

industrial growth) was important for promoting and defending democracy. For the struggle for democracy is best seen as a struggle for *power*, for workers need democracy as means to both protect and advance their interests in the course of industrialisation.[12]

Evidence from the West and elsewhere can certainly sustain this view. Liberal democratic institutions, declining social inequalities, a flourishing civil society, a widening policy consensus, a secular public and bureaucratic ideology and the extension and institutionalisation of civil and human rights have seldom *preceded* economic development based on industrialisation and urbanisation. Indeed, the foundations of most modern advanced industrial economies were laid under non-democratic or highly limited democratic conditions – as in Britain (1750–1850), much of Western Europe,[13] and especially Bismarckian Germany and Meiji Japan. Moreover, most major post-1960 'success' stories of economic growth in the Third World—Brazil, South Korea, Taiwan and, more recently, Thailand and Indonesia—have not occurred under conditions remotely approximating continuous and stable democracy: quite the opposite. Even in the developmentally successful democratic societies—such as Botswana, Malaysia and Singapore—*de facto* one-party rule has been the norm for 30 years. In all these societies there are now powerful internal demands for the establishment or extension of democratic practices, and dominant parties in *de facto* one-party democracies are under pressure. This has served to underline the view that democracy is a consequence of development and hence sustain the earlier arguments in modernisation theory.

An even stronger argument is that the 'premature' introduction of democracy may actually hamper development in its early stages when there is '. . . a cruel choice between rapid (self-sustained) expansion and democratic processes',[14] and when there is the greatest need for effective state action or direction.[15] This is because early stages of development require capital accumulation for infrastructure and investment before advanced welfare systems or high wages can be afforded. Democratic systems, the argument goes, are likely to curtail processes of accumulation in favour of consumption.

[12] D. Rueschemeyer *et al.*, *Capitalist Development and Democracy*, Cambridge: Polity, 1992, pp. 140, 77.

[13] C. Tilly, 'Reflections on the history of European state-making', in C. Tilly (ed), *The Formation of National States in Western Europe*, Princeton NJ: Princeton University Press, 1975, pp. 3–83; and C. Tilly, 'Western state-making and theories of political transformation', *ibid*, pp. 601–638.

[14] J. Bhagwati, *The Economics of Underdeveloped Countries*, London: Weidenfeld and Nicolson, 1966, p. 204.

[15] I. Adelman & C. Morris, *Society, Politics and Economic Development*, Baltimore MD: Johns Hopkins Press, 1967; S. A. Hewlett, 'Human rights and economic inequalities – tradeoffs in historical perspective', *Political Science Quarterly*, 94, 1979, pp. 453–473: and J. Cotton, 'From authoritarianism to democracy in South Korea', *Political Studies*, XXXVII, 1989, pp. 244–259.

Even from the earliest days, these arguments have not gone unchallenged[16] and have stimulated a debate which has regularly been reviewed.[17] The critics argue that democracy and development are both compatible and functional for each other. If there is a trade-off between development and democracy, they claim, a slightly lower rate of growth is an acceptable price to pay for a democratic polity, civil liberties and a good human rights record. And they point out that there have been many more non-democratic than democratic regimes which at various times have had dismal or disastrous developmental records, such as Romania, Argentina, Haiti, Ghana, Myanmar, Peru, Ethiopia and Mozambique. In these, neither growth nor liberty have prospered. And, as Table 26.1 shows, while some Third World democracies have had stagnant or even negative average annual growth records over the last 25 years, others have performed very respectably when compared with non-democratic states.

These data suggest that, contrary to confident current claims, there is no

Table 26.1 Selected average annual rates of growth of GNP per capita: 1965 to 1990 (%)

Democratic regimes:		*Non-democratic regimes:*	
Jamaica	−1.3	Zaire	−2.2
Trinidad and Tobago	0.0	Nigeria	0.1
Venezuela	−1.0	Zambia	−1.9
Senegal	−0.6	Libya	−3.0
India	1.9	South Korea	7.1
Sri Lanka	2.9	Taiwan	7.0
Malaysia	4.0	Indonesia	4.5
Costa Rica	1.4	Brazil	3.3
Botswana	8.4	China	5.8
Mauritius	3.2	Algeria	2.4
Singapore	6.5	Thailand	4.4

Source: World Bank, *World Development Report 1992*, New York, Oxford University Press, 1992 and Council for Economic Planning and Development, *Taiwan Statistical Data Book*, Taipei, Republic of China, 1992.

[16] W. McCord, *The Springtime of Freedom*, New York: Oxford University Press, 1965: R. E. Goodin, 'The development-rights trade-off: some unwarranted economic and political assumptions', *Universal Human Rights*, 1, 1979, pp. 31–42; A. Kohli, 'Democracy and development', in J. P. Lewis & V. Kallab (eds) *Development Strategies Reconsidered*, New Brunswick, NJ: Transaction Books, 1986, pp. 153–182; and R. Sklar, 'Developmental democracy', in R. Sklar & C. S. Whitaker (eds), *African Politics and Problems of Development*, Boulder CO: Lynne Riener, 1991, pp. 285–312.

[17] S. P. Huntington & J. I. Dominguez, 'Political development', in F. I. Greenstein & N. W. Polsby (eds), *Handbook of Political Science*, Vol. 3, Reading: Addison-Wesley, 1975, pp. 1–114. See pp. 47–66; L. Sirowy & A. Inkeles, 'The effects of democracy on economic growth and inequality: a review', *Studies in Comparative International Development*, 25(1), 1990, pp. 126–157; and L. Diamond, 'Economic development and democracy reconsidered', *American Behavioral Scientist*, 35(4/5), 1992, pp. 450–499.

necessary relationship between democracy and development nor, more generally, between any regime type and economic performance. Rather, a more complex conclusion is appropriate. This starts from the hard fact that *both* democratic and non-democratic Third World regimes have been able to generate high levels of economic development, although there are fewer formal democracies among them (Botswana, Malaysia, Singapore and Mauritius). However, there are many more non-democratic regimes with both poor developmental achievements and appalling human rights records. Finally and emphatically, *no* examples of good or sustained growth in the developing world have occurred under conditions of uncompromising economic liberalism, whether democratic or not. From Costa Rica to China and from Botswana to Thailand, the state has played an active role in influencing economic behaviour and has often had a significant material stake in the economy itself.

Crucially, then, it has *not* been regime type but the kind and character of the *state* and its associated politics that have been decisive in influencing developmental performance. This in turn highlights the primacy of *politics*, not simply governance, as a central determinant of development. In short, the combination of democratic politics and economic liberalism has rarely been associated with the critical early breakthroughs from agrarianism to industrialism, now or in the past.[18]

The evidence suggests that it is far from clear that economic liberalisation will generate development and raise welfare across the board in the Third World. But will democratisation make it any more sure? Or together will they institute a period of political turbulence, democratic reversal and economic decline?

Prognosis

The often appalling human rights record in much of the Third World has been well documented over the years.[19] There can therefore be few who do not welcome the rise in the number of formally democratic states in the world, especially if this leads to an improvement in human rights records, which democracy usually brings.[20] The percentage of formally democratic states in the world increased from 25% in 1973 to 45% in 1990 to 68% in 1992.[21] Nonetheless, the crucial question remains: will these new or born-again democracies survive and will they promote growth?

[18] A. Gerschenkron, *Economic Backwardness in Historical Perspective*, Cambridge, MA: Harvard University Press, 1962.

[19] M. Nowack & T. Swinehart, (eds), *Human Rights in Developing Countries, 1989 Yearbook*, Kehl, West Germany: N. P. Engel, 1989.

[20] C. Humana, *World Human Rights Guide*, London: Pan, 1987.

[21] S. P. Huntington, *The Third Wave. Democratization in the late Twentieth Century*. Norman, OK: University of Oklahoma Press, 1991, p. 26; and Freedom House, *Freedom in the World. Political Rights and Civil Liberties, 1991–1992*. New York, Freedom House, 1992.

I want to suggest that the remainder of the present decade will yield a very mixed pattern in which only a relatively few of the new democracies will survive. But each outcome will depend on the way in which the continuing costs and benefits of economic liberalisation are distributed, on the one hand, and the way in which democratic openings will be used by the excluded or oppressed to correct past inequalities or new hardships, on the other. That is to say, it is not the *combination* of liberal economies and democratic politics that will determine developmental outcomes, but the *manner of their interaction*. Under many circumstances such interactions will constitute explosive and anti-developmental mixtures. Such a prognosis, however, depends on a view of the conditions which ensure democratic survival.

Conditions for democratic politics

The evidence from previous breakdowns in the twentieth century[22] suggests some important conditions for enduring democratic politics.

First, the geographical, constitutional and political legitimacy of the state is essential. Without this, democratic politics will be fatally threatened. Coup, counter-coup, secession and civil war are the consequences in a state whose fundamental organisational framework is not accepted. This has commonly been the case in Africa and is right now occurring in many newly-independent states of the former Soviet Union and, ferociously, in what was once Yugoslavia.

Second, democratic politics in any legitimate polity presupposes a secure and broadly-based consensus about the rules of the political game. Democracy requires all-round loyalty to the democratic process itself, whatever its form and particulars. Losing parties in elections must accept the outcome and embrace the status of 'loyal opposition'.[23] They must not defect to anti-democratic forces as happened in Mrs Aquino's troubled post-Marcos administration, nor threaten to use democracy to suspend it (as the FIS did in Algeria in 1991/2), or resort to violence, as after the 1992 Angolan elections. As Przeworski has pointed out, where democracy is emerging from right-wing authoritarianism (as occurred in Latin America and, now, in South Africa) the right-wing political forces that are associated with the old regime must now attach themselves to democratic values. But equally the left (for instance the ANC in South Africa) must commit itself to demobilise and turn to ballot-box organisation in its struggle for power.[24] Without these reciprocal movements towards playing by the rules of the game, democracy can neither work nor be secure.

Third, sustaining democratic politics also depends on governmental restraint

[22] J. J. Linz, 'The breakdown of democratic regimes: crisis, breakdown and re-equilibration', in J. J. Linz & A. Stepan (eds), *The Breakdown of Democratic Regimes*. Baltimore, MD: Johns Hopkins, 1978, pp. 3–124; and Huntington, *op cit*, note 21.

[23] Linz, *op cit*, note 22.

[24] A. Przeworski, 'Democracy as a contingent outcome of conflict', in J. Elster & R. Slagstad (eds), *Constitutionalism and Democracy*. Cambridge: Cambridge University Press, 1988, pp. 59–80. See p. 73.

on the extent of policy change undertaken by the winning party or parties. That is to say, democratically elected governments (especially minority governments or those with slim legislative support and dubious mandates for major reforms) will threaten democratic politics if they pursue contentious policies which seriously and adversely affect other major interests – as President Allende discovered in Chile in the early 1970s.[25]

Another way of making this point is to adopt Przeworski's insightful contribution and say that democracy is best understood as a system of power under which no group can guarantee that its interests will automatically or always prevail. It is a system under which everyone must subject his or her interests 'to competition and uncertainty'. But although outcomes are therefore in theory 'open' and indeterminate, democracy will only work when '. . . relevant political forces can find institutions that would provide reasonable guarantees that their interests would not be affected in a highly adverse manner in the course of democratic competition'.[26] That is to say, no group will commit itself to the democratic process if it feels that losing an election will mean that it will be wiped out. Concretely, for example, the democratic class compromise between capital and labour in a society such as Venezuela, has until recently meant that '. . . capitalists accepted democratic institutions as a means for workers to make effective claims to improve their material conditions, while workers accepted the private appropriation of profit by capitalists' or, in short, '. . . workers accept capitalism and capitalists accept democracy'[27] The compromise may also be between regions, ethnic groups or even religious groups (though the latter are much harder to organise as principles rather than interests are often involved). It is a key function of institutional arrangements to protect these compromises. Victorious parties have to recognise that there are other legitimate interests or ideals that need to be respected, brought into negotiation over policies and not simply side-lined, attacked or suppressed. This is why the 'winner-takes-all' approach is hardly conducive to democratic politics.

The profound dilemma which this last point raises is that non-consensual and non-democratic measures may often be essential in the early stages of developmental sequences in laying the foundations for growth – and also sustainable democracy in the long run. Land reform is a good example, since it is widely recognised, even by the World Bank[28] that this may be an important condition for agricultural and rural development. However, precisely because it is resisted by powerful interests with intimate connections to the state, land reform in many developing countries (democratic or not) has either failed or been sabotaged.[29]

[25] A. Valenzuela, 'Chile', in Linz & Stepan *op cit*, Part IV, note 22, pp. 1–133.

[26] Przeworski *op cit*. note 24, p. 64.

[27] J. McCoy, 'The state and the democratic compromise in Venezuela'. *Journal of Developing Societies*, IV, 1988, pp. 85–104. See pp. 85–86.

[28] World Bank, *World Development Report 1991*, New York: Oxford University Press 1991.

[29] B. J. Kerkvliet, 'Land reform in the Philippines since the Marcos coup', *Pacific Affairs*, 47, 1974, pp. 286–304; and R. J. Herring, 'Zulfikar Ali Bhutto and the "eradication of feudalism" in Pakistan', *Comparative Studies in Society and History*, 21(4), 1979, pp. 519–557.

In short, and uncomfortably, democratic politics is seldom the politics of radical economic change. Yet radical change may be precisely what is required at key points in developmental processes, and this is one of the key reasons why the relationship between democracy and development is so tense. Democratic politics is, rather, the politics of accommodation, compromise and the centre. Given a diversity of interests in society, this is inevitable. For this reason, democracy is improbable in highly polarised societies, whether divided by income, class, ethnicity, religion or culture. Whether liberal or socialist, the logic of democratic politics is necessarily consensual, conservative and incrementalist. For many that is its virtue: for others, its vice.

A fourth condition for democratic politics is a rich and pluralistic civil society.[30] Civil society consists of those cultural, political or economic '. . . areas of social life . . . which are organised by private or voluntary arrangements between individuals and groups outside the *direct* control of the state'[31] These may be youth groups, trade unions or business organisations, as well as consumer or interest groups through which people seek to exercise pressure and restraint on the state and thereby strengthen the assumptions and practices of democratic self-management in complex societies.[32] However, one feature of many Third World societies is that the institutions of civil society, to the extent that they exist, have been penetrated and 'captured' by dominant one-party states and thus transformed into agencies of the regime.

Fifth, stable democracy also requires that there be no serious threat to the authority and power of the state[33] such as private armies – as in Peru, Colombia, Northern Ireland, the Philippines and Somalia. Where such armies were born and nurtured in the struggle for democracy itself, their subsequent disbanding or incorporation in the armed forces of the new state will be crucial (as occurred in Zimbabwe and will be necessary in South Africa).

Sixth, with important exceptions (such as Malaysia and Mauritius), democratic politics has not been easy to sustain in societies divided by regional tensions or ethnic, cultural or religious pluralism, especially where this is compounded by real or perceived material inequalities, as in Nigeria, Cyprus, Fiji and Indonesia. Democratic breakdown in such societies is likely until political or institutional solutions are found, or until a unifying and largely secular national consensus emerges which reduces plural differences to secondary importance.[34]

[30] L. Diamond & J. J. Linz, 'Introduction: politics, society and democracy in Latin America', in L. Diamond *et al. Democracy in Developing Countries*. Vol. 4. Latin America, Boulder, CO: Lynne Riener, 1989, pp. 1–58; and Rueschemeyer *et al. op cit*, note 12, p. 6.

[31] D. Held, *Models of Democracy*, Cambridge: Polity, 1987, p. 281.

[32] R. A. Dahl, *A Preface to Economic Democracy*, Cambridge: Polity, 1985; and D. Held, 'The contemporary polarization of democratic theory: the case for a third way', in A. Leftwich (ed), *New Developments in Political Science*, Aldershot, Edward Elgar, 1990, pp. 8–23.

[33] Linz, *op cit*, note 22.

[34] Huntington & Dominguez, *op cit*, note 17; and K. A. Bollen & R. W. Jackman, 'Economic and noneconomic determinants of political democracy in the 1960's, *Research in Political Sociology*, I, 1985, pp. 27–48.

Finally, deepening economic crises which sharpen already acute social inequalities have commonly constituted the background against which abrupt political change has occurred[35] whether from authoritarian rule to democracy or in the reverse direction. Such inequalities do not help democracy and are generally far sharper in the Third World than in the industrial world.[36]

Those currently pushing for democracy as a component of good governance appear oblivious to how few of these conditions for democratic endurance exist in the Third World and elsewhere, and what their implications for democracy might be.

. . . .

Conclusions

If correct, this bleak scenario suggests a number of conclusions. First, from a developmental point of view, the general but simplistic appeal for better 'governance' as a condition of development is virtuous but naive. For an independent and competent administration is not simply a product of 'institution building' or improved training, but of *politics*. And if the politics do not give rise to the kind of state which can generate, sustain and protect an effective and independent capacity for governance, then there will be no positive developmental consequences.

Second, as shown above, there are good theoretical reasons why, in the prevailing conditions in many Third World countries, democracy is unlikely to be the political form which can generate such a state or system of governance: quite the opposite. Furthermore, historical evidence shows that faith in the economic and political liberalism of the minimal state as the universally appropriate *means* of development is deeply flawed. Successful modern transformative episodes of economic development, from the 19th century to the present, have almost always involved both a strong *and* an active state to help initiate, accelerate and shape this process.[37] Bismarckian Germany, Japan after the Meiji revolution, Turkey under Ataturk, the Soviet Union after 1917, twentieth-century Sweden, China for the first quarter century after the revolution, as well as Taiwan, Korea, Thailand and post-independence Mauritius, Singapore, Malaysia and Botswana after 1960 all provide abundant evidence of this.

[35] C. Tilly, 'Does modernization breed revolution?', *Comparative Politics*, 5, 1973, pp. 425–47.

[36] UNDP (United Nations Development Programme), *Human Development Report 1992*, New York: Oxford University Press, 1992, Tables 17 and 28.

[37] Gerschenkron, *op cit*, note 18; B. Supple, 'The state and the Industrial Revolution, 1700–1914', in C. M. Cipolla (ed), *The Fontana Economic History of Europe*, London: Collins, 1973 pp. 301–353; and S. Haggard, *Pathways from the Periphery*, Ithaca, NY: Cornell University Press, 1990.

Third, as their diversity in space and time illustrates, such transformations have depended less on regime type or policy orientation than on the character and capacity of the state, whether democratic or not. In short, it has been *politics and the state* rather than governance or democracy that explains the differences between successful and unsuccessful developmental records.

Fourth, what is required, then, is not necessarily a democratic state (though it would be highly desirable if it could also be that), but a *developmental state*. By this I mean a state whose political and bureaucratic élite has the genuine developmental determination and autonomous capacity to define, pursue and implement developmental goals.[38] For contestation about the nature and direction of economic development, the distribution of the costs and benefits of transformation, the institutional forms of government and the character of the state appropriate to this, is contestation about *power*. Hence it is not simply a managerial question, as the World Bank's literature on governance asserts, but a political one. For all processes of 'development' express crucially the central core of *politics*: conflict, negotiation and cooperation over the use, production and distribution of resources.[39]

Fifth, given the recent demise and implausible prospects of the state 'socialist path' to development,[40] successful developmental states are likely to pursue 'market-friendly' strategies.[41] But to be developmentally effective, and to avoid the turbulence suggested above, such strategies will need to include far-reaching programmes of economic empowerment in order to bring the economically marginalised more fully into productive economic activity. The means for this will vary widely, but will include land reform, such as occurred in Korea,[42] the extension of property rights,[43] training, job creation and imaginative forms of credit.

Such measures require not simply less government but both better government[44] and *stronger* government. To make market-friendly strategies work, developmental states in many societies will also need to liberate the poor, especially the rural poor, from the continued domination of traditional landed élites and anti-developmental oligarchs who both oppose empowerment and often stand in the way of development and democracy. All this is

[38] Johnson, C., *MITI and the Japanese Miracle*, Stanford, Stanford University Press, 1982; Leftwich, *op cit*, note 8; and R. Wade, *Governing the Market: Economic Theory and the Role of Government in East Asian Industrialization*, Princeton NJ: Princeton University Press, 1990.

[39] A. Leftwich, *Redefining Politics. People, Resources and Power*, London: Methuen, 1983.

[40] C. Clapham, 'The collapse of socialist development in the Third World', *Third World Quarterly*, 13(1), 1992, pp. 13–26.

[41] World Bank, *op cit*, note 28, p. 1.

[42] E. Lee, 'Egalitarian peasant farming and rural development: the case of South Korea', *World Development*, 7, 1979, pp. 493–517.

[43] H. De Soto, *The Other Path. The Invisible Revolution in the Third World*, New York: Harper & Row, 1990.

[44] R. Sandbrook, 'Taming the African Leviathan', *World Policy Journal*, VII(4), 1990, pp. 672–701.

therefore not just a matter of 'governance', as the technicist illusion would have us believe: it is a matter of *politics*.

Without a developmental state, democratic or not, no contemporary developing society is likely to achieve developmental breakthrough. And without such states, democratic market-friendly strategies will sooner or later break up on the rocks of their own internally-generated economic inequalities and escalating political strife, especially in 'premature' democracies. This is why current Western insistence on markets, governance and democracy as the keys to development seems both so ideological and naive.

Such conclusions are unfashionable and uncomfortable. But if eliminating the continuing offence of poverty and misery is the *real* target, then unlimited liberal democracy and unrestrained economic liberty may be the last thing the developing world needs as it whirls towards the 21st century.

Note

I am grateful to Barry Gills, Geoffrey Hawthorn, David Held, Peter Larmour and John Peterson for comments on earlier drafts of this paper. However, I alone am responsible for the arguments developed here.

---

27 Nancy Scheper-Hughes, 'Everyday Violence: Bodies, Death, and Silence'

Excerpts from: N. Scheper-Hughes, *Death Without Weeping: The Violence of Everyday Life in Brazil*, chapter 6. Berkeley: University of California Press (1992)

Violence and the Taken-for-Granted World

> The tradition of the oppressed teaches us that the 'state of emergency' in which we live is not the exception but the rule. We must attain to a conception of history that is in keeping with this insight . . . One reason why Fascism has a chance is that in the name of progress its opponents treat it as a historical norm.
>
> Walter Benjamin (cited in Taussig, 1989, p. 64)

Writing about El Salvador in 1982, Joan Didion noted in her characteristically spartan prose that 'the dead and pieces of the dead turn up everywhere, everyday, as taken-for-granted as in a nightmare or in a horror movie' (1982, p. 9). In *Salvador* there are walls of bodies; they are strewn across the landscape, and they pile up in open graves, in ditches, in public restrooms, in bus stations, along the sides of the road. 'Vultures, of course, suggest the presence of a body. A knot of children on the street suggest the presence of a body'. Some bodies even turn up in a place called Puerto del Diablo, a well-known tourist site described in Didion's inflight magazine as a location 'offering excellent subjects for color photography.'

It is the anonymity and the routinization of it all that strikes the naive reader as so terrifying. Who are all these *desaparecidos* – the unknown and the 'disappeared' – both the poor souls with plucked eyes and exposed, mutilated genitals lying in a ditch and those unidentifiable men in uniform standing over the ditches with guns in their hands? It is the contradiction of wartime crimes against ordinary peacetime citizens that is so appalling. Later we can expect the unraveling, the recriminations, the not-so-guilty confessions, the church-run commissions, the government-sponsored investigations, the arrests of tense and unyielding men in uniform, and finally the optimistic reports – Brazil, Argentina (later, perhaps even El Salvador) *nunca mais*. Quoth the raven, '*Nunca mais*.' After the fall, after the aberration, we expect a return to the normative, to peacetime sobriety, to notions of civil society, human rights, the sanctity of the person (Mauss's *personne morale), habeas corpus*, and the unalienable rights to the ownership of one's body.

But here I intrude with a shadowy question. What if the disappearances, the piling up of civilians in common graves, the anonymity, and the routinization of violence and indifference were not, in fact, an aberration? What if the social spaces before and after such seemingly chaotic and inexplicable acts were filled with rumors and whisperings, with hints and allegations of what could happen, especially to those thought of by agents of the social consensus as neither persons nor individuals? What if a climate of anxious, ontological insecurity about the rights to ownership of one's body was fostered by a studied, bureaucratic indifference to the lives and deaths of 'marginals,' criminals and other no-account people? What if the public routinization of daily mortifications and little abominations, piling up like so many corpses on the social landscape, provided the text and blueprint for what only appeared later to be aberrant, inexplicable, and extraordinary outbreaks of state violence against citizens?

In fact, the 'extraordinary' outbreaks of state violence against citizens, as in Didion's *Salvador*, during the Argentine 'Dirty War' (Suarez-Orozco, 1990), in Guatemala up through the present day (Paul, 1988; Green, 1989), or in the harshest period following the Brazilian military coup of 1964 (Dassin, 1986) entail the generalizing to recalcitrant members of the middle classes what is, in fact, normatively practiced in threats or open violence against the poor, marginal, and 'disorderly' popular classes. For the popular classes every day is, as Taussig (1989) succinctly put it, 'terror as usual.' A state of emergency occurs when the violence that is normally contained to that social space suddenly explodes into open violence against the 'less dangerous' social classes. What makes the outbreaks 'extraordinary,' then, is only that the violent tactics are turned against 'respectable' citizens, those usually shielded from state, especially police, terrorism.

If, in the following ethnographic fragments, I seem to be taking an unduly harsh and critical view of the 'state' of things in Brazil, let me hasten to say at the outset that I view this interpretation as generalizable to other bureaucratic states at a comparable level of political-economic 'development' and in a different form to those characterized by a more 'developed' stage of industrial capitalism such as our own. Violence is also 'taken for granted' and routinized in parts of our police underworld operating through SWAT team attacks on suspected crack houses and crack dealers in inner-city neighborhoods. And state terrorism takes other forms as well. It is found in the cool jargon of nuclear weapons researchers, our own silent, yet deadly, technicians of practical knowledge. Carol Cohn (1987) penetrated this clean, closed world and returned with a chilling description of the way our nuclear scientists have created a soothing and normalizing discourse with which to discuss our government's capacity for blowing populations of bodies to smithereens. 'Bio-power,' indeed.

. . . .

Citizenship and justice in Brazil

We tend to think of the Western political traditions and concepts of democracy, citizenship, and the modern state – as well as the necessary preconditions for their existence – as universally shared among modern nations. But as the recent events in Eastern Europe indicate – especially the difficulty with which newly liberated citizens are attempting to 'reclaim the public' and recreate civil society following the 'fall' of repressive and totalitarian communist regimes – the concepts of democracy, equality, and civil society may have very specific and different cultural and historical referents. In Eastern Europe the relationship between civil society and the state was perceived not in terms of collaboration and consensus but rather in terms of mutual hostility and antagonism.

In Brazil the political traditions of republican democracy and equality, influenced by both the French and American revolutions, have always been mediated by traditional notions of hierarchy, privilege, and distinction. The Brazilian constitution, like the American constitution, was adopted before slavery was abolished, and by the end of the nineteenth century the public sphere had been constituted exclusively for a very small, elite group. Liberty and democracy became the exclusive preserves of the dominant minority, those educated and landed men (and, later, women) of breeding, culture, and distinction. Civil liberties and human rights were cast as 'privileges' and 'favors' bestowed by superiors on subordinates within relations structured by notions of personal honor and loyalty. 'Favors' included everything from personal protection, material goods, jobs, and status to the right to vote. Consequently, up through the first half of the twentieth century in the Northeast, votes and elections were controlled by a few local big men and their clients.

Roberto da Matta pointed out that although Brazilian law is based on liberal and democratic principles of universalism and equality, its practice often diverges from theory and it 'tends to be applied in a rigorous way only to the masses who have neither powerful relatives nor important family names.' He went on to state that 'in a society like Brazil's universal laws may be used for the exploitation of labor rather than for the liberation of society.' Those who are wealthy or who have political connections can always manage to 'slip under or over legal barriers' (1984, p. 233).

Brazil's system of criminal justice is a 'mixed system' containing elements of both the American and the European civil law tradition (Kant 1990). Contrary to the American system, there is no common-law tradition whereby precedents and jury verdicts can actually participate, in conjunction with the legislature, in making the law. And in addition to many modern, egalitarian, and individual rights protected by the Brazilian criminal justice system (such as the right to counsel and to *ampla defesa* – that is, the right to produce any possible evidence on equal footing with the prosecution), there are other, more traditional, and less liberal traditions. First among these is the tradition of progressing from a position of 'systematic suspicion,' rather than from an assumption of innocence, and, relatedly, the judge's 'interrogation' of the accused relying on information produced by prior police investigations that are 'inquisitorial' in nature. In the words of one police chief interviewed by Roberto de Lima Kant, police interrogations entail 'a proceeding *against everything* and *everyone* to find out the *truth* of the facts' (1990, p. 6). Within this inquisitorial system, 'torture becomes a legitimate – if unofficial – means of police investigation for obtaining information or a confession'. In all, 'Brazilian criminal proceedings are organized to show a gradual, step by step, ritual of progressive incrimination and humiliation, the outcome of which must be either the confession or the acquittal. The legal proceedings are represented as a punishment in themselves' (*ibid.*).

Within this political and legal context, one can understand the *moradores*' awesome fear of the judicial system and their reluctance to use the courts to redress even the most horrendous violations of their basic human rights. And, as Teresa Caldeira (1990) noted, the first stirrings of a new political discourse on 'human rights,' initiated by the progressive wing of the Catholic church and by leftist political parties in Brazil in the late 1970s and early 1980s and fueled in part by the international work of Amnesty International, was readily subverted by the Right. Powerful conservative forces in Brazil translated 'human rights' into a profane discourse on special favors, dispensations, and privileges for criminals. Worse, the Brazilian Right played unfairly on the general population's fears of an escalating urban violence. These fears are particularly pronounced in poor, marginalized, and shantytown communities. And so, for example, following a 1989 presidential address broadcast on the radio and over loudspeakers in town announcing much-needed proposed prison reforms in

Brazil, the immediate response of many residents of the Alto[1] seemed paradoxical. Black Zulaide, for example, began to wail and wring her hands: 'Now we are finished for sure,' she kept repeating. 'Even our president has turned against us. He wants to set all the criminals free so that they can kill and steal and rape us at will.' It seemed to have escaped Black Zulaide that her own sons had at various times suffered at the hands of police at the local jail and that the prison reform act was meant to protect *her* class in particular. Nevertheless, Zulaide's fears had been fueled by the negative commentary of the police, following the broadcast, on the effects these criminal reforms would have on the people of Bom Jesus but especially on those living in 'dangerous' *bairros* such as the Alto do Cruzeiro and needing the firm hand of the law to make life minimally 'safe.'

Similarly, Teresa Caldeira offered two illustrations of right-wing ideological warfare that equated the defense of human rights with the defense of special privileges for criminals. The first is from the 'Manifesto of the Association of Police Chiefs' of the state of Sao Paulo, which was addressed to the general population of the city on October 4, 1985. The manifesto takes to task the reformist policies of the then-ruling central-leftist political coalition, the PMDB:

> The situation today is one of total anxiety for you and total tranquility for those who kill, rob, and rape. Your family is destroyed and your patrimony acquired with much sacrifice is being reduced. . . .. How many crimes have occurred in your neighborhood and how many criminals were found responsible for them? . . . The bandits are protected by so called human rights, something that the government considers that you, an honest and hard working citizen, do not deserve. (1990, p. 6).

Her second example is taken from an article published on September 11, 1983, in the largest daily newspaper of Sao Paulo, *A Folha de São Paulo*, written by an army colonel and the state secretary of public security:

> The population's dissatisfaction with the police, including the demand for tougher practices . . . originates from the trumped up philosophy of 'human rights' applied in favor of bandits and criminals. This philosophy gives preference to the marginal, protecting his 'right' to go around armed, robbing, killing, and raping at will.

Under the political ideology of favors and privileges, extended only to those who behave well, human rights cannot logically be extended to criminals and marginals, those who have broken, or who simply live outside, the law. When this negative conception of human rights is superimposed on a very narrow definition of 'crime' that does not recognize the criminal and violent acts of the powerful and the elite, it is easy to see how everyday violence against the poor is routinized and defended, even by some of the poor themselves.

[1] Editor's footnote: The *Alto* referred to in this extract is the Alto do Cruzeira, a hillside shantytown in north-east Brazil, where Nancy Scheper-Hughes lived and worked in 1964–66 and to which she returned as an 'anthropologist' for fourteen months between 1982 and 1989.

Mundane surrealism

In Mario Vargas Llosa's novel *The Real Life of Alejandro Mayta*, the Peruvian narrator comments on the relations of imagination to politics and of literary fiction to history:

> Information in this country has ceased to be objective and has become pure fantasy – in newspapers, radio, television, and in ordinary conversation. To report among us now means either to interpret reality according to our desires or fears, or to say simply what is convenient. It is an attempt to make up for our ignorance of what is going on – which in our heart of hearts we understand as irremediable and definitive. Since it is impossible to know what is really happening, we Peruvians lie, invent, dream, and take refuge in illusion. Because of these strange circumstances, Peruvian life, a life in which so few actually do read, has become literary. (1986, p. 246)

The magical realism of Latin American fiction has its counterparts in the mundane surrealism of ethnographic description, where it is also difficult to separate fact from fiction, rumor and fantasy from historical event, and the events of the imagination from the events of the everyday political drama. The blurring of fiction and reality creates a kind of mass hysteria and paranoia that can be seen as a new technique of social control in which everyone suspects and fears every other: a collective hostile gaze, a human panopticon is created. But when this expresses itself positively and a state of alarm or a state of emergency is produced . . . the shocks reveal the disorder in the order and call into question the 'normality of the abnormal,' which is finally shown for what it really is.

Peacetime crimes

> The peoples' death was as it had always been:
> as if nobody had died, nothing,
> as if those stones were falling
> on the earth, or water on water . . .
> Nobody hid this crime.
> This crime was committed
> in the middle of the Plaza.
>
> Pablo Neruda (1991, pp. 186–7)

What makes the political tactic of disappearance so nauseating – a tactic used strategically throughout Brazil during the military years (1964–1985) against suspected subversives and 'agitators' and now applied to a different and perhaps an even more terrifying context (i.e., against the shantytown poor and the economic marginals now thought of as a species of public enemy) – is that it does not occur in a vacuum. Rather, the disappearances occur as part of a larger context of wholly expectable, indeed even anticipated, behavior.

Among the people of the Alto, disappearances form part of the backdrop of everyday life and confirm their worst fears and anxieties—that of losing themselves and their loved ones to the random forces and institutionalized violence of the state.

The practices of 'everyday violence' constitute another sort of state 'terror', one that operates in the ordinary, mundane world of the *moradores*[2] both in the form of rumors and wild imaginings and in the daily enactments of various public rituals that bring the people of the Alto into contact with the state: in public clinics and hospitals, in the civil registry office, in the public morgue, and in the municipal cemetery. These scenes provide the larger context that makes the more exceptional and strategic, politically motivated disappearances not only allowable but also predictable and expected.

'You gringos,' a Salvadorian peasant told an American visitor, 'are always worried about violence done with machine guns and machetes. But there is another kind of violence that you should be aware of, too. I used to work on a hacienda. My job was to take care of the dueño's dogs. I gave them meat and bowls of milk, food that I couldn't give my own family. When the dogs were sick, I took them to the veterinarian. When my children were sick, the dueño gave me his sympathy, but no medicine as they died' (cited in Chomsky, 1957: p. 6)

Similarly, the *moradores* of the Alto speak of bodies that are routinely violated and abused, mutilated and lost, disappeared into anonymous public spaces—hospitals and prisons but also morgues and the public cemetery. And they speak of themselves as the 'anonymous,' the 'nobodies' of Bom Jesus da Mata. For if one is a 'somebody,' a *fildalgo* (a son of a person of influence), and a 'person' in the aristocratic world of the plantation *casa grande*, and if one is an 'individual' in the more open, competitive, and bourgeois world of the new market economy (the *rua*), then one is surely a nobody, a mere *fulano-de-tal* (a so-and-so) and João Pequeno (little guy) in the anonymous world of the sugarcane cutter (the *mata*).

Moradores refer, for example, to their collective invisibility, to the ways they are lost to the public census and to other state and municipal statistics. The otherwise carefully drafted municipal street map of Bom Jesus includes the Alto do Cruzeiro, but more than two-thirds of its tangle of congested, unpaved roads and paths are not included, leaving it a semiotic zero of more than five thousand people in the midst of the bustling market town. CELPE, the state-owned power and light company, keeps track, of course, of those streets and houses that have access to electricity, but the names the company has assigned to identify the many intersecting *bicos, travessas*, and *ruas* of the Alto do not conform to the names used among the *moradores* themselves. The usual right of the 'colonizer' to name the space he has claimed is not extended to the marginal settlers of the Alto do Cruzeiro.

The people of the Alto are invisible and discounted in many other ways. Of

2 *Moradores*: inhabitants, residents, dwellers, squatters, tenants.

no account in life, the people of the Alto are equally of no account in death. On average, more than half of all deaths in the *município* are of shantytown children under the age of five, the majority of them the victims of acute and chronic malnutrition. But one would have to read between the lines because the death of Alto children is so routine and so inconsequential that for more than three-fourths of recorded deaths, the cause of death is left blank on the death certificates and in the ledger books of the municipal civil registry office. In a highly bureaucratic society in which triplicates of every form are required for the most banal of events (registering a car, for example), the registration of child death is informal, and anyone may serve as a witness. Their deaths, like their lives, are quite invisible, and we may as well speak of their bodies, too, as having been disappeared.

The various mundane and everyday tactics of disappearance are practiced perversely and strategically against people who view their world and express their own political goals in terms of bodily idioms and metaphors. The people of the Alto inhabit a world with a comfortable human shape, a world that is intimately embodied. I have already suggested that the *moradores* of the Alto 'think' the world with their bodies within a somatic culture. At their base community meetings the people of the Alto say to each other with conviction and with feeling, 'Every man should be the *dono* [owner] of his own body.' Not only their politics but their spirituality can be described as 'embodied' in a popular Catholicism, with its many expressions of the carnal and of physical union with Jesus, with His mother Mary, and with the multitude of saints, more than enough for every day of the year and to guide every human purpose. There is a saint for every locale, for every activity, and for every part of the body. And the body parts of the saints, splintered into the tiniest relics, are guarded and venerated as sacred objects.

Embodiment does not end with death for the people of the Alto. Death is itself no stranger to people who handle a corpse with confidence, if not with ease. ('When you die, Dona Nanci,' little Zefinha used to say affectionately, 'I'm going to be the first to eat your big legs,' the highest compliment she could think of to pay me.) On the death of a loved one, a local photographer will often be called to take a photo of the adult or child in her or his coffin. That same photo will be retouched to erase the most apparent signs of death, and it will become the formal portrait that is hung proudly on the wall. The deceased continue to appear in visions, dreams, and apparitions through which they make their demands for simple pleasures and creature comforts explicit. As wretched *almas penadas*, 'restless souls' from purgatory, the dead may request food and drink or a pair of shoes or stockings to cover feet that are cold and blistered from endless wandering. Because the people of the Alto imagine their own souls to have a human shape, they will bury an amputated foot in a tiny coffin in the local cemetery so that later it can be reunited with its owner, who can then face his Master whole and standing 'on his own two feet.'

Against these compelling images of bodily autonomy and certitude is the reality of bodies that are simultaneously discounted and preyed on and sometimes mutilated and dismembered. And so the people of the Alto come to imagine that there is nothing so bad, so terrible that it cannot happen to them, to their bodies, because of sickness (*por culpa de doença*), because of doctors (*por culpa dos médicos*), because of politics and power (*por culpa da política*), or because of the state and its unwieldy, hostile bureaucracy (*por culpa da burocracia*).

I am not going so far as to suggest that the fears of mutilation and of misplacement of the body are not shared with other social classes of Brazil, which also 'privilege' the body in a culture that prides itself on its heightened expressions and pleasures of the sensual. What is, however, specific to the marginal classes of the Alto do Cruzeiro is a self-conscious sort of thinking with and through the body, a 'remembering' of the body and of one's 'rights' in it and to it. The affluent social classes take for granted these rights to bodily integrity and autonomy to the extent that they 'go without saying.' The police oppressors know their victims all too well, well enough to mutilate, castrate, make disappear, misplace, or otherwise lose the bodies of the poor, to actualize their very worst fears. It is the sharing of symbols between the torturer and the tortured that makes the terror so effective.

The unquestionability of the body was, for Wittgenstein, where all knowledge and certainty began. 'If you do know that here is one hand', he began his last book, *On Certainty*, 'we'll grant you all the rest' (1969, p. 2e). And yet Wittgenstein himself, writing this book while he was working with patients hospitalized during the war, was forced to reflect on the circumstances that might take away the certainty of the body. Here, in the context of *Nordestino* life, I am exploring another set of circumstances that have given a great many people grounds to lose their sense of bodily certitude to terrible bouts of existential doubt – 'My God, my God, what ever will become of us?' – the fear of being made to vanish, to disappear without a trace.

It is reminiscent of the situation described by Taussig with reference to a similar political situation in Colombia: 'I am referring to a state of doubleness of social being in which one moves in bursts between somehow accepting the situation as normal, only to be thrown into a panic or shocked into disorientation by an event, a rumor, a sight, something said, or not said – something even while it requires the normal in order to make its impact, destroys it' (1989, p. 8). The intolerableness of the situation is increased by its ambiguity. Consciousness moves in and out of an acceptance of the state of things as normal and expectable – violence as taken for granted and sudden ruptures whereby one is suddenly thrown into a state of shock (*susto, pasmo, nervios*) – that is endemic, a graphic body metaphor secretly expressing and publicizing the reality of the untenable situation. There are nervous, anxious whisperings, suggestions, hints. Strange rumors surface.

Selected references

Caldeira, T. 1990: The experience of violence: order, disorder, and social discrimination in Brazil. Mss., Department of Anthropology: University of California, Berkeley.

Chomsky, N. 1985: *Turning the tide: United States intervention in Central America and the struggle for peace*. Boston: South End.

Cohn, C. 1987: Sex and death in the rational world of defense intellectuals. *Signs* 12 (4), 687–718.

Da Matta, R. 1984: On carnival, informality, and magic: a point of view from Brazil. In Bruner, E. (ed.), *Text, play and story*. Washington, DC: American Ethnological Society.

Dassin, J. (ed.) 1986: *Torture in Brazil: A report by the Archdiocese of São Paulo*. New York: Random House.

Didion, J. 1982: *Salvador*. New York: Pocket Books.

Green, L. 1989: The realities of survival: Mayan widows and development aid in rural Guatemala. Paper presented at the meeting of the American Anthropological Association, Washington, DC, 15 November.

Kant, R. de Lima 1990: Criminal justice: a comparative approach, Brazil and the United States. Mss., Kellogg Institute: University of Notre Dame.

Llosa, Mario Vargas 1986: *The real life of Alejandro Mayta*. New York: Vintage.

Neruda, P. 1991: *Canto General*: Berkeley: University of California Press.

Paul, B. 1988: The operation of a death squad in a Lake Atitlan community. In Cammack, R. (ed.), *Harvest of violence: The Mayan Indians and the Guatemalan crisis*. Norman: Univerisity of Oklahoma Press.

Suarez-Orozco, M. 1990: Speaking of the unspeakable: toward a psychological understanding of responses to terror. *Ethos* 18, 353–83.

Taussig, M. 1989: Terror as usual. *Social Text* (Fall–Winter), 3–20.

Wittgenstein, L. 1969: *On certainty*. New York: Harper and Row.

INDEX

Most references are to development, which is therefore generally omitted as a qualifier. Spellings have been standardized to those most frequently used in the text eg *labour* instead of *labor* and *urbanization* instead of *urbanisation*.